Lecture Notes in Physics

Edited by J. Ehlers, München, K. Hepp, Zürich
R. Kippenhahn, München, H. A. Weidenmüller, Heidelberg
and J. Zittartz, Köln
Managing Editor: W. Beiglböck, Heidelberg

89

Microscopic Optical Potentials

Proceedings of the Hamburg Topical Workshop
on Nuclear Physics
Held at the University of Hamburg,
Hamburg, Germany, September 25–27, 1978

Edited by H. V. von Geramb

Springer-Verlag Berlin Heidelberg GmbH 1979

Editor
H. V. von Geramb
Universität Hamburg
Jungiusstraße 9
D–2000 Hamburg 13

ISBN 978-3-540-09106-6   ISBN 978-3-540-35542-7 (eBook)
DOI 10.1007/978-3-540-35542-7

Originally published by Springer-Verlag Berlin Heidelberg New York in 1979

2153/3140-543210

PREFACE

The Workshop on Microscopic Optical Potentials held at the University
of Hamburg, September 25-27, 1978, was arranged on an international level
to unite the various theoretical activities in the understanding of elastic and limited inelastic nuclear scattering with a potential model.
This volume contains selected topics covered in the contributions and
discussions. The contents deal with all projectile masses ranging from
nucleons to heavy ions, and projectile energies ranging typically from
10 MeV to pion threshold. Whenever possible, emphasis was placed on microscopic theories which link the nucleon nucleon interaction with aspects
of the nuclear many body problem as far as they concern the evaluation
of complex optical potentials. For composite projectiles, this put the
semiphenomenological folding models in the foreground, together with the
analyses of experimental angular distributions. The thereby often shown
precision fits defined new standards in precision and theoretical reproduction power. Experimental work per se is not included, and when, then
only in a few selected topics.

The guide used in arranging the contents of the proceedings was the projectile mass. The volume begins with the reviews on the nucleon nucleus
optical potential and then gradually increases the projectile mass until
we reach in the second half the aspects of heavy ion potentials. Towards
the end, some special topics regarding the mainstream of the potential
theories are included.

On this occassion we should like to thank Mrs. Larsen for her valuable
secretarial help in organizing the Workshop and the preparation of this
manuscript, as well as Mrs. Berghaus for her assistance in the workshop
preparations.

We acknowledge the support of the Workshop by the *Bundesministerium für
Forschung und Technologie* and the *Universität Hamburg*.

Hamburg, 1978   H.V. von Geramb

LIST OF CONTENTS

LIST OF PARTICIPANTS

ALONS, P.W.F., Vrije University, Holland

ANDERS, B., University of Hamburg, Germany

AUSTIN, S.M., Michigan State University, USA

BALTZ, A.J., Brookhaven National Laboratory, USA

BARRETT, R.C., University of Surrey, England

BAUR, G., KFA Jülich, Germany

BENDISCIOLI, G., INFN Pavia, Italy

BIESBOER, F., Vrije University, Holland

BRIEVA, F.A., University of Oxford, England

BRINK, D.M., University of Oxford, England

BRISSAUD, I., Physique Nucléaire, Orsay, France

BRÜCKMANN, H., University of Hamburg, Germany

CEULENEER, R., University of Mons, Belgium

COHLER, M.D., King's College, London, England

DE LEO, R., University of Bari, Italy

DJALOEIS, A., KFA Jülich, Germany

DUHM, H., University of Hamburg, Germany

EPPEL, D., University of Hamburg, Germany

ESCUDIE, J.L., CEN-Saclay, France

FICK, D., University of Marburg, Germany

FINCKH, E., University of Erlangen-Nürnberg, Germany

FRIESE, J., University of Hamburg, Germany

FÜHRER, T., University of Hamburg, Germany

GERAMB, H.V., University of Hamburg, Germany

GIANNINI, M., University of Genova, Italy

GILS, H.J., KFZ Karlsruhe, Germany

GOEKE, K., KFA Jülich, Germany

GRALLA, S., University of Hamburg, Germany

GRAW, G., University of München, Germany

GRIDNEV, K., University of Leningrad, UdSSR

HALLFARTH, G., University of Hamburg, Germany

HEFTER, E.F., TU Hannover, Germany

HEIMLICH, F., University of Hamburg, Germany

HODGSON, P.E., University of Oxford, England

IOANNIDES, A.A., University of Surrey, England

JACKSON, D.F., University of Surrey, England

KARBAN, O., University of Birmingham, England

KORF, G., University of Hamburg, Germany

KREWALD, S., KFA Jülich, Germany

KRÖGER, M., University of Hamburg, Germany

KROTSCHECK, E., University of Hamburg, Germany

LANDOWNE, S., University of München, Germany

LARA, P., University of Hamburg, Germany

LEEB, H., Atominstitut, Vienna, Austria

LEJEUNE, A., University of Liége, Belgium

LEZOCH, P., University of Hamburg, Germany

LINDNER, A., University of Hamburg, Germany

LINDSTRÖM, G., University of Hamburg, Germany

LOMBARD, R., CEN-Saclay, France

LOVE, W.G., University of Georgia, USA

MACKINTOSH, R.S., Daresbury Laboratory, England

MADSEN, V., Oregon State University, USA

MAHAUX, Cl., University of Liége, Belgium

MAJKA, Z., KFZ Karlsruhe, Germany

McCARTHY, I.E., Flinders University of South Australia, Australia

MICHELETTI, S., University of Milan, Italy

MICKLINGHOFF, M., University of Hamburg, Germany

MÜLLER, K.H., TH Darmstadt, Germany

NGO, H., Physique Nucléaire, Orsay, France

NORDLAND, O., University of Hamburg, Germany

OSTERFELD, F., KFA Jülich, Germany

PETROVICH, F.L., Florida State University, USA

PIGNANELLI, M., University of Milan, Italy

PLASTINO, A., LaPlata, Argentina

PLISCHKE, P., University of Hamburg, Germany

PUT, L.W., University of Groningen, Netherlands

QURESHI,I., University of Surrey, England

REINHARD, P.G., University of Mainz, Germany

RÖSEL, F., University of Basel, Switzerland

ROMAN, S., University of Birmingham, England

SCHEID, W., University of Giessen, Germany

SCHWARZ, A., University of Hamburg, Germany

SCOBEL, W., University of Hamburg, Germany

SINHA, B., Bhabha Institute, Bsmbay, India

STANCU, F., University of Liége, Belgium

STRAUSS, W., University of Hamburg, Germany

STROHBUSCH, U., University of Hamburg, Germany

STUMM, J., University of Erlangen-Nürnberg, Germany

TANG, Y.C., University of Minnesota, USA

TARRATS, A., CEN-Saclay, France

TIELENS, A., University of Hamburg, Germany

TRAUTMANN, D., University of Basel, Switzerland

TROST, H., University of Hamburg, Germany

TUNGATE, G., MPI Kernphysik, Heidelberg, Germany

ÜBERALL, T., University of Hamburg, Germany

VAN GIAI, N., Physique Nucléaire, Orsay, France

VAN HIENEN, J.F.A., Vrije University, Holland

VAN HALL, P.J., Eindhoven University of Technology, Holland

VINH MAU, N., Physique Nucléaire, Orsay, France

VLACHODIMITROPOULOS, P., University of Hamburg, Germany

WEIGEL, M., University of München, Germany

WENDLERy W., University of Hamburg, Germany

WICK, K., University of Hamburg, Germany

WIKTOR, S., Institute of Nuclear Physics, Cracow, Poland

WORZECK, J., University of Hamburg, Germany

ZABOLITZKY, J.G., Ruhr-University, Bochum, Germany

ZARUBIN, P., University of Leningrad, UdSSR

NUCLEAR MATTER APPROACH TO THE NUCLEON-NUCLEUS OPTICAL MODEL

C. Mahaux

Institut de Physique, Université de Liège, Belgium

1. Introduction

It can be demonstrated quite generally that a single-particle operator exists which, when introduced in the one-body Schroedinger equation, yields the elastic part of the full many-channel wave function.[1,2] This operator is the *generalized* optical-model potential;[3,4] it is complex, has a complicated non-locality, a wild energy dependence and is therefore mainly of formal interest. The practical usefulness of the optical model derives from the empirical finding that many experimental data can be fairly well fitted with an optical-model potential (OMP) which has a simple radial shape, and depends smoothly on energy and on target mass number.[5] Our survey addresses itself to this OMP and is based on the three properties just mentioned.

(a) The fact that the OMP depends smoothly on mass number implies that one can gain useful information from the theoretical study of the large target limit, i.e. of *nuclear matter*.

(b) The fact that its radial shape is simple suggests that this shape mainly reflects a dependence upon matter density. Consequently, it is feasible to investigate the essential features of the OMP in a finite nucleus by studying nuclear matter at various densities, and by applying a *local density approximation*.

(c) The fact that the OMP depends smoothly on energy indicates that it is possible to construct a unified theoretical approach which would apply at negative, low and intermediate energies. This requires the use as input of a *realistic* nucleon-nucleon interaction, which accounts for free nucleon-nucleon scattering up to several hundreds MeV.

At *negative energy*, the OMP essentially reduces to the shell-model potential, with an additional imaginary part which attaches a finite mean free path to each nucleon. It is thus intimately related to the energies and widths of hole states created in pick-up or in knock-out processes,[6,7] and to the validity of the shell model itself. At *low energy*, the OMP for nucleons plays a prominent part in the interpretation and in the theoretical prediction of most cross sections : it enters in the analysis of elastic and total cross sections; of

direct, precompound and compound processes; of radiative capture; ...
Moreover, it is the building stone for most calculations of the OMP
for composite projectiles. At *intermediate energy*, the OMP remains an
essential tool, but its empirical properties are not yet well-establi-
shed since the new generation of cyclotrons has only recently become
operative. Hence, a theoretical guideline at intermediate energy would
be very helpful; the approach described here is able to fill the missing link between the descriptions of the OMP at low energy via nuclear
reaction theory [8] and at high energy via the multiple scattering
theory.[9,10] The latter uses as input the cross sections for collisions between free nucleons rather than a nucleon-nucleon potential.

The content of Sect. 2 is the following. Many technical difficulties simplify in the large target limit. However, a problem arises
since one can no longer take as initial state an incoming projectile
located at large distance. We briefly describe how an OMP can nevertheless be defined in this case. We outline the interest and the main
limitations of the nuclear matter approach. We also then deal with
approximation schemes for the calculation of the OMP in nuclear matter.
For a realistic nucleon-nucleon interaction, it appears appropriate to
use a low-density expansion,[11,12] in the spirit of the Bethe-Brueckner theory for the binding energy of nuclear matter.[13]

The leading term of this expansion is the Brueckner-Hartree-Fock
approximation, for which we present a few numerical results in Sect. 3
in the case of nuclear matter. We also give estimates of the relative
importance of some higher order terms.

In Sect. 4, we discuss several ways of constructing the OMP in
finite nuclei from the OMP in nuclear matter. We compare their respective merits and drawbacks.

Some prospects of the nuclear matter approach to the nucleon-
nucleus OMP are outlined in Sect. 5.

2. Nuclear matter approach

2.1. Justification and limitations

Accurate fits to the elastic scattering cross sections can be obtained only if the OMP parameters are changed from target to target in
a somewhat erratic way.[14] This reflects the influence of nuclear
structure effects. For instance, it is expected that the strength of
the imaginary part of the OMP is smaller than average for doubly-closed shell nuclei, but there exists only a very dim empirical confirma-
tion of this feature.[15] It is fair to say that one has not yet been

able to establish a direct relationship between the empirical deviations from the average set of OMP parameters and specific nuclear
structure properties. Information on these matters may ultimately emerge from theories of the OMP which explicitly involve a detailed description of the low-lying excited states of the target.[16,17] These
theories are thus complementary to the nuclear matter approach dealt
with here. In its present state, the latter can yield information only
on an *average* set of OMP parameters. Any detailed agreement between
the corresponding theoretical cross section and an experimental cross
section must therefore be considered as *accidental*, or else must involve the adjustment of some parameters. In the latter case, the interpretation of the agreement with experiment of the adjusted OMP requires
extreme   caution since the variability of some quantities may amount
to mock up processes in a manner that corresponds to imposing unphysical constraints on the calculated OMP.[18]

There exist several sets of average OMP parameters.[5,14] This corresponds to the fact that the experimental data seem to be sensitive
chiefly to only a few combinations of these parameters. In particular,
it appears that the scattering cross sections mainly depend on the volume integrals per nucleon (J_V/A and J_W/A) of the real part (V)
and of the imaginary part (W) of the OMP :[19-21]

$$J_V/A = A^{-1} \int d^3r \, V(r) \quad .$$

(2.1)

This is illustrated by the small size of the vertical lines in Figs. 1
and 2, which represent standard deviations from the mean value. Note,
however, that these deviations are significantly larger ($\approx$ 30 %) in
the case of the imaginary part than in the case of the real part ($\approx$
10 %) . This partly reflects the larger sensitivity of W to the target structure, and also to the projectile energy.

The quantities shown in Figs. 1 and 2 are typical examples of observables that the nuclear matter approach should aim at reproducing.
In addition, this theory may also provide valuable information on some
characteristics of the *average* OMP which are difficult or impossible to
isolate in the empirical analyses, e.g. its non-locality, its Coulomb
correction, its symmetry, spin-orbit and spin-spin components, its detailed geometry, or on the average properties of the effective interaction to be used in the study of inelastic nucleon scattering. In its
present state, the nuclear matter approach cannot, however, include the
collectivity effects related to the finiteness of the nucleus and to
spurious center-of-mass motion. Moreover, all detailed calculations to-

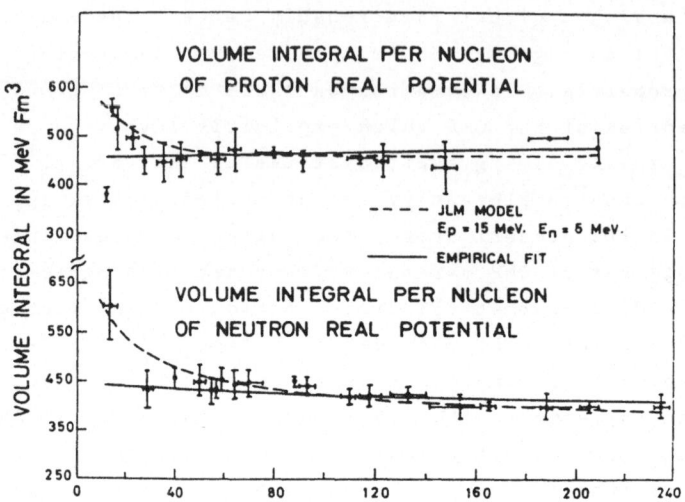

Fig. 1. Dependence on mass number of the volume integral per nucleon of the real part of the OMP, for protons with energy 10 < E < 20 MeV and for neutrons with energy 1 < E < 10 MeV . The dashes represent the theoretical values[22] obtained from the Brueckner-Hartree-Fock approximation. From Ref. 23.

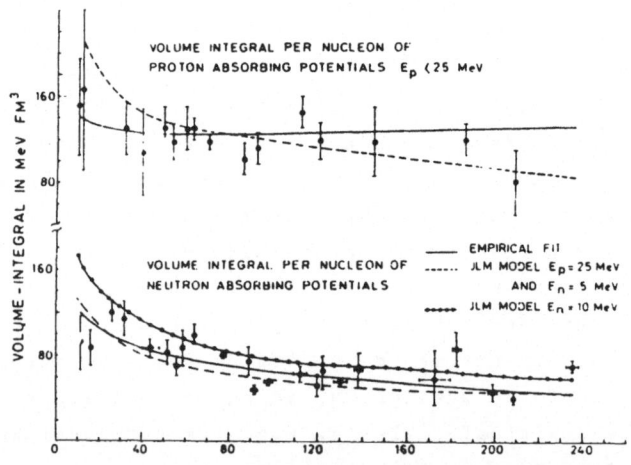

Fig. 2. Same as Fig. 1, for the imaginary part of the OMP. From Ref.24.

date are based on the Brueckner-Hartree-Fock approximation whose accuracy is limited to at best ten per cent (Sect. 4). This reinforces our previous warning that a detailed agreement between individual experimental and theoretical cross sections is bound to be accidental.

2.2. Definition of the generalized optical-model potential

The optical-model wave function $\psi_E(\vec{r})$ should give the probability amplitude of finding at location $\vec{r}$ the incoming nucleon with energy E on top of the target ground state Φ . Hence, our problem consists in studying the propagation of a nucleon inside the nuclear medium, and this can be accomplished with greater ease in an infinite system.

Let Φ denote the normalized ground state wave function of nuclear matter with density ρ . In the free Fermi gas model, Φ is approximated by a Slater determinant of plane waves where all momenta smaller than the Fermi momentum k_F are occupied while those larger than k_F are empty. Henceforth, the letters j , ℓ , m , ... are reserved to momenta smaller than k_F (*hole* states), the letters a , b , c , ... to momenta larger than k_F (*particle* states) and the letter k to an arbitrary momentum.

The true (correlated) ground state of nuclear matter differs from a free Fermi gas. The number of nucleons (with given spin and isospin) with momentum k is plotted in Fig. 3.

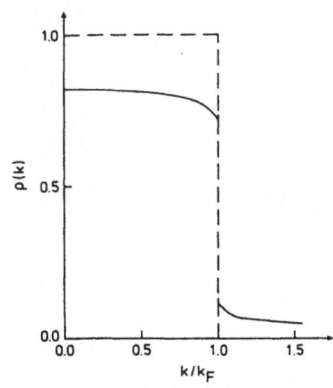

Fig. 3. The full curves represent the momentum distribution in the ground state of nuclear matter. The dashes correspond to the free Fermi gas model. From Ref. 12.

Let us create a nucleon at time t=0 and at location $\vec{r}$ on top of the nuclear matter ground state. The probability amplitude for finding this nucleon at location $\vec{r}'$ and at time t > 0 on top of the ground state is measured by the one-body Green function

$$G(|\vec{r}-\vec{r}'|;t) = -i \langle\Phi| a(\vec{r}',t) a^{\dagger}(\vec{r},0)|\Phi\rangle \quad . \tag{2.2}$$

A Fourier transformation over the variables $|\vec{r}-\vec{r}'|$ and/or t yields the Green functions $G(k,t)$ or $G(k,E)$. We note the correspondence

$$k \leftrightarrow |\vec{r}-\vec{r}'| \quad , \quad E \leftrightarrow t \quad . \tag{2.3}$$

Clearly, the one-body Green function contains the information relevant to the optical model. It can be shown that the *generalized* optical-model potential (Sect. 1) in nuclear matter can be identified with the mass operator

$$M(k;E) = V(k;E) + i W(k;E) \tag{2.4}$$

which is related to the Green function by $(\hbar = 1)$

$$G(k;E) = [E - \frac{k^2}{2m} - M(k;E)]^{-1} \quad . \tag{2.5}$$

A Fourier transformation over the momentum variable k yields the non-local *and* energy-dependent operator

$$M_{\rho}(|\vec{r}-\vec{r}'|;E) \leftrightarrow M_{\rho}(k;E) \quad , \tag{2.6}$$

which as indicated depends on the density ρ.

In a finite system, the mass operator is in general a complicated function of $\vec{r}$ and $\vec{r}'$:

$$M(\vec{r},\vec{r}';E) = V(\vec{r},\vec{r}';E) + i W(\vec{r},\vec{r}';E) \quad . \tag{2.7}$$

When introduced in the Schroedinger equation

$$-\frac{1}{2m} \nabla^2 \psi_E + \int d^3r' M(\vec{r},\vec{r}';E) \psi_E(\vec{r}') = E \psi_E(\vec{r}) \quad , \tag{2.8}$$

it yields the projection $\psi_E(\vec{r})$ of the full many-channel wave function on the target ground state.[2,25]

2.3. Definition of the OMP

The structure of the generalized optical-model potential is very poorly known and is anyhow too complicated to be used in phenomenological analyses. The practical OMP is simpler; usually it is local and energy-dependent, sometimes non-local but energy-independent.

In order to see how to derive a theoretical expression for this OMP, one has to recall that the optical model assumes that the probability amplitude for finding a nucleon with momentum k after a time t decreases *exponentially* with t . Hence, the corresponding Green function is given by

$$G_{O.M.}(k;t) = -i \exp(-i e(k) t) \exp(t/2 \overline{W}(k)) , \qquad (2.9)$$

where $\overline{W}$ (< 0) is the imaginary part of the OMP and e(k) is the projectile energy. A Fourier transformation over t yields

$$G_{O.M.}(k;E) = [E - e(k) - \frac{i}{2} \overline{W}(k)]^{-1} . \qquad (2.10)$$

Equations (2.5) and (2.10) show that the optical model approximates the one-body Green function by its pole approximation. Indeed, one can write Eq. (2.5) in the form

$$G(k;E) = R(k) [E - e(k) - \frac{i}{2} \overline{W}(k)]^{-1} + \text{background} ; \qquad (2.11)$$

the real part of the pole is given by[12]

$$e(k) = \frac{k^2}{2m} + V(k;e(k)) , \qquad (2.12)$$

and its imaginary part by

$$\overline{W}(k) = R(k) W(k;e(k)) . \qquad (2.13)$$

The residue

$$R(k) \simeq (\frac{\overline{m}}{m})^{-1} = \{1 - \frac{\partial V(k,E)}{\partial E}\}^{-1} \qquad (2.14)$$

measures to what extent the optical model is valid, since one would have R(k) = 1 if it would be exact (Eq. (2.10)).

Equation (2.12) shows that one obtains the OMP from the generalized optical-model potential M(k;E) by relating the projectile energy and momentum :

$$E = \frac{k^2}{2m} + V(k;E) . \qquad (2.15)$$

If one uses (2.15) to express k in terms of E , one gets the *energy-dependent local* OMP

$$M(k(E);E) = M_L(E) . \qquad (2.16)$$

If one uses (2.15) to express E in terms of k , one gets the *energy-independent non-local* OMP

$$M(k,e(k)) = M_{NL}(k) \leftrightarrow M_{NL}(|\vec{r}-\vec{r}'|) . \qquad (2.17)$$

Henceforth, we shall often omit the subscripts L and NL , and shall use the argument of M (M(k) or M(E)) to indicate whether we deal with the non-local or with the local forms of the OMP.

We note that the imaginary part of (2.16) and of (2.17) should be multiplied by R(k) (eq. (2.13)); we shall not write this explicitly, unless useful.

2.4. Simple local density approximation

The quantities M(E) and M(k) defined in eqs. (2.16) depend on the density of nuclear matter :

$$M_L^{(\rho)}(E) \quad , \quad M_{NL}^{(\rho)}(k) \quad . \tag{2.18}$$

In a finite nucleus, the density $\rho(r)$ is not uniform. The simplest way of constructing the OMP in a finite nucleus consists in using the following *local density approximation* : At location r the OMP is the same as in an infinite medium whose density would be the local value $\rho(r)$. Equations (2.16) and (2.17) then yield two theoretical approximations for the OMP in a finite nucleus :

$$M_L^{(\rho(r))}(E) = M_L(r;E) \tag{2.19}$$

$$M_{NL}^{(\rho(r))}(|\vec{r}-\vec{r}'|) = M_{NL}(r;|\vec{r}-\vec{r}'|) \quad . \tag{2.20}$$

These OMP are *phase equivalent* in the sense that they yield approximately the same elastic scattering phase shifts. The prescription (2.15)-(2.20) for constructing equivalent potentials is the local energy approximation of Perey and Saxon[26] and of Frahn.[27] Although the wave functions computed from M_L and from M_{NL} are the same at large distance, they differ *inside* the potential well by the correction factor (2.29).[28-30]

2.5. Non-locality and energy dependence

The energy dependence of the real part of the local OMP (2.16) is measured by the *effective mass* m* , which is defined by

$$\frac{m^*}{m} = 1 - \frac{d}{dE} V_L(E) \quad . \tag{2.21}$$

It can easily be checked that an equivalent expression for m* is

$$\frac{m^*}{m} = \{1 + \frac{m}{k} \frac{d}{dk} V_{NL}(k)\}^{-1} \quad . \tag{2.22}$$

Equations (2.21) and (2.22) enable one to find the energy dependence of the local equivalent of a non-local field.

The energy dependence of the real part of the *generalized* optical-model potential $M(k;E)$ measured by the E-mass $\bar{m}$ (see Eq. (2.14))

$$\frac{\bar{m}}{m} = 1 - \frac{\partial}{\partial E} V(k,E) \quad . \tag{2.23}$$

Similarly, the non-locality of the *generalized* optical-model dependence is measured by the k-mass $\tilde{m}$ [12]

$$\frac{\tilde{m}}{m} = \{1 + \frac{m}{k} \frac{\partial}{\partial k} V(k,E)\}^{-1} \quad . \tag{2.24}$$

The quantities m^{*}, $\tilde{m}$ and $\bar{m}$ are related by[12]

$$\frac{m^{*}}{m} = \frac{\tilde{m}}{m} \cdot \frac{\bar{m}}{m} \quad . \tag{2.25}$$

The effective mass m^{*} is the only one which can be obtained from phenomenological analyses, since these involve the OMP (2.19) or (2.20) rather than the generalized optical-model potential. Hence, one must resort to the theory if one wants to disentangle the non-locality from the energy-dependence of this generalized optical-model potential.

Early investigations[31,32] were based on the subtracted dispersion relation

$$V(k,E) = V(k,E_o) + \frac{1}{\pi} (E-E_o) \int dE' \frac{W(k,E')}{(E'-E_o)(E'-E)} \quad . \tag{2.26}$$

They indicated that *at low energy* the E-mass $\bar{m}$ is larger than the bare mass m

$$\frac{\bar{m}}{m} > 1 \quad , \tag{2.27}$$

while the empirical value of m^{*} in the nuclear interior is

$$\frac{m^{*}}{m} \approx 0.68 \quad . \tag{2.28}$$

This shows that at low energy the energy dependence of the OMP mainly reflects the non-locality of the mean field.

The distinction between the non-locality and the energy dependence of the generalized optical-model potential is not only of formal interest. For instance, it is necessary in order to calculate the non-locality correction factor

$$(\frac{\tilde{m}}{m})^{\frac{1}{2}} \tag{2.29}$$

by which the amplitude of the wave function *inside* the non-local potential well is reduced as compared to the wave function computed from the local equivalent potential.

It is also through relation (2.25) that one has been able to demonstrate the existence of a plateau in the energy dependence of $V_L(E)$ for weakly bound states (E small and negative).[12,33] This has a number of interesting implications.[34]

2.6. Approximation schemes in nuclear matter

Early attempts[35] to calculate the generalized optical-model potential $M(k,E)$ had been based on the infinite set of equations which couple the one-, two-, ...-body Green functions. Several versions of this approach exist, which depend on the way of truncating the set of equations.[36] The most accurate version is quite difficult to apply in the case of a realistic nucleon-nucleon interaction.[37-39]

If the nucleon-nucleon potential v would be weak, one could use perturbation theory, whose first order term is the *Hartree-Fock* approximation

$$M^{(HF)}(k) = \sum_{j < k_F} \langle \vec{k}, \vec{j} | v | \vec{k}, \vec{j} \rangle_A \quad , \qquad (2.30)$$

where the index A refers to antisymmetrization. The ket $|\vec{k}, \vec{j}\rangle$ denotes a product of two normalized plane waves with momenta k and j, respectively. The Hartree-Fock approximation is represented by the diagram shown on the left-hand side of Fig. 4. There, a horizontal dashed

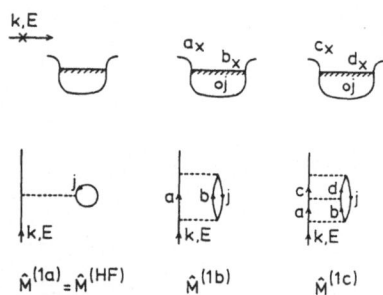

Fig. 4. First three terms of the subseries of the perturbation expansion that is summed in the Brueckner-Hartree-Fock approximation (graph (a) in Fig. 5). From Ref. 8.

line corresponds to the nucleon-nucleon potential v , a downward going line to a hole state and an upward going line to a particle state.

Since a realistic nucleon-nucleon interaction is not weak, one cannot use straightforward perturbation theory. In the Bethe-Brueckner theory for the *binding energy* of nuclear matter, one sums all graphs of the perturbation expansion which have the same number of hole lines. This rearrangement is argued to be appropriate because nuclear matter has a small density.[13] The rate of convergence of this *low-density* or *hole-line* expansion is measured by the depletion of the Fermi sea in the correlated ground state, i.e. by the *smallness parameter*

$$\kappa = 1 - \rho(0.75 \, k_F) \quad . \tag{2.31a}$$

A typical value for κ is (see Fig. 3)

$$\kappa \simeq 0.2 \tag{2.32a}$$

in the case of hard core interactions, and

$$\kappa \simeq 0.15 \tag{2.32b}$$

for soft core potentials. In the case of *binding energy* calculations, we have argued that the self-consistent condition (2.39) given below minimizes the (three-hole line) correction which is proportional to κ , so that the main correction is of order κ^2 (renormalization).

A similar low-density expansion can be derived for the generalized optical-model potential. Its leading term is the Brueckner-Hartree-Fock (BHF) approximation. It gives the sum of the subseries of the one-hole line graphs, of which the first three terms are shown in Fig. 4. On the basis of Eqs. (2.31)-(2.32), one could fear that its accuracy is only about 20 %. However, we shall argue in Sect. 3.3 that *the smallness parameter in that case is given by*

$$\bar{\kappa} = \kappa - 3 \, \kappa^2 \tag{2.31b}$$

rather than by κ . In the examples (2.32a) and (2.32b), one has

$$\bar{\kappa} \simeq 0.08 \quad . \tag{2.32c}$$

Hence, the accuracy of the BHF approximation should be about 10 %. Its expression is given by

$$M^{(BHF)}(k,E) = \sum_{j<k_F} \langle \vec{k}, \vec{j} | g[E+e(j)] | \vec{k}, \vec{j} \rangle_A \quad . \tag{2.33}$$

Here, the operator $g[w]$ is the *reaction matrix*; it is the solution of the following integral equation :

$$g[w] = v + v \sum_{a,b>k_F} \frac{|\vec{a}, \vec{b}\rangle \, \langle \vec{a}, \vec{b}|}{w - e(a) - e(b) + i\delta} \, g[w] \quad . \tag{2.34}$$

By comparing Eqs. (2.30) and (2.33), one sees that $g[E+e(j)]$ plays the role of a complex effective interaction. It is somewhat complicated, since it is a non-local and complex operator. Moreover, it depends on a parameter (w) which in practice is often related to the momenta of the two interacting nucleons, but which can also depend on the momenta of other nucleons (propagation off the energy shell).

The BHF approximation does not include the contributions to the perturbation expansion which are represented by graphs with two- and more hole-lines. Two of these terms are shown in Fig. 5. When combined with other two-hole line diagrams, these give rise to graphs (b) and (c) of Fig. 6 where the BHF approximation is represented by diagram (a). A wiggly line corresponds to a g-matrix.

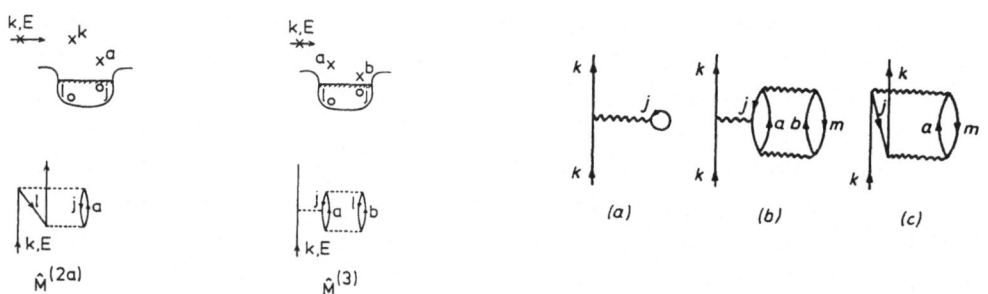

Fig. 5. Two terms of the perturbation expansion which are the progenitors of the graphs (b) and (c) in Fig. 6. Their physical meaning is sketched on the top of the figure. From Ref. 8.

Fig. 6. Graphical representation of a few terms of the low-density expansion for the OMP : Graph (a) represents the Brueckner-Hartree-Fock approximation, graph (b) the renormalization correction and graph (c) the correlation term. From Ref. 40.

It is convenient to write Eq. (2.34) in the abbreviated form

$$ g[w] \ = \ v + v \ \frac{Q}{w - e + i\delta} \ g[w] \quad , \tag{2.35} $$

where the Pauli operator Q allows only *particles* to appear as intermediate states :

$$ Q \ = \ \sum_{a,b>k_F} |\vec{a},\vec{b}> <\vec{a},\vec{b}| \quad . \tag{2.36} $$

When writing Eqs. (2.33) and (2.34), we purposely used the notation (2.12) for the nucleon energies. In principle, however, one could write in (2.33) and (2.34)

$$ e(h) \ = \ \frac{h^2}{2m} + U(h) \quad , \tag{2.37} $$

where U(h) is an arbitrary auxiliary potential. Kidwai and Rook[41,42] used for instance

$$ U(h) \ = \ 0 \quad (all \ h) \tag{2.38} $$

in their evaluation of the BHF approximation. However, a number of arguments[11,12,43] have been given in favour of the choice (2.12) :

$$U(h) = V(h, e(h)) \quad (all \quad h) \quad . \tag{2.39}$$

This is a *self-consistent* condition, since the auxiliary potential U(h) is needed in the calculation of V(k,E) . In practice, the requirement (2.39) is thus quite time-consuming. The justification of the choice (2.39) is firmly established for h smaller than or somewhat larger than k_F , but somewhat less well-founded for large values of h .[43]

3. <u>Numerical results in nuclear matter</u>

3.1. <u>Introduction</u>

Most calculations have been limited to the BHF approximation (2.34). They are surveyed in Sect. 3.2, where we discuss the computational procedures, the real part of the OMP, and its imaginary part. In Sect. 3.3, we discuss the magnitude of the higher order terms represented in Fig. 6 by the graphs (b) and (c). Section 3.4 is devoted to the small components of the OMP, namely its symmetry, Coulomb, spin-orbit and spin-spin components. We draw a few conclusions in Sect. 3.5.

3.2. <u>Brueckner-Hartree-Fock approximation</u>

3.2.1. *Computational procedures*

Accurate and reasonably fast methods have been devised for calculating the BHF approximation (2.33) in the bound state case $E+e(j) \ll e(a) + e(b)$.[13,44-47] Their extension to the scattering problem, $E+e(j) > 2 e(k_F)$, is relatively straightforward, but it is not easy to combine the requirements of accuracy and of rapidity.

Practically all methods share the same starting point : One defines a correlated two-nucleon wave function by the relation

$$v \, \Psi_w(\vec{k}, \vec{j}) = g[w] |\vec{k}, \vec{j}> \quad . \tag{3.1}$$

Three procedures have been used to solve Eq. (2.34).

(a) In the method of *Brueckner and Gammel*,[48] the wave function Ψ_w is expanded in partial waves. The radial components are solutions of the following integral equation, where q and P denote relative and total momenta, while J , ℓ , ℓ' and S are angular momentum quantum numbers :

$$u_{\ell,\ell',w}^{J,S,P}(qr) = j_{\ell,w}(qr)\,\delta_{\ell\ell'} + \sum_{\ell''} \int r'^2\, F_{\ell',w}^{P}(q,r,r')\; v_{\ell'\ell''}^{JS}(r')$$

$$u_{\ell,\ell',w}^{J,S,P}(qr')\,dr' \quad . \tag{3.2}$$

The kernel F must be calculated by performing an integration over $t = |\vec{a}-\vec{b}|$: this is the lenghthiest part of the calculation. Once F is stored, Eq. (3.2) can be solved by discretization and matrix inversion.

The latter operation is relatively easy only if the nucleon-nucleon interaction has a hard core. Even in this case, it is difficult to carry out with speed and accuracy for low partial waves if v(r) is very deep near the hard core radius, as in the Hamada-Johnston potential.[49]

The method of Brueckner and Gammel has been extended to the scattering problem by the Liège group.[50] The accuracy of the program has been tested by comparison with other codes which apply to the case E+e(j) << e(a) + e(b) ,[51] and by computing the free nucleon-nucleon phase shifts. It is somewhat better than 5 % in the case of Reid's hard core interaction,[52] and somewhat worse than 5 % for the low partial waves of the Hamada-Johnston interaction.

(b) *Brieva and Rook*[53] use an extension of the reference spectrum method.[54] They approximate the kernel F by an algebraic expression which enables (3.2) to be solved in a simple way. The form of this algebraic expression is suggested by the two limiting situations E+e(j) << e(a) + e(b) , and Q = 1 , U(h) = 0 . Two parameters appear in the expression and are adjusted in order to obtain a fair (5 %) fit to F(r,r') for $|\vec{r}-\vec{r}'|$ < 1 fm , and for each value of w , P , k_F , q .

The reliability of this method is difficult to evaluate on the basis of the information given in Ref. 53. I believe that it should not be better than 10 %. Indeed, it essentially reduces to the standard reference spectrum method for bound states, and this is known to "give quite wrong results" (Ref. 13, page 112). The reason is that a fit to the kernel F with 5 % accuracy does not yield precise values for u(qr) in Eq. (2.32). Moreover, values of F for $|\vec{r}-\vec{r}'|$ > 1 fm in Eq. (3.2) still contribute appreciably, especially to the imaginary part of u(qr) ; this domain is not well rendered by the algebraic parametrization of F .

Finally, it appears that the expression used in Ref. 53 for the Pauli operator Q is incorrect for $|\vec{a}+\vec{b}|$ > 2 k_F .[50] This may affect the conclusions drawn in Ref. 53 on the importance of large partial waves as well as on the values of the OMP at intermediate energies (and also at low energy via the self-consistent condition (2.39)).

We conclude that it would be useful to make a critical comparison between the results obtained from the methods of Refs. 50 and 53, in the case of Reid's hard core interaction. It would also be of interest to compare the results obtained from the method of Ref. 53 with other procedures in the case of the binding energy of nuclear matter.

One of the advantages of the method of Brieva and Rook[53] is that it applies to soft core as well as to hard core nucleon-nucleon potentials. In the latter case, it offers a slightly[55] better treatment of the hard core contribution that the method of Brueckner and Gammel.

(c) Grangé and Lejeune[56] recently extended to the scattering problem the method of *Haftel and Tabakin*[45,57] which amounts to write Eq. (3.1) in momentum representation and to solve the equation corresponding to (3.2) by matrix inversion. This procedure is quite accurate but is restricted to soft core interactions, such as Reid's.[56]

A new computational method has been developed by Legindgaard[47] for bound states. There one evaluates corrections to the reference spectrum approximation by the inversion of a small matrix and appears quite promising. It should be rather straightforward to extend it to the scattering case.

In conclusion, the accuracy of the available codes for the calculation of the BHF approximation is not better than five or ten per cent. This is an additional (see Sects. 2.1, 2.6) reason for viewing as accidental any *detailed* agreement between theoretical and experimental cross sections.

The numerical results presented in the rest of Sect. 3 have been obtained from the procedure of Brueckner and Gammel, and in the case of Reid's hard core nucleon-nucleon interaction.

3.2.2. *Real part of the OMP*

We showed in Sect. 2.3 that there exist two equivalent forms of the OMP.

The non-local potential $M_{NL}(k)$ is defined by Eqs. (2.12),(2.17) and,in the case of the BHF approximation, by Eq. (2.33). It is represented by the dashes in Fig. 7, for the Fermi momentum $k_F = 1.35 \text{ fm}^{-1}$ which corresponds to the density in the inner region of a finite nucleus. The full curve shows the Gaussian phenomenological parametrization adopted by Perey and Buck[28] in their analysis of elastic scattering data at low energy, namely

$$V_{NL}(|\vec{r}-\vec{r}'|) \quad \propto \quad \exp(- |\vec{r}-\vec{r}'|^2/\beta^2) \quad . \tag{3.3}$$

In Fig. 7, the calculated non-locality range is equal to

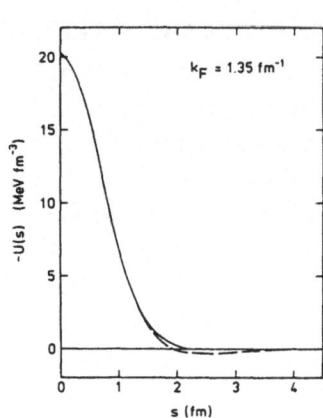

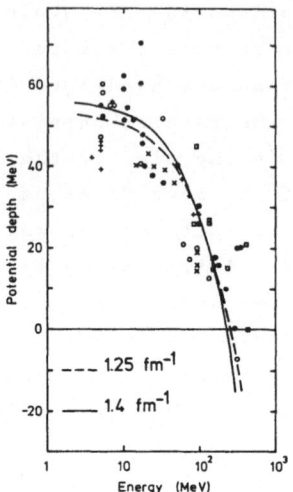

Fig. 7. Dependence on $s = |\vec{r}-\vec{r}'|$ of (minus) the real part of the non-local OMP (Eq. (2.17)). The full curve represents the empirical Gaussian parametrization of Perey and Buck;[28] the dashes show the results of the BHF approximation. From Ref. 12.

Fig. 8. Energy dependence of (minus) the real part of the OMP. The dots, crosses and squares represent empirical depths.[58] The curves show the BHF approximation, for $k_F = 1.40$ fm^{-1} and 1.25 fm^{-1}, respectively. From Ref. 59.

$$\beta = 1.02 \text{ fm} ; \tag{3.4}$$

this is somewhat larger than the phenomenological value 0.85 fm of Perey and Buck.[28] However, (3.4) gives the global non-locality range, *averaged* over *all* momenta. The computed non-locality range at low energy is 0.84 fm . The small deviation between the theoretical curve and the Gaussian approximation for large k partly corresponds to the property that the calculated (and the empirical) potential depth changes sign at high momentum, which is not included in the parametrization (3.3).

This change of sign is apparent in Fig. 8 which shows the energy dependence of $-V_L(E)$. We note that it occurs at a lower energy for $k_F = 1.40$ fm^{-1} than for $k_F = 1.25$ fm^{-1} . It is also apparent that the energy dependence becomes close to logarithmic for E > 100 MeV . This is in keeping with the dispersion relation (2.26), because the empirical $W(k,E')$ approaches a constant for $E' \rightarrow \infty$.[60]

In Fig. 9, the computed depth is compared with phenomenological values for ^{40}Ca and ^{58}Ni . We see that the agreement between the BHF approximation and experiment is significantly better for the self-consistent choice (2.39) for the auxiliary potential U(h) (full curve) than for the choice (2.38) which had been adopted by Rook[42] (dashes).

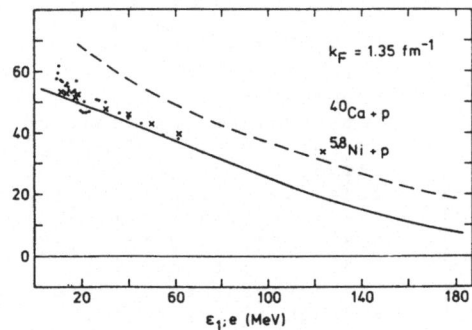

Fig. 9. The dots and crosses represent empirical depths, for protons on ^{40}Ca and ^{58}Ni . The full and the dashed curves show the BHF approximation for the choices (2.39) and (2.38), respectively, of the auxiliary potential U . From Ref. 12.

In principle, this auxiliary potential is an arbitrary parameter and the value of the *sum* of the low-density series should be independent of U(h) . However, the *rate of convergence* of the series depends on U(h) . In the case of *binding energy* calculations, it appears that stability with respect to the choice of U(h) is reached if one includes the three-hole line graphs.[61-63] A similar study has not yet been carried out for the *mass operator*. Various formal arguments[11,12,43] indicate that in this case the rate of convergence of the low-density expansion is considerably worsened if one deviates from the self-consistent choice (2.39). This is supported by the fact that the BHF approximation is very sensitive to the choice of U , and is in much better agreement with the data for the self-consistent choice (Fig. 9). The latter is even more stringent in the case of the imaginary part of the OMP (see Fig. 26 of Ref. 12).

The agreement between empirical and calculated depths shown in Fig. 9 should not be overrated. Indeed, the phenomenological depth depends on the radius assumed for the well. Moreover, higher order corrections to the BHF approximation are certainly not negligible (Sect. 3.3).

3.2.3. *Imaginary part of the OMP*

In Fig. 10, we compare empirical and calculated values of the imaginary part of the OMP. The dots, crosses and squares have been compiled from phenomenological OMP;[58] they represent the value of −W at the center of the nucleus if one assumes pure volume absorption. The dependence upon the assumed radial shape of W(r) is one of the origins of the large scatter of the empirical points. Another major reason is that the imaginary part W is much more sensitive to nuclear structure effects than the real part V of the OMP.

The fact that the curves calculated for three densities intersect indicates that the local density approximation will yield surface ab-

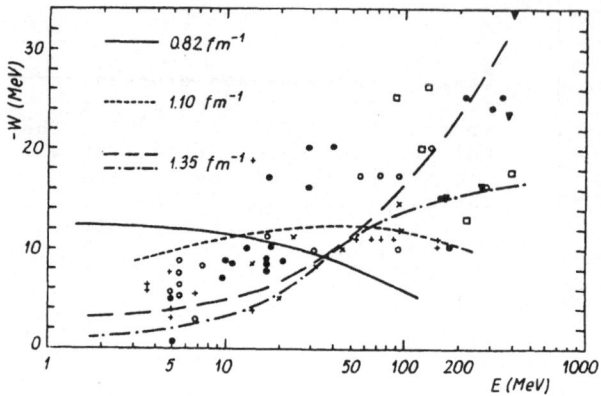

Fig. 10. Dependence on energy of the imaginary part of the OMP. The dots, crosses and squares are empirical depths compiled by Passatore.[58] The full curve, the short and the long dashes represent the BHF approximation for k_F = 0.82 fm^{-1}, 1.10 fm^{-1} and 1.35 fm^{-1}, respectively. The dash-and-dots include a rough estimate of the graphs (b) and (c) of Fig. 6. From Ref. 40.

sorption below E   50 MeV and volume absorption above that energy.

Finally, we mention that the curves shown in Fig. 10 include the factor R(k) (Eqs. (2.13),(2.14)) which has apparently been omitted by Brieva and Rook.[53,64] This factor is smaller than unity (Fig. 11).

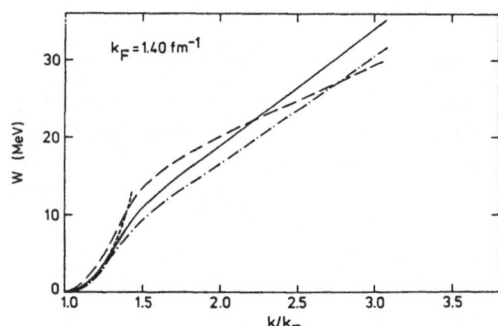

Fig. 11. Momentum dependence of minus the imaginary part of the OMP, for k_F = 1.4 fm^{-1}. The full curve shows the quantity $W(k; e(k))$, and the dash-and-dots the quantity $\overline{W}(k)$ (Eq. (2.12)). The long dashes include an estimate of graph (c) of Fig. 6. The short dashes represent the asymptotic behaviour $W(k) \propto (k - k_F)^2$. From Ref. 65.

Its omission is not inconsistent with the accuracy of the BHF approximation, but adds to our repeated warning that the overall accuracy cannot be claimed to be better than about 10 %, especially in the case of the imaginary part W .

3.2.4. Negative energies

The smooth behaviour of $V_L(E)$ when E varies from negative (bound states) to positive (scattering states) energies is nicely illustrated by Fig. 12. There, the depth $V_L(E_j)$ for $-60 < E_j < 130$ MeV was fitted to the observed energy $E_j < 0$ of the bound single-particle state or to the elastic scattering cross sections at the energy $E_j > 0$, for N=Z nuclei with $12 \leqslant A \leqslant 40$. However, a closer look at this figure suggests the existence of a plateau near the Fermi energy : the average depth seems to be almost independent of E for

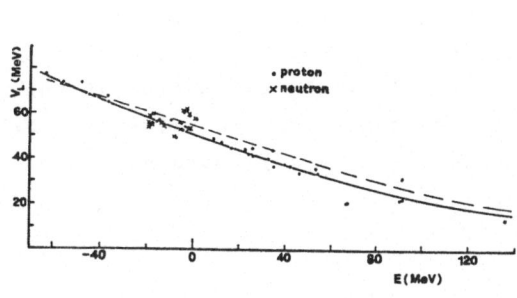

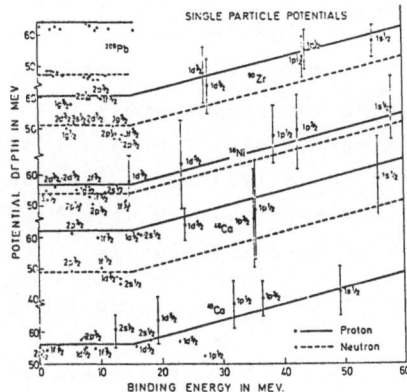

Fig. 12. The dots and the crosses represent empirical depths for N=Z nuclei with A<40. The full curve is
drawn through these points. The long
dashes show the BHF approximation.
From Refs. 8, 66.

Fig. 13. Dependence on the binding energy (-E) of the depth
of the OMP that fits the observed energies (dots, crosses) of
bound single-particle states.
From Bear and Hodgson.[67]

-20 < E < 0 MeV . This plateau is exhibited in Fig. 13, taken from the
recent analysis by Bear and Hodgson[67] for 40 ⩽ A ⩽ 90 .

It is in nice agreement with the Brueckner-Hartree-Fock approximation. Indeed, detailed calculations[12,33] show that the effective
mass m^{*} has a narrow maximum in the vicinity of the Fermi surface
(Fig. 14). Equation (2.21) shows that this corresponds to a plateau in

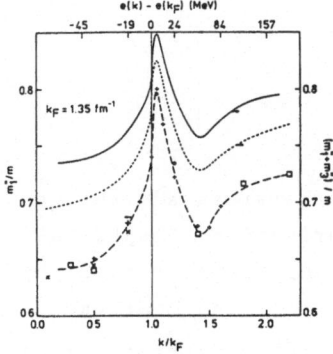

Fig. 14. The long dashes represent the
dependence on momentum (lower scale) and
on energy (upper scale) of the effective
mass m^{*}/m in the BHF approximation.
From Ref. 12.

the energy dependence of $V_{L}(E)$. This plateau is even more pronounced
at lower density.[12] The analysis shows that this phenomenon is entirely due to the E-mass $\bar{m}$ (Eq. (2.23)), i.e. to the true dependence
of the non-local *generalized* optical-model potential. This has a number of implications, as we discuss in Ref. 34.

3.3. Higher order corrections

We first give a simple estimate which illustrates our previous warning concerning the limited accuracy of the BHF approximation. If this approximation would be exact, it should yield saturation at the value $k_F = 1.35$ fm^{-1} for the Fermi momentum, with the value -16 MeV for the Fermi energy $e(k_F)$.[68] The value of $e(k_F)$ obtained from the BHF approximation is [12]

$$e^{(BHF)}(k_F = 1.35 \text{ fm}^{-1}) = -24.5 \text{ MeV} \quad , \tag{3.5}$$

which is 8.5 MeV too small (Fig. 15). Thus, the potential energy

$$U^{(BHF)}(k_F) = -62.3 \text{ MeV}$$

is too deep by 15 % at the Fermi surface. The size of this error is in keeping with the value (2.32a) of the smallness parameter.

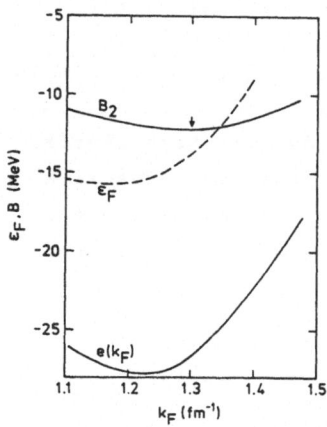

Fig. 15. Density dependence of the average binding energy per nucleon B_2 in the BHF approximation, and of the corresponding Fermi energy $e(k_F)$. The dashes include the effect of the renormalization graph (b) of Fig. 6. From Ref. 12.

If one aims at a better accuracy, one must therefore add the contribution of higher order terms to the BHF approximation. Graphs (b) and (c) of Fig. 6 are of particular interest because they carry a physical meaning. Diagram (b) accounts for the fact that the momentum state $j < k_F$ is partly empty in the *correlated* ground state. From Eq. (2.31a) we find that graph (b) is approximately equal to

$$M^{(b)}(k,E) \simeq -\kappa M_1(k,E) \quad . \tag{3.6}$$

However, this contribution should also be included in the self-consistent condition (2.39). In the case of the *real* part of the OMP, this reduces the correction due to graph (b) and yields approximately (see below)

$$V^{(BHF)}(k,E) + V^{(b)}(k,E) \simeq (1 + \kappa)^{-1} V^{(BHF)}(k,E) \quad . \tag{3.7a}$$

The dashes in Fig. 15 represent the corresponding Fermi energy

$$\varepsilon_F = \frac{k_F^2}{2m} + (1 + \kappa)^{-1} V^{(BHF)}(k_F, \varepsilon_F) \quad . \tag{3.7b}$$

We see that it is approximately 10 MeV *larger* than the BHF value e(k_F) .

Graph (c) of Fig. 6 expresses the fact that the presence of the projectile nucleon k "blocks" the admixture of the two particle-two hole line configurations (ak , $j^{-1}m^{-1}$) in the *correlated* ground state whence the name "correlation graph" given to (c). Its contribution has not yet been evaluated accurately for realistic interactions. Curve 3 of Fig. 16 shows the value obtained by Sartor[70] for a semi-realistic interaction. It is in fair agreement with earlier and cruder estimates.[71-73] We see that this contribution is large for k < k_F but almost negligible for k > k_F . When added to the BHF value, it yields the short dashes in Fig. 17. However, the self-consistency condition (2.39) should also include this contribution; this brings the BHF approximation down to the long dashes. Figure 17 shows that the OMP at *positive* energy is hardly affected by the correlation graph (c) of Fig. 6.

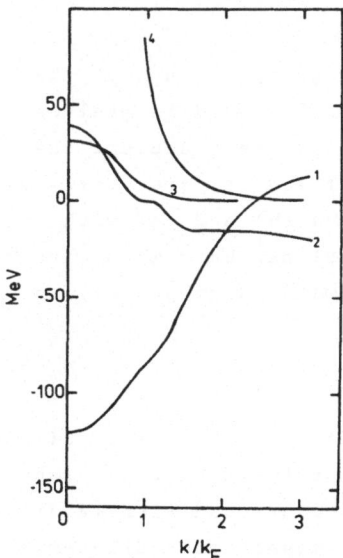

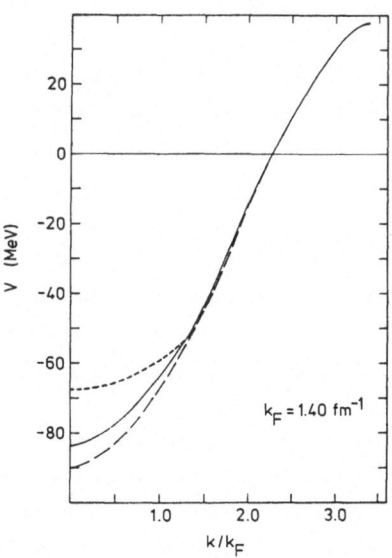

Fig. 16. Curve 1 represents the real part of the OMP in the BHF approximation for the interaction of Hamman and Ho-Kim.[69] Curve 3 shows the value of the core polarization graph (c); curve 4 is 10 times curve 3. Curve 2 is the imaginary part of the OMP; it arises from $M^{(c)}$ for k< k_F and from $M^{(BHF)} = M^{(a)}$ for k>k_F. From Ref. 70.

Fig. 17. The full curve shows the real part of the OMP in the BHF approximation (graph (a) of Fig. 6).The short dashes are obtained by adding graph (c) of Fig. 6. The long dashes show the BHF value when graph (c) is added to graph (a) in the self-consistent condition (2.39). From Ref. 65.

The contribution of the renormalization graph (b) of Fig. 6 is more worrisome in view of Eqs. (3.6) and (2.32a),(2.32b). These show that the sum (a) + (b) is approximately 15 or 20 % smaller, in absolute value, than the BHF approximation. This is confirmed by Fig. 18, where the dash-and-dots represent the contribution of the sum of graphs (a), (b) and (c) of Fig. 6 to the depth of the OMP in the inner region of the nucleus. A twenty per cent error is about two times larger than the 10 % accuracy at which we aim.

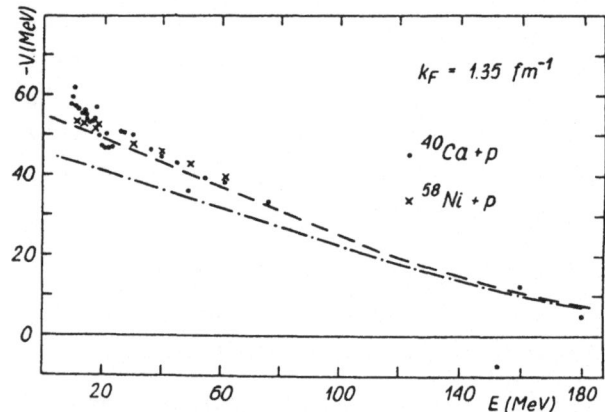

Fig. 18. The dots and crosses are the same as in Fig. 9, and the dashes reproduce the BHF approximation (full line in Fig. 9). The dash-and-dots represent the sum of the real parts of graphs (a), (b) and (c) of Fig. 6. From Ref. 40.

We believe that the situation is not so grim, because of the following main reason. Graphs (b) and (c) of Fig. 6 should themselves be renormalized.[7] In other words, one should take into account the fact that the hole lines j and m in graphs (b) and (c) are partly empty. This renormalization amounts to combine graphs (b) and (c) with three- and four-hole line graphs, just as (a) and (b) can be combined to renormalize (a). Hence, the value of the *renormalized* graphs (b) and (c) is

$$M_R^{(b)} + M_R^{(c)} \simeq (1 - \kappa)^2 [M^{(b)} + M^{(c)}] \quad , \qquad (3.8)$$

where the index R on the left-hand side refers to the renormalization.

In addition, one should take into account the fact that graphs (b) and (c) should be included in the self-consistent condition (2.39). In the case of the *real* part of $M^{(BHF)}$, this amounts to adding $-\kappa [V^{(b)} + V^{(c)}]$ to $M^{(a)}$, whence the factor $(1 + \kappa)^{-1}$ instead of $(1 - \kappa)$ in Eq. (3.7a).

Gathering these results, we find the following result for the sum of graphs (a) and (b) of Fig. 6, including renormalization and self-consistency effects :

$$V^{(a)}(k,E) + V_R^{(b)}(k,E) \simeq [1 + \kappa (1 - \kappa)^2]^{-1} V^{(a)}(k,E) \quad . \qquad (3.9a)$$

Using the parameter $\bar{\kappa}$ of Eq. (2.31b), we can write

$$V^{(a)}(k,E) + V_R^{(b)}(k,E) \simeq (1 - \bar{\kappa})\, V^{(a)}(k,E) \quad . \qquad (3.9b)$$

We conclude that the effect of renormalization and of core polarization is repulsive and is roughly equal to 8 % of the real part of the BHF approximation. It decreases with decreasing density, so that the modification of the volume integral per nucleon (2.1) should even be somewhat smaller.[40]

We now turn to the imaginary part of the OMP. The effect of graph (c) is only indirect, via the self-consistent condition (2.39) (see long dashes in Fig. 11). That of graph (b) can be deduced from Eq. (3.6). The dash-and-dots in Fig. 10 show a rough estimate of the sum $W^{(a)} + W^{(b)} + W^{(c)}$, including self-consistency effects but not taking into account the renormalization of graphs (b) and (c). The latter should reduce the difference between the long dashes and the dash-and-dots in Fig. 10 by a factor

$$(1 - \kappa)^2 \simeq 0.6 \quad . \qquad (3.10)$$

Graph (c) of Fig. 6 gives the leading contribution to the imaginary part of the OMP below the Fermi momentum. It is represented by curve 2 in Fig. 16, and is related to the spreading width of deeply bound single-particle states.[70,74-76]

3.4. Small components

In my opinion, one of the main *practical* interests of a calculable theory of the OMP is that it yields information on properties which are difficult to study experimentally, such as for instance the non-locality and the small components. Among the latter, the spin-spin, the symmetry and Coulomb components have a meaning in nuclear matter; the spin-orbit component has been studied by Brieva and Rook.[77] The spin-spin component has been evaluated in Ref. 78; its relationship with experiment is difficult to establish.

The symmetry component $U_1(N-Z)/A$ has been studied in detail in Ref. 79. There, two effects have been included which are omitted in Ref. 64 and which partly cancel one another :

(a) The non-locality of the mean field leads to a sizeable decrease of the calculated value.[80]

(b) The dependence on neutron excess of the Pauli operator Q (Eq. (2.36)). This enlarges the calculated symmetry term.

The resulting value of U_1 is shown in Fig. 19, in the BHF ap-

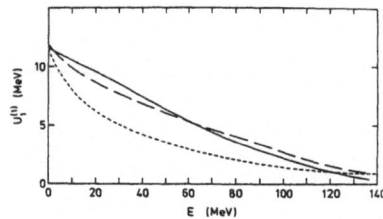

Fig. 19. Dependence on energy of the real part of the symmetry component of the OMP, for k_F = 1.35 fm^{-1} (full curve), 1.10 fm^{-1} (long dashes) and 0.82 fm^{-1} (short dashes), respectively. From Ref. 79.

proximation. It is in fair agreement with the empirical value ($\simeq$ 14 MeV) determined from low-energy *neutron* scattering; I believe that this agreement would be further improved if one would give a better treatment of the non-locality correction (a). It is significantly smaller than the value ($U_1 \simeq$ 24 MeV) usually determined from *proton* scattering. However, it was shown in Ref. 79 that the latter discrepancy can be explained in terms of the Coulomb correction. The latter arises from the non-locality of the OMP. The calculation of Ref. 79 shows that this Coulomb correction is underestimated by about 20 %. This appears to be corroborated by recent experiments.[81]

3.5. Conclusions

The accuracy of the BHF approximation to the OMP in nuclear matter is limited to about 10 %, because of purely computational problems and because of the higher order terms in the low-density expansion. In particular, diagrams (b) and (c) of Fig. 6 reduce the absolute value of the real and of the imaginary parts by 10 %. However, no estimate exists for the size of the other two-hole graphs.[11]

4. Numerical results in finite nuclei

4.1. Introduction

The construction of the OMP in a finite nucleus from the nuclear matter results essentially amounts to using one or another version of the local density approximation (LDA) for computing a complex effective interaction between nucleons.[22,64] This raises a number of problems, some of which are examined in the next sections. Here, we want to emphasize a few general points.

(a) Some *static* effects of the non-uniformity of $\rho(r)$ can be taken into account. However, one has not included in the nuclear matter approach the *dynamical* effects associated with surface vibrations.

(b) The LDA disregards *center-of-mass corrections*. In particular, the corresponding imaginary part includes the spurious contribution of a fictitious channel where the center-of-mass of the target is excited.[82] According to Hughes, Fallieros and Goulard,[83] this may lead to a large (≃ 60 %) overestimate of |W| at *low* energy and for *light* nuclei.

(c) The OMP is sensitive to shell effects.[84] These are not contained in the LDA. It was emphasized in Sect. 1 that the nuclear matter approach can only yield information on the OMP *averaged* over energy and mass number. In my opinion, any detailed comparison between one specific experimental cross section and a theoretical prediction is of limited interest.

(d) It is sometimes argued that the LDA is justified only at high energy, when the wave length of the projectile is short compared to the domain over which the density changes appreciably. We believe that this view is too pessimistic. Rather, it appears plausible that the validity of the LDA hinges upon the property that the *density dependence* of the effective interaction $\hat{v}(\rho)$ is small.

More specifically, $\hat{v}(\rho)$ should vary little over a distance equal to the range t of the effective interaction :

$$\left| t \, \frac{d\rho}{dr} \, \frac{d}{d\rho} \, [\hat{v}(\rho)] \right| \ll \left| \hat{v}(\rho) \right| \quad . \tag{4.1}$$

This condition is most stringent at the half-density radius. In Ref. 22, the corresponding limit for the range t was found equal to

$$t \ll 2.4 \text{ fm} \quad . \tag{4.2}$$

The specific limit (4.2) corresponds to the LDA described in Sect. 4.3. We believe that the condition (4.1) is less stringent if one uses the LDA of Eisen, Day, Brieva and Rook[64,85] (Sect. 4.4), because the density dependence of the corresponding effective interaction is weaker. In any case, it appears dangerous to attach any significance to the outcome of the nuclear matter approach if it displays wiggles in the radial form factor, unless these wiggles can be ascribed a sound physical origin. We return to this in Sect. 4.4.

4.2. Simple_local_density_approximation

In Eqs. (2.19) and (2.20), we defined the "simple" LDA. It is assumed that the OMP at a given location r in the nucleus, where the density is equal to $\rho(r)$, takes the same value as in a uniform medium with that local value of the density. While this simple LDA is crude, it is able to yield semi-quantitative conclusions on the global properties of the OMP : depth, energy dependence, non-locality, small components, main features of the form factors, etc. Its main drawback can easily be corrected (Sect. 4.3).

The top of Fig. 20 shows the matter density distribution in ^{208}Pb.

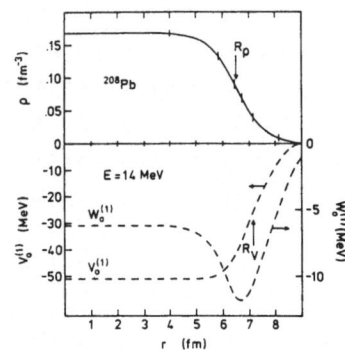

Fig. 20. Matter density distribution in ^{208}Pb (top) and values of the real and of the imaginary parts of the OMP at 14 MeV (bottom), as computed from the BHF approximation and from the simple LDA of Sect. 4.2. From Ref. 79.

Here and below, the Fermi distribution given by Negele[86] was used. The short vertical straight lines indicate the densities at which the OMP was computed by the Liège group.[22,79] The corresponding values of the OMP shown at the bottom of Fig. 20 are obtained in the framework of the BHF approximation and of the simple LDA. We see that the real part of the OMP has a Woods-Saxon shape. In contrast, the imaginary part is surface peaked.

One expects from Fig. 10 that the imaginary part changes from surface to volume at about 50 MeV. This is confirmed in Fig. 21, in the example A = 170. The corresponding real part of the OMP is shown in Fig. 22, which exhibits a remarkable feature : the change of sign

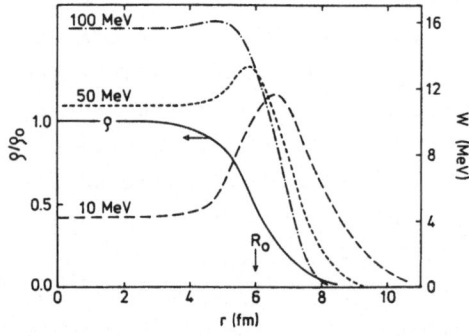

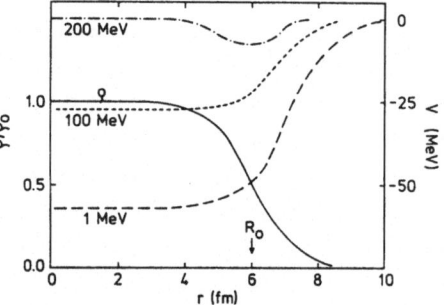

Fig. 21. The full curve shows the matter density distribution for A = 170; ρ_0 denotes the central density. The long dashes, the short dashes and the dash-and-dots represent (minus) the imaginary part of the OMP, at 10, 50 and 100 MeV, respectively, as calculated in the framework of the BHF approximation and of the simple LDA of Sect. 4.2.

Fig. 22. Analogous to Fig. 21, in this case for the real part of the OMP at 1 MeV (long dashes), 100 MeV (short dashes) and 200 MeV (dash-and-dots).

occurs at a lower energy in the inner region than in the outer region of the nucleus.[8,12,40,87,88] This is a consequence of the intersect of the two theoretical curves in Fig. 8. The existence of this wine-bottle bottom shape is supported by the empirical analysis of total and elastic cross sections.[89] This phenomenon deserves scrutiny in view of the recent availability of intermediate energy proton beams. It is therefore presently being studied in detail by Lejeune and my-self.[90] In particular, we included higher order partial waves and fully confirmed its existence. It can be understood on the basis on rather general arguments based on the comparison between the BHF and the impulse approximations.[34,90]

We insisted in Sect. 2.1 on the fact that the experimental data are sensitive to the volume integral per nucleon of the OMP, rather than to the individual parameters. Hence, a comparison between theoretical and empirical OMP should primarily bear on this quantity. Figures 1 and 2 show that the BHF approximation and the simple LDA are quite successful.

One could also investigate more detailed properties of the OMP. For instance, it appears that the scattering cross sections are sensitive to the root mean square radius of the real and of the imaginary parts of the OMP. In this respect, the simple LDA fails : it systematically underestimates the experimental value by as much as 1 fm . *Examples* are shown in Figs. 23-26 where the full lines correspond to the simple LDA.

Let us first discuss the real part of the OMP (Figs. 23 and 25). The theoretical volume integrals are in very good agreement with the empirical values. This is also the case for the improved LDA described in Sect. 4.3 : the corresponding J_V/A cannot be graphically distinguished from the full curves.

In contrast, the values of J_V/A computed by Brieva and Rook[64] in the framework of the simple LDA discussed here, or of their folded effective interaction procedure (Sect. 4.4) falls significantly below the results of the Liège group (and below the empirical values). It would be surprizing if this discrepancy were due to the fact that these authors adopted the Hamada-Johnston rather than Reid's interaction (and more partial waves, but this works in the other direction). Rather, I believe that this discrepancy reflects an inaccuracy in their numerical code (Sect. 3.2.1), or the fact that they seem to have used as input a *theoretical* matter distribution density calculated from a Hartree-Fock model,[86] without corrections for center-of-mass motion and renormalization. These theoretical densities are not quite realis-

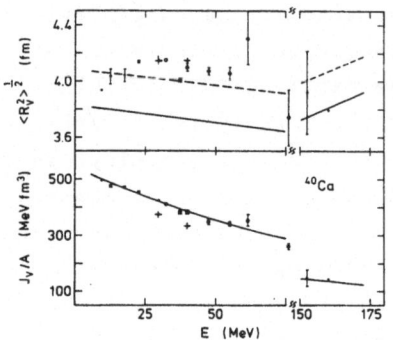

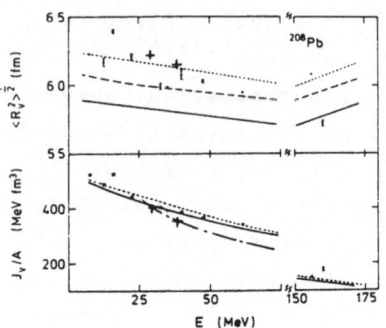

Fig. 23. Root mean square radius (upper part) and volume integral per nucleon (lower part) of the real part of the OMP for proton scattering by ^{40}Ca . The dots are empirical values. The full curve corresponds to the simple LDA of Sect. 4.2, the long dashes to the improved LDA of Sect. 4.3. The crosses are theoretical values obtained by Brieva and Rook[64] from their folded effective interaction. Adapted from Ref. 22.

Fig. 25. Same as Fig. 23, for ^{208}Pb . In addition, the short dashes correspond to the assumption that the root mean square radius of the neutron distribution exceeds that of the proton distribution by 0.23 fm . The dash-and-dots interpolate between the values computed by Brieva and Rook at 30.3 and 40 MeV (crosses) and at 100.4 MeV . Adapted from Ref. 22.

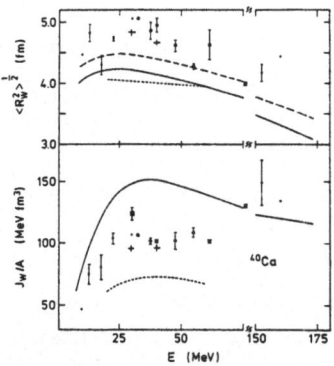

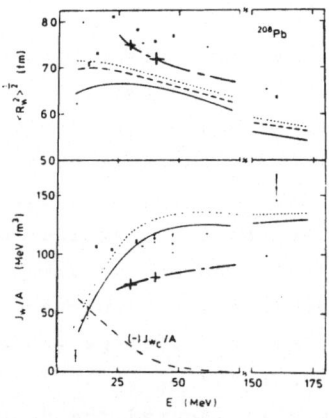

Fig. 24. Same as Fig. 23, for the imaginary part of the OMP. In addition, the short dashes represent the theoretical values computed by N. Vinh Mau and Bouyssy[17] from the random phase approximation. Adapted from Ref. 22.

Fig. 26. Same as Fig. 25, for the imaginary part of the OMP. Adapted from Ref. 22.

tic (see Ref. 86 and Sect. 4.4). The discrepancy could also arise from the fact that Brieva and Rook's expression involves cancellations between large terms with opposite signs, and is thus sensitive to numerical inaccuracies.

The volume integral J_W of the calculated[22] imaginary part of the OMP is in good agreement with the empirical values for heavy nuclei, see an example in Fig. 26. In contrast, it is significantly too large for light nuclei, see an example in Fig. 24. Here also, we note that the theoretical numbers obtained by Brieva and Rook[64] are considerably smaller than those of the Liège group. I believe that the explanation of this discrepancy lies in the reasons enumerated above. In particular, it seems that the simple LDA used by Brieva and Rook should be identical to the one given here only inasmuch as they would perform no average over angular momentum, or over the momentum $\vec{j}$ of the target nucleon. The delicate balance between the exchange and the direct contributions in their expression for the OMP may strongly enhance the inaccuracy necessarily implied by these averages (Sect. 4.4).

The fact that our theoretical volume integral J_W is too large for the doubly magic nuclei leads us to critically examine the following drawbacks of the nuclear matter approach :

(a) The BHF value is an overestimate (see Fig. 10). This cannot explain why the discrepancy is largest for light nuclei.

(b) The dynamical finiteness effects are omitted. For instance, the theory does not include vibrational excited states of the target. In fact, this omission would probably lead to an underestimate of the empirical value.

(c) In our version of the low-density expansion, there exists *no gap* between single-particle energies of the occupied (hole) and of the unoccupied (particle) orbits (Eq. (2.39)). This version probably overcompensates the omission mentioned under (c) that it is supposed to mock up.[12] This error is particularly true at low density, and is thus more important in light nuclei. However, it is difficult to conceive that it is still important at several tens MeV.

(d) The spurious channel corresponding to the center-of-mass motion is implicitly included in the nuclear matter approach (Sect. 4.1). This is a plausible explanation for the too large value of the calculated J_W .

(e) The empirical value of J_W is smaller than average for doubly-closed shell nuclei.[15,91] This is a natural consequence of the smaller density of particle-hole excited states. This should be particularly true for light nuclei. Hence, it is not surprizing that the theory yields too large values of J_W/A for these nuclei. This explanation would also be compatible with the fact that the BHF approximation works well on the average (Fig. 2).

4.3. Improved local density approximation

The main failure of the approximation described in Sect. 4.2 resides in the fact that it yields too small root-mean square radii. It can be seen that this is a predictable consequence of the crudeness of the simple LDA.[22,34]

Let us for simplicity consider the Hartree-Fock approximation for a central effective nucleon-nucleon interaction $\hat{v}$. In nuclear matter, the real part of the OMP then reads

$$M_\rho^{(HF)}(k) = \rho \int \hat{v}(s) \; [1 - \frac{1}{4} j_o(k_\rho s) \; S_\rho(s)] \; d^3s \quad , \tag{4.3}$$

where

$$S_\rho(s) = 3(k_F s)^{-3} \; [\sin(k_F s) - k_F s \; \cos(k_F s)] \tag{4.4}$$

is the Slater function. The simple LDA of Sect. 4.2 yields

$$M^{(LDA)}(r,k) = \rho(r) \int \hat{v}(s) \; [1 - \frac{1}{4} j_o(k_{\rho(r)} s) \; S_{\rho(r)}(s)] \; d^3s \quad . \tag{4.5}$$

In a finite nucleus, the Hartree-Fock approximation is given by
$(\vec{s} = \vec{r} - \vec{r}')$

$$M^{(HF)}(r,k) = \int \rho(r') \; \hat{v}(s) \; [1 - \frac{1}{4} j_o(k_{\rho(r)} s) \; S_{\rho(r)}(s)] \; d^3s \quad , \tag{4.6}$$

provided that one uses the Slater and the local momentum approximations.

Comparing Eqs. (4.5) and (4.6), we see that they are identical only for a *zero-range* effective interaction. Hence, the simple LDA of Sect. 4.3. misses the effect of the *range* of the effective interaction on the value of the real part of the OMP in a *finite* nucleus.

In Ref. 22, a range t was introduced in a phenomenological way by assuming a Gaussian form factor :

$$[1 - \frac{1}{4} j_o(k_\rho s) \; S_\rho(s)] \; \hat{v}(s) = \hat{v}_o (t \; \sqrt{\pi})^{-3} \; \exp(- s^2/t^2) \quad . \tag{4.7}$$

Then, Eq. (4.3) gives

$$\hat{v}_o = M_\rho(k)/\rho = M^{(LDA)}(r',k)/\rho(r') \quad . \tag{4.8}$$

Equations (4.6),(4.7) and (4.8) suggest the following *improved* LDA :

$$M(r,k) = (t \; \sqrt{\pi})^{-3} \int M^{(LDA)}(r',k) \; \exp(- |\vec{r} - \vec{r}'|^2/t^2) \; d^3r' \quad . \tag{4.9}$$

The long dashes in Figs. 23-26 are obtained from Eq. (4.9), with the *fitted* value t = 1.2 fm for the range of the effective interaction (4.7). We see that the agreement with the empirical root mean square radii is now greatly improved, without affecting the volume integrals per nucleon.

Theoretical elastic scattering cross sections associated with this improved LDA have been computed by Lejeune and Hodgson.[92] Since the nuclear matter approach can be meaningful only for the *average* OMP, these authors appropriately studied a wide range of energies and nuclei. Examples are shown by the full curves in Fig. 27.

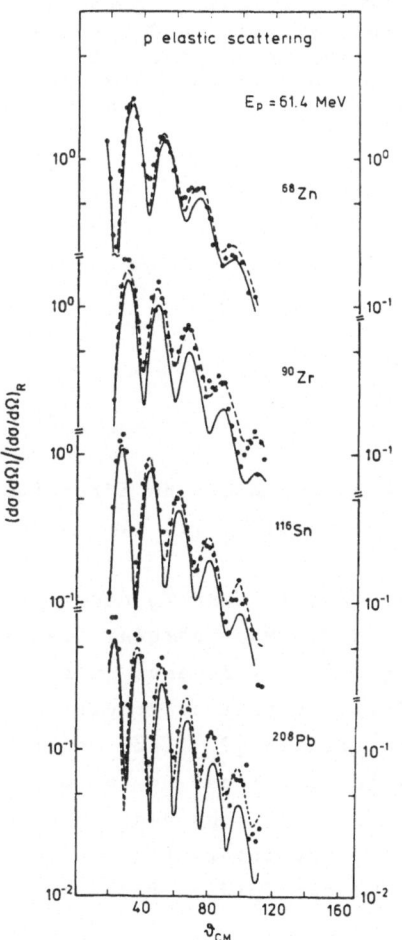

In order to gain a feeling on the accuracy of the BHF approximation and of the improved LDA, Lejeune and Hodgson multiplied the strengths of the real and of the imaginary parts of the OMP by adjustable factors R_V and R_W . This yields the dashed curves in Fig. 27. The fitted values of R_V range from 0.93 (30 MeV protons on ^{12}C) to 1.12 (61 MeV protons on ^{208}Pb) , which is the expected uncertainty of the BHF approximation. Except for 5 MeV neutrons on ^{40}Ca (R_W = 0.38) , the values of R_W range from 0.60 (30 MeV protons on ^{12}C) to 1.02 (61 MeV protons on ^{68}Zn) . These results illustrate the fact that the calculated imaginary part of the OMP is approximately 20 % too large for medium-weight and heavy closed shell nuclei, and 30 to 40 % too large for light closed shell nuclei.

As expected from the discussion in Sect. 4.2, Brieva and Rook[64] find that the strength of their computed imaginary part is too *small* by 54 % in the case of 30 MeV protons on ^{40}Ca , while Lejeune and Hodgson[92] find that it is too *large* by 28 %.

Fig. 27. Cross sections for the scattering of 61.4 MeV protons. The full curves are computed from the BHF approximation and the improved LDA of Sect. 4.3. The dashes are obtained by multiplying the strength of the OMP by renormalization factors. From Ref. 92.

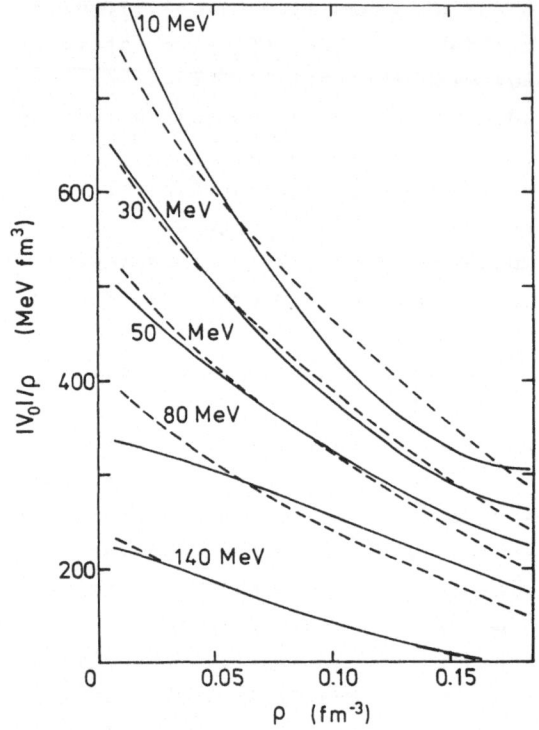

Fig. 28. The full curves represent the density dependence
of the strength (4.8) of the
effective interaction; the
dashes correspond to the parametrization (4.11). From
Ref. 22.

The improved LDA amounts to defining the strength of the effective
interaction by Eq. (4.8), and to attach a range to it by Eq. (4.7). We
note three main points :

(a) *The strength of the effective interaction is density-dependent.*
This dependence is exhibited in Fig. 28 and has *two main origins.* The
first one is that the reaction matrix g (Eq. (2.34)) depends on the
density, via the Pauli operator Q and the self-consistent condition
(2.39). The second one is that the effective strength $\hat{v}_o$ incorporates
the effect of *exchange* : the Slater function and the local momentum
k_ρ are both functions of the density. In other words, the density
dependence shown in Fig. 28 is that of an effective interaction which
should be used in the calculation of the OMP by the folding formula

$$V(r) \quad = \quad \int \rho(r') \; \hat{v}(|\vec{r}-\vec{r}'|) \; d^3r' \quad , \tag{4.10}$$

without including explicitly the exchange term.

This should be kept in mind for instance when using a density-
dependent interaction for computing the OMP between heavy ions[85,93,94)]
or for nucleon inelastic scattering,[95,96)] or when comparing this density dependence with that of an effective interaction used in the cal-
·culation of a nucleon OMP by a folding procedure with exchange.[64,97,98)]

In particular, the density dependence of the latter should be weaker than that shown in Fig. 28 and which roughly corresponds to (dashes in Fig. 28, ρ in fm^{-3})

$$\hat{v}_o = F(E) \ (1 - 2 \ \rho^{2/3}) \quad . \tag{4.11}$$

Hence, we expect that the condition (4.1), which must be fulfilled for the LDA to be applicable, is less stringent than (4.2) suggests if exchange effects are included explicitly.

(b) *The strength of the effective interaction is energy-dependent.* Part of this energy dependence originates from the exchange contribution. Part of it is due to the energy dependence and to the non-locality of the reaction matrix itself, which renders the direct Hartree term energy dependent and non-local.

No problem arises when using an energy-dependent effective interaction in optical-model calculations. However, it would be very difficult to adopt it in a fully self-consistent calculation, where the density distribution $\rho(r)$ would be computed from $\hat{v}(r)$ rather than used as input.[99] Indeed, the corresponding bound single-particle states are not orthogonal, except in light nuclei where orthogonality is provided by angular momentum quantum numbers. It would be of interest to study to what extent the energy dependence of the effective interaction could be replaced by a non-locality.[98]

(c) *The strength of the effective interaction is complex.* This should be the case if the effective interaction is used in a folding calculation of the complex OMP. The folding procedure is based on the analogy with the Hartree-Fock approximation. Hence, the justification of using a complex effective interaction in a folding formula apparently requires that one should be able to write the imaginary part of the OMP in the form (4.10). This may perhaps be achieved by using the unitarity-like relation[11,100]

$$W(k,E) = -\pi \sum_{j<k_F} \sum_{a,b>k_F} \delta[E+e(j)-e(a)-e(b)] \ |<\vec{k},\vec{j}|g[E+e(j)]|\vec{a},\vec{b}>_A|^2 \ . \tag{4.12}$$

While the sum over $j < k_F$ seems to lead to the desired form, the Fourier transform of the integrand in (4.12) is quite different from that of Re $<\vec{k},\vec{j}|g[E+e(j)]|\vec{k},\vec{j}>_A$ which appears in the real part of the OMP. In particular, there exists no reason to expect the radial form factor of the imaginary part of the effective interaction to be similar to that of the real part, as assumed for simplicity in Ref. 64.

4.4. Folded microscopic interaction

The approach of Day, Eisen, Friedman, Brieva, Rook and Von Geramb[64,77,85,101,102] is described elsewhere in these proceedings. Therefore, we shall not discuss it in detail. Its basic aim is to construct a local effective interaction which would have the same matrix elements as the g-matrix, at least between the relevant single-particle states. This is analogous to the procedure used by Negele[86] and others in density-dependent Hartree-Fock calculations of ground state properties.

The effective interaction is essentially defined by (see Eq.(3.2))

$$\hat{v}_E^{ST}(r) = \frac{\sum\limits_{j<k_F} \sum\limits_{\ell'} j_\ell(qr) \, v_{\ell\ell'}^{JS}(r) \, u_{\ell,\ell',E+e(j)}^{J,S,P}(r)}{\sum\limits_{j<k_F} j_\ell^2(qr)} \quad , \qquad (4.13)$$

where for simplicity we drop any explicity reference to an average over J and ℓ , and where T refers to the isospin. While Eisen, Day and Friedman[85,101] only used the real part of Eq. (4.13), Brieva and Rook[64] applied it to the imaginary part as well.

Since the radial dependence of $\hat{v}_\ell^{JS}$ is quite complicated, Brieva, Rook and Von Geramb[102] constructed its Fourier transform $\tilde{v}_E^{ST}(\tilde{k})$ and fitted the real and the imaginary parts of this quantity with a sum of five Gaussians. This sum can then be transformed back into r-space to yield a smooth effective interaction $\bar{v}_E^{ST}(r)$.

In Fig. 29, we show the OMP calculated by Brieva and Rook for the scattering of 30.3 MeV protons by ^{40}Ca .[64] The wiggles and the long-range of the form factor of $W(r)$ as calculated from the simple

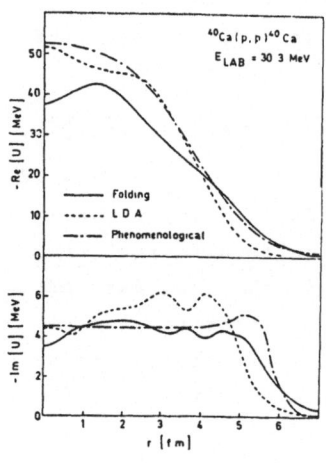

Fig. 29. Real (top) and imaginary (bottom) parts of the OMP for the scattering of 30.3 MeV protons by ^{40}Ca . The dash-and-dots represent a phenomenological OMP.[103] The theoretical curves have been computed by Brieva and Rook from the BHF approximation, with the simple LDA of Sect. 4.2 (short dashes) or the folded interaction method of Sect. 4.4 (full curves). From Ref. 64.

LDA of Sect. 4.2 are quite surprizing. I believe that they reflect either an inaccuracy of the BHF code (Sect. 3.2.1), or the fact that the input matter density is unrealistic, or else to the use of averages in Eq. (4.13). The full curves are computed from the OMP given by Eq. (4.6), with $\hat{v}$ defined by Eq. (4.13). Here again, I have some doubts on the accuracy and meaningfulness of the calculated imaginary part.

We now mention a few open problems and remarks concerning this folded effective interaction procedure.

(a) As compared to the improved LDA of Sect. 4.3, it presents the merit of introducing no adjusted parameter, and the drawback of involving a complicated effective interaction which requires a lengthy tabulation.

(b) Partly because of the appearance of sums over j on the right-hand side of Eq. (4.14), the *local* effective interaction $\hat{v}$ does not have the same matrix elements as the reaction matrix g. This is not important for the real part of $\hat{v}$, but may considerably affect its *imaginary* part which is very sensitive to threshold effects at low energy.

(c) It might be preferable to construct the effective interaction $\hat{v}$ in such a way that it has approximately the same matrix elements as g between Woods-Saxon harmonic oscillator single-particle wave functions.[96]

(d) It would be useful to analyze what is the relevant range of values for $\underset{\sim}{k}$ in the Fourier transform $\underset{\sim}{v}$. This is possible by using the techniques recently developed by Goldfarb.[104]

(e) The starting energy $E+e(j)$ that appears on the right-hand side of Eq. (4.14) is not the same as in nuclear matter.[86,105] This should be taken into account in all versions of the LDA, especially with respect to the imaginary part of $\hat{v}$.

(f) The meaning of the imaginary part of $\hat{v}$ should be clarified, especially with respect to its use in the analysis of inelastic scattering.[95,102,106]

(g) When assessing the success of the theory in rendering inelastic scattering cross sections, one should take into account the non-locality correction factor (2.29).

(h) The method of Brieva and Rook[53] for computing the reaction matrix gives a better treatment of the contribution of the inner region of the hard core than the method of Brueckner and Gammel[48] (Sect. 3.2.1). However, the latter method replaces this contribution by an effective delta-potential, which may be more practical in numerical applications of $\hat{v}$.

(i) The factor $R(k)$ (Eq. (2.13)) was omitted in Refs. 53, 64, 102.

(j) One could use a better approximation than the Slater function (4.4) for the mixed density matrix,[107] or use a microscopic description of exchange.

(k) It would be useful to fit $\tilde{v}(\tilde{k})$ by a sum of Fourier transforms of Yukawas, where the one with the longest range would correspond to a truncated OPEP interaction. This is important if one constructs the OMP between light ions from the effective interaction $\tilde{v}(r)$.[85,101]

(l) It would be convenient to use a definition of $\hat{v}$ more closely analogous to Negele's,[86] in order to avoid large cancellations between the direct and exchange terms. These might indeed lead to spurious wiggles in the calculated OMP.

5. Outlook

In the present survey, we have had a tendency to be quite critical, in keeping with the spirit of a workshop. This should not conceal the many successes of the nuclear matter approach. It appears to be able to render the properties of the *average* OMP with an accuracy of the order of 10 %, except for the imaginary part of light targets. It also seems to provide a promising way of constructing the effective interaction for the analysis of inelastic nucleon scattering. We mainly concentrated our discussion on low energy, but the negative and intermediate energy domain are of comparable theoretical and practical interest.[34]

The main open problems in the case of nuclear matter concern the accuracy of the Brueckner-Hartree-Fock codes (Sect. 3.2.1), the justification and the implementation of the self-consistent condition (2.39) and the evaluation of the two-hole line diagrams (Sect. 3.3).

In the case of finite nuclei, the main problem consists in finding an appropriate local density approximation. In particular, the imaginary part of the calculated OMP in light nuclei contains a spurious contribution due to the excitation of the center-of-mass of the target, and it is also quite sensitive to nuclear structure effects. The construction and the use of a *complex* effective interaction involve conceptual and practical problems; some of these are listed at the end of Sect. 4.4.

We gratefully acknowledge stimulating discussions with B. Day, J.-P. Jeukenne, A. Lejeune, W.G. Love and R. Satchler.

References

1. H. Feshbach, Ann.Rev.Nucl.Sci. 8, 49 (1958)
2. J.S. Bell and E.J. Squires, Phys.Rev.Letters 3, 96 (1959)
3. H. Feshbach, Ann.Phys. (N.Y.) 5, 357 (1958); 19, 287 (1962)
4. C. Mahaux and H.A. Weidenmüller, Shell-Model Approach to Nuclear Reactions (North-Holland, Amsterdam, 1969)
5. P.E. Hodgson, Nuclear Reactions and Nuclear Structure (Clarendon Press, Oxford, 1971)
6. D.H.E. Gross and R. Lipperheide, Nucl.Phys. A150, 449 (1970)
7. C.A. Engelbrecht and H.A. Weidenmüller, Nucl.Phys. A184, 385 (1972)
8. J.-P. Jeukenne, A. Lejeune and C. Mahaux, Proceedings of the International Conference on the Interactions of Neutrons with Nuclei, Lowell, edited by E. Sheldon (National Technical Information Service, Springfield, 1976), Vol. 1, p. 451
9. M.L. Goldberger and K.M. Watson, Collision Theory (Wiley & Sons, New York, 1964)
10. A.L. Fetter and K.M. Watson, in Advances in Theoretical Physics, edited by K.A. Brueckner (Academic Press, New York, 1965)
11. J. Hüfner and C. Mahaux, Ann.Phys. (N.Y.) 73, 525 (1972)
12. J.-P. Jeukenne, A. Lejeune and C. Mahaux, Phys.Reports 25C, 83(1976)
13. H.A. Bethe, Ann.Rev. Nucl.Sci. 21, 93 (1971)
14. C.M. Perey and F.G. Perey, Atomic Data and Nuclear Data Tables 17, 1 (1976)
15. G. Eder, H. Leeb and H. Oberhummer, J.Phys.G 3, L127 (1977)
16. C.L. Rao, M. Reeves III and G.R. Satchler, Nucl.Phys. A207, 182 (1973)
17. N. Vinh Mau and A. Bouyssy, Nucl.Phys. A257, 189 (1976)
18. G.M. Lerner and E.F. Redish, Nucl.Phys. A193, 565 (1972)
19. G.W. Greenlees, G.J. Pyle and Y.C. Tang, Phys.Rev. 171, 1115 (1968)
20. F.G. Perey, Nuclear Spectroscopy and Reactions, edited by J. Cerny (Academic Press, New York, 1974), part B, p. 137
21. D.C. Agrawal and P.C. Sood, Phys.Rev. C11, 1854 (1975)
22. J.-P. Jeukenne, A. Lejeune and C. Mahaux, Phys.Rev. C16, 80 (1977)
23. S. Kailas and S.K. Gupta, Phys.Rev. C17, 2236 (1978)
24. S. Kailas and S.K. Gupta, Phys.Letters 71B, 271 (1977)
25. R. Lipperheide, Nucl.Phys. 89, 97 (1966)
26. F.G. Perey and D.S. Saxon, Phys.Rev. 10, 107 (1964)
27. W.E. Frahn, Nucl.Phys. 66, 358 (1965)
28. F.G. Perey and B. Buck, Nucl.Phys. 32, 353 (1962)
29. F.G. Perey, Phys.Rev. 131, 745 (1963)
30. F.G. Perey and A.M. Saruis, Nucl.Phys. 70, 225 (1965)
31. M. Bertero and G. Passatore, Z.Naturforsch. 28a, 519 (1973)
32. G. Passatore, in Nuclear Optical Model Potential (Springer Verlag, 1976)
33. J.-P. Jeukenne, A. Lejeune and C. Mahaux, Phys.Letters 59B, 208 (1975)
34. C. Mahaux, Proceedings of the Summer Institute in Nuclear Theory, Banff, Canada, 1978 (to be published by Plenum Press)
35. A.S. Reiner, Phys.Rev. 138, B389 (1965)
36. M. Weigel and G. Wegmann, Fortschr.Phys. 19, 451 (1971)
37. Q. Ho-Kim and F.C. Khanna, Ann.Phys. (N.Y.) 86, 233 (1974)
38. C. Marville, Ph.D. Thesis (Liège, 1978), unpublished
39. C. Marville and P. Haensel, Z.Physik A284, 83 (1978) and references contained therein
40. J.-P. Jeukenne, A. Lejeune and C. Mahaux, Nukleonika 20 n2, 181 (1975)
41. H.R. Kidwai and J.R. Rook, Nucl.Phys. A169, 417 (1971)
42. J.R. Rook, Nucl.Phys. A222, 596 (1974)
43. A. Lejeune and C. Mahaux, Nucl.Phys. A295, 189 (1978)
44. G. Dahll, E. Østgaard and B. Brandow, Nucl.Phys. A124, 481 (1969)

45. M.I. Haftel and F. Tabakin, Nucl.Phys. A158, 1 (1970)
46. D.W.L. Sprung, Advances in Nuclear Physics, edited by M. Baranger
    and E. Vogt (Plenum Press, New York, 1972) 5, 225 (1972)
47. W. Legindgaard, Nucl.Phys. A297, 429 (1978)
48. K.A. Brueckner and J.L. Gammel, Phys.Rev. 109, 1023 (1958)
49. T. Hamada and D. Johnston, Nucl.Phys. 34, 382 (1962)
50. J.-P. Jeukenne, A. Lejeune and C. Mahaux, Phys.Rev. C10, 1391 (1974)
51. P.J. Siemens, Nucl.Phys. A141, 225 (1970)
52. R.V. Reid, Ann.Phys. (N.Y.) 50, 411 (1968)
53. F.A. Brieva and J.R. Rook, Nucl.Phys. A291, 299 (1977)
54. H.A. Bethe, B.H. Brandow and A.G. Petschek, Phys.Rev. 129, 225
    (1963)
55. E. Østgaard, Phys.Rev. 168, 1139 (1968)
56. P. Grangé and A. Lejeune, to be published
57. P.K. Banerjee and D.W.L. Sprung, Can.J.Phys. 49, 1899 (1971)
58. G. Passatore, Nucl.Phys. A95, 694 (1967)
59. J.-P. Jeukenne, A. Lejeune and C. Mahaux, Proceedings of the 12th
    International Winter Meeting on Nuclear Physics, Villars (Switzer-
    land), 1974
60. G. Passatore, Nucl.Phys. A110, 91 (1968)
61. B.D. Day, Rev.Mod.Phys. (in press)
62. C. Mahaux and R. Sartor, to be published
63. C. Mahaux, in Proceedings of the Conference on Recent Progress in
    Many-Body Theories (Trieste, Oct. 2-7, 1978) (North-Holland, in
    press)
64. F.A. Brieva and J.R. Rook, Nucl.Phys. A291, 317 (1977)
65. J.-P. Jeukenne, A. Lejeune and C. Mahaux, unpublished
66. M.M. Giannini and G. Ricco, Ann.Phys. (N.Y.) 102, 458 (1976)
67. K. Bear and P.E. Hodgson, Preprint Oxford (1978)
68. N.M. Hugenholtz and L. Van Hove, Physica 24, 363 (1958)
69. T.F. Hamman and Q. Ho-Kim, Nuovo Cim. 64B, 356 (1969);
    Q. Ho-Kim and R. Provencher, Nuovo Cim. 14A, 633 (1973)
70. R. Sartor, Nucl.Phys. A267, 29 (1976)
71. K.A. Brueckner and D.T. Goldman, Phys.Rev.117, 207 (1960);
    K.A. Brueckner, J.L. Gammel and J.T. Kubis, Phys.Rev. 118, 1438
    (1960)
72. H.S. Köhler, Phys.Rev. 137, B1145 (1965)
73. H.S. Köhler, Nucl.Phys. A170, 88 (1971)
74. C. Mahaux, in Physics of Medium-Light Nuclei, edited by P. Blasi
    & R.A. Ricci (Editrice Compositori, Bologna, 1978), p. 365
75. R. Sartor, Nucl.Phys. A289, 329 (1977)
76. H. Orland and R. Schaeffer, Nucl.Phys. A299, 442 (1978)
77. F.A. Brieva and J.R. Rook, Nucl.Phys. A297, 206 (1978)
78. J. Dabrowski and P. Haensel, Can.J.Phys. 52, 1768 (1974)
79. J.-P. Jeukenne, A. Lejeune and C. Mahaux, Phys.Rev. C15, 10 (1977)
80. J. Dabrowski, Phys.Letters 8, 90 (1964)
81. J. Rapaport, Phys. Letters 70B, 141 (1977)
82. G.E. Brown, private warning
83. T.A. Hughes, S. Fallieros and B. Goulard, Nuclei and Particles
    1, 93 (1971)
84. E. Colombo, R. de Leo, J.L. Escudié, E. Fabrici, S. Micheletti,
    M. Pignanelli and F. Resmini, Proceedings of the 1977 Tokyo Con-
    ference, J.Phys.Soc.Japan 44, Suppl. p. 543 (1978)
85. Y. Eisen and B.D. Day, Phys.Letters 63B, 253 (1976)
86. J.W. Negele, Phys.Rev. C1, 1260 (1970)
87. J.-P. Jeukenne, A. Lejeune and C. Mahaux, Journ.Physique 35,
    Suppl. 11, C5-7 (1974)
88. J.-P. Jeukenne, A. Lejeune and C. Mahaux, in Proceedings of the
    International Conference on Nuclear Self-Consistent Fields
    (Trieste, 1975), edited by G. Ripka and M. Porneuf (North-Holland,
    Amsterdam, 1975) p. 155

89. L.R.B. Elton, Nucl.Phys. 89, 69 (1966)
90. A. Lejeune and C. Mahaux, to be published
91. A. Lejeune, private communication
92. A. Lejeune and P.E. Hodgson, Nucl.Phys. A295, 301 (1978)
93. G.R. Satchler and W.G. Love, Phys.Letters 65B, 415 (1976)
94. F. Petrovich, D. Stanley and J.J. Bevelacqua, Bull.Am.Phys.Soc. 21, 973 (1976)
95. D.E. Bainum, R.W. Finlay, J. Rapaport, J.D. Carlson and W.G. Love, Phys.Rev. C16, 1377 (1977)
96. G. Bertsch, J. Borysowicz, H. McManus and W.G. Love, Nucl.Phys. A284, 399 (1977)
97. B. Sinha, Phys.Reports 20C, 1 (1975)
98. B. Sinha and S.A. Moszkowski, Nucl.Phys. A302,237 (1978)
99. C.B. Dover and Nguyen Van Giai, Nucl.Phys. A177, 559 (1971); ibid. A190, 373 (1972)
100. G.L. Shaw, Ann.Phys.(N.Y.) 3, 509 (1959)
101. Y. Eisen, B. Day and E. Friedman, Phys.Letters 56B,313 (1975)
102. H.V. Von Geramb, F.A. Brieva and J.R. Rook, preprint (1978)
103. R.S. Mackintosh and L.A. Cordero, Phys.Letters 68B, 213 (1977)
104. L.J.B. Goldfarb, Nucl.Phys. A301, 497 (1978)
105. X. Campi and D.W.L. Sprung, Nucl.Phys. A194, 401 (1972)
106. G.R. Satchler, Z.Physik 260, 209 (1973)
107. X. Campi and A. Bouyssy, Phys.Letters 73B, 263 (1978).

NUCLEAR STRUCTURE APPROACH TO THE NUCLEON-NUCLEUS OPTICAL MODEL

N. Vinh Mau
Division de Physique Théorique*, Institut de Physique Nucléaire
91406 ORSAY Cedex - France

In the many body theory, several microscopic approaches to the optical model potential have been developed. If the optical potential is defined as a one body potential which is able to reproduce the scattering wave functions, Bell and Squires[1] have shown its identification with the mass operator of the one particle Green function. In section I of this paper, the mass operator is calculated by summing up two classes of diagrams of the perturbation theory : the two particle ladder diagrams which lead to the Brueckner-Hartree-Fock potential and the particle-hole bubble diagrams which take into account two step processes with excitation of the target nucleus. Both contributions to the optical potential are discussed simultaneously with an attempt to compare their relative contributions. In section II the contribution of particle-hole diagrams is considered with some emphasis on the possible choices of the effective nucleon-nucleon interaction. Some results on nucleon-alpha and nucleon-calcium 40 potentials are presented. In section III a generalization of the approach of section II is developed for alpha-nucleus scattering and the complex optical potential is calculated for α-^{40}Ca scattering. Section IV is devoted to some general conclusions.

I. CALCULATION OF THE MASS OPERATOR

The mass operator is defined by the following integral equation :

$$G_1(x,x_0) = G_1^{(o)}(x,x_0) - \int G_1^{(o)}(x,x') M(x',x'') G_1(x'',x_0) d^4x' d^4x'' \qquad (1)$$

where $G_1(x,x_0)$ is the one-particle Green function at points x and x_0 of the four dimension space ($x = \vec{r}, t$) and $M(x',x'')$ is the mass operator. $G_1^{(o)}$ is the Green function of a free nucleon.

The optical potential is thus defined as [1] :

---

*Laboratoire associé au C.N.R.S.

$$V(\vec{r},\vec{r}',E) = M(\vec{r},\vec{r}',E)$$

$$= \int_{-\infty}^{+\infty} e^{iE(t-t')} M(x,x') d(t-t') \qquad (2)$$

The mass operator can be calculated in two equivalent frameworks. One uses the many-body Green functions equations, the other uses the perturbative expansion of M in terms of nucleon-nucleon interaction. In the first framework where only two body correlations are taken into account, M is obtained as a sum of three kinds of terms represented by diagrams a), b) and c) of figure 1 [2]. We know from nuclear structure studies that the leading term of G_2, the particle-particle Green function, is obtained by a resummation of all

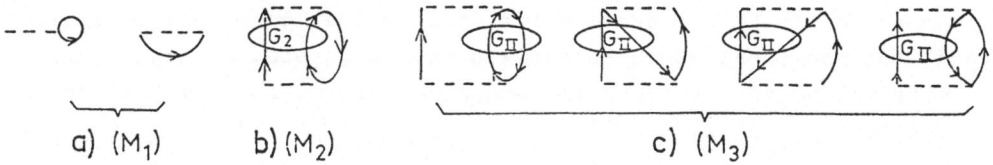

a) (M_1) b) (M_2) c) (M_3)

Fig.1 - Diagrams contributing to the mass operator

ladder graphs while the leading term of G_{II}, the particle-hole propagator, comes from a resummation of all bubble diagrams. Let us call M_1, M_2 and M_3 the three terms corresponding to diagrams a), b) and c) of fig.1 respectively. The mass operator is thus given by the following expression :

$$M = M_1 + M_2 + M_3 - 2 M^{(2)} \qquad (3)$$

$M^{(2)}$ is the second-order contribution to M, in perturbation theory. Indeed this contribution is counted once in M_2 and twice in M_3 therefore the summation of all terms in fig.1 would involve three times the second-order perturbative diagrams and it implies the subtraction of $2M^{(2)}$.

The expression (3) of the mass operator, or equivalently of the optical potential, contains both the Brueckner-Hartree-Fock approach and the nuclear structure approach to the optical model potential.

Indeed, the potential of eq.(3) has three approximated forms :

1) The Hartree-Fock potential : neglecting two-body correlations we get

$$M(\vec{r},\vec{r}',E) \approx M_1(\vec{r},\vec{r}',E)$$

or
$$V(r,r',E) \simeq V_{HF}(\vec{r},\vec{r}') \tag{4}$$

$$= \int d^3r_1 \; v(\vec{r}-\vec{r}_1)\rho(\vec{r}_1)\delta(\vec{r}-\vec{r}') - v(\vec{r}-\vec{r}')\rho(\vec{r},\vec{r}')$$

$v(\vec{r}_1-\vec{r}_2)$ is the nucleon-nucleon interaction and $\rho(\vec{r})$ and $\rho(\vec{r},\vec{r}')$ respectively the diagonal and non diagonal densities of the target ground state. To this approximation, the optical potential is real, static but non local. Such potentials have suggested a large number of works taking into account the exchange term or not, with different interactions and densities[3].

2) The Brueckner-Hartree-Fock potential : neglecting M_3 but including M_2 obtained by summation of ladder diagrams in the particle-particle Green function, we get [2] the Brueckner-Hartree-Fock potential :

$$M = M_1 + M_2$$

or
$$V(\vec{r},\vec{r}',E) \approx V_1(\vec{r},\vec{r}',E) = V_{BHF}(\vec{r}\;\vec{r}')E$$

$$= \sum_i n_i \int d^3r_1 \; d^3r_2 \; \varphi_i^*(r_1)[g(\vec{r},\vec{r}_1;\vec{r}'\;\vec{r}_2,E+\varepsilon_i) - \\ - g(\vec{r},\vec{r}_1;\vec{r}_2\;\vec{r}',E+\varepsilon_i)]\varphi_i(\vec{r}_2) \tag{5}$$

φ_i and ε_i are the individual particle wave functions and energies, n_i the occupation number of state i and $g(\vec{r},\vec{r}',E)$ obeys the Bethe-Goldstone integral equation and is the Brueckner - g matrix. This potential is complex, non local and energy dependent and calculations of $V_1(r,r',E)$ [4,5] will be reported in this Symposium by C. Mahaux and F. Brieva. In eq.(5) there is no double-counting in second-order therefore no subtraction.

3) The "nuclear structure" potential : assuming that we are
working with a weak effective interaction, M_2 is approximated
by its lowest term which is of second-order in v and the contribution
of M_3 is included. We thus get the optical potential as :

$$V(\vec{r},\vec{r}\,',E) \cong V_2(\vec{r},\vec{r}\,',E)$$

$$V_2(\vec{r},\vec{r}\,',E) = M_1 + M_3 - M^{(2)} \tag{6}$$

In this approach, there is still a double counting in
second order term because of complete antisymmetrization in the particle-hole diagrams of fig.1. $V_2(\vec{r},\vec{r}\,',E)$ may be called the nuclear
structure approach to the optical potential since it takes into
account intermediate channels corresponding to excitations of the
target nucleus states described in the RPA. In terms of RPA amplitudes,
$c_{ij}^{(n)}$ ((ij) represent the particle-hole configuration and n a particular RPA state of energy E_n), M_3 becomes

$$M_3(\vec{r},\vec{r}\,',E) = \sum_n \sum_{\substack{ij \\ i'j'}} \sum_\lambda c_{i'j'}^{(n)} \cdot \frac{{}_A\langle 0|v|ij,\lambda\rangle_A \; {}_A\langle i'j',\lambda|v|0\rangle}{E - E_n - \varepsilon_\lambda + i\delta} \tag{7}$$

λ represents the particle in intermediate state with
energy ε_λ, $\sum_\lambda$ means a summation over discrete states and an integration over continuum states, the index A refers to antisymmetrization
between the particle λ and the particle in the nucleus.

If we neglect exchange terms in M_3 and introduce Ψ_n, the
RPA wave function of the nucleus in state n, M_3 reduces to :

$$M_3^{NA}(\vec{r},\vec{r}\,',E) = \sum_{n\neq 0} \frac{\langle 0|v|\Psi_n\lambda\rangle\langle\Psi_n\lambda|v|0\rangle}{E - E_n - \varepsilon_\lambda + i\delta} \tag{8}$$

This expression together with eq.(6) where the subtraction
of $M^{(2)}$ disappears because we do not take antisymmetrization into
account in M_3, leads to the usual nuclear theory approach to optical
potential. The potential V_2, eqs.(6) and (7) is complex, non local and
energy dependent. Because of the energy denominator in M_3, the
contribution to the imaginary part of V_2 comes only from states n
with an excitation energy E_n lower than the incident energy E. We
therefore expect a strong energy dependence of the absorptive potential
for low incident energy. The real potential has a smother energy
dependence because its main contribution comes from the Hartree-Fock

term and because all RPA states may contribute to the real part of M_3.

4) <u>Comparison between V_1 and V_2</u> : up to second - order in v, V_1 eq.(5) and V_2 eq.(6) are identical. They differ only by higher order terms.

In terms of V_1, the optical potential of eq.(3) can be written as :

$$V = V_1 + (M_3 - 2M^{(2)})$$ (9)

When E, the incident energy, is much larger than the energies of the collective RPA states, the difference $E - E_n$ is not sensitive to a variation of E_n in a range of some MeV and closure can be performed over $c_{ij}^{(n)}$ coefficients in eq.(7) to give the relation :

$$M_3 \xrightarrow[E > E_{min}]{} 2 M^{(2)}$$ (10)

Thus V_1 is a very good approximation to V, eq.(9), for E larger than some minimum energy, E_{min} which depends on the target nucleus but is expected to be around 50-100 MeV. When the target nucleus has no collective states, $M_3 - 2 M^{(2)}$ is always negligible and V_1 is a good approximation to V whatever is the energy.

At low energies, M_3 is different from $2 M^{(2)}$ (an example is given in reference[2]) and we expect that V_1 will give an underestimation of the optical potential mainly for its imaginary part.

On the other hand, the validity of V_2 which takes into account the contribution of collective states relies on the assumption of a weak effective nucleon-nucleon interaction. Such interactions have been used successfully in nuclear structure studies and they can be used with some confidence for calculating the optical potential of low energy nucleon scattering only.

Therefore one could think of these two approaches as complementary in the sense that one ($V \approx V_2$) should be more adequate at low energy but is expected to fail at high energy while the other ($V \approx V_1$) should give a good representation of the optical potential at high energy but neglects large nuclear effects at low energy.

However the results obtained for V_1 [4,5] show good agreement with experimental data at low energy as well as at high energy and do not need any additive contribution from the collectivity of nuclear states.

II. CALCULATION OF THE OPTICAL POTENTIAL IN THE NUCLEAR
    STRUCTURE APPROACH

In this section we first discuss the possible choices of the effective nucleon-nucleon interaction. Then we present some numerical results for n-α and n-^{40}Ca potentials.

Let us write :

$$V(\vec{r},\vec{r}',E) \simeq V_2(\vec{r},\vec{r}',E) = - U - iW \qquad (11)$$

with
$$- U = M_1 + Re(M_3 - M^{(2)})$$
$$- \bar{W} = Im (M_3 - M^{(2)})$$

To calculate U and W we need v, the effective nucleon-nucleon potential, the target ground state density and the RPA amplitudes.

The choice of the effective nucleon-nucleon interaction is somewhat ambiguous. In the present state of the calculations v has been chosen among three classes of effective interactions :

1. In our first calculations[2,6] we were mainly interested by the imaginary potential for which the RPA amplitudes and energies are important input. Therefore v was chosen so that it reproduces well the spectrum of the target nucleus in RPA calculations. Two such forces have been used for ^{40}Ca - target : a zero range force[2] which leads to simpler numerical calculations and a finite range force adjusted by Gillet-Sanderson. The corresponding RPA amplitudes built with harmonic oscillator wave functions for particles and holes as well as the harmonic oscillator ground state density are used to derive U and W. The same type of force has also been used by Slanina and Mc Manus[7].

2. M_3, eqs.(7-8), contains the inelastic scattering form factors and if we think of optical potential in terms of reaction theory, the effective interaction should be able to reproduce the experimental

form factors and cross sections. Satchler[8] has shown that the Gillet-Sanderson force leads for p-^{40}Ca inelastic scattering to too low cross sections but he gets very nice agreement with the Reichstein-Tang interaction. Therefore in a more recent calculation, this force was used for the calculation of matrix elements in M_3 while the RPA amplitudes and energies were kept the same as before as well as M_1. With the same kind of argument Satchler et al. [9] used experimental informations on form factors to calculate M_3 eq.(8).

3. From the point of view of many body theory, V_2 should be calculated in a self-consistent way, that is to say we should solve the Hartree-Fock problem with an effective interaction, then use the same interaction to calculate the RPA states and M_3. In this scheme, v should be chosen to reproduce first the individual particle spectrum. We know that, in order to get good agreement for Hartree-Fock states, a density dependence has to be introduced in the effective interaction. This density dependence comes, in the local density approximation, from the summation of ladder diagrams. Therefore with a density-dependent force suitably determined for Hartree-Fock problem, the first term of V_2, eq.(6), should reproduce well the potential V_1, eq.(5), at least at low energy. Therefore with such force, one has to care about double counting and use eq.(9) rather than eq.(6). For n-^{208}Pb scattering, the potential V_2 has been calculated with a Skyrme interaction by V. Bernard and N. Van Giai[10] and the results will be reported in this Symposium. The Skyrme interaction has the great advantage of its simplicity but is phenomenological and only simulate the contribution of ladder diagrams. More fundamental density-dependent forces for Hartree-Fock problem have been derived by Campi-Sprung[11] from the G-matrix calculated in nuclear matter with the Reid soft-core nucleon-nucleon interaction. The comparison between M_1 calculated with Campi-Sprung interactions and the potentials of Jeukenne et al. and Brieva et al. would be very interesting at least at low energy. This calculation has been done for n-α potential[12].

We now discuss briefly the results obtained for the n-^{40}Ca potential calculated with effective interactions of type 1) and 2) and for n-α potentials obtained with interactions of type 1) and 3) with and without density dependence.

A) n-^{40}Ca potentials : we discuss only the local potentials equivalent to the non local ones in the usual approximation of Perey-Saxon. In the case of a zero range force the results have been published long ago and we only give in fig.2 the imaginary potential W at E = 14 MeV compared to its second order

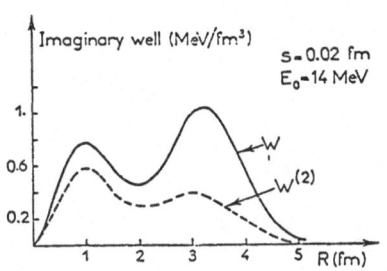

contribution. One sees that, at low energy at least, the collectivity of RPA states brings more absorption mainly at the surface. The difference between these two potentials should in principle be added to the potential V_1. In fig.3 are plotted U for the Gillet-Sanderson force (GSI) of type 1) and W for GSI and the Reichstein-Tang interaction (RTI) of type 2) at 30 MeV. The RTI gives stronger absorption in agreement with the work of Satchler[8] on inelastic scattering cross sections. H. Ngô [13] has calcula-

Fig.2 - E_o =14 MeV. The contribution of all RPA 1^-, 3^- and 5^- states to W(s,R) (——) is compared to corresponding second order contribution, $W^{(2)}$(s,R) for s = 0.02 fm.

ted the elastic scattering cross sections with the two potentials of fig.3. The comparison with experimental data shows that U is too weak by 10 % about while W for RTI is good within 5 %.

Figures 4 to 7 show the mean square radius and volume integrals of our potentials U and W as a function of energy compared to the results of Jeukenne, Lejeune and Mahaux and to the experimental or phenomenological values. The agreement for the radii is quite good for both finite range potentials. The general

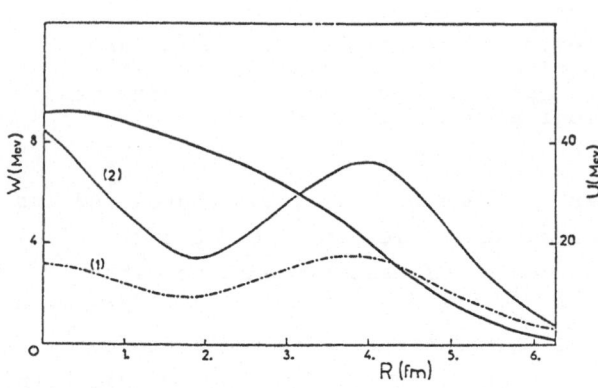

Fig.3 - the n-^{40}Ca potential for E_n = 30.3 MeV —— U calculated with GSI, ---- W calculated with RTI, -·-·- W calculated with GSI.

trend of the energy dependence is also well reproduced but the volume integrals are too small for all energies and always smaller than in Jeukenne et al calculations.

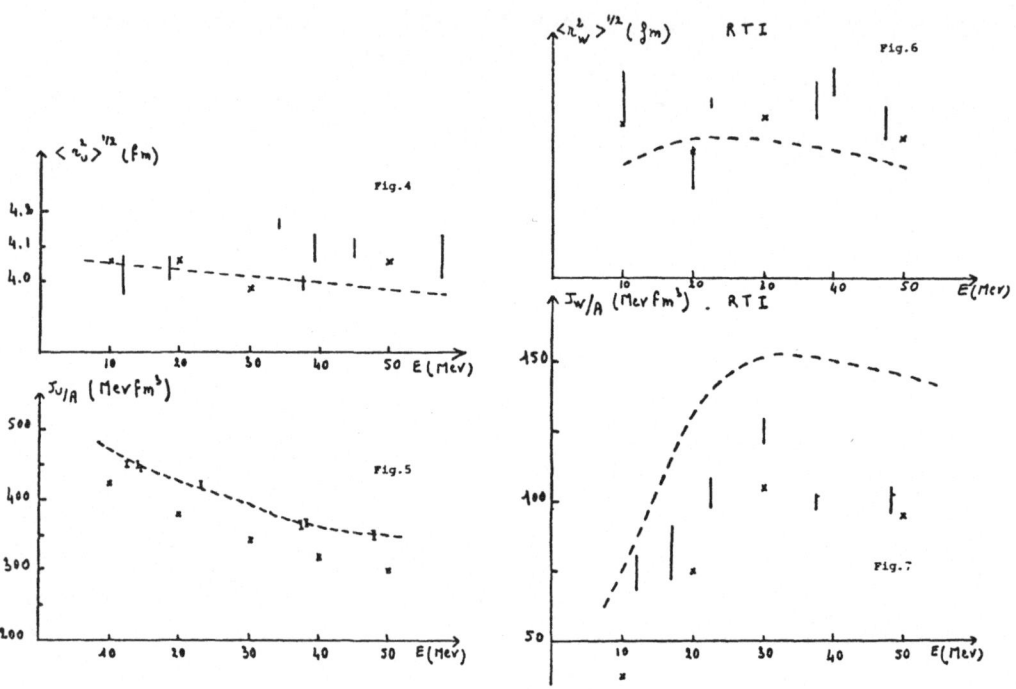

Figs. 4 to 7 - Energy dependence of the mean square radius and volume integral per particle of the α-^{40}Ca real (U) and imaginary (W) potentials : x represent our results, ---- the results of ref.4), I the phenomenological values.

B) n-α_potentials : the α-particle has no collective states and its first excited state lies at 20 MeV about. Therefore for an energy in the center of mass system lower than 20 MeV, the optical potential is purely real. If we use an effective interaction fitted to the bound state of the α-particle the potential must be well represented by the first term M_1 of eq.(9). M_1 has been calculated with a Brink-Boeker interaction first [14], using the alpha density calculated with the same interaction by Brink and Boeker [15]. This interaction is density independent and then does not take into account the summation of ladder diagrams. The n-α potential has also been calculated with the Campi-Sprung force G-0 of type 3) mentioned above, and with a gaussian α-density fitted on Hartree-Fock density calculated by Campi.

This potential should be a good approximation to V_1 at least at low energy. For comparison the calculation has also been performed with a Serber interaction.

All the calculated potentials are strongly non local with a large non locality range compared to the range of the potential. Therefore the derivation of a local equivalent potential is somewhat ambiguous. This problem has been studied in details [12] but for our present purpose it is sufficient to consider the results obtained in the Perey-Saxon approximation. In fig.8 the three potentials are

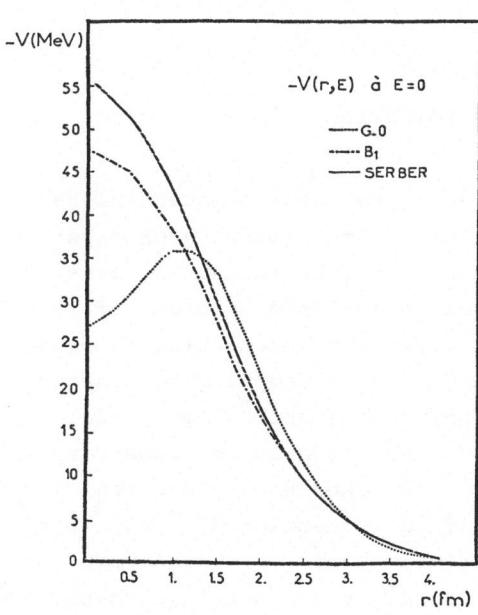

plotted for Serber, Bl (Brink-Boeker) and G0 (Campi-Sprung) interactions. G0 leads to a much weaker potential at short distances but a stronger one at the surface compared to the two others which can be fitted by gaussian form factors [14]. The energy dependence is shown for G0 and for 3 values of r in fig.9. It depends on the distance and is much larger at short distances than at the surface. It is also true for Bl and Serber interaction. It is thus difficult to compare it with the energy dependence of phenomenological potentials. M. Lassaut [12] has calculated the scattering phase shifts. She gets good agreement with data at low energy but the best agreement is obtained for G0.

Fig.8 - The nucleon-alpha interaction at zero energy for three effective nucleon-nucleon interactions.

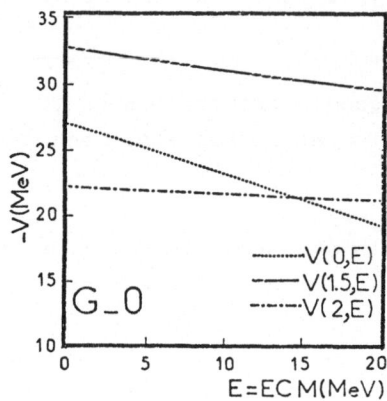

The potential
calculated with G0
can be considered as
an application of the
nuclear matter
approach to the opti-
cal model potential.

Fig.9 - energy dependence of the n-α
potential for three values of r :
r = 0, 1.5 and 2 fm.

III. CALCULATION OF THE ALPHA-NUCLEUS POTENTIAL

Our knowledge of the effective alpha-nucleus potential has
progressed these recent years because of the multiplicity of experiments on alpha elastic or inelastic scattering by various targets
and at various energies and their detailed analysis in terms of
optical model. The derivation of this effective interaction in many
body theory is an interesting problem and it is tempting to generalize the previous approaches to nucleon-nucleus optical potential
for α-nucleus potential. The alpha is considered as an elementary
particle and the fundamental interaction is then the α-nucleon
effective potential which has been studied in section II. The main
difference between α and nucleon scattering comes from the fact
that alpha particle and nucleons in the target are distinguishable
so that we do not have any antisymmetrization terms in the expansion
of M. The alpha-nucleon potential is weak enough so that we only
need calculate V_2 eq.(6) :

$$V_\alpha \overset{\sim}{=} V_2(\alpha) = M_{1\alpha} + M_{2\alpha} \tag{12}$$

where $M_{1\alpha}$ is the usual folding model real potential with no exchange
term. $M_{2\alpha}$ is given by eq.(8) where λ represents now the alpha
particle in intermediate states and v the alpha-nucleon interaction.

v is basically a non local potential but because of numerical difficulties it will be approximated by its   equivalent local potential
in the calculation of $M_{2\alpha}$. Detailed calculations of both terms $M_{1\alpha}$
and $M_{2\alpha}$ are given in reference 17) and here we briefly summarize
our approximations. $M_{1\alpha}$ is easily calculated either with the non
local or the local alpha-nucleon potential. The calculation of $M_{2\alpha}$
is more complicate but we are first interested by the general properties of the imaginary part of the optical potential which is quite
poorly known. Therefore we make simplifying assumptions as follows :

   a) the α-particle wave function can be approximated by W.K.B.
solutions corresponding to the potential $M_{1\alpha}$ calculated previously

   b) E_n , the alpha-particle energy, is large enough so that all
n states with a strong probability of excitation are included in the
summation and contribute to $Im\Delta V$.

   c) E_n, the excitation energy of state n, can be replaced by an
average value $\bar{E}_n$. Because of assumptions a) and b) the dependence
of $M_{2\alpha}$ upon E_n comes through the quantities $E_\alpha - E_n - M_{1\alpha}$ and
$E_\alpha - E_n$ which we replace by average values corresponding to $\bar{E}_n$.
Changing $\bar{E}_n$. is equivalent to change E_α and in the following numerical results we shall put $\bar{E}_n = 0$

   d) ψ_o , the ground state wave function, is a Slater determinant

   e) the α-nucleon interaction is local and gaussian as found with
B1 or Serber interaction. Let us write $v_\alpha(r) = V_o e^{-r^2/\mu^2}$ (13) . With
these assumptions, the only dependence of $M_{2\alpha}$ upon n comes from
the wave functions ψ_n and a closure relation over n is used and
leads to :

$$M_2(\alpha) = \bar{f}[<\psi_o| \sum_{i=1}^{A} v_\alpha(i) \sum_{j=1}^{A} v_\alpha(j) |\psi_o> - <\psi_o| \sum_{i=1}^{A} v_\alpha(i) |\psi_o><\psi_o| \sum_{i=1}^{A} v_\alpha(i) |\psi_o>]$$

$$(14)$$

where $\bar{f}$ is a complex function[17] depending on E_α. With assumption d),
$M_{2\alpha}$ is expressed in terms of $\rho_A(r)$ and $\rho_A(\vec{r}-\vec{r}', \frac{\vec{r}+\vec{r}'}{2})$ which are
respectively the diagonal and non diagonal ground state density of
the target nucleus. With approximation e) and the potential of
eq. (13), eq.(14) becomes :

$$M_{2\alpha}(s,X) = \bar{f} v_o^2[e^{-s^2/2\mu^2} \int e^{-2|\vec{X}-\vec{X}'|^2/\mu^2} \rho_A(X')d^3X'$$
$$- \frac{1}{4} \int e^{-|\vec{s}-\vec{s}'|^2/2\mu^2} e^{-2|\vec{X}-\vec{X}'|^2/\mu^2} \rho_A^2(s',X')d^3s'd^3X']$$

$$(15)$$

s and X are the relative and centre of mass coordinates. This term is non local and depends upon energy.

In this simple model, $M_{2\alpha}$ is easy to calculate. Because $\overline{f}$ is a complex function, $M_{2\alpha}$ has a real part which adds to $M_{1\alpha}$ and an imaginary part which leads to the absorptive part of the optical potential. It is clear from eq.(15) that both terms vary very smoothly with A. Therefore one finds that our imaginary optical potentials will not manifest any large variation from nucleus to nucleus.

Our model is only valid when the α-particle energy is large enough and is not able to reproduce the differences observed at backwards angles between the cross sections of alpha elastic scattering at low energy on Ca^{40} and Ca^{44} for example. But it tells us that these differences should disappear when the incident energy increases, accordingly to experimental data.

As we assume that all inelastic channels are opened at the energies considered in this work, our imaginary potential has no energy dependence due to the opening of new channels and its variation with energy is that one of the function $\overline{f}$ which depends on energy through the quantity $E_\alpha - M_{1\alpha}$. As $M_{1\alpha}$ is a deep potential, we expect a smooth energy dependence for the imaginary potential.

We have considered that in intermediate step, the alpha stays an alpha and have therefore excluded all contributions from α-break-up channels. Nevertheless by using closure over the target states we have implicity included channels with break-up of the target nucleus like (α,αp), (α,αn), (α,αd) ... channels.

The calculation of the optical potential for $α-^{40}Ca$ elastic scattering has been performed with the α-nucleon interaction derived from B1 Brink and Boeker interaction. The local density of calcium is calculated with harmonic oscillator wave functions (b_o = 2.08 fm) and the non local density taken from Campi and Bouyssy's work[18].

The local optical potential is calculated with the Perey-Saxon approximation and is :

$$V_\alpha(R) = U_\alpha(R) + i\, W_\alpha(R)$$

where $U_\alpha(R) = (M_{1\alpha} + \text{Re } M_{2\alpha})_{\text{local}}$

$W_\alpha(R) = (\text{Im } M_{2\alpha})_{\text{local}}$

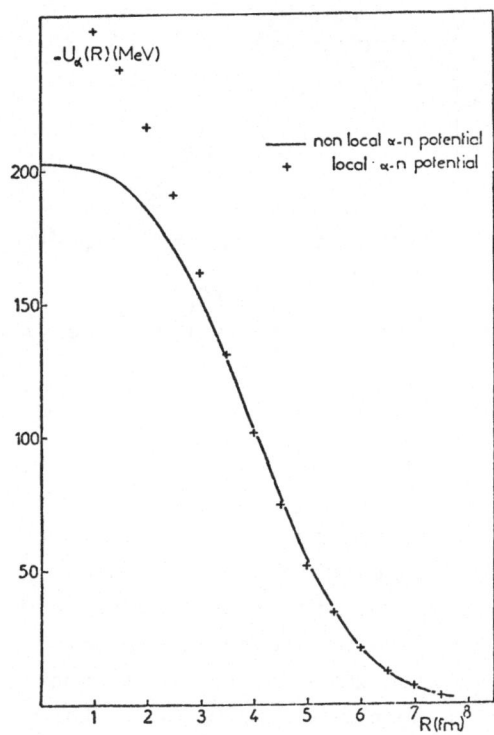

Fig.10 - the α-^{40}Ca real potential at zero energy calculated with the non local($-$) and local ($+$) n-α potentials.

Re $M_{2\alpha}$ is always negligible[17] and $U_\alpha(R)$ is only given by the first term $M_{1\alpha}$. On fig.10 we have shown the potentials $U_\alpha^1(R)$ and $U_\alpha^2(R)$ calculated with the non local α-nucleon potential derived from B1 interaction and with its local equivalent. $U_\alpha^1(R)$ is shallower than $U_\alpha^2(R)$ at short distances but both are very similar for $R \geqslant 4$ fm. U_α^1 can be fitted by a Saxon-Woods squared potential, in agreement with recent phenomenological analysis[19].

The energy dependence of $U_\alpha^1(R)$ is linear as shown in fig.11 and hardly depends on the distance. It can be parametrized as :

$$U(R,E_\alpha) = [1 - \alpha(R)E_\alpha]U(R)$$

with $\alpha(0) \approx 0.0021$ MeV^{-1}.

On fig.12 is plotted the imaginary potential, $W(R)$, calculated for E_α = 50, 75, 100 and 150 MeV. $W(R)$ has the shape of a Saxon-Woods potential with a radius R_W larger than the radius of the real potential. When we parametrized it as a Saxon-Woods potential we get a radius $R_W \simeq 5.8$ fm and a diffuseness parameter $a_W \simeq 0.65$ fm at E_α = 50 MeV. When the energy increases, the strength and a_W increase slowly while R_W seems to decrease slightly.

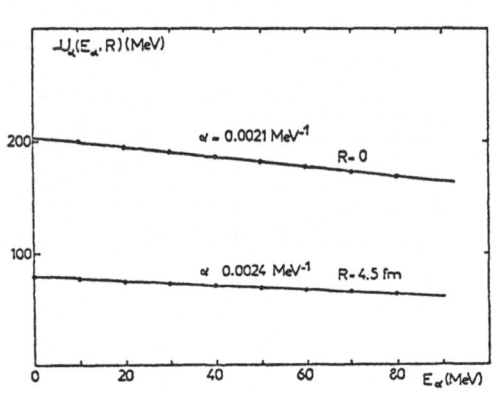

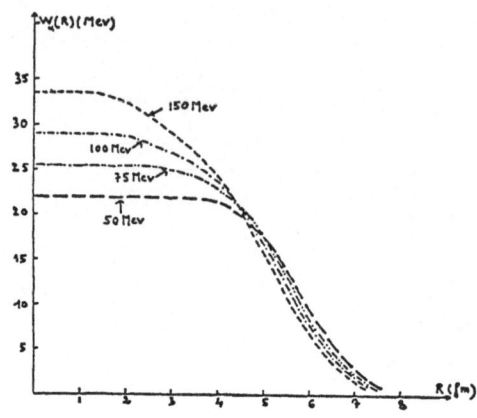

Fig.11 - energy dependence of the α-^{40}Ca real potential calculated with the non local α-nucleon interaction.

Fig.12 - the α-^{40}Ca imaginary potential for several α-energies.

IV. CONCLUSION

We have discussed the validity of the nuclear structure approach to the nucleon-nucleus optical model potential and shown that it is essentially a low energy approach. The calculated nucleon-alpha and nucleon-calcium 40 potentials have some sensitivity on the choice of the effective nucleon-nucleon interaction. An attempt to derive the alpha-nucleus potential within the same approach has been proposed. With simplifying assumptions it provides some global informations on the absorptive part of the α-^{40}Ca potential and could be of some help for the analysis of experimental data.

The author would like to thank her co-workers on the studies, some of them still unpublished, described in section II : A. Bouyssy, H. Ngô and M. Lassaut.

1) J.S. Bell and E.J. Squires, Phys. Rev. Lett. 3 (1959) 96
2) N. Vinh Mau and A. Bouyssy, Nucl. Phys. A257 (1976) 189
3) See for references the Review by B. Sinha, Physics Reports 20C (1975)
4) C. Mahaux, this Symposium - J.P. Jeukenne, A. Lejeune and C. Mahaux, Phys. Reports (1976)
5) F.A. Brieva, this Symposium - F.A. Brieva and J.R. Rook, Nucl. Phys. A291 (1977) 299 and 317
6) A. Bouyssy, H. Ngô and N. Vinh Mau, to be published
7) Slanina, Ph. D. Thesis, Michigan State University (1969)
8) G.R. Satchler, Z. Physik 260 (1973) 209

9) C.L. Rao, M. Reeves and G.R. Satchler, Nucl. Phys. $\underline{A207}$ (1973) 482

10) V. Bernard, Thèse de 3è Cycle, Orsay (1978) - V. Bernard and N. Van Giai, this Symposium and to be published

11) X. Campi and D.W.L. Sprung, Nucl. Phys. $\underline{A194}$ (1972) 401

12) M. Lassaut, Thèse de 3è Cycle, Orsay (1978) to be published

13) H. Ngô, to be published

14) M. Lassaut and N. Vinh Mau, Phys. Lett. $\underline{70B}$ (1977) 1

15) D.M. Brink and E. Boeker, Nucl. Phys. $\underline{A91}$ (1967) 1

16) X. Campi, Thèse, Orsay (1973) - X. Campi and D.W.L. Sprung Nucl. Phys. $\underline{A194}$ (1972) 401

17) N. Vinh Mau, Phys. Lett. $\underline{71B}$ (1977) 5 and Second Louvain-Krakow Seminar - Louvain (1978)

18) X. Campi and A. Bouyssy, Phys. Lett. $\underline{73B}$ (1978) 263

19) Louvain-Zagreb-Mons-Krakow-Munich Collaboration, to be published to Phys. Rev. C.

ON THE GENERAL THEORY OF THE NUCLEON OPTICAL POTENTIAL

M. K. Weigel

Sektion Physik der Universität München, 8046 Garching

Abstract. The aim of this contribution is to obtain some insight in the structure of the so-called "generalized optical potential" GOP for nucleons. We shall follow the pioneering work of Bell and Squires, who connected the optical model problem with the many-body Green's function (GF) theory. The benefit of this approach is the possibility to utilize modern many-body techniques together with known numerical results in the investigation of the GOP. We will concentrate on the general features of the optical potential and discuss briefly some approximation schemes feasible for the calculation of the GOP. In the discussion of numerical results, we restrict ourselves to investigations connected with dispersion relations.

I. Introduction

In the theoretical formulation of the concept of the microscopic optical potential, one may roughly distinguish two main methods (Pa 76):

a) Derivation from the many-channel scattering theory

The basis of this method is the projection P onto the elastic channel subspace (Fe 58, 62), which leads to a "Schrödinger" equation for the projected wave function $P\psi_E$ with an energy-dependent, non-local optical potential of the following form $(Q: = 1 - P; H = T + V)$

$$v^{opt} = <o|V|o> + <o|VQ(E - QHQ + i\eta)^{-1} QV|o> \qquad (I.1)$$

One can also use the time-dependent formulation (Co 58) or directly the Lippmann-Schwinger equation for the T-matrix (Fet 65; Ke 59), respectively.

b) <u>Investigation of the generalized optical potential (GOP) by utilizing the Greens-function many-particle theory</u>

It was shown first by Bell and Squires (Be 59), that the so-called "generalized optical potential", which describes exactly - apart from recoil - the elastic nucleon scattering can be identified with the so-called effective single-particle potential (or "irreducible mass operator") in the equation of motion for the single-particle propagator.

The investigation in the GF-scheme combines many features necessary in a modern and extensive investigation of the GOP for instance: 1) The use of the second quantization formulation guarantees a correct treatment of the Pauli principle . 2) The GOP can be divided in a simple manner into an energy-independent (folding term) and an energy-dependent ("resonant") part. 3) The connection to modern nuclear structure approaches - as for example RPA-calculations of different kinds - can be directly established, which opens the possibility to use nuclear structure results in the calculation of the GOP. It seems, that in the low-energy region coherent excitations of the target influence strongly the GOP. For larger energy-ranges the mean energy-behaviour can be obtained by taking the particle-particle correlations into account (ladder approximation). 4) One can obtain special decompositions of $V^{opt}(E)$, which makes it - for instance - possible to trace the contributions of inelastic, pick-up and knock-out processes to the imaginary part of $V^{opt}(E)$. 5) Since the GF obbey spectral representations with known structure, one can obtain the analytic properties of the GOP in a unique manner, and so deduce a dispersion relation between the

real and imaginary part of v^{opt}. 6) The normalization condition
for the scattering states of the non-hermitian and energy-dependent GOP can be formulated. Furthermore it is possible to
generalize the optical theorem for the GOP.

For these reasons, we will use the GF-formulation in our presentation. The main guide lines are demonstrated in the following scheme:

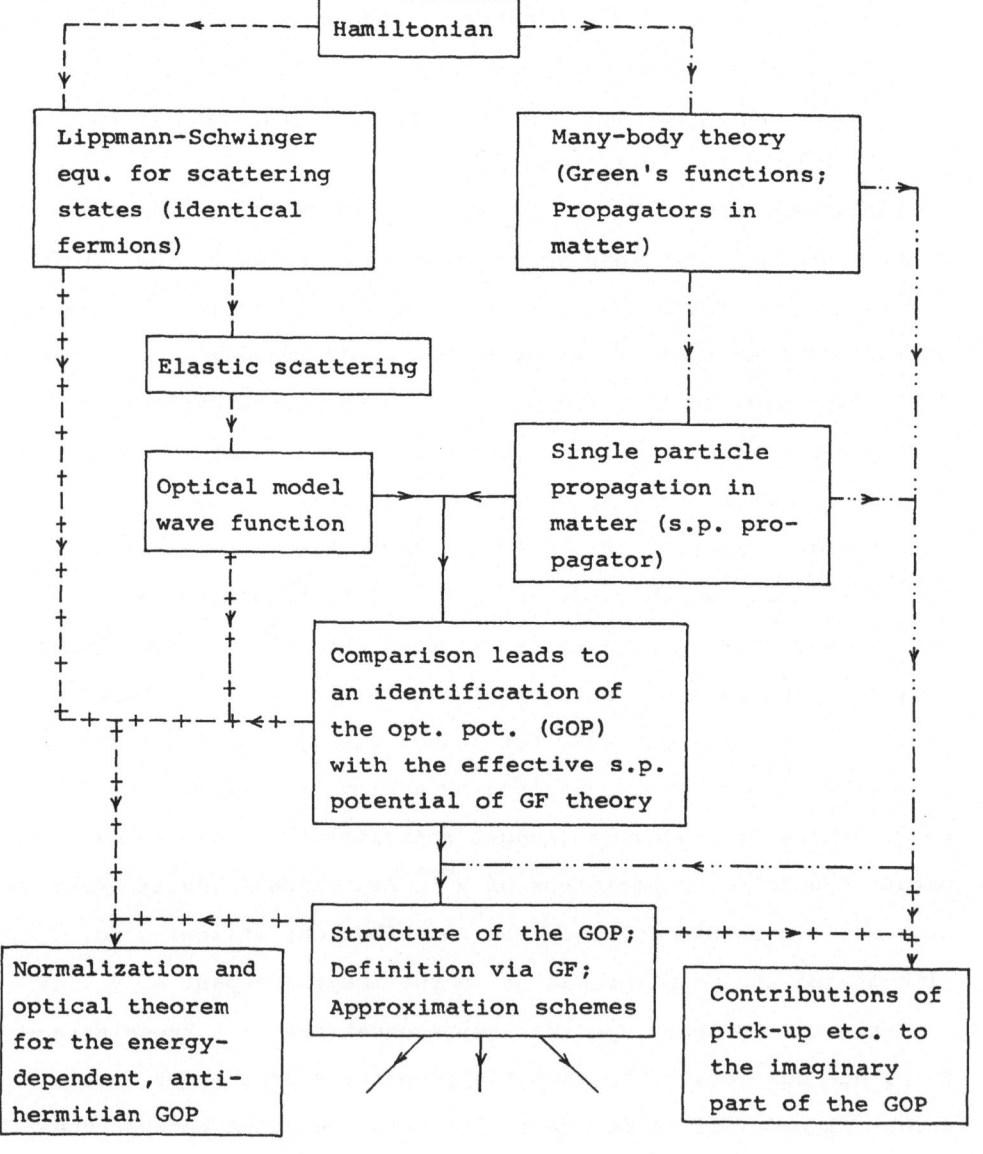

We will shortly describe the motivation for pursing the optical model problem in this way.

The dynamics and statistics of the many-body system is described by the hamiltonian written in second quantization. A formal procedure for the description of the optical model problem is given on the left side (broken arrows ----→). First one formulates the Lippmann-Schwinger equation for scattering states (for identical fermions) and selects in the next step the elastic scattering case. By defining the optical model wave function according to the standard interpretation a formal expression for this function is obtainable. As pointed out by Bell and Squires the optical potential should be closely related to the propagation of a particle in matter. Problems of such kind can be investigated in the framework of the Green's function formulation of the many-body problem. With the given hamiltonian one can therefore study the many-body structure (-·-·-→). For the GOP one is interested in the propagation of a particle on the background of the N-particle (target) system described by the so-called one-particle Green function (GF). For this GF one can derive with the help of Heisenberg's equation of motion for the creation and annihilation operators of the particles an equation for the single-particle propagator, in which the influence of the matter onto the single-particle propagation is incorporated into a so-called effective single-particle potential. A direct comparison between the optical model wave function and the one-particle GF shows, that the first can be identified with a certain (mathematical or physical, respectively) limit of the s.p. GF. This comparison (indicated by full arrows ——→) leads to an identification of the GOP with the effective s.p.-potential of GF theory. Furthermore one obtains now the "Schrödinger" equation for the elastic channel. The main advantage of this procedure is the fact, that the GF-many-body theory has been investigated in greater detail, and we know much about the general formal struc-

ture and possible approximation schemes, respectively. Furthermore
many numerical calculations have been performed. This knowledge can
now be used directly in the investigation of the GOP (indicated by
··-··-··→ arrows). So we obtain a definition of the GOP in the GF-
scheme; we can read off the general (analytic) structure and furthermore we have many approximation schemes available for the GOP (or
effective s.p.-potential, respectively). In the sketch we have shown
some additional problems, which are tractable in the GF-formalism
(-+-+→ arrows). Since they are not essential for the general structure, we will not discuss them further (see ref. E 76). Furthermore,
we will postpone the investigation with respect to the detailed
structure etc. of the GOP ($\swarrow \downarrow \searrow$) until we have obtained some insight
into the structure of the scattering problem and the GF-theory, respectively. In the next section we shall briefly discuss some needed
GF-theory and the scattering problem. Then we are going to investigate the GOP in more detail. In the last section, we will shortly
sketch the use of dispersion relations. We present only
results relevant to this last topic since for the other approximation,
the outcome of numerical calculations has been presented by several
other authors of this conference.

II. Scattering Theory and Optical Model Wave Function

The many-body system is described by the hamiltonian

$$H = \sum_{\alpha\beta} (\frac{p^2}{2m})_{\alpha\beta} a_\alpha^\dagger a_\beta + \frac{1}{2} \sum_{\substack{\alpha\beta \\ \gamma\delta}} v_{\alpha\beta\gamma\delta} a_\alpha^\dagger a_\beta^\dagger a_\delta a_\gamma = T + V, \qquad (II.1)$$

with the usual anticommutator relations for the annihilation and creation (Schrödinger) operators, respectively. The scattering states for
the N+1-particle system with an incoming particle with momentum $\vec{p}_i$
($\varepsilon_i = \frac{p_i^2}{2m}$) and the target system in the exact groundstate of H ($H|0> = E_N^0|0>$) is given by (Na 60, Vi 67)

$$|\Omega^{(+)}> := |\vec{P}_i^{(+)}; \; 0> = (a^{\dagger}_{\vec{P}_i} + (\varepsilon_i + E_N^O - H + i\eta)^{-1} [V, a^{\dagger}_{\vec{P}_i}]) |0>$$

$$= \lim_{\eta \to 0} \frac{i\eta}{\varepsilon_i + E_N^O - H + i\eta} a^{\dagger}_{\vec{P}_i} |0> \qquad (II.2)$$

$$= \lim_{t \to -\infty} e^{iHt} e^{-i(E_N^O + \varepsilon_i)t} a^{\dagger}_{\vec{P}_i} |0>$$

The three formulations are equivalent. The last expression - for instance - is due to the Lehmann-Zimmermann-Symanzik treatment of scattering problems (Vi 67). The first term in (II.2) describes the incoming plane wave; the scattering term is determined by the structure of the so-called "doorway-states", which emerge formally by evaluating the commutator:

$$[V, a^{\dagger}_{\vec{P}_i}]_- = \sum_{\alpha\beta\delta} v_{\alpha\beta i\delta} \, a^{\dagger}_{\alpha} a^{\dagger}_{\beta} a_{\delta} \qquad (II.3)$$

$\vec{P}_i$ x

As we shall see in the discussion of the optical model, the last structure will - as in the inelastic scattering - also determine the behaviour of the optical potential.

According to the standard interpretation the optical model wave function is given by ($r := \vec{r}, s_r, t_r; \; p := \vec{P}_i, s_i, t_i$):

$$\varphi_p^{(+)}(r) := <0|a_r|\Omega^{(+)}>, \qquad (II.4)$$

which can be expressed according to (II.2) by the following (Abelian) limit:

$$\varphi_p^{(+)}(r) = \lim_{t \to +\infty} e^{i\varepsilon_i t} \; <0|a_r \; 1 \; ^{-i(H-E_N^0)t} \; a_p^\dagger|0>$$

$$\text{(II.5)}$$

$$= \lim_{\eta \to 0} i\eta <0|a_r \; (\varepsilon_i - (H-E_N^0) + i\eta)^{-1} a_p^\dagger|0>$$

III. Green's Functions

a) General Definition

The GF (2n-point functions; propagators) are defined as the ex-spectation values (in the true groundstate $|0>$ of the N-particle system = target nucleus) of n time-ordered (Heisenberg) annihilation and creation operators, respectively.

$$G^{(n)}_{1\ldots n,\; 1'\ldots n'} := (-i)^n <0|T\; a_{\alpha 1}(t_1)\ldots a_{\alpha_n}(t_n)\; a^\dagger_{\alpha_{n'}}(t_{n'})\ldots$$
$$a^\dagger_{\alpha_{1'}}(t_{1'})|0>$$

$$\text{(III.1)}$$

They describe - for instance - the propagation of particles, holes, particle-pairs etc. on the background of the many-particle system. The essential information about the system is contained in the GF and all relevant quantities of the many-body system can be obtained from them by trivial operations. The equation of motion is obtainable by utilizing Heisenberg's equation.

b) Single-Particle-Propagator

For the nucleon optical potential we have to consider the one-particle Green function, which has according to (III.1) the following form ($\tau := t_\lambda - t_\mu$):

$$iG_{\lambda\mu}(\tau) := \begin{cases} <0|a_\lambda e^{-i(H-E_N^0)\tau} a_\mu^\dagger|0> , & \tau > 0 \\ \\ -<0|a_\mu^\dagger e^{i(H-E_N^0)\tau} a_\lambda|0> , & \tau < 0 \end{cases}$$

$$\text{(III.2a)}$$

or after Fouriertransform with respect to τ:

$$G_{\lambda\mu}(\omega) = <0| (a_\lambda (\omega - H + E_N^O + i\eta)^{-1} a_\mu^\dagger + a_\mu^\dagger (\omega + H - E_N^O - i\eta)^{-1} a_\lambda)|0>$$

$$= \oint_{|N+1>} <0|a_\lambda|N+1> (\omega - E_{N+1} + E_N^O + i\eta)^{-1} <N+1|a_\mu|0>$$

$$+ \oint_{|N-1>} <0|a_\mu^\dagger|N-1> (\omega + E_{N-1} - E_N^O - i\eta)^{-1} <N-1|a_\lambda|0> \qquad (III.2b)$$

$$= \oint_{\nu\varepsilon(N\pm1)} \frac{c_\lambda^{N\pm1} c_\mu^{N\pm1}{}^*}{\omega - \omega_\nu}$$

From the last expression one can read off (explicitly) the Lehmann or spectral representation of the single-particle propagator

$$G_{\lambda\mu}(\omega) = G_{\lambda\mu}^{(+)}(\omega) + G_{\lambda\mu}^{(-)}(\omega) , \qquad (III.3)$$

with

$$G_{\lambda\mu}^{(\pm)}(\omega) = \int_{-\infty}^{+\infty} d\omega' \frac{A_{\lambda\mu}^{(\pm)}(\omega')}{\omega - \omega' + i\eta} \qquad (III.4)$$

For nonrelativistic systems with a given hamiltonian the spectral function A is defined explicitly as

$$A_{\lambda\mu}^{(\pm)}(\omega) = \oint_{\substack{\nu\varepsilon(N+1) \\ (N-1)}} c_\lambda^{N\pm1} c_\mu^{N\pm1}{}^* \delta(\omega - \omega_\nu) , \qquad (III.5)$$

For the higher order GF one can obtain the corresponding Lehmann representations in an analogue manner (Ko 62).

The equation of motion (Dyson-equation) for the 2-point function has the following form:

$$\sum_\mu \{\omega\delta_{\alpha\mu} - (\frac{p^2}{2m})_{\alpha\mu} - V_{\alpha\mu}(\omega)\} G_{\mu\beta}(\omega) = \delta_{\alpha\beta} , \qquad (III.6)$$

where the effect of the medium onto the particle propagation has been taken into account by the so-called "effective single-particle potential" $V_{\alpha\beta}(\omega)$. It has two parts

$$V_{\alpha\beta}(\omega) = V_{\alpha\beta}^{HF} + V_{\alpha\beta}^{R}(\omega) \quad . \tag{III.7}$$

V^{HF} is hermitian and energy-independent and can be obtained by the Hartree-Fock procedure (with the true groundstate $|0>$):

$$V_{\alpha\beta}^{HF} := \sum_{\sigma\mu} v_{\alpha\sigma\beta\mu} <0|a_{\sigma}^{\dagger}a_{\mu}|0> \quad . \tag{III.8}$$

The second energy-dependent and non-hermitian (resonant) part V^R is (implicitly) defined by

$$\sum_{\mu} V_{\alpha\mu}^{R}(\omega) \ G_{\mu\beta}(\omega) :=$$

$$-i \sum_{\rho\sigma\mu} v_{\alpha\rho\sigma\mu} (G_{\sigma\mu\beta\rho}^{(2)}(\omega) - G_{\sigma\beta}(\omega)G_{\mu\rho}(\omega) + G_{\sigma\rho}(\omega)G_{\mu\beta}(\omega)) \quad . \tag{III.9}$$

In the last relation the time arguments of $\sigma\rho\mu$ in $G^{(2)}$ are equal and describe the propagation of a correlated 2p-1h or 2h-1p triple, respectively. The shell-model correlations have been subtracted by the last two terms.

Optical model: The connection between the optical model and the GF can be established by means of the limiting procedure of (II.5). One gets by comparison of (II.5) and (III.1,2)

$$\varphi_p^{(+)}(r) = \lim_{\tau \to +\infty} e^{i\varepsilon_i\tau} \ iG_{rp}(\tau)$$

$$= \lim_{\eta \to 0} i\eta \ G_{rp}(\varepsilon_i) \tag{III.10}$$

If the limit is performed in equ. (III.6), one gets the "Schrödinger" equation for the optical model wavefunction (Σ means summation or integration, respectively)

$$\sum_{\mu} \{\epsilon_i \delta_{\alpha\mu} - (\frac{p^2}{2m})_{\alpha\mu} - V_{\alpha\mu}(\epsilon_i)\}\varphi_{\mu,p}^{(+)} = 0 \quad , \tag{III.11}$$

and the generalized optical potential GOP has to be identified with the effective s.p.-potential taken at $\omega = \epsilon_i$:

$$V_{\alpha\beta}^{opt}(\epsilon_i) = V_{\alpha\beta}^{HF} + V_{\alpha\beta}^{R}(\epsilon_i) \tag{III.12}$$

IV. Generalized Optical Potential (Structure and Approximations)

According to the discussion in the introduction the main motivation for choosing the method of Bell and Squires is the close connection with the GF-theory, which has been investigated in many respects in the last years. In this way we obtain some detailed insight in the structure of the GOP. A possible procedure is demonstrated in the graph shown on the next page:.

From the GF-theory we know several methods for the investigation of the effective single-particle potential. We are going to discuss two main methods, which in principle allow the complete calculation of the s.p. potential, namely the (renormalized) perturbation expansion and the so-called "doorway state RPA" (indicated by full arrows ———→). From both methods one can obtain either by partial summation of the perturbation expansion (Ma 76) or by using the so-called weak coupling limit the two main approximation schemes, which have been used in the past in the calculation of the optical potential (—·—·—·→). These approximations can also be obtained directly from the definition of the GF and other approximation schemes (—+—+—→). With the so-called phonon contributions of the target system one takes the influence of virtual coherent excitations of the target onto the particle propagation into account, which seems to be impor-

Structure and Approximation Schemes for the GOP

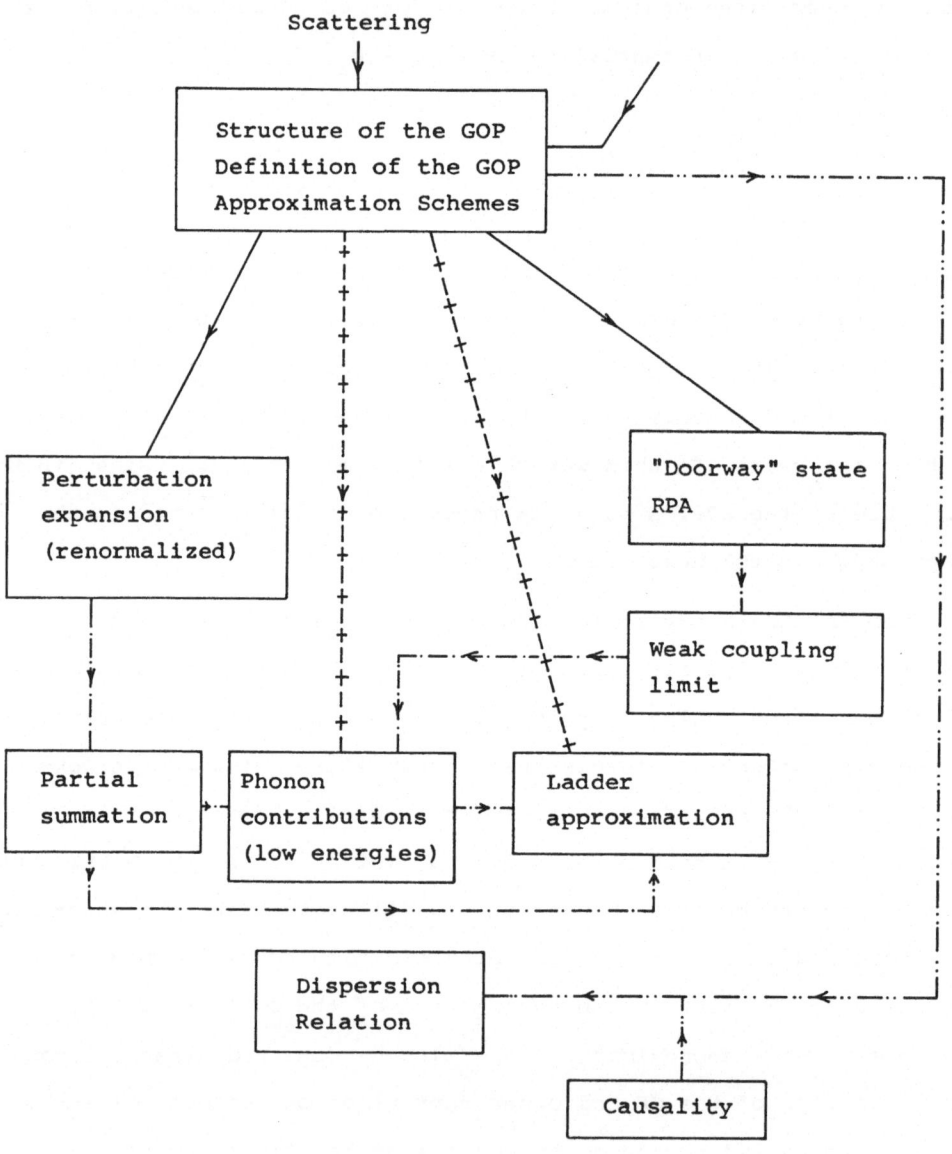

tant in the low energy region (see contribution of Prof. Vinh Mau).
In the ladder approximation one considers p-p (or h-h) correlations.
In this approximation it is possible to obtain the total optical potential in a kind of folding procedure, where the free interaction
has to be replaced by the effective scattering matrix in medium (see
contributions of Profs. Mahaux, v. Geramb, Brieva and Rook, respectively). This method seems to be capable to reproduce the optical potential in a wide energy-range within 10-15 %. A slightly different
approach is the use of dispersion relations, which can be deduced
directly from the analytic structure of the GOP (– ·· – ·· – ··→). (We have
given only a selection of methods, for other treatments see for
example Em 71, Mat 76, Ab 63, Ba 62, We 71, Ma 59, Fe 71, Zh 65,
Mi 67.)

a) Perturbation Theory of V:

The perturbation expansion of the single-particle potential V is
often formulated with the aid of diagrams. If we use the following
picture dictionary:

$$v_{1'2'12} := \quad \begin{matrix} 2' & 2 \\ \square \\ 1' & 1 \end{matrix} \quad = \quad (\quad \begin{matrix} 2',2 \\ \{ \\ 1',1 \end{matrix} \quad + \text{ exch. for local potentials}) \qquad \text{(IV.1)}$$

and

$$G_{\lambda\mu}(\omega) = \quad \underset{\lambda \quad \mu}{\overset{\omega}{\longleftarrow}} \quad , \qquad \text{(IV.2)}$$

we obtain for the first graph of v^R the following structure:

$$\frac{i}{2} v^R_{1'1}(\omega) := \quad = \quad \left(\quad + \quad \right) \qquad \text{(IV.3)}$$

$$= \quad (\underset{\beta\alpha\delta}{\Sigma} \frac{v_{1'\beta\alpha\delta} \, v_{\alpha\delta1\beta}}{\omega+\varepsilon_\beta-\varepsilon_\alpha-\varepsilon_\delta+i\eta} + \text{ exch.}),$$

which easily can be translated in explicit expressions.

A neat and fast way to obtain the whole series in a systematic way is due to Brenig and Wagner (Br 63). One formulates the iterative scheme with the aid of an external source:

$$Q := \Sigma \; a_\alpha^\dagger \, (t_1) \; q_{\alpha\beta}(t_1,t_2) \; a_\beta(t_2) \quad , \tag{IV.4}$$

which is put equal to zero after the expansion. By means of Dirac's perturbation theory (with respect to q) or Schwinger's principle one obtains the following coupled system (summation over all doubly occuring variables):

$$V_{1'1} = -2i \; v_{1'm1n} \; G_{nm} + iv_{1'man} \; G_{ar} \frac{\delta V_{r1}}{\delta_{qnm}}$$

$$\frac{\delta G_{1'1}}{\delta_{q2'2}} = G_{1'2'} \; G_{21} + G_{1k} \frac{\delta V_{ka}}{\delta_{q2'2}} \; G_{a1} \tag{IV.5}$$

$$= (-G^{(2)}_{1'212'} + G_{1'1} \; G_{22'}) \quad ,$$

which can be iterated starting with $\frac{\delta V}{\delta q} = 0$. (The advantage of this formulation is the use of the full (physical) propagator G from the start).

One essential point of the series - valid for <u>any</u> treatment of V^R - is the <u>non-occurence</u> of <u>only one</u> s.p. propagator in the intermediate states (for this reason V^R is also called the irreducible mass operator). This structure implies, that no resonances of V^R are located at the energies of the <u>compound N+1 - particle system</u>. Such a resonances would also contradict the Dyson-equation (III.6) for $G_{\alpha\mu}(\omega)$.

b) <u>PPH - GF and Mass Operator</u>

A more detailed analysis by Ethofer and Schuck (ES 69) and Winter (Wi 72, 75) shows, that the total resonant part of V^R can be (impli-

citly) defined by insertion of the correlated p-p-h (triple)-propaga-
tor into the intermediate states. This can be seen by inserting the
equation of motion for $G^{(2)}$ into the definition (III.9) of V^R and the
use of the Dyson-equation (III.6) (E 76). Explicitly one obtains the
symmetric expression:

$$V^R_{1'1}(t_{1'} - t_1) =$$

$$\Sigma \, v_{1'\beta\gamma\delta} \, \underbrace{(G^{(3)}_{\beta'\gamma\delta,\gamma'\delta'\beta} - G^{(2)}_{\gamma\delta\beta\mu} \, G^{-1}_{\mu\gamma} \, G^{(2)}_{\nu\beta'\gamma'\delta'})}_{=: \, R_{\beta'\gamma\delta,\gamma'\delta'\beta}(t_{1'} - t_1)} \, v_{\gamma'\delta'1\beta'} \qquad (IV.6)$$

In relation (IV.6) the propagator R describes the propagation of a
correlated (p-p-h) triple (due to the instantanous interaction v
the time arguments of $\beta'\gamma'\delta'$ ($a^\dagger_{\gamma'}(t_1) a^\dagger_{\delta'}(t_1) a_\beta(t_1)$) and $\beta\gamma\delta(t_{1'})$,
respectively, are equal). The second term removes the resonances of
the compound N+1-system ("non-doorway states"). (If one would take
$G^{(3)}$ alone in (IV.6), one obtains directly the T-matrix without the
shape-elastic part.)

The Bethe-Salpeter-equation for R is rather complicated. If we
compare the problem with the propagation of three particles in vacuo
(Faddeev problem) - which is known to be difficult - we encounter for
the triple propagation an even more complicated situation due to the
possibility of (retarded) interactions via p-h-pairs, for instance:

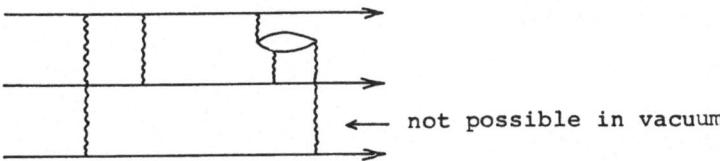

One can derive three equivalent versions of the Bethe-Salpeter equation, depending on the chosen channel-subsummations. The formal structure of the BS-equations for R reads as follows:

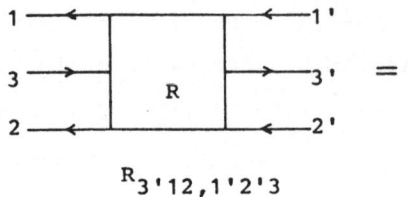

$$R_{3'12,1'2'3}$$

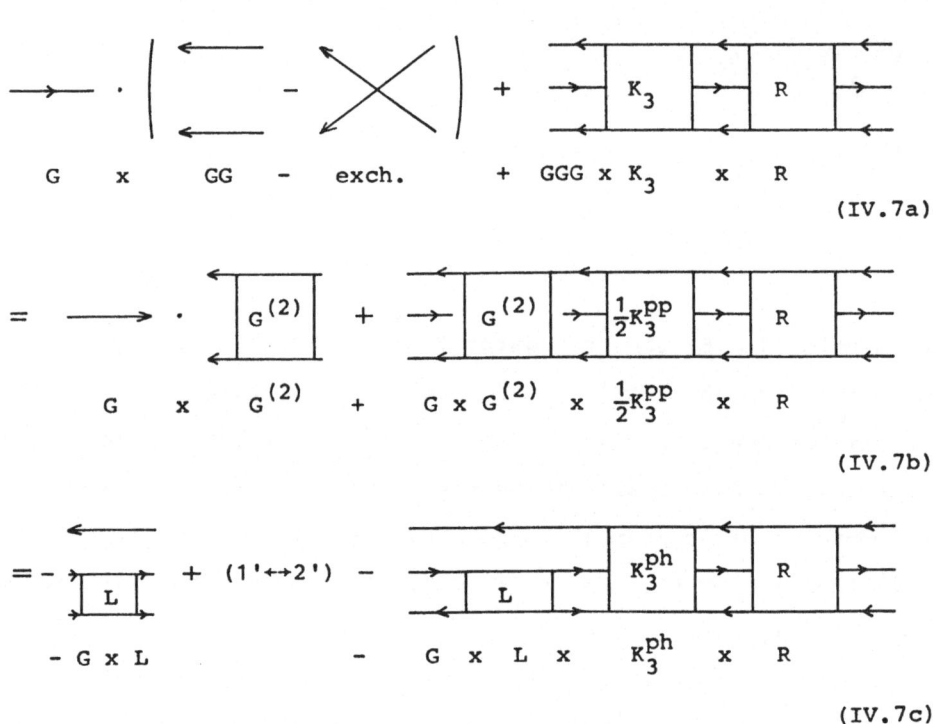

$$= \qquad \cdot \left(\begin{array}{c} \longleftarrow \\ \\ \longrightarrow \end{array} - \times \right) + \quad K_3 \quad R \tag{IV.7a}$$

G x GG - exch. + GGG x K_3 x R

$$= \qquad \cdot G^{(2)} + G^{(2)} \quad \tfrac{1}{2}K_3^{pp} \quad R \tag{IV.7b}$$

G x $G^{(2)}$ + G x $G^{(2)}$ x $\tfrac{1}{2}K_3^{pp}$ x R

$$= - L + (1' \leftrightarrow 2') - L \quad K_3^{ph} \quad R \tag{IV.7c}$$

- G x L $\qquad$ - G x L x K_3^{ph} x R

Here, $G^{(2)}$ denotes $G_{12,1'2'}$ (p-p-channel) and L the so-called linear response function $L = G_{121'2'} - G_{11'}G_{22'}$ (p-h-channel). The effective three-body forces K differ in the various channels (for details see Wi 75, E 76, ES 69).

Due to

$$V^R(\omega) = v \, R(\omega) \, v \tag{IV.6b}$$

$R(\omega)$ determines uniquely the energy-dependence of $V^R(\omega)$. Since $R(\omega)$ is composed of certain GF, it obbeys a spectral representation of similiar form as the 2-point GF (see III.3-5):

$$R_{123,1'2'3'}(\omega) = R_{\cdots}^{(+)}(\omega) + R_{\cdots}^{(-)}(\omega) \qquad (IV.8)$$

with

$$R_{\cdots}^{(\pm)}(\omega) = \int_{-\infty}^{+\infty} d\omega' \frac{S_{\cdots}^{(\pm)}(\omega')}{\omega'-\omega\pm i\eta} \qquad (IV.9)$$

and the spectral functions

$$S_{123,1'2'3'}^{(\pm)}(\omega) = \sum_{\nu\epsilon\left\{\begin{matrix}2p1h\\2h1p\end{matrix}\right\}} \rho_{233'}^{\nu} \rho_{1'2'1}^{\nu*} \delta(\omega-e_{\nu}) \qquad (IV.10)$$

The imaginary part of v^R is directly given by the spectral functions

$$Im\ v_{1'1}^{R}(\omega) = \pi \sum v_{1'\delta\beta\gamma} (S_{\delta'\beta\gamma,\beta'\gamma'\delta}^{(+)}(\omega) - S_{\cdots}^{(-)}) v_{\beta'\gamma'1\delta'} \qquad (IV.11)$$

The explicit form for the antisymmetric triple amplitudes ρ was given by Winter (Wi 72, 75), who formulated the "doorway"-RPA eigenproblem for the amplitudes ρ and the energies e_{ν}. (The direct solution of the equation of motion for the propagators is always more complicated as the corresponding eigenproblem!) This can be achieved with the standard procedure of the GF-theory by inserting the spectral decomposition into the Bethe-Salpeter equation for the triple propagator (integrated over the redundant energy-variables) and taking the limit $\omega \rightarrow e_{\nu}$. The eigenproblem has the standard form, which reads in the quasiparticle approximation as follows:

$$(e_{\nu}1 - \tilde{H}_S - K_3)\rho^{\nu} = 0 \qquad (IV.12)$$

with

$$(IV.13)$$

$$\tilde{H}_S = \begin{pmatrix} (\varepsilon_{p_1}+\varepsilon_{p_2}+\varepsilon_{h_3})\delta_{p_1p_1'}\delta_{p_2p_2'}\delta_{h_3h_3'} & 0 \\ 0 & -(\varepsilon_{h_1}+\varepsilon_{h_2}+\varepsilon_{p_3})\delta_{h_1h_1'}\delta_{h_2h_2'}\delta_{p_3p_3'} \end{pmatrix},$$

$$K_3 = \begin{pmatrix} K_{h_3'p_1p_2,p_1'p_2'h_3} & K_{p_3'p_1p_2,h_1'h_2'h_3} \\ \\ K_{h_3'h_1h_2,p_1'p_2'p_3} & K_{p_3'h_1h_2,h_1'h_2'p_3} \end{pmatrix} \qquad (IV.14)$$

$$\rho^\nu = \begin{pmatrix} \rho^\nu_{p_1'p_2'h_3'} \\ \\ \rho^\nu_{h_1'h_2'p_3'} \end{pmatrix} \qquad (IV.15)$$

The three-body interaction K_3 can be split into two contributions.1)A symmetrically Faddeev part ($K^{pp}_{12,1'2'}\delta_{33'}+...$), which contains the irreducible p-p- and p-h-two-body-interactions (K^{pp} and K^{ph}), which ought to be used in the p-p- and p-h-RPA, respectively. (Calculation of the eigenstates of the N±2-particle system and the exited states of the N-particle system.) 2) The second part is a true three-body interaction and can be defined in term of Feynman graphs.

c) Weak-Coupling Limit

So far, in most cases the so-called weak-coupling limit has been used in numerical calculations. Expressed in the language of the triple propagator, such approximations take only the disconnected part of R into account. The first term of (IV.7a) would correspond to the term given in (IV.3). If one considers the unconnected terms of (IV.7c) one obtains a theory like that of Vinh Mau (splitting of the triple propagation into physical phonons plus particle), which leads to following contributions ($H|N> = E_N|N>$)

$$V^R_{1'1}(\omega) = \qquad\qquad\qquad\qquad (IV.16)$$

$$= \left(2 \sum_{|N> \neq |0>} \sum_{|N+1>} \left\{ v_{1'\beta\gamma\delta} \frac{<0|a_\beta^\dagger a_\delta|N><0|a_\gamma|N+1><N+1|a_{\gamma'}|0><N|a_\delta^\dagger, a_{\gamma'}|0>}{\omega - (E_N + E_{N+1}) + i\eta} \right. \right)$$

$$\left. x \quad v_{\gamma'\delta'1\beta} \right\}$$

$$+$$

$$= (2 \Sigma v \frac{<N|a^\dagger a|0><N-1|a|0>...}{\omega + (E_N + E_{N-1}) - i\eta} v)$$

Decomposition according to the first term of (IV.7b) gives the following result $(H|N\pm2> = E_{N\pm2}|N\pm2>)$; virtual transition to an eigenstate of the N+2-system plus one hole):

$$v_{1'1}^R (\omega) = \qquad\qquad\qquad\qquad (IV.17)$$

$$= (\Sigma v \frac{<0|aa|N+2><0|a^\dagger|N-1><N-1|a|0><N+2|a^\dagger a^\dagger|0>}{\omega - (E_{N+2} - E_{N-1}) + i\eta} v)$$

$$+ \text{"exch."}$$

The physical interpretation of these graphs is obvious. The great advantage of such approximations is the possibility to use the outcome of nuclear structure calculations (for instance the p-h-RPA for $|N>$ gives $<0|a^\dagger a|N>$ and E_N etc.) in the determination of the optical potential. In this manner one reaches a more unique description of several nuclear quantities. The ladder approximation is a special case of (IV.17), in which the pair correlation is calculated by replacing the irreducible p-p-interaction K^{pp} by the free nucleon-nucleon force (see contribution of Prof. Mahaux). This approximation implies (refinements are possible), that two particles in matter can interact only via the instantanous two-body interaction (interaction via p-h-creation is not included). The influence of the medium enters only via the Pauli principle, which leads to a density dependent force. The advantage of this method is the folding structure of the <u>total</u> GOP of the following form (T denotes the effective scattering matrix in matter).

$$V_{1'1}(\omega) = \begin{array}{c} \boxed{T} \\ 1' \quad 1 \end{array}$$

$$\approx \sum_{\mu} T_{1'\mu 1\mu}(\omega + \varepsilon_\mu) n_\mu$$

(IV.18)

V. Dispersion Relations

From the analytic structure of the GOP and the causality condition one can easily deduce the following dispersion relation (Fe 58, 62; Li 66; Pa 67, 68).

$$\langle \vec{r}| \left(\text{Re } V^{opt}(\varepsilon) - V^{HF} - \sum_{\nu} \frac{\lambda_\nu}{\varepsilon - \varepsilon_\nu} - \frac{P}{\pi} \int_{\varepsilon_s}^{\infty} d\varepsilon' \frac{\text{Im} V^{opt}(\varepsilon')}{\varepsilon' - \varepsilon} \right) |\vec{r}'\rangle = 0, \qquad (V.1)$$

which holds for non-singular forces. Since the convergence of the principal-value integral is slow, one can also use subtracted versions of the dispersion relations, for instance (ε_o fixed):

$$\langle \vec{r} | \{ Re(V^{opt}(\varepsilon) - V^{opt}(\varepsilon_o) + \sum_\nu \lambda_\nu \frac{\varepsilon - \varepsilon_o}{(\varepsilon - \varepsilon_\nu)(\varepsilon_o - \varepsilon_\nu)}$$

$$- \frac{\varepsilon - \varepsilon_o}{\pi} P \int_{\varepsilon_s}^{\infty} d\varepsilon' \frac{Im V^{opt}(\varepsilon')}{(\varepsilon' - \varepsilon)(\varepsilon' - \varepsilon_o)} \} | \vec{r}' \rangle = 0. \tag{V.2}$$

The generalization for singular forces (hard-core with radius c) can also be obtained (Wi 78, Ga 76) and leads to a relation with the following general structure:

$$Re \, V(\varepsilon) = Re \, V^{hc}(\varepsilon) + \text{HF-correlation term} \, (\ldots \int_c^\infty drv(r) \ldots)$$

$$+ \frac{P}{\pi} \int d\varepsilon' \, \frac{Im(V(\varepsilon') - V^{hc}(\varepsilon'))}{\varepsilon' - \varepsilon} \quad . \tag{V.3}$$

The dispersion relations allow the calculation of the energy-dependent real part of the optical potential with the knowledge of the imaginary part, which often is easier to determine (see for instance (IV.11)). Furthermore, due to the properties of the principal value, one needs the correct behaviour of the imaginary part mainly in the region of stronger energy-dependence. Since the dispersion relation can be obtained with quite general assumptions, one can in principle either test a specific treatment of the optical potential or use the dispersion relation directly for the calculation of the real part, respectively. (For an extensive discussion see Pa 76).

VI. Discussion

The outcome of numerical calculations for the GOP in the weak-coupling limit has been reported in greater detail in several contributions. Therefore we are going to concentrate the discussion on the use of dispersion relations for the optical potential. We have chosen two examples, which may illustrate the situation. The first two figures - taken from ref. (Ga 76) - show the energy- and momentum dependence of the real - and imaginary part of the mass operator in nuclear matter. (The bars indicate the on-energy-shell values.) The calculation of the real and imaginary part was performed in a kind of ladder approximation (Ga 76) <u>not</u> using the dispersion relation. The figures exhibit a typical dispersion relation behaviour, that the maximum of $ImV(p,\omega)$ nearly coincides with the point of inflexion of Re $V(p,\omega)$. We may therefore conclude, that the direct use of the dispersion relation with a calculated Im $V(\omega)$ is a useful tool to obtain the real part of the potential.

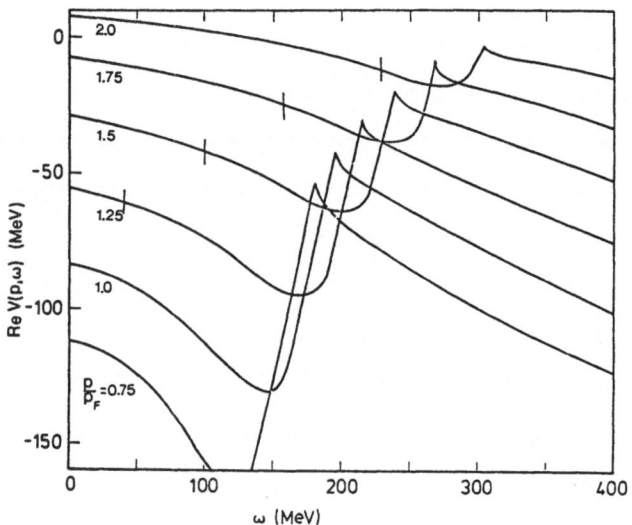

Fig. 1 The real part of the mass operator in nuclear matter versus ω
for different values of p/p_F (Ga 76). The bars indicate the
on-energy values

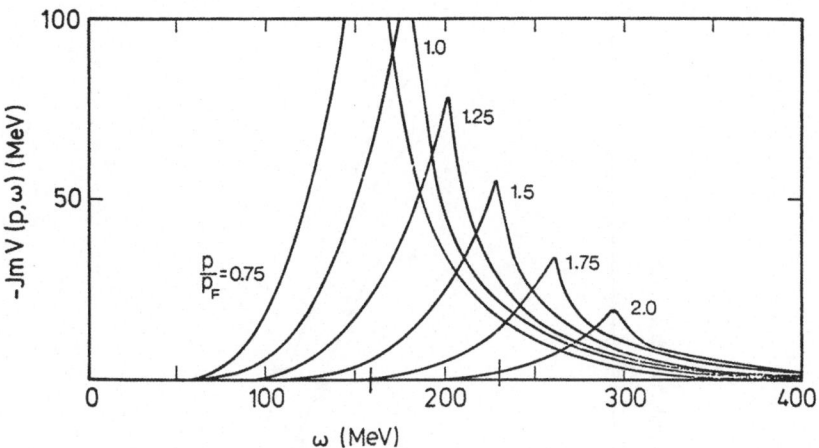

Fig. 2 The imaginary part of the mass operator in nuclear matter

versus ω for different values $p_{/p_F}$ (Ga 76).

In the next figures - taken from ref. EW 76 - we show the outcome of
a dispersion relation method by using a calculated imaginary potential
(for a detailed discussion see EW 76). In the first figure the energy-dependence of the depth of Re V(E, r = 0) in comparison with the
experimental values is shown. It seems, that the energy-dependence is
correct, but the absolute values are too great (A). If one uses the
renormalized version - fixed at 400 MeV - one obtains the curve B. In
the next figure we have explicitly shown the total - and dispersive
potential.   It can be seen, that the dispersive part is even for
smaller energies important and dominates for higher energies. The
Hartree-Fock-term reduces for higher energies to a small constant
Hartree potential above 300 MeV.

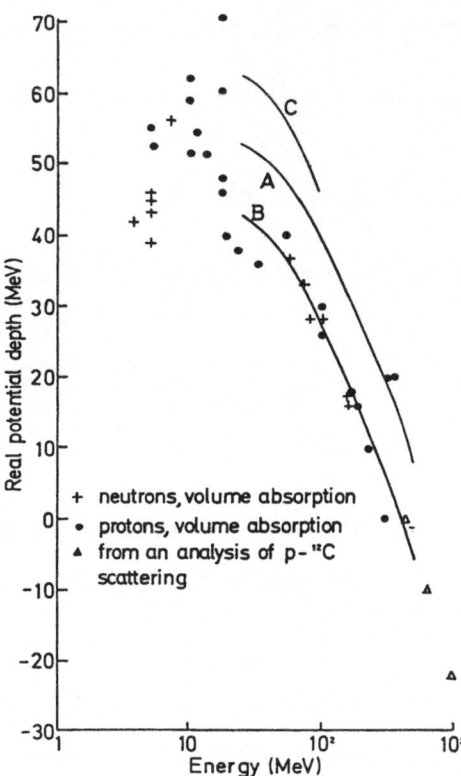

Fig. 3 Energy dependence of the total equivalent local real potential
$V^{opt}(r = 0, \varepsilon)$ compared with phenomenological data (EW 76). A,
real part obtained with the unsubtracted dispersion relation;
B, real part renormalized with the subtracted dispersion relation (ε_o = 400 MeV); C, real potential depth in the Greenless approximation (Ro 74).

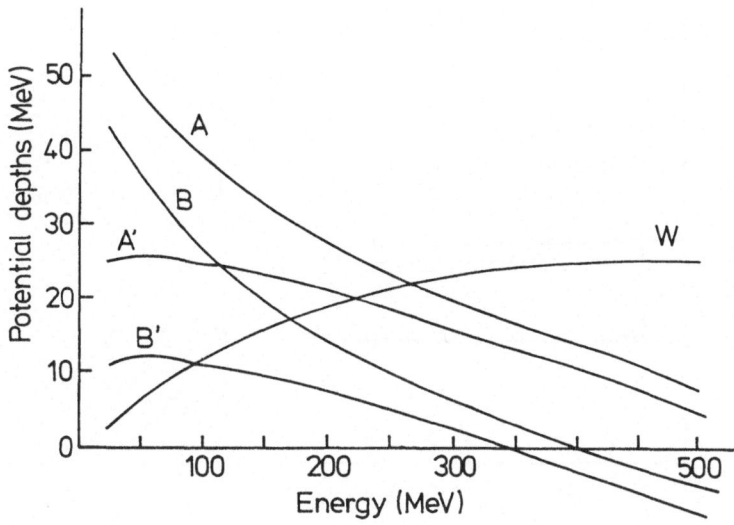

Fig. 4   Energy variation of the different local potential depths; for
A and B, respectively, see fig. 3. A' and B' show the dispersive contributions for A and B. W denotes the imaginary part
of the optical potential.

The radial dependence of V^{opt} for Au (A = 197) is shown in the last
figures for several energies (obtained with the unsubtracted dispersion relation; the outcome for the subtracted dispersion relation can
be obtained by using fig. 4). For the total potential one gets with
respect to energy- and radial dependence on the average the exspected
behaviour (Ho 71).

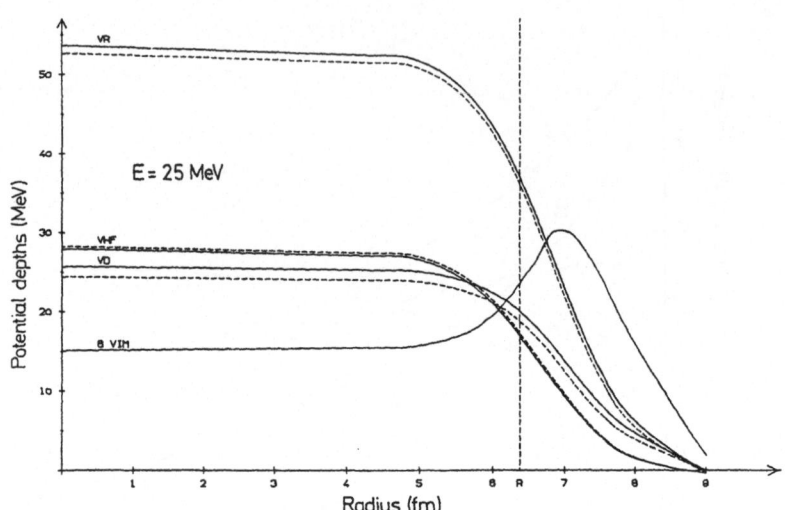

Fig. 5   Radial dependence of the (non-renormalized) equivalent local

optical potential for incident energy E = 25 MeV (VR, real

part; VD, dispersive contribution, VHF, Hartree-Fock term;

VIM, imaginary potential). The renormalized potential is ob-

tainable by use of fig. 4 (factor 0.81). The full (broken)

curves correspond to the nonlocal (local) approach for the

imaginary part.

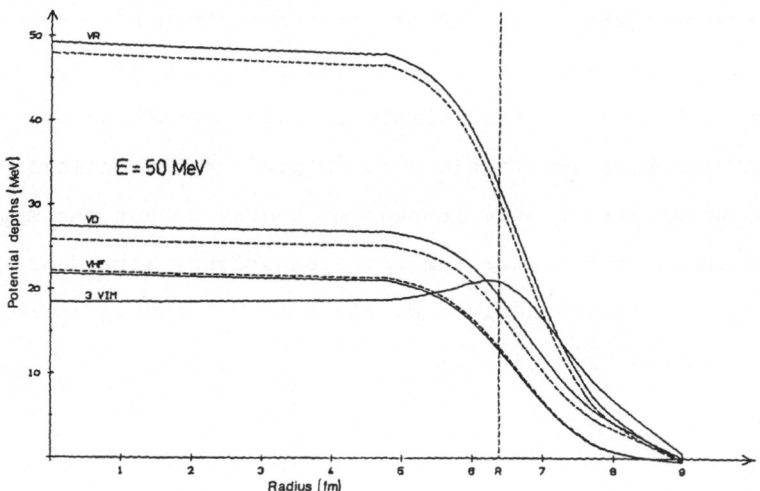

Fig. 6 Optical potential as a function of the radial distance for incident energy E = 50 MeV (for notations see fig. 5).

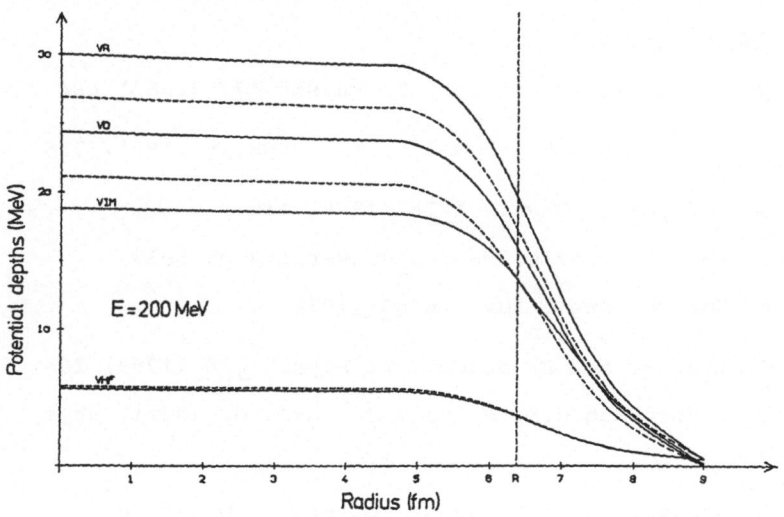

Fig. 7 Optical potential as a function of the radial distance for incident energy E = 200 MeV (for notations see fig. 5).

The curves show again, that the dispersive contribution is very important and plays the essential role for higher energies. For this reason, one can not exspect, that simple folding procedures with energy-independent two-body interactions can reproduce the optical potential in a wide energy-range. Only in smaller energy-ranges one may simulate in a phenomenological manner the right behaviour, since one has in certain regions simultanously a decrease of V^{HF} and an increase of V^{Dis}, respectively (see fig. 4).

References:

Ab 63   A.A. Abrikosov, L. Gorkov and I. Dzyaloshinsky; Methods of Quantum Field Theory in Statistical Physics (Pergamon 1963)

Be 59   J.S. Bell and E.J. Squires; Phys. Rev. Lett. $\underline{3}$ (1959) 96

Ba 62   G. Baym and L.P. Kadanoff; Quantum Statistical Mechanics (Benjamin 1962)

Br 63   W. Brenig and H. Wagner; Z. Physik $\underline{173}$ (1963) 484

Co 58   F. Coester and H. Kümmel; Nucl. Phys. $\underline{9}$ (1958) 225

E 76   G. Eckart; Z. Physik $\underline{A278}$ (1976) 145

         G. Eckart; thesis, Munich university pp 5-39

Em 71   K. Emrich; Nucl. Phys. $\underline{A160}$ (1971) 1

ES 69   S. Ethofer and P. Schuck; Z. Physik $\underline{228}$ (1969) 264

EW 76   G. Eckart and M.K. Weigel; J. Phys. G. (Nucl. Phys.) $\underline{2}$ (1976) 487

         G. Eckart; thesis, Munich university pp 40-76

Fe 58   H. Feshbach; Ann. Phys. $\underline{5}$ (1958) 357

Fe 62   H. Feshbach; Ann. Phys. $\underline{19}$ (1962) 287

Fet 65   A. Fetter and K. Watson; Adv. Theor. Phys. $\underline{1}$ (1965) 115

Fe 71   A.L. Fetter and J.D. Walecka; Quantum Theory of Many-Particle Systems (McGraw-Hill 1971)

Ga 76   H. Gall and M.K. Weigel; Z. Physik $\underline{A276}$ (1976) 45

Ho 71   P. Hodgson; Nuclear Reactions and Nuclear Structure
         (Clarendon 1971)

Ke 59   A.K. Kerman, H. McManus and R.M. Thaler; Ann. Phys. $\underline{8}$ (1959)
         551

Ko 62   D.H. Kobe; Ann. Phys. $\underline{19}$ (1962) 448

Li 66   R. Lipperheide; Nucl. Phys. $\underline{89}$ (1966) 97

Mat 76   R.D. Mattuck; A guide to Feynman diagrams in the many-body
         problem (McGraw-Hill 1976)

Ma 59   P.C. Martin and J. Schwinger; Phys. Rev. $\underline{115}$ (1959) 1342

Mi 67   A.B. Migdal; Theory of Finite Fermi Systems (Wiley 1967)

Pa 67   G. Passatore; Nucl. Phys. $\underline{A95}$ (1967) 694; Nucl. Phys. $\underline{A110}$
   68   (1968) 91

Pa 76   G. Passatore in Nuclear Optical Model Potential in Lecture
         Notes in Physics (Springer 1976) pp 1-19, pp 177-203

Na 60   N. Naminiki; Prog. Theor. Phys. $\underline{23}$ (1960) 629

Ro 74   J.R. Rook; Nucl. Phys. $\underline{A222}$ (1974) 596

Vi 67   F. Villars; in Fundamentals in Nuclear Theory (Vienna 1967;
         IAEA) pp 269-332

We 71   M.K. Weigel and G. Wegmann; Fortschritte d. Phys. $\underline{19}$ (1971)
         451

Wi 72   J. Winter; Nucl. Phys. $\underline{A194}$ (1972) 535

Wi 75   J. Winter; thesis, Munich university 1975

Wi 78   J. Winter; Fortschritte d. Phys. $\underline{26}$ (1978) 29

Zh 65   F.A. Zhivopistsev; Sov. J. Nucl. Phys. $\underline{1}$ (1965) 429

Nucleon scattering from nuclei with nuclear matter t-matrices

F.A. Brieva

Nuclear Physics Laboratory, Oxford

1. Introduction

The presence of strong interactions between two free nucleons makes the usual perturbation theory unacceptable for attempting a microscopic description of nuclear structure and the scattering of nucleons from nuclei. In order to overcome this problem, it has been realized for many years that the introduction of effective interactions, derived from a free internucleon force, is a first step towards achieving this goal. Two different situations can be envisaged: firstly, the calculation of the effective interaction between bound nucleons [SIE 70,NE 70,SP 71] where a high degree of sophistication has been reached and very encouraging results obtained and secondly, the extension to the case in which one nucleon is unbound, namely elastic and inelastic scattering. This later problem has remained open and the developments have followed a rather phenomenological character, except perhaps in the high energy region where the impulse approximation is valid. Only recently, consistent efforts towards the calculation of the optical model potential starting from a realistic internucleon force [JE 74,JE 76,BRI 77] have given some insight on the effective interaction for nucleon scattering.

Since the calculation of effective interactions for bound nucleons has been reviewed elsewhere (see, for example, [NEM 71]), we shall restrict ourselves to a discussion of the calculation of effective forces to be used in the description of elastic and inelastic nucleon scattering. A first approximation to the true effective interaction in nuclei, t-matrix, which is generally complex in the scattering situation under study, is to evaluate it in nuclear matter within the framework of Brueckner's theory [BRU 58,HU 72] . The implicit hypothesis is that the true complex t-matrix, which would include any specific effects of the excitation spectrum for the particular nucleus we consider, can be approximately replaced by the nuclear matter t-matrix at the local density and that this procedure includes excitation effects in an average way. Further, though not necessarily adequate, simplifications enable a local representation of this t-matrix in coordinate space to be obtained. Thus, the resulting effective interaction depends on the internucleon separation distance as well as on the density of the target nucleus and the energy of the incident nucleon [BRI 77].

The extension of the nuclear matter t-matrix for the case of
finite nuclei is usually made through the local density approximation
[BE 68] and it is the point of view adopted here. The most direct
situation where these complex effective forces can be applied is the
calculation of the optical model potential for elastic nucleon
scattering from nuclei. This is obtained, in first order, by folding
the effective internucleon force with the nuclear matter density
describing the target nucleus and including exchange effects. Such
an approach may be regarded as complementary to that of Jeukenne et al.
[JE 77] who calculated the optical model directly in nuclear matter
and then obtained the potential for finite nuclei by using an
improved version of the local density approximation. Similar folding
prescriptions have been used previously (see [SIN 75] and references
therein) to calculate the real part of the optical potential while
leaving the absorptive component to be determined phenomenologically.

The folding model has also been extended [MA 76,BRI 78a] to
describe inelastic nucleon scattering from strongly deformed nuclei
and within the framework of the rotational model. In this case, the
deformed nucleon-nucleus optical model potential takes into account
the intrinsic nuclear degrees of freedom while the rotational degrees
of freedom are explicitly treated through the coupled-channel
formalism [GL 67].

A last set of applications to the effective interactions are
the calculation of inelastic transitions within the distorted wave
t-matrix approximation. Some encouraging results have recently been
reported [BRI 78b] and further studies can be found in [GER 78a].

In general, by using the nuclear matter t-matrix, we hope to
have a consistent parameter-free description of nucleon scattering
from nuclei and over a wide range of energies. In particular, if
this approach gives a satisfactory agreement with the experimental
data, we can test the reliability of the standard phenomenological
models. Further, if the accuracy of the theory allows it, we can
attempt to extract structure information. At the same time we must
try to establish the limitations of the present approach. In fact,
it should not be surprising that the theory fails to describe
scattering from light nuclei where the plane wave and local density
approximations are inadequate as well as when the properties of a
particular nucleus are strongly coupled to the reaction process,
resonances, collective effects, channels opening, etc. The high

energy region, say incident nucleon energy above 100 MeV, is also difficult to test mainly due to the lack of extensive experimental data. This problem prevents an investigation of the need for higher order terms in the multiple scattering series which might be expected at high energies.

2. Effective interactions

We shall assume that the interaction between two nucleons is well described by the transition operator $t(\omega)$ (or reaction matrix $g(\omega)$) which sums all the ladder diagrams [BRU 58, JE 76] and is a solution of the Bethe-Goldstone integral equation [Be 57]

$$ t(\omega) = V + V \sum_{a,b > k_F} \frac{|\vec{a},\vec{b}\rangle\langle\vec{a},\vec{b}|}{\omega - e(a) - e(b) + i\eta} t(\omega) , \qquad (2.1) $$

where ω is the starting energy, V is the free internucleon potential, $\vec{a}, \vec{b}$ are the momenta in the intermediate states, k_F is the Fermi momentum and $e(k)$ is the single-particle energy. Following [Je 76, Le 78a] we take ($\hbar = 1$)

$$ e(k) = \frac{k^2}{2m} + Re\, M(k, e(k)) , \qquad \text{for all k,} \qquad (2.2) $$

with $M(k, e(k))$ the leading term of the mass operator in the Brueckner-Hartree-Fock approximation [Hu 72],

$$ M(k, E) = \sum_{p < k_F} \langle \vec{p}\,\vec{k}| t(E + e(p))|\vec{p}\,\vec{k}\rangle_A . \qquad (2.3) $$

In eq.(2.2), the abbreviation Re refers to the real part of the mass operator while in eq.(2.3) the subindex A indicates that antisymmetrized matrix elements of the transition operator must be taken.

The operator equation (2.1) defines a two-nucleon correlated wavefunction $|\psi\rangle$ by

$$ V|\psi_\omega\rangle = t(\omega)|\phi\rangle , \qquad (2.4) $$

where $|\phi\rangle$ is a plane wave characterized by the relative momentum of the nucleon pair. From equations (2.1) and (2.4), $|\psi_\omega\rangle$ satisfies a Bethe-Goldstone integral equation,

$$ |\psi_\omega\rangle = |\phi\rangle + G(\omega) V|\psi_\omega\rangle , \qquad (2.5) $$

with G the two-nucleon propagator defined by eq.(2.1). The calculation of the two-nucleon correlated wavefunction has been

reported elsewhere [Je 74,BRI 77] . Then, the calculation of the
t-matrix elements and the mass operator, eq.(2.3), is straightforward.
Here we shall assume that the correlated wavefunction can be
calculated and proceed to define an effective interaction.

There are several ways of defining a coordinate representation
for the effective forces [NEM 71]. We require the effective
interaction to be local in coordinate space, that is depending only
on the internucleon distance, and that it reproduces the leading term
of the mass operator in nuclear matter when used in the Born
approximation [BRI 77]. This is satisfied by averaging over the
momentum of the bound nucleon. Thus we obtain in states of L, S, J
quantum numbers [BRI 77],

$$
t_{L'L}^{JS}(r;k_F,E) = \frac{\sum\limits_{\vec{p}<k_F}\sum\limits_{L''}\frac{1}{k_o^2}\mathcal{J}_{L'}(k_o r)\, V_{LL''}^{JS}(r)\, \overline{v}_{LL'',\alpha}^{JS}(r)}{\sum\limits_{\vec{p}<k_F}\frac{1}{k_o^2}\mathcal{J}_{L'}(k_o r)\mathcal{J}_{L}(k_o r)} \quad ,(2.6)
$$

where $k_o = \frac{1}{2}|\vec{k}-\vec{p}|$ is the relative momentum of the nucleon pair,
$\mathcal{J}_L(k_o r) = (k_o r)j_L(k_o r)$ with $j_L(x)$ the spherical Bessel function of
order L, $V_{LL''}^{JS}(r)$ are the reduced matrix elements of the realistic
internucleon force, $\overline{v}_{LL'',\alpha}^{JS}(r)$ is the radial part of the correlated
wavefunction with L" allowing for the tensor coupling in the free
internucleon force and α representing the dependence on E, k_F, $\vec{k}$
and $\vec{p}$.

This averaging procedure ensures that the weighted average
diagonal t-matrix element is reproduced correctly but no particular
element is correct [SP 71]. A similar approach has been used in
structure calculations [SI 70,NE 71] though carried a step further
and averaging over the momentum of the other interacting nucleon.
This is not possible when one of the nucleons is initially unbound
and therefore the dependence on its momentum remains explicit.
Using eq.(2.2), the momentum dependence can be expressed as an energy
dependence which seems more suitable for actual applications. An
interesting consequence of eq.(2.6) is that the momentum average
gives a part of the density dependence of the interaction besides the
dependence coming explicitly from the Pauli principle. This reflects
a particular assumption on the weighting procedure which seems
reasonable for heavier nuclei but not so clear for light nuclei. A
way of avoiding these approximations would be to insist in defining
a non-local, density and energy dependent effective interaction but,

at present, this alternative approach remains open.

Once the $t^{JS}_{LL'}(r;k_F,E)$ matrix elements are known, it is starightforward to calculate the different components (central, spin-orbit, tensor, etc.) of the effective interaction [BRI 78c]. Thus, for example, the central component is given by

$$t^{LS}_c (r; k_F, E) = \frac{\sum_J (2J+1)\, t^{JS}_{LL}(r; k_F, E)}{(2L+1)(2S+1)} \quad , \tag{2.7}$$

and the spin-orbit component by

$$t^{L,S=1}_{S.O.}(r; k_F, E) = \frac{\sum_J (2J+1)\, \langle \vec{S}\cdot\vec{L}\rangle\, t^{JS}_{LL}(r; k_F, E)}{2L(L+1)(2L+1)} \quad , \tag{2.8}$$

with

$$\langle \vec{S}\cdot\vec{L}\rangle = \tfrac{1}{2}\left[J(J+1) - L(L+1) - 2 \right] \quad . \tag{2.9}$$

In many applications it is convenient to introduce a further simplification. We define an L-independent effective interaction for given spin-isospin states,

$$t^{ST}(r; k_F, E) = \frac{\sum_L (2L+1)\, t^{LS}(r; k_F, E)\, W_L(r; k_F, E)}{\sum_L (2L+1)\, W_L(r; k_F, E)} \quad , \tag{2.10}$$

with

$$W_L(r; k_F, E) = \sum_{p<k_F} \frac{1}{k_0^2}\, j_L^2(k_0 r) \quad , \tag{2.11}$$

and where T refers to the total isospin of the nucleon pair and the sum over L is over even or odd values so as to have a totally antisymmetric state.

The t^{ST} effective interactions are complex and depend on density and energy. Calculations of the central and spin-orbit components of t^{ST} have been performed [BRI 77, BRI 78c] starting from the Hamada-Johnston internucleon force [HA 62]. The smooth dependence of the interaction on energy and density has allowed a reasonable parametrization of it as a linear combination of Gaussian form factors to be obtained,

$$t^{ST}(r; k_F, E) = \sum_{i=1}^{5} A_i(k_F, E)\, e^{-b_i r^2} \quad , \tag{2.12}$$

89

both for the real and imaginary parts of the central interaction.
The $A_i(k_F,E)$ coefficients have been tabulated for different values of
the Fermi momentum between $k_F = 0.6$ and 1.4 fm^{-1} and many energies in
the 5 - 150 MeV range.

As an illustrative example of the central components of the
interaction, we present in fig.1 the Fourier transforms of t_c^{ST},

$$t_c^{ST}(k;k_F,E) = \frac{4\pi}{k} \int_0^\infty t_c^{ST}(r;k_F,E) \sin(kr)\, r\, dr \quad, \quad (2.13)$$

for $k_F = 1$ fm^{-1} and corresponding to an incident energy E = 30 MeV.

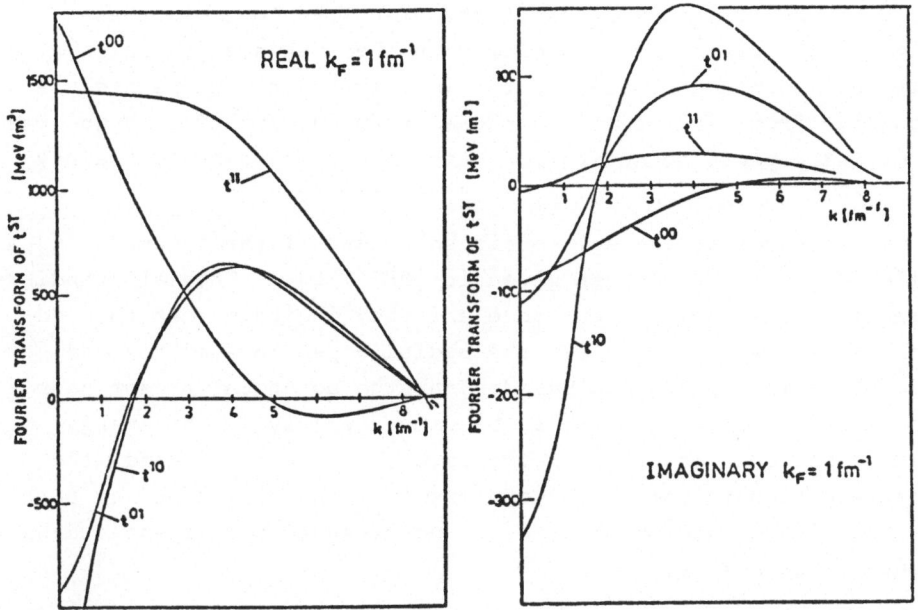

Fig.1 Fourier transform of the real and imaginary part of the
effective interaction t_c^{ST}; $k_F = 1$ fm^{-1} and E = 30 MeV.
From [BRI 78b].

The general features of the calculated forces are: for the
real part of the interaction, the even components are attractive in
the low momentum region and becoming eventually repulsive for
$k \gtrsim 2$ fm^{-1}. The odd components are strong and repulsive. The
behaviour in the low momentum region corroborates phenomenological
findings [BRI 78b], except perhaps for the singlet odd components.
This is not crucial due to the difficulties in determining empirically
the odd components of the interaction because of the strong

cancellation between the direct and exchange contributions. The density and energy dependence is rather small but changes do exist when the density decreases and/or the energy increases. One important feature is that the density dependence is smaller than that obtained from effective interactions between bound nucleons which makes the latter doubtful to use in a scattering situation. For the imaginary part of the force, the even components are also "attractive" in the low momentum region and changing sign for $k > 2$ fm^{-1}. However, the ratio between the singlet even and triplet even interactions is different from that found for the corresponding real components. The odd components are relatively small, a feature that will make exchange effects less important for the absorption. One important character-istic of these imaginary parts of the effective force is their strong density dependence, specially in the even states. Such density dependence decreases as the incident energy increases. Also, it is clear from fig.1 that the radial form factors of both the real and imaginary parts do not follow each other very closely, mainly those connected with the odd part of the forces.

With respect to the spin-orbit component of the interaction, a detailed discussion has been given in [BRI 78c] . The main findings concern the real part of the force and they indicate that the odd component follows very closely the radial dependence of the odd spin-orbit component in the Hamada-Johnston potential except near and in the hard core region where it becomes repulsive. The energy and density dependence was found negligible. On the other hand, the even interaction is somewhat different from the corresponding part in the realistic internucleon force and presents a noticeable density and energy dependence.

There are, at present, two main limitations in the calculation of the effective forces. They refer to the calculation of the interaction for very low values of the density ($k_F < 0.6$ fm^{-1}) and the inclusion of proton and neutron density differences. Such corrections will have eventually to be included for a better description of the effective interaction.

3. Applications of the effective interactions

The most direct application of the effective forces calculated from a free internucleon force is the calculation of the optical model potential for nucleon scattering from nuclei. In this respect, we shall present results using the folding model for elastic scattering and inelastic scattering from deformed nuclei. Another interesting set of applications is related to the microscopic calculation of inelastic transitions.

3.1 Folding model

If we consider the first term in a multiple scattering series [JO 63], the nucleon-nucleus optical potential can be written as the sum of a local direct term and a non-local exchange term [OW 70] , namely

$$M(\vec{r}_1,\vec{r}_1',E) = \delta(\vec{r}_1-\vec{r}_1') \sum_n \int \phi_n^*(\vec{r}_2) \, t_D(|\vec{r}_1-\vec{r}_2|, \rho, E) \, \phi_n(\vec{r}_2) \, d\vec{r}_2$$

$$+ \sum_n \phi_n^*(\vec{r}_1) \, t_{EX}(|\vec{r}_1-\vec{r}_1'|, \rho, E) \, \phi_n(\vec{r}_1') \,, \qquad (3.1)$$

where $\vec{r}_1$ and $\vec{r}_2$ refer to the incident and a bound nucleon respectively, $\phi_n(\vec{r}_2)$ is the bound-state single-particle wavefunction with n representing the appropriate quantum numbers and t_D and t_{EX} are the direct and exchange internucleon effective interactions. In this approximation to the optical potential, the source of non-locality lies only in the exchange term due to the assumption on the locality of the effective force. A more rigorous approach would have both the direct and exchange components of the potential non-local. The new feature of eq.(3.1) is that it gives simultaneously both the real and imaginary parts of the optical potential.

In order to relate the calculated optical potential to the phenomenological ones [HO 71] it is convenient to define a local equivalent optical potential, U, by

$$U(\vec{r}_1,E) \, \Psi(\vec{r}_1) = \int M(\vec{r}_1,\vec{r}_1',E) \, \Psi(\vec{r}_1') \, d\vec{r}_1' \,, \qquad (3.2)$$

where $\Psi(\vec{r}_1)$ is the scattering wavefunction of the incident nucleon. The nucleon optical potential can then be written in the standard form

$$U(\vec{r}_1,E) = U_c(\vec{r}_1,E) + U_{s.o.}(\vec{r}_1,E) \,, \qquad (3.3)$$

where the central component is given by [BRI 77]

$$U_c(\vec{r}_1,E) \equiv -V(\vec{r}_1,E) - i\,W(\vec{r}_1,E)$$

$$= \int \rho(\vec{r}_2) \, t_D(|\vec{r}_1-\vec{r}_2|, \rho, E) \, d\vec{r}_2 \quad +$$

$$\int \rho(\vec{r}_1,\vec{r}_2) \, t_{EX}(|\vec{r}_1-\vec{r}_2|, \rho, E) \, j_0(k\,|\vec{r}_1-\vec{r}_2|) \, d\vec{r}_2 \,, \qquad (3.4)$$

with ρ the target density and the spin-orbit component, for a spherical matter distribution, by [BRI 78c]

$$U_{s.o.}(\vec{r}_1, E) \equiv - \left[V_{s.o.}(\vec{r}_1, E) + i \, W_{s.o.}(\vec{r}_1, E) \right] \vec{l}_1 \cdot \vec{\sigma}_1$$

$$= - \frac{1}{3} \pi \left[B^D(\rho, E) - R^{EX}(\rho, E) \right] \frac{1}{r_1} \frac{\partial \rho(r_1)}{\partial r_1} \quad , \quad (3.5)$$

with

$$B^D(\rho, E) = \int t^D_{s.o.}(x, \rho, E) \, x^4 \, dx \quad , \quad (3.6)$$

and

$$R^{EX}(\rho, E) = \frac{3}{k} \int t^{EX}_{s.o.}(x, \rho, E) \, x^3 \, dx \quad . \quad (3.7)$$

In eqs. (3.4) and (3.5), a local momentum approximation for the exchange term has been used in order to factorize out the incident nucleon wavefunction $\Psi(\vec{r}_1)$ in eq.(3.2). The local momentum is calculated self-consistently (eq.(2.2)),

$$\frac{k^2}{2m} = E + V(\vec{r}_1, E) \quad . \quad (3.8)$$

This kind of approximation seems reasonable to reproduce the global features of the cross-section and polarization though in detail there are discrepancies with the results obtained with the non-local potential [GEO 78].

We also need a prescription to use the effective interactions in the local density approximation. A popular choice is to replace the nuclear matter density ρ_{NM} by

$$\rho_{NM} \longrightarrow \rho \left(\tfrac{1}{2} (\vec{r}_1 + \vec{r}_2) \right) \quad , \quad (3.9)$$

when evaluating the forces in the calculation of the different components of the optical potential. Alternative prescriptions do exist [NEM 71]. They make little difference in the real part of the optical potential while the absorption is more sensitive to this choice. In any case, this problem is intrinsic to the local density approximation and a weakness of the present approach.

In this first stage of microscopic calculations of the optical potential some further approximations have been used and we refer to [BRI 77] and [BRI 78c] for them. As an example, we mention the approximation of the single particle mixed density $\rho(\vec{r}, \vec{r}')$ in eq.(3.4) by the first term of an expansion proposed in [NE 72],

$$\rho \left(\vec{r}, \vec{r}\,' \right) \approx \rho \left(\tfrac{1}{2} \left(\vec{r} + \vec{r}\,' \right) \right) \frac{3}{s\, k_F} \, j_1 \left(s\, k_F \right) \quad , \qquad (3.10)$$

with $s = | \vec{r} - \vec{r}\,' |$ and the Fermi momentum k_F given by

$$k_F^3 = \frac{3}{2} \, \pi^2 \, \rho \left(\tfrac{1}{2} \left(\vec{r} + \vec{r}\,' \right) \right) \qquad (3.11)$$

Also, Coulomb corrections for proton scattering from nuclei have been included.

To illustrate the results obtained for the optical potential from the model already described, we present in fig.2 the real and imaginary parts of the central component (V and W) and the real part of the spin-orbit potential ($V_{S.O.}$) for proton elastic scattering from ^{40}Ca.

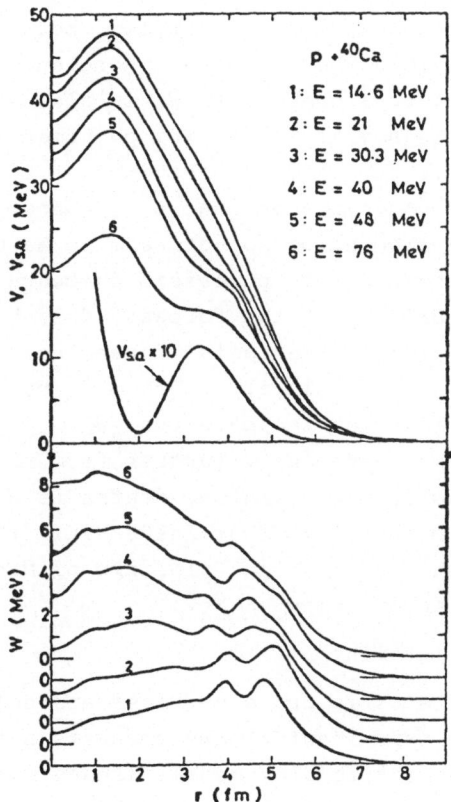

Fig.2 Radial and energy dependence of the optical potential for elastic scattering of protons from ^{40}Ca. From [BRI 78d].

The effective forces were calculated from the Hamada-Johnston potential and the matter density to describe ^{40}Ca was taken from [NE 70]. For the real part of the central component, V, we observe from fig.2 that the radial shape departs substantially from the usual Saxon-Woods parameterization so popular in phenomenological analysis. This implies that caution must be taken when trying to extract and interpret small components of the potential through the empirical analysis. The energy dependence of the potential form factor is due to the density and energy dependence of the effective force and the presence of exchange effects. The real part of the spin-orbit component, $V_{S.O.}$, is on the contrary only slightly energy dependent. As shown in fig.2, it peaks well inside the central component and the strength and shape hardly change in the energy range shown. This $V_{S.O.}$ component presents far less fluctuations, both in strength and radial parameterization, than that obtained phenomenologically. The imaginary part of the central component, W, is also shown in fig.2. In general it follows the same trend as the phenomenological absorption, that is, a surface peaked behaviour at lower energies while the volume term becomes predominant as the energy increases. The oscillations in the absorption do not seem to have any physical meaning and they are rather due to numerical inaccuracy in the original calculations [BRI 77] of the forces. This has been checked recently by Geramb [GER 78b]. The radial dependence of $W_{S.O.}$ has not been included in fig.2 for clarity. The main results have been discussed in [BRI 78c] and they indicate that $W_{S.O.}$ has different sign from $V_{S.O.}$, it peaks appreciably further from the nuclear centre than $V_{S.O.}$ and the ratio of the potential depths is $W_{S.O.}/V_{S.O.} \approx -0.05$ at E=20 MeV and of the order of -0.2 at E=200 MeV. The change of sign does not agree with the empirical findings of Mackintosh and Kobos [MA 78] in the region below 45 MeV.

3.2 Elastic nucleon scattering

The calculation of elastic cross sections and polarization and their comparison to the experimental ones presents a severe test of the effective forces and the folding model already discussed. Several results are already available using this approach [BRI 78d] or an alternative one developed by Jeukenne et al. [Je 77,Le 78b]. The important point is that essentially no parameters enter in the former calculations. Here, we shall take the proton plus ^{40}Ca system to illustrate the results obtained. The experimental data (for references, see van Oers [OE 71] or [BRI 78d]) cover the 15-76

MeV range. The theoretical results and the data are shown in fig.3
for the differential elastic cross sections and in fig.4 for the
polarization. We observe a reasonable agreement between theory and
experiment over this wide energy range. A better fit could eventually
be obtained by slightly modifying the strength of the different
components of the potential but, at the present stage, it seems more
appropriate to study the validity of the theory than to obtain the
best fits to the data.

Fig.3   Ratio-to-Rutherford
         differential cross
         sections for elastic
         proton scattering
         from ^{40}Ca at
         different energies.
         From [BRI 78d].

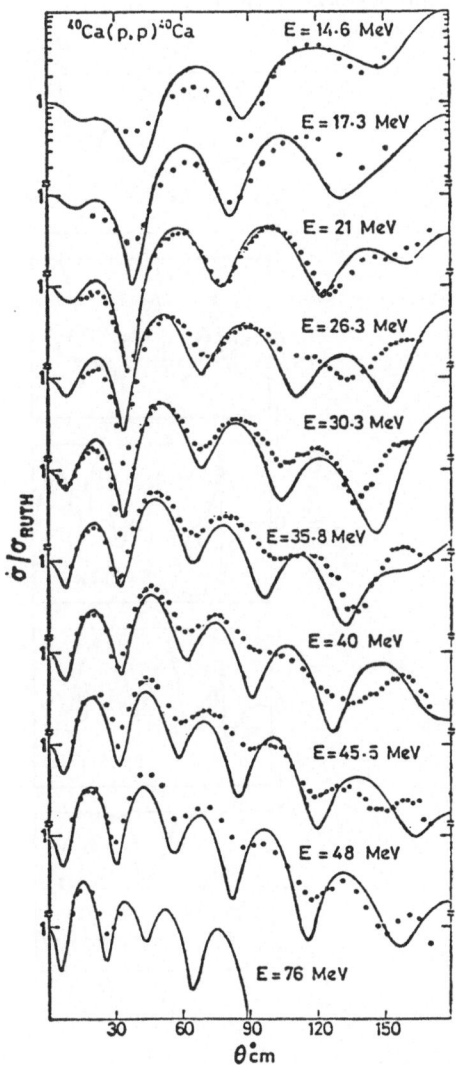

The cross sections for this reaction vary smoothly with energy, except around 16 MeV where the (p,n) reaction threshold for ^{40}Ca lies. Such a trend is, on average, well reproduced. There seems to be an exception around 26 MeV where the calculated potential does not give the correct oscillatory behaviour in the cross section at backward angles. The reason for such anomaly is probably the existence of strongly coupled channel resonance effects which are not included in the present approach to the optical potential. Similar anomaly for proton elastic scattering from ^{40}Ca has been detected by Kobos and Mackintosh [KO 78] in their detailed phenomenological analyses.

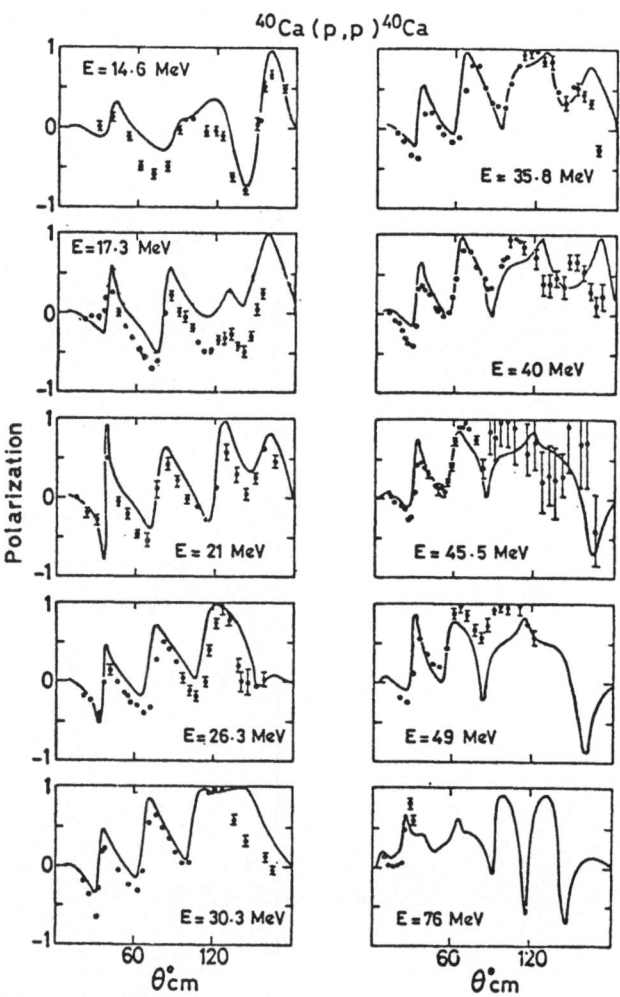

Fig. 4   Polarization for proton elastic scattering from ^{40}Ca at different energies. From BRI 78d .

The results obtained for the polarization (fig.4) are also very
satisfactory, except perhaps at 17.3 MeV where the oscillations of
the polarization are nicely reproduced but the magnitude is shifted.
The most interesting aspect regarding these results is that they
were obtained using an essentially constant spin-orbit term, both in
radial form factor and strength, over all the energy range. This
contrast with the phenomenological analysis of van Oers [OE 71] where
the radial parameters and strength vary greatly from one energy to
another. Therefore, it can be argued that in the standard
parameterization of the optical potential, the spin-orbit term is
trying to mock up small components of the potential indeed present in
the central part or components that have not been considered at all.
In this context, the ℓ-dependent optical potentials [KO 78] may help
to elucidate the problem.

A similar quality of fit can be obtained for nucleon elastic
scattering from heavier nuclei [LE 78b,BRI 78d,GER 78a]. However,
for proton scattering from lighter nuclei, say ^{16}O,the theoretical
cross sections and polarizations deviate considerably from the
experimental ones. This should not be surprising since the plane
wave and local density approximation used throughout are questionable.
Still, further work is required specially that related to the
several approximations used in the present model before more concrete
statements can be made.

3.2 Inelastic nucleon scattering from deformed nuclei.

The folding model finds an interesting field of applications in
the description of inelastic nucleon scattering from strongly deformed
nuclei [MA 76,BR 78a]. Based on the hypothesis that in a strongly
deformed nucleus all excitation levels have the same intrinsic
structure and differ only by different rates of rotations of the
system [GL 67], then we can calculate the nucleon-nucleus optical
potential by folding the effective internucleon interaction with the
matter density of the target nucleus and including exchange effects
in the body-fixed frame. This takes into account, in an average way,
the intrinsic nuclear degrees of freedom. Then, the degrees of
freedom associated with the rotation can be treated explicitly
through the coupled channel formalism [GL 67].

The calculation of the coupling interaction in the coupled-
channel formalism involves the knowledge of the nucleon-nucleus
optical potential, U, referred to the laboratory-fixed frame. If we
accept the validity of the adiabatic rotational model, then this

potential depends essentially on $\vec{r}_1$, the coordinate describing the
position of the incident particle, $\vec{\Omega}$, representing the collective
coordinates and E, the energy of the incident particle. To calculate
such an interaction, we write the following expansion for $U(\vec{r}_1, \vec{\Omega}, E)$
in the laboratory-fixed frame,

$$U(\vec{r}_1, \vec{\Omega}, E) = \sum_{\lambda, \mu} U_\lambda(r_1, E) \, Y_{\lambda\mu}(\hat{r}_1) \, D^\lambda_{\mu 0}(\vec{\Omega}) \, , \quad (3.12)$$

where $Y_{\lambda\mu}(\hat{r})$ is a spherical harmonics function and $D^\lambda_{\mu 0}(\vec{\Omega})$ is the
rotation matrix. The radial form factors in eq.(3.12) can be
extracted from the nucleon-nucleus potential calculated in the body-
fixed frame, $U(\vec{r}_1', E)$,

$$U_\lambda(r_1, E) = \int U(\vec{r}_1', E) \, Y_{\lambda 0}(\hat{r}_1') \, d\hat{r}_1' \quad . \quad (3.13)$$

In eq.3.13, the optical potential $U(\vec{r}_1', E)$ is calculated from the
folding model discussed earlier. This is consistent with the
adiabatic approximation where we neglect the excitation energies and
therefore we have essentially an elastic scattering situation as
viewed from the body-fixed frame.

Several applications using this approach have been made BRI 78a
Here, we present as an example the analysis of inelastic proton
scattering from [154]Sm at 50.8 MeV. To generate the optical potential,
the proton and neutron density were taken proportional and
parameterized as deformed Fermi distributions [MA 77, BR 78 a]. The
parameters characterizing the density were taken to be those extracted
from inelastic electron scattering measurements [CO 76]. The
comparison of the angular distributions from the 0^+, 2^+ and 4^+ states
of the ground state rotational band of [154]Sm with the experimental
data [WO 75] is presented in fig.5. The agreement is satisfactory.
Further, it can be appreciated that the present model improves on
the description of some detailed aspects of the cross section with
respect to some phenomenological analyses [WO 76]. One reason for
this is that the calculated form factors, like in the case of
elastic scattering discussed earlier, differ from the standard
parameterization and mainly as a consequence of the density dependence
of the interaction and the presence of exchange effects.

One advantage of this parameter-free description of nucleon
inelastic scattering from deformed nuclei is the possibility of
investigating directly the nuclear density deformation. Some work
done in this direction[BRI 78a] suggests, for example, that in [154]Sm

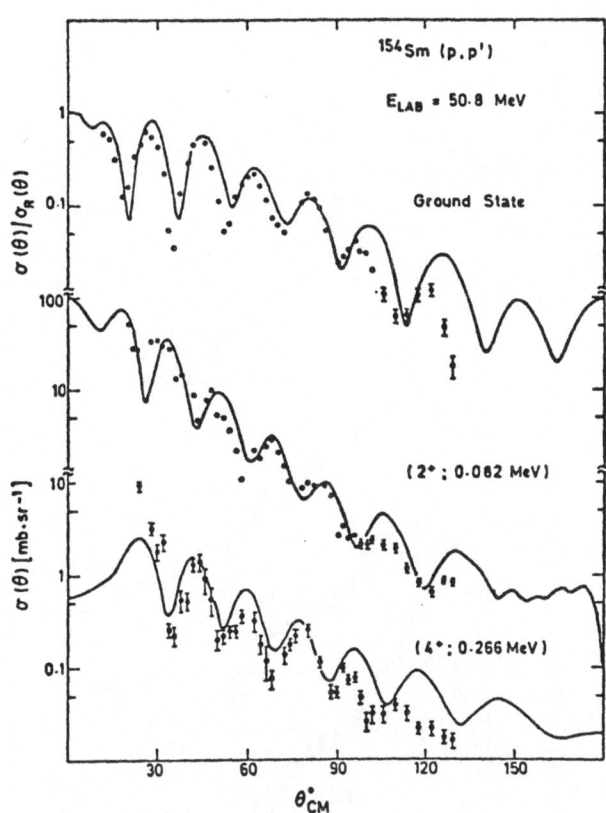

Fig.5 Scattering cross sections for 50.8 MeV protons from the
0⁺ ground state, 2⁺ and 4⁺ levels in ¹⁵⁴Sm.
From [BRI 78a].

the proton density deformation parameters are consistent with those
extracted from inelastic electron scattering and that the neutron
density deformation parameters are not likely to be more than around
15% smaller than those for the proton density.

A last remark is necessary with regard to inelastic proton
scattering from nuclei in the s-d shell. Again, as in the case of
elastic scattering from light nuclei, the model fails to give a
satisfactory fit to the data.

3.4   Inelastic transitions

The generality of the effective interactions calculated in
nuclear matter can be further tested by calculating inelastic
transitions to weakly coupled states and using the distorted wave
t-matrix approximation [DO 66, GER 71]. Some reservation, however,
should be expressed since the average procedures involved in the

definition of the effective forces imply that the off-shell properties
of the t-matrix are not necessarily correct and that some error is
made when selecting valence orbits only close to the Fermi surface,
as in fact is more or less done, in the calculation of the inelastic
transitions. Nevertheless, we expect that such problems are not
critical for the reproduction of the global features of the cross
sections.

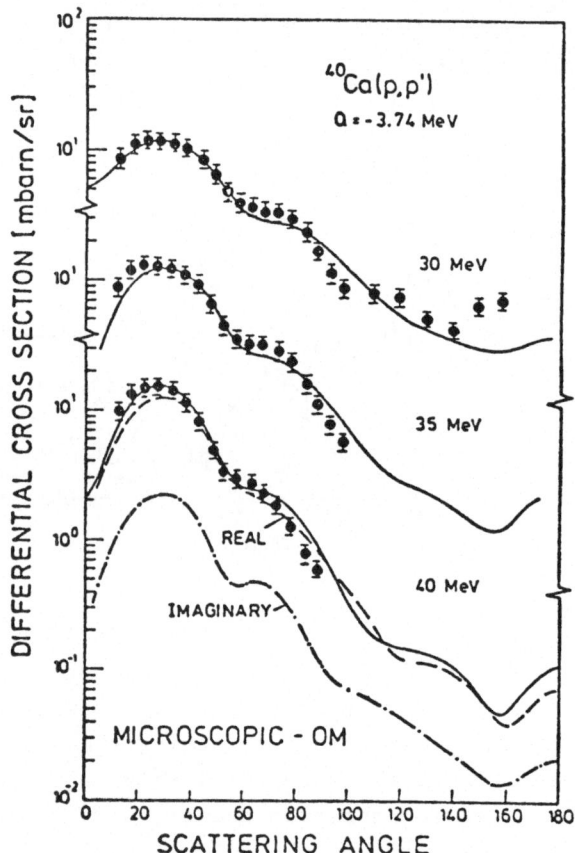

Fig.6 Differential cross sections for the inelastic scattering
of 30, 35 and 40 MeV protons from ^{40}Ca to the J ,
Q=3⁻, 3.74 MeV state.   From [BRI 78b].

As an example, we show in fig.6 the differential cross sections
for the inelastic scattering of 30, 35 and 40 MeV protons from ^{40}Ca
to the J^{π}, Q=3⁻, 3.74 MeV state [GR 72,BRI 78b]. The transition
density was constructed from an RPA vector which contained all
forward and backward going amplitudes limited within the s,d to f,p
shells. It reproduces reasonably the shape of the experimental
transition density while for the reproduction of the experimental

B(E3) value a renormalization of 1.3 is required. The optical potential for the entrance and exit channels were also calculated using the folding model discussed in sect.3.1. Thus we have a fully microscopic description of the inelastic transition. It was found that the theory underestimates systematically the cross sections for about 40% which is 10% larger than the electromagnetic renormalization. This later value sets more or less the limits within which the parameter free calculation reproduces the experimental data.

In general, all the features obtained by the microscopic calculation are in good agreement with the phenomenological findings. For example, the odd state forces yield large direct and exchange amplitudes but cancel to a great extent, leaving the even interactions to be the major contribution to the cross section. Also, the contribution of the imaginary part of the interaction to the transition amplitude is typically 30% of the real component. They interfere constructively at forward angles and destructively beyond a scattering angle around 100°. Overall, the presence of the imaginary component of the force yield cross sections with more pronounced maxima and minima. Further applications of the effective interactions to the calculation of inelastic transitions are reported by Geramb [GER 78a].

4. Summary

We have presented here a wide variety of results which show that a reasonable description of nucleon scattering data can be obtained starting from a free internucleon potential. This is achieved by calculating complex effective internucleon force which takes into account some many-body effects. Two main approximations have been used in the calculation of the interaction: firstly, the nuclear matter hypothesis which makes necessary the use of a local density approximation when trying to apply these interactions to the finite nucleus situation. In this respect, we expect that the average properties of the true effective interaction for the physical situation we consider are well represented. Secondly, the requirement that the effective interaction is local in coordinate representation. This simplifies the applicability of the effective forces although off-shell properties of the t-matrix are not necessarily well reproduced and it introduces additional density dependence.

To calculate the nucleon-nucleus optical potential we make use of the nuclear matter effective forces and the folding model.

Many-particle scattering effects are neglected. Moreover, a number of further approximations have been used. They refer to the approximation of exchange effects in order to extract a local optical potential, approximation of the mixed density matrix, etc. which may have some influence when calculating fine details of the cross sections and should therefore be improved upon. Nonetheless, the gross features of the experimental data are well reproduced.

The general trend of both theory and phenomenological models agrees well. However, the theoretical results seem to contain far more structure, specially in the calculated optical potentials and form factors, than phenomenology does. If this structure matters, it is possible to question some of the empirical findings and their interpretation.

Failure to reproduce the experimental data has been detected for nucleon scattering from light nuclei and when the properties of a nucleus are strongly coupled to the reaction process. This is not surprising since the approximations involved in the theory become less reliable under the conditions before mentioned. However, further studies are required before reaching a more definite conclusion.

References

BE 57   H.A. Bethe and J. Goldstone, Proc.Roy.Soc.A238(1957)551.

BE 68   H.A. Bethe, Phys.Rev.167(1968)879.

BRI 77   F.A. Brieva and J.R. Rook, Nucl.Phys.A291(1977)299,317.

BRI 78a   F.A. Brieva and B.Z. Georgiev, Nucl.Phys.,to be published.

BRI 78b   F.A. Brieva, H.V. Geramb and J.R. Rook, Phys.Lett.B, to
            be published.

BRI 78c   F.A. Brieva and J.R. Rook, Nucl.Phys.A297(1978)206.

BRI 78d   F.A. Brieva and J.R. Rook, Nucl.Phys.,to be published.

BRU 58   K.A. Brueckner and J.L. Gammel, Phys.Rev.109(1958)1023.

CO 76   T. Cooper et al., Phys.Rev.C13(1976)1083.

DO 66   L.R. Dodd and K.R. Greider, Phys.Rev.146(1966)675.

GEO 78   B.Z. Georgiev and R.S. Mackintosh, Phys.Lett.73B(1978)250.

GER 71   H.V. Geramb and K.A. Amos, Nucl.Phys.A163(1971)337.

GER 78a   H.V. Geramb, contribution to these proceedings.

GER 78b   H.V. Geramb, private communication.

GL 67   N.K. Glendenning, Proc.Int.School of Phys. Enrico Fermi,
            course 40 (1967), ed. M.Jean (Academic Press, NY,1969).

GR 72   C.R. Gruhn et al., Phys.Rev.C6(1972)415.

HA 62   T. Hamada and I.D. Johnston, Nucl.Phys.34(1962)382.

HO 71   P.E. Hodgson, Nuclear reactions and nuclear structure,
        (Clarendon Press, Oxford, 1971).

HU 72   J. Hüfner and C. Mahaux, Ann. of Phys.73(1972)525.

JE 74   J.P. Jeukenne, A. Lejeune and C. Mahaux, Phys.Rev.C10
        (1974)1391.

JE 76   J.P. Jeukenne, A. Lejeune and C. Mahaux, Phys.Reports 25C
        (1976)83.

JE 77   J.P. Jeukenne, A. Lejeune and C. Mahaux, Phys.Rev.C16
        (1977)80.

JO 63   P.B. Jones, The optical model in nuclear and particle
        physics, (Interscience publishers,1963).

KO 68   A.M. Kobos and R.S. Mackintosh, preprint, Daresbury
        Laboratory, No. DL/NSF/P79.

LE 78a A. Lejeune and C. Mahaux, Nucl.Phys.A295(1978)189.

LE 78b A. Lejeune and P.E. Hodgson, Nucl.Phys.A295(1978)301.

MA 76   R.S. Mackintosh, Nucl.Phys.A266(1976)379.

MA 77   R.S. Mackintosh, Rep.Prog.Phys.40(1977)731.

MA 78   R.S. Mackintosh and A.M. Kobos, J.Phys.G4(1978)L135.

NE 70   J.W. Negele, Phys.Rev.C1(1970)1260.

NE 72   J.W. Negele and D. Vautherin, Phys.Rev.C5(1972)1472.

NEM 71 J. Németh in The structure of nuclei, International
        course on nuclear theory, Trieste, 1971.

OE 71   W.T.H. van Oers, Phys.Rev.C3(1971)1550.

OW 70   L.W. Owen and G.R. Satchler, Phys.Rev.Lett 25 (1970)1720.

SIE 70 P.J. Siemens, Nucl.Phys.A141(1970)225.

SIN 75 B. Sinha, Phys.Reports 20C(1975)1.

SP 71   D.W.L. Sprung and P.K. Banerjee, Nucl.Phys.A168(1971)273.

WO 75   P.B. Woollam et al., Nucl.Phys.A179(1972)657.

Effective Nuclear Matter Interactions
Applied to Finite Nuclei

H.V. von Geramb
University of Hamburg, Germany

F.A. Brieva and J.R. Rook
Nuclear Physics Laboratory, Oxford, England

In the last years several attempts have been made to understand and
compute the elastic and inelastic nucleon scattering from complex nuclei
when starting from a realistic internucleon potential for free scattering /1,2,3/. A reaction theory which aims towards this goal has been developed with model constraints of direct reactions and a local density
approximation (LDA) in the extension from infinite nuclear matter to finite nuclei. Good agreement with experiment was obtained for the calculation of the mass operator in nuclear matter and then the optical potential with an improved version of the LDA.

Our approach is based on the calculation of the internucleon effective interaction in nuclear matter and its representation in coordinate
space as a superposition of central, tensor and spin orbit potentials.
The extension to finite nuclei is achieved with an LDA for the effective
interaction, used as an antisymmetrized folding model or DWTA for the
computation of the optical potential and inelastic transitions /3,5/.
In the DWTA calculation we may start with distorted waves from the microscopic optical model potential. The calculations thus achieve the status
of being free of adjusted parameters when the input is limited to the
Hamada-Johnston internucleon potential /6/, experimental density distributions /7/ for the target and shell model transition densities.

The purpose of this contribution is to study the range of validity
and implications of the theory /3,8/ by making a comparison with experiment. The topics covered are as follows: We summarize the salient features of the theory for the optical model and the two nucleon t-matrix
in infinite nuclear matter. Further, we outline the coordinate space representations of the t-matrix. The application distinguishes the study of
the optical model potentials, elastic channel differential cross section
analyses and total reaction cross sections.

1. The Optical Model in Infinite Nuclear Matter and the t-Matrix

A nucleon, in transversing infinite nuclear matter, feels an average complex potential, depending on its momentum k, its energy E and the
density k_F of the medium. In lowest order of the low density expansion

/9/ the potential is related to the diagonal Brückner reaction matrix element

$$U(k_F,k,E) = \sum_{|\vec{p}|<k_F} <\vec{p}\vec{k}|t(\omega)|\vec{p}\vec{k}>_A \quad . \tag{1}$$

The starting energy

$$\omega = E(k) + \varepsilon(p) \tag{2}$$

is defined selfconsistently /9/ for all particles below and above the Fermi energy, which is taken to be ($\hbar=1$)

$$\varepsilon(p) = p^2/2m + \text{Re}\left[U(k_F,p,\varepsilon)\right] \quad . \tag{3}$$

The operator $t(\omega)$ satisfies a Bethe-Goldstone integral equation

$$t(\omega) = V + VG^{(+)}(\omega)t(\omega) \tag{4}$$

with

$$G^{(+)}(\omega) = \frac{Q(\vec{q}_1,\vec{q}_2)}{\omega-\varepsilon(q_1)-\varepsilon(q_2)+i\delta} \quad . \tag{5}$$

The Pauli operator Q in this equation has the properties

$$Q(\vec{q}_1,\vec{q}_2)|\vec{q}_1,\vec{q}_2> = \begin{cases} q_1q_2 & \text{if } q_1 \text{ and } q_2 > k_F \\ 0 & \text{else.} \end{cases} \tag{6}$$

The Fermi momentum k_F is simply related to the nuclear matter density of the isotropic, symmetric and noninteraction Fermi gas by

$$\rho_{NM} = \frac{2}{3\pi^2} k_F^3 \quad . \tag{7}$$

If $t(\omega)$ operates on the *uncorrelated* (unperturbed plane waves) two body wave function, $\phi_{kq}\equiv$ with the relative wave function of $|kq>$, we obtain the *correlated* two body wave function, viz.

$$\psi_{kq} = \phi_{kq} + G(\omega)t(\omega)\phi_{kq} = \phi_{kq} + G(\omega)V\psi_{kq} \tag{8}$$

and therefrom the identity

$$t\phi_{kq} = V\psi_{kq} \quad . \tag{8a}$$

Equation (8) is familiar from the formulation of the scattering of two free nucleons from each other. In principle, standard techniques are available to solve this equation straightforward on a computer, even with complicated forms, i.e. when containing a hard core in V. The hidden Pauli operator, however, and the selfconsistency requirement for the single particle energies prohibit any straightforward solution and it is desirable to use approximations. Nuclear matter reaction matrix calculations are the guide /10,11/.

In binding energy calculations we always have $\omega<2\varepsilon(k_F)$ and therefore lack any singularity in the kernel of $G(\omega)$. In the asymptotic region no phase shifts occur and the correlated wave function heals towards the uncorrelated plane waves.

In the other extreme, at very high energies with $k \gg k_F$, the Pauli operator can be put equal to unity and the single particle energies relate simply to its momentum, viz.
$$\varepsilon(k) = \frac{k^2}{2m}$$

and eq.(8) becomes the free particle Lippmann-Schwinger equation. Intuitively, these two extreme situations lead us to use a combined approach suitable for nuclear matter as well as for free scattering. Brieva and Rook /3/ first suggested utilizing the advantages of the *reference spectrum method* (RSM)/10/ in a generalized version.

The RSM approximates the real spectrum of eq.(3) by

$$\varepsilon^{RSM}(q) = \frac{q^2}{2m} + \text{constant} \quad . \tag{9}$$

Following Brieva and Rook's suggestion to decompose the propagator of eq.(5) yields

$$G^{(+)}(\omega) = G_R + G_F^{(+)} \tag{10}$$

with
$$G_R = \frac{-m\, c_R}{q^2 + \gamma^2} \tag{11}$$

and
$$G_F^{(+)} = \frac{m\, c_F}{Q_0^2 - q^2 + i\delta} \quad . \tag{12}$$

Q_0 is the pole position in $G^{(+)}(\omega)$, whereas c_R, c_F and γ are optimized to satisfy eq.(10). This separation of $G^{(+)}(\omega)$ into a Green's function without pole, G_R, and into one which permits scattering, leads us to write eq.(8) as coupled integral equations

$$\Lambda_{kq} = \phi_{kq} + G_R V \psi_{kq}$$
$$\psi_{kq} = \Lambda_{kq} + G_F^{(+)} V \psi_{kq} \quad . \tag{13}$$

The here occuring $\psi_{kq} = \psi_{kq}^{(RSM)}$ should be a reasonable first approximation to the correlated wave function in eqs.(8,8a). Λ_{kq} is an auxiliary function. Just as the advantage of the reference spectrum method lies in transforming the Bethe-Goldstone integral equation into a differential equation, the selected functional form of G_R and $G_F^{(+)}$ in eqs.(11,12) permits us to do the same for the system of equations, eq.(13). This yields

$$(\Delta - \gamma^2)\left[\Lambda_{kq} - \phi_{kq}\right] = m\, c_R V \psi_{kq}$$
$$(\Delta + Q_0^2)\left[\psi_{kq} - \Lambda_{kq}\right] = m\, c_R V \psi_{kq} \quad . \tag{14}$$

Such a system of differential equations is readily solved by numerical quadruture and partial wave decomposition. At present, numerical solutions do not go beyond this coupled system of equations, and thus the reference spectrum method. Improvements are probably needed, and may be pursued following /12/.

At present, numerical results are limited within $k_F = 0,6 \div 1,4 \, \text{fm}^{-1}$, which corresponds to covering 10-100% of the finite nuclei densities. Towards even lower densities, the virtual deuteron formation (actually in both channels T=0 and 1) introduces severe numerical problems. This shows some relation to other optical model work /13/ and requires further detailed study. The major uncertainty in the imaginary optical potentials also has its origin in this very low density region $k_F < 0,5 \, \text{fm}^{-1}$.

2. R-Space Representation of the Operator t(ω)

We follow the suggestion of Siemens /14/ in his approach to generate a local energy and density dependent t-operator in coordinate space. As shall be delineated in the following sections, this approach has the advantage of being more universally applicable and yields generally better results /1,4,15/. The effective operator comprises of central, spin-orbit and tensor components

$$t(r) = \sum_{ST} t_0^{ST}(r) \cdot P^S \cdot P^T + \sum_T t_1^{\,T}(r)(1 \cdot S) \cdot P^T + \sum_T t_2^{\,T}(r) r^2 S_{12} \cdot P^T = \sum_k (R_k(r) \cdot S_k) \ . \tag{15}$$

With partial waves and some Racah algebra we extract the tensor amplitudes /16/

$$\langle L||R_k||L'\rangle\langle ST||S_k||ST\rangle = \sum_J \hat{k}\hat{J} \, (-)^{J-L'-S} (\hat{T})^{1/2} (LSJT|t|L'SJT) \begin{Bmatrix} L & S & J \\ S & L' & k \end{Bmatrix} \tag{16}$$

with matrix elements for t

$$(LSJT|t(\vec{k},\vec{k})|L'SJT) \tag{17}$$

$$= \frac{\displaystyle\sum_{L''} \int_{|\vec{p}|<k_F} d^3p \left| j_L(\tfrac{1}{2}|\vec{k}-\vec{p}| \cdot r) v_{LL''}^{J,ST}(r) \ U_{L'L''}^{J,ST}(r,k_F,\vec{k},p) \right|}{\displaystyle\int_{|\vec{p}|<k_F} d^3p \left| j_L(\tfrac{1}{2}|\vec{k}-\vec{p}|r) \ j_{L'}(\tfrac{1}{2}|\vec{k}-\vec{p}|r) \right|} \ ,$$

In the central components $S_0 = P^S P^T$ and

$$R_0 = R_0^{(LL,ST)}(r) = \frac{\sum_J (LSJT|t|LSJT)}{\hat{S} \cdot \hat{L}} \ . \tag{18}$$

the rank one tensor is limited to the functional form of a spin-orbit potential with

$$S_1 = (\vec{S}_1 + \vec{S}_2) \cdot P^T = \tfrac{1}{2}(\vec{\sigma}_1 + \vec{\sigma}_2) \cdot P^T \ . \tag{19}$$

and
$$R_1 = R_1^{(LL,T)}(r) \cdot \vec{1} = \frac{\sum_J \hat{J} <\vec{1} \cdot \vec{S}> (L1JT|t|L1JT)}{2\hat{L} \cdot L(L+1)} \cdot \vec{1} \quad . \tag{20}$$

This is only defined for $L \geq 1$ and $(\vec{1} \cdot \vec{S}) = 1/2 \left[J(J+1) - L(L+1) - 2 \right]$

For the tensor operator we adopt

$$S_2 = \left[\vec{S}_1 \times \vec{S}_2 \right]_2 \cdot P^T \tag{21}$$

and

$$R_2 = (\frac{8\pi}{15})^{1/2} R_2^{(LL',T)}(r) \cdot r^2 Y_2(\hat{r}) \quad . $$

With this functional form, the tensor force is readily obtained from

$$r^2 R_2^{(LL'T)}(r) = \frac{5\sqrt{6} \sum_J \hat{J}(-)^{J+1} (L1JT|t|L'1JT) \left\{ \begin{smallmatrix} L & 1 & J \\ 1 & L' & 2 \end{smallmatrix} \right\}}{<LL'oo|2o> \cdot \sqrt{\hat{L} \cdot \hat{L}'}} \quad . \tag{22}$$

To further simplify these interactions from their 1-dependence we evaluate the averages with

$$(e^{i\vec{k} \cdot \vec{r}} |R_k| e^{i\vec{k} \cdot \vec{r}}) = (e^{i\vec{k} \cdot \vec{r}} | \sum_{LL'} R^{(LL'ST)} O_k | e^{i\vec{k} \cdot \vec{r}}) \tag{23}$$

$$= t_k^{ST}(r) \; (e^{i\vec{k} \cdot \vec{r}} |O_k| e^{i\vec{k} \cdot \vec{r}}) \quad . $$

The effective interaction defined in relative spin and isospin states (S,T) and odd/even character obtains the form

central interaction
$$t_o^{ST}(r) = \frac{\sum_{L \geq o} \varepsilon(L+S+T+1) \hat{L} \; R_o^{(LL,ST)} W_{LL}}{\sum_{L \geq o} \varepsilon(L+S+T+1) \hat{L} \; W_{LL}} \tag{24}$$

with $\quad \varepsilon(k) = \begin{smallmatrix} o & k = odd \\ 1 & k = even \end{smallmatrix}$ and $\quad X = \begin{smallmatrix} 1 \\ Y_{2o}(1/2(\widehat{k-p})) \end{smallmatrix}$ for $\begin{smallmatrix} k = 0,1 \\ k = 2 \text{ (tensor)} \end{smallmatrix}$

$$W_{LL'} = \int_{|\vec{p}| < k_F} X \, d^3p \left[j_L(1/2|\vec{k}-\vec{p}|r) \; j_{L'}(1/2|\vec{k}-\vec{p}|r) \right] \quad . \tag{25}$$

spin orbit interaction
$$t_1^T(r) = \frac{\sum_{L \geq 1} \varepsilon(L+T) \cdot \hat{L} \cdot R_1^{(LL,T)}(r) W_{LL}}{\sum_{L \geq 1} \varepsilon(L+T) \cdot \hat{L} \cdot W_{LL}} \quad . \tag{26}$$

and the *tensor interaction* $\tag{27}$
$$t_2^T(r) = \frac{\sum_{LL'} \varepsilon(L+T) (\hat{L}\hat{L}') \quad i^{L-L'} <LL'oo|2o>^2 R_2^{(LL'T)} W_{LL'}}{\sum_{LL'} \varepsilon(L+T) (\hat{L}\hat{L}') \quad i^{L-L'} <LL'oo|2o>^2 W_{LL'}} \quad . $$

3. Folded Optical Model Potential

The nucleon-nucleus optical potential for finite nuclei can be generated from the effective interaction $t(\omega)$ with a local density approximation. This procedure assumes that at each point in the nucleus the
value of the optical model is well approximated with the optical model
of infinitely extended nuclear matter calculated with the local quantum
numbers. The equivalent local central optical potential in LDA /1/ assumes the form

$$U^{(LDA)}(\vec{r},E) = \rho(\vec{r}) \int d^3x \left[t^D(x,k_F(\vec{r}),E) + \right.$$

$$\left. \frac{3}{xk_F(r)} j_1(xk_F(r)) t^{EX}(x,k_F(\vec{r}),E) j_o(k(x)\cdot x) \right]$$

(28)

where

$$k_F(r) = \left[\frac{3\pi^2}{2} \rho(\vec{r}) \right]^{1/3}$$

(29)

and

$$k(r) = \sqrt{2m(E-U^{LDA}(r)-V_{coul}(r))} \quad .$$

This expression contains the Slater expansion of the mixed density and
t^D and t^{EX} are linear combinations of t^{ST} for the direct and exchange
terms. The properties and shortcomings of this OMP have been discussed
extensively /1/. An improved version of this LDA has been applied recently /4/ and good agreement with experimental data was obtained.

Another improvement on the finite nucleus OMP is obtained when
starting from the general formulation of a folding expression of the
effective interaction with finite nuclear matter distributions /17/.
It yields a nonlocal in r-space and energy dependent potential

$$U(\vec{r},\vec{r}';E) = \delta(\vec{r}-\vec{r}') \sum_n \int \phi_n^*(\vec{s}) t^D(\vec{r},\vec{s};E) \phi_n(\vec{s}) d^3s$$

$$+ \sum_n \phi_n^*(\vec{r}) t^{EX}(\vec{r},\vec{r}';E) \phi_n(\vec{r}') \quad .$$

(30)

Using the local momentum approximation for the exchange term, a local
energy dependent potential results

$$U(\vec{r};E) = \int U(\vec{r},\vec{r}';E) d^3r'$$

(31)

$$= \int \delta(\vec{r}-\vec{r}') \{ \int \rho(s) t^D(\vec{r},\vec{s};E) d^3s + \rho(\vec{r},\vec{r}') t^{EX}(\vec{r},\vec{r}') j_o(k|\vec{r}-\vec{r}'|) \} d^3r' \quad .$$

For the effective interaction we adopt the idea of the LDA and use the
translational invariant form of the infinite nuclear matter interaction

$$t^{D,EX}(\vec{r}_1,\vec{r}_2;E) = t^{D,EX}(|\vec{r}_1-\vec{r}_2|, k_F(\frac{\vec{r}_1+\vec{r}_2}{2});E) \quad .$$

(32)

The linear combinations of t^{ST} differentiate between the incident and target nucleon nature. For incident protons it reads

$$t_{p(p)}^{D,EX} = t_{n(n)}^{D,EX} = \frac{1}{4} \, (t_o^{01} \pm 3t_o^{11}) \tag{33a}$$

$$t_{p(n)}^{D,EX} = t_{n(p)}^{D,EX} = \frac{1}{8}(3t_o^{10} + t_o^{01} + t_o^{00} + 3t_o^{11}) \; . \tag{33b}$$

With single particle wave functions of a phenomenological shell model or Hartree-Fock calculation, the densities in eq.(31) are easily gener-ated and the local optical model potential generated.

In case the evaluation of densities appears cumbersome, we may use experimental densities from electron scattering instead. The foremost problem is then connected with the mixed density occuring in the exchange integral.

The expansion of $\rho(\vec{r},\vec{r}\,')$ around the diagonal density has long been known and various versions are applied today. The simplest and least ac-curate in finite nuclei is Slater's expansion

$$\rho(\vec{r},\vec{r}\,') = \rho(\frac{\vec{r}+\vec{r}\,'}{2}) \cdot \frac{3}{\hat{k} \cdot s} \, j_1 \, (ks) \tag{34}$$

with $s=|\vec{r}-\vec{r}\,'|$ and $\hat{k}=k_F(\frac{\vec{r}+\vec{r}\,'}{2})$, eq.(29). A better approximation is due to Negele and Vautherin /18/. More recently Campi and Bouyssy /19/ proposed an expansion which maintains the functional form of the Slater expansion, but with

$$\hat{k}(R) = \left| \frac{5}{3\rho(R)} \, (\tau(R) - \frac{1}{4}\nabla^2\rho(R)) \right|^{1/2} \; . \tag{35}$$

In figs. 1 and 2 we compare the $\hat{k}$ values for ^{16}O and ^{40}Ca. Most

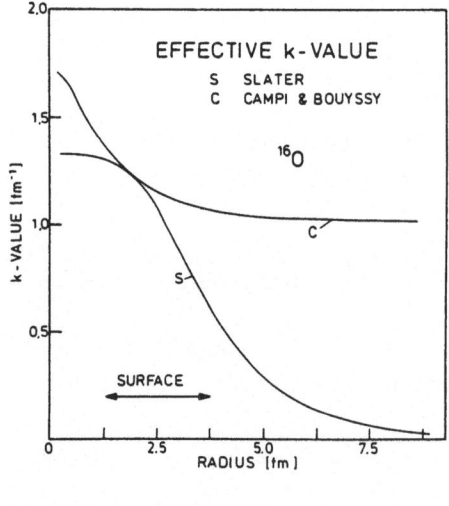

FIGURE 1

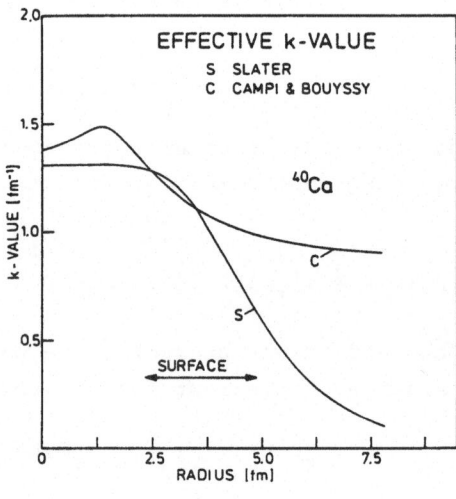

FIGURE 2

obvious is the difference in the asymptotic values associated with eqs. (34) and (35). The optical potentials computed with the two approximations differ typically 10-15%, particularly in the surface region. In fig.3 we compare the two results for 30.3 MeV protons and neutrons scattered from ^{40}Ca. The major changes are seen in the real potentials where exchange amplitudes are very large. The imaginary potential alters by less than 2%.

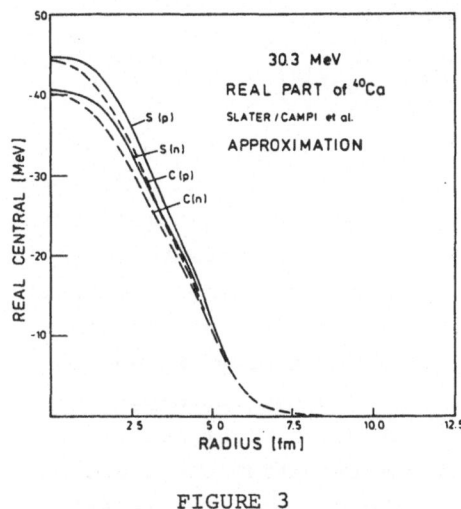

FIGURE 3

In eq.(34) the center of mass LDA is used. In G-matrix calculations, other forms have also been chosen /20/ with computational advantages. The arithmetic mean

$$\rho_A = \frac{1}{2} \left| \rho(\vec{r}_1) + \rho(\vec{r}_2) \right| \qquad (36)$$

or the geometric mean

$$\rho_G = \left| \rho(\vec{r}_1) \; \rho(\vec{r}_2) \right|^{1/2} \qquad (37)$$

are two examples of some of the many possible alternatives. Since the nuclear forces and the effective interaction $t(\omega)$ have short ranges, no great differences occur. We have verified this for the real OMP where changes are restricted to below 5%. The imaginary potential is more strongly influenced due to the exponential density dependence of the imaginary parts of $t(\omega)$. At present we register it as a deficiency, but not the foremost problem.

4. Optical Model Results

The nucleon nucleus OMP is based on the nuclear matter t-matrices computed with the Hamada-Johnston internucleon potential. The effective interaction $t(\omega)$ is now available for 25 energies ranging E=10÷200 MeV, and 8 different densities k_F=0,6÷1,4 fm^{-1}. The diagonal densities are computed with *two parameter* (2PF) or *three parameter* (3PF) *Fermi distributions* /21/ or Hartree Fock densities /22/.

The full expression for the optical potential of a proton then reads

$$U_p(\vec{r},E) = \int d^3s \rho(|\vec{r}+\vec{s}|) \{Zt^D_{p(p)}(s,k_F(|\vec{r}+\tfrac{\vec{s}}{2}|),\tilde{E}) + Nt^D_{p(n)}(s,k_F(|\vec{r}+\tfrac{\vec{s}}{2}|),\tilde{E})\}$$

$$+ \int d^3s \rho(|\vec{r}+\tfrac{\vec{s}}{2}|) 3 \cdot (\hat{k}(|\vec{r}+\tfrac{\vec{s}}{2}|)s)^{-1} j_1(\hat{k}(|\vec{r}+\tfrac{\vec{s}}{2}|s) \cdot j_o(k(r)s) \cdot \qquad (38)$$

$$\{Zt^{EX}_{p(p)}(s,k_F(|\vec{r}+\tfrac{\vec{s}}{2}|),\tilde{E}) + Nt^{EX}_{p(n)}(s,k_F(|\vec{r}+\tfrac{\vec{s}}{2}|),\tilde{E})\}$$

with the quasiclassical approximation for Coulomb effects

$$\tilde{E} = E - V_{coul}(r) \tag{39}$$

and
$$k(r) = \sqrt{2m(E-U_p(r)-V_{coul}(r))} \ . \tag{40}$$

The selfconsistency condition in k(r) was met by integrating from large radii inward in small steps (Δ=0,1 fm) and setting

$$k(r) = \sqrt{2m(E-2U_p(r+\Delta)+U_p(r+2\Delta)-V_{coul}(r))} \ . \tag{41}$$

With figs. 4 to 9 we display the features of the microscopic OMP for ^{58}Ni. Its density is parameterized as 2PF (r_o=1.09 fm and a=0,54 fm). The real central potential, fig.4, resembles closely the phenomenological Woods Saxon shape at lower energies and deteriorates towards higher energies into a volume and a surface term. For the imaginary potential, fig.5, we obtain at lower energies the pronounced surface absorption with a volume part which then dominates in the high energy region. In fig.6 the potential components from direct and exchange are separately displayed. The repulsive nature of *direct* is at first sight surprising, but we soon realize that it rests solely on the repulsive odd state interactions which have a large strength but a short range. The relative sign between *direct* and *exchange* for different relative spin and isospin states, $(-)^{S+T+1}$, yields then the required cancellations. *In toto*, attractive potentials with a depth around -40 MeV are obtained. The energy dependence displayed in fig.4 is coming largely from the energy dependence

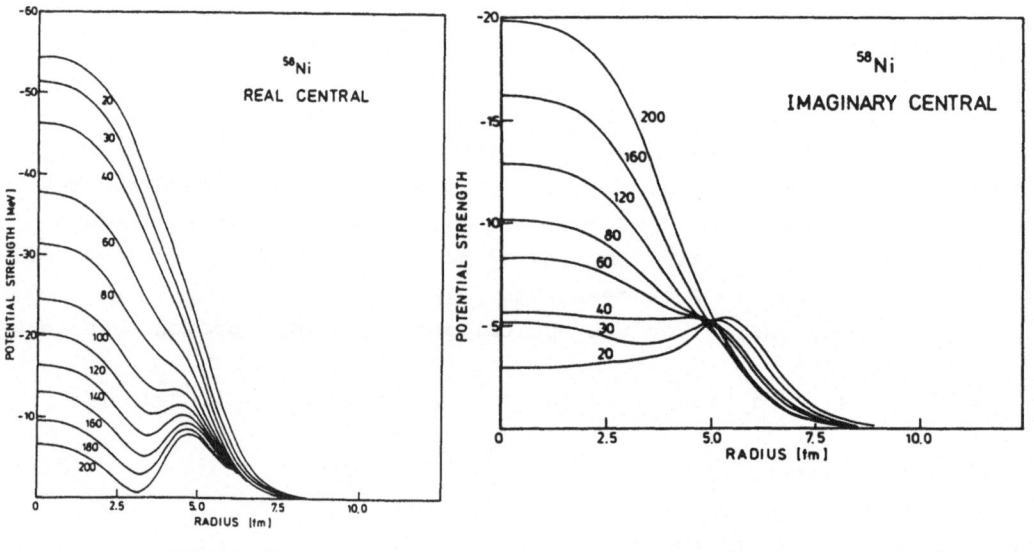

FIGURE 4   FIGURE 5

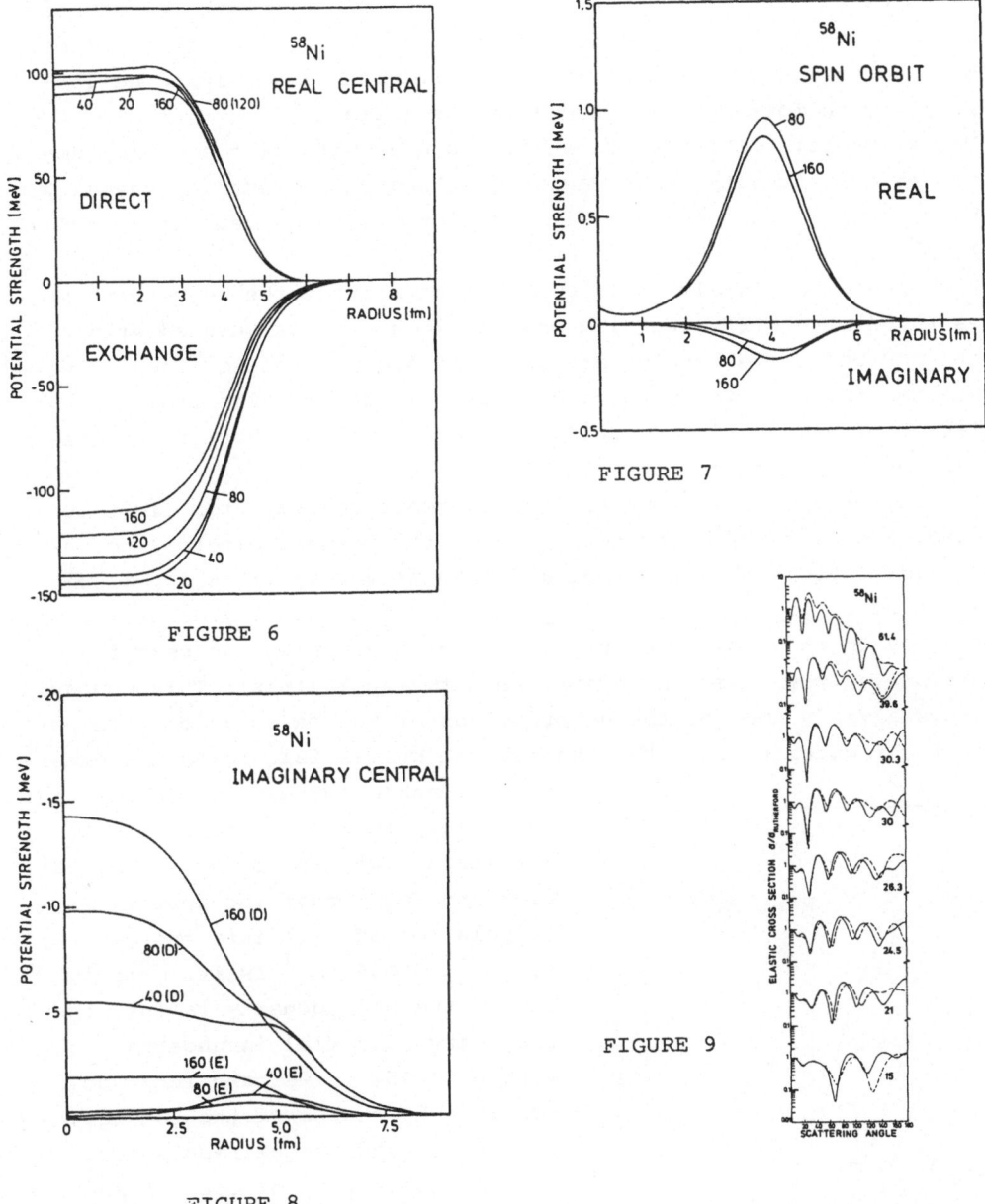

FIGURE 6

FIGURE 7

FIGURE 8

FIGURE 9

of exchange. The imaginary potential always has an absorptive character, in direct as well as in exchange contributions. The interesting inter-play between surface and volume absorption as a function of energy comes purely from the density dependence of the imaginary effective interaction.

Since in this calculation no explicit shell effects or effects of collective motion (long range correlations) have been included, it appears to be an interesting task to unify the picture with results obtained in the more structure oriented OMP /23/. The latter calculations attribute the surface potential to the excitation of low lying collective states with form factors peaked in the surface.

The spin-orbit potential is surface peaked with little energy dependence in its strength. The imaginary spin-orbit potential generally remains very small and obtains only some importance at very high energies >150 MeV.

With these potentials, E=15÷60 MeV (and without the adjustment procedures /3,8/) the differential cross sections were calculated with a standard OMP code. The results are shown in fig.9 as solid lines. The dashed lines in this figure represent best fit OMP results from the compilation of Perey and Perey /24/. The angular distributions of the polarization compares equally well /8/.

A study of optical potentials for elements ranging from light to heavy nuclei also shows close consistency with phenomenological potentials. Figs.10 to 14 show the real and imaginary potentials for a selected set of isotopes whose density was simply parameterized as 2PF with r_o=1,09 fm and a=0,54 fm. Figure 14 again compares the best fit potentials /24/ with the microscopic results (full lines). This comparison shows already some of the shortcomings of the calculations. In particular for lighter nuclei, the theoretical curves fall below the experiment (dashed curve). This shows that our absorption is typically 10-15% too strong. We have investigated this fact and found that the applied extrapolation of t(ω) into the density range k_F=0÷0,6 fm^{-1} is its origin. Beyond the half density radius, the imaginary potential is uncertain to within 10-15% if we take into consideration the choice of the LDA eqs. /34,38,39/ and the extrapolation.

A recent work of Coulter and Satchler /13/ and Ioannides and Johnson /25/ makes us suspect that the deuteron formation (pick-up channel) is a very important ingredient of the imaginary surface potential. At the present time we can add to this problem only our observations

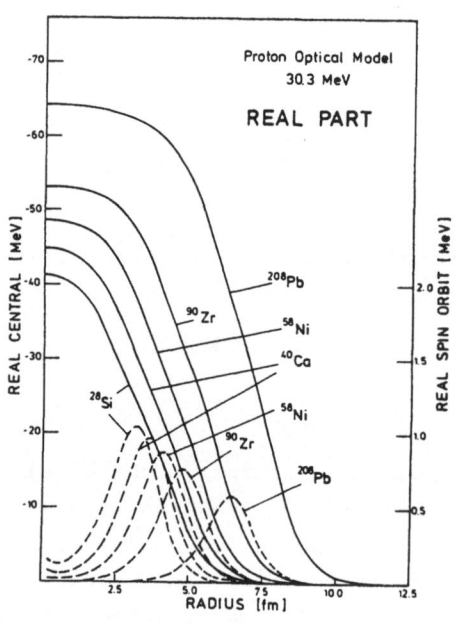

FIGURE 10

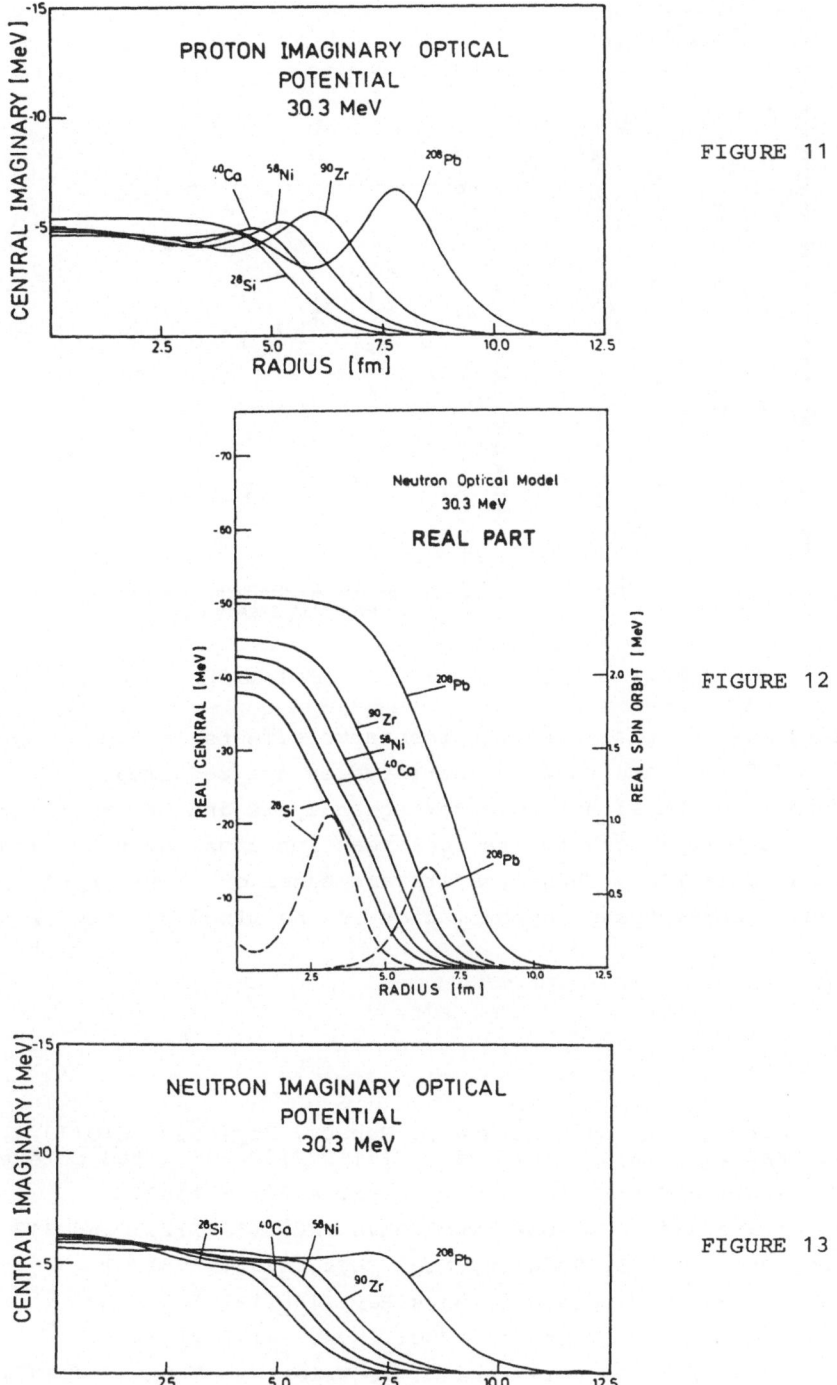

FIGURE 11

FIGURE 12

FIGURE 13

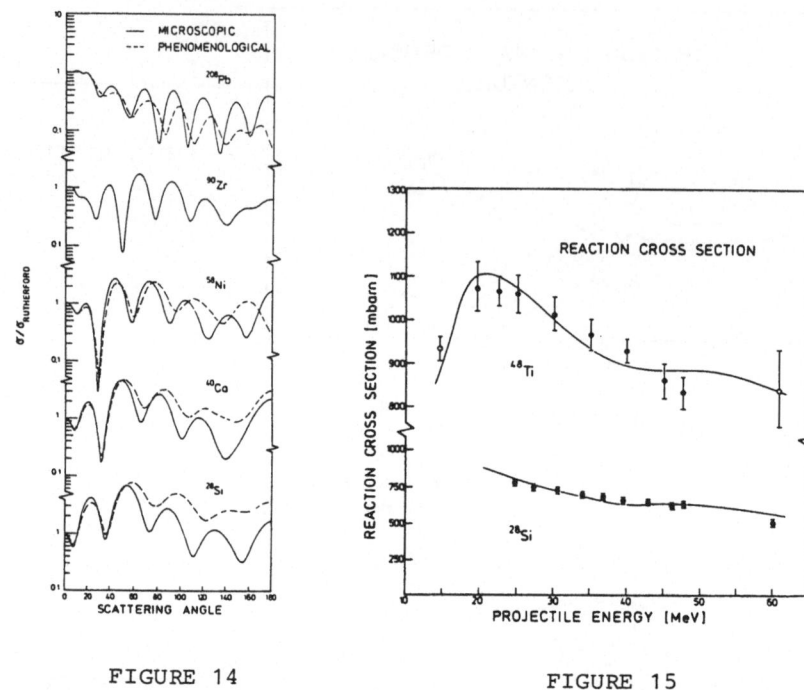

FIGURE 14   FIGURE 15

that the eq.(14) displayed strong resonance effects in the low density region $k_F < 0,5$ fm^{-1} and that further studies are required.

To conclude the elastic scattering analysis and its application, we complare in fig.15 the theoretical reaction cross section with recent experimental data /26/. These results are based on a 2PF (r_o=1 fm, a=0,5 fm) and the Campi-Bouyssy approximation in the mixed density expansion eq.(34).

References

/1/ J.P. Jeukenne, A. Lejeune and C. Mahaux, Phys.Rev. C10(1974)1391; Phys.Rep. 25(1976)83; Phys.Rev. C16(1977)80 and cited references

/2/ H.R. Kidwai and R.J. Rook, Nucl.Phys. A169(1971)417

/3/ F.A. Brieva and R.J. Rook, Nucl.Phys. A291(1977)299 and 317

/4/ A. Lejeune and P.E. Hodgson, Nucl.Phys. A295(1978)301

/5/ L.R. Dodd and K.R. Greider, Phys.Rev. 146(1966)675

/6/ T. Hamada and I.D. Johnston, Nucl.Phys. 34(1962)382

/7/ C.W. DeJager, H. De Vries and C. De Vries, Nucl.Data 14(1974)479

/8/ F.A. Brieva and R.J. Rook, Nucl.Phys. A297(1978)206

/9/ J. Hüfner and C. Mahaux, Ann.Phys. (NY) 73(1972)525

/10/ H.A. Bethe, B.H. Brandow and A.G. Petschek, Phys.Rev. <u>129</u>(1963)225

/11/ D.W.L. Sprung, Adv.Nucl.Phys. Vol.5, ed. by M. Baranger and E.Vogt, Plenum (1972)

/12/ W. Legindgaard, Nucl.Phys. <u>A297</u>(1978)429

/13/ P.W. Coulter and G.R. Satchler, Nucl.Phys. <u>A293</u>(1977)269

/14/ P.J. Siemens, Nucl.Phys. <u>A141</u>(1970)225

/15/ F.A. Brieva and R.J. Rook, Oxford preprint (1978)

/16/ A. de Shalit and I. Talmi, Nucl.Sehll Theory, Acad.Press (1963)

/17/ L.W. Owen and G.R. Satchler, Phys.Rev.Lett. <u>25</u>(1970)1720

/18/ J.W. Negele and D. Vautherin, Phys.Rev. <u>C5</u>(1972)1472 and Phys.Rev. <u>C11</u>(1975)1031

/19/ X. Campi and A. Bouyssy, Phys.Lett. <u>73B</u>(1978)263
A. Tielens, Diplomarbeit, Hamburg (1978)

/20/ C.W. Wong and S.A. Moszkowski, Nucl.Phys. <u>A239</u>(1975)209 and cited references

/21/ C.W. De Jager, H. De Vries and C. Ce Vries, Nucl.Data <u>14</u>(1974)479

/22/ J.W. Negele, Phys. Rev. <u>C1</u>(1970)1260

/23/ N. Vinh Mau, Theory of Nuclear Structure (IAEA),Vienna,(1970)931
N. Vinh Mau and A. Bouyssy, Nucl.Phys. <u>A257</u>(1976)189

/24/ C.M. Perey and F.G. Perey, Nucl.Data <u>17</u>(1976)1

/25/ A.A. Ioannides and R.C. Johnson, Phys,Rev. <u>C17</u>(1978)1331

/26/ T.N. Nasr et al., Can. J. Phys., <u>56</u>(1978)56
W.F. McGill et al., Phys.Rev. <u>C10</u>(1974)2237

EFFECTS OF PARTICLE-VIBRATION COUPLING ON THE HARTREE-FOCK POTENTIAL

V. Bernard and Nguyen Van Giai
Division de Physique Théorique[+], Institut de Physique Nucléaire,
91406 ORSAY Cedex, France

1. INTRODUCTION

The microscopic description of many properties of nuclear
ground states in terms of an average field has reached a good quantitative level with the appearance of numerous Hartree-Fock (HF) calculations using effective interactions. It is established by now that
these interactions determined either from purely phenomenological
considerations[1,2] or from more theoretical grounds[3,4] must be density dependent. This density dependence leads to calculated total
binding energies and single particle spectra which are in satisfactory agreement with experiment.

The comparison between the HF and experimental single particle spectra can be extended to the continuum region, and it has been
shown that the HF potential is very similar to the empirical optical
potential at low energies[5]. On the other hand, there are successful
attempts to calculate the optical model potential in terms of the
t-matrix of a nucleon pair scattering in the medium and using the
local density approximation[6,7]. In fact, density dependent effective
interactions can be viewed as parametrizations of the Brueckner
G-matrix, and the corresponding HF potentials are therefore related
to the optical potentials of the t-matrix approach.

However, the HF approximation does not include the effects
of long range correlations of the particle-hole type which would
bring in correction terms to the real part of the potential and contributions to the absorptive potential. These effects have been
studied in refs.8,9) for the case of ^{40}Ca. In this work we present
a calculation of the real and imaginary parts of the optical potential V_{opt} in which the HF term and the correction terms are derived
from an effective interaction of the Skyrme type. A more detailed
account can be found in refs. 10,11). In section 2 we give the
outline of the method. In section 3 we discuss the results for the

---

[+]Laboratoire associé au C.N.R.S.

absorptive potential in the nucleus ^{208}Pb whereas the real part of the potential is discussed in section 4.

2. METHOD OF CALCULATION

It is convenient to use the Green's function formalism for the study of the mass operator Σ or its Fourier transform V_{opt}. This formalism provides a systematic framework for the expansion of V_{opt} in terms of many body correlations and it includes automatically the exchange effects due to the identity of the incident and target nucleons. This approach has been used by N. Vinh Mau[12] who showed that if one makes the following assumptions : i) three-body Green's functions can be replaced by suitable combinations of one- and two-body Green's functions ; ii) particle-particle propagators are treated in the ladder approximation and particle-hole propagators in the bubble (RPA) approximation ; iii) single particle propagators can be described by HF propagators, then the complex optical potential can be written as :

$$V_{opt}(\vec{r},\vec{r}';E) = V_{HF} + \Delta V$$
$$\Delta V = V_{RPA} + V_{pp} - 2V^{(2)} \tag{1}$$

where V_{HF} is the HF potential, V_{RPA} and V_{pp} are respectively correction terms due to correlations of the particle-hole and particle-particle type. The quantity $V^{(2)}$ is just the second order contribution to the optical potential, and the subtraction term $-2V^{(2)}$ in eq.(1) takes care of double counting[12]. The potential V_{HF} is non local, real and does not depend on the energy if one assumes an energy-independent interaction whereas ΔV is complex and depends on the projectile energy E. We shall not write explicitly the spin and isospin variables in order to keep simple notations.

In the case of the Skyrme interaction the calculation of V_{HF} is straightforward because of the simple analytical structure of the force[1]. The HF field is determined by a local potential $U(\vec{r})$ and an effective mass $m^*(\vec{r})$ which describes the special type of non-locality of the Skyrme-HF field. The quantities $U(\vec{r})$ and $m^*(\vec{r})$ depend on the self-consistent nuclear density $\rho(\vec{r})$ and kinetic energy density $\tau(\vec{r})$. The numerical application presented in this work has been performed with the set of force parameters SIII[2].

The contribution of the collective particle-hole states is

$$V_{RPA}(\vec{r},\vec{r}';E) = \frac{-1}{2\pi}\, \tilde{v}(\vec{r})\tilde{v}(\vec{r}') \int_{-\infty}^{+\infty} G_1(\vec{r},\vec{r}';E-E') G_{RPA}(\vec{r},\vec{r}';E')\,dE' \qquad (2)$$

where G_1 is the HF single particle propagator :

$$G_1(\vec{r},\vec{r}';E) = -i\left(\sum_h \frac{\varphi_h(\vec{r})\varphi_h^{*}(\vec{r}')}{\varepsilon_h - E - i\eta} + \sum_p \frac{\varphi_p(\vec{r})\varphi_p^{*}(\vec{r}')}{\varepsilon_p - E + i\eta}\right) \qquad (3)$$

and G_{RPA} is the particle-hole propagator calculated in the RPA :

$$G_{RPA}(\vec{r},\vec{r}';E) = \sum_N \langle\Psi_o|\psi^{+}(\vec{r})\psi(\vec{r})|\Psi_N\rangle\langle\Psi_N|\psi^{+}(\vec{r}')\psi(\vec{r}')|\Psi_o\rangle \qquad (4)$$

$$\left[\frac{1}{E_N - E - i\eta} + \frac{1}{E_N + E - i\eta}\right]$$

The quantities φ_h and ε_h (φ_p and ε_p) are the HF wave functions and energies corresponding to occupied (unoccupied) orbitals whereas the $|\Psi_N\rangle$ are RPA states with energies E_N, $|\Psi_o\rangle$ being the correlated ground state. In eq.(2) $\tilde{v}(\vec{r})$ is the strength of the antisymmetrized particle-hole interaction $\tilde{v}_{ph}$ derived from the Skyrme force :

$$\tilde{v}_{ph}(1,2) = \tilde{v}(\vec{r}_1)\,\delta(\vec{r}_1-\vec{r}_2) \qquad (5)$$

The strength $\tilde{v}(r)$ depends on the distance r because of the density dependence of the Skyrme force. Also, $\tilde{v}_{ph}$ contains momentum dependent terms which we neglect when calculating V_{RPA}. They are however taken into account in the calculation of the propagator G_{RPA} appearing in eq.(2).

The second order contribution $V^{(2)}$ is given by an expression similar to eq.(2) where G_{RPA} is replaced by the unperturbed particle-hole propagator G_{ph} :

$$G_{ph}(\vec{r},\vec{r}';E) = \sum_{p,h}\left(\frac{\varphi_p(\vec{r})\varphi_h^{*}(\vec{r})\varphi_p^{*}(\vec{r}')\varphi_h(\vec{r}')}{\varepsilon_p - \varepsilon_h - E - i\eta}\right.$$

$$\left. + \frac{\varphi_h(\vec{r})\varphi_p^{*}(\vec{r})\varphi_h^{*}(\vec{r}')\varphi_p(\vec{r}')}{\varepsilon_p - \varepsilon_h + E - i\eta}\right) \qquad (6)$$

The computation of the particle-particle contribution V_{pp} would require the knowledge of the two-particle propagator calculated in the ladder approximation. If the HF term V_{HF} had been derived from a G-matrix then V_{pp} or at least its real part would be already contained in V_{HF} since the G-matrix sums up the ladder diagrams appearing in V_{pp}. We shall not include the real part of V_{pp} in our calculation since a link can be established between HF calculations using a G-matrix or an effective interaction of the Skyrme type[13]. Furthermore, the imaginary part of V_{pp} will be calculated in lowest order, i.e. we use $\text{Im } V_{pp} \simeq \text{Im } V^{(2)}_{pp}$ in eq.(1).

In order to obtain the propagators G_{RPA} we have solved the RPA integral equation

$$G_{RPA} = G_{ph} + G_{ph} \tilde{v}_{ph} G_{RPA} \tag{7}$$

Using the method of refs. 14,15) this can be done in coordinate space. The effects of the continuum on the RPA states are therefore taken into account. The particle-hole interaction $\tilde{v}_{ph}$ is defined as the second derivative of the HF energy density with respect to the density ρ[14]. For each L-multipole the function $G^{(L)}_{RPA}(E)$ is computed up to an energy E_{max} such that a large fraction (70-90 %) of the linear energy-weighted sum rule m_1 is exhausted. The results are then used to cast G_{RPA} in the form (4) with a finite, generally small number of terms in the sum. Some of the energies E_N can be complex when the corresponding states are resonances with a finite escape width. The convolution integral in eq. (2) can be calculated analytically with the aid of eq. (4). This procedure insures that for each L-multipole the most important collective states, i.e. those contributing most to the sum rule m_1 are included in the calculation of V_{opt}. In the results for ^{208}Pb we shall present in the next sections we have included the contributions of natural parity states corresponding to L = 0,2,3,4,5 for isoscalar modes and L = 0,1,2 for isovector modes.

3. THE ABSORPTIVE POTENTIAL

The non-local potential $\Delta V(\vec{r},\vec{r}')$ has been calculated by making a multipole expansion in the variables $\vec{r}$ and $\vec{r}'$. It is however useful to define ΔV as a function of $\vec{R} = (\vec{r} + \vec{r}')/2$ and

$\vec{s} = \vec{r} - \vec{r}'$. This can be done easily since ΔV is practically indepen-dendent of the relative angle between $\vec{R}$ and $\vec{s}$. In this section we discuss the properties of the quantity $W(R,s) \equiv Im \ \Delta V$.

The calculation of ΔV has been performed for neutrons incident on ^{208}Pb with energies ranging from 5 to 30 MeV. An energy averaging procedure has been used with an averaging interval I = 3MeV. The results are not sensitive to this choice as long as the interval I remains large enough. In figs. 1 and 2 are summarized the results obtained for $W(R,s)$.

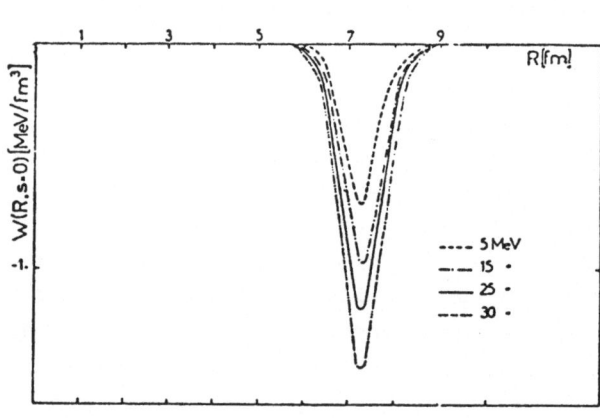

Fig.1 : The R-dependence of W(R,s)

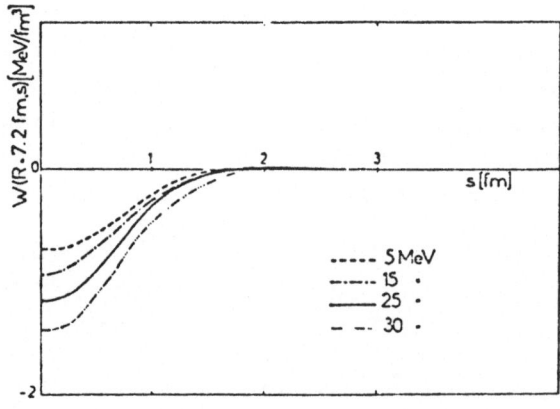

Fig.2 : The s-dependence of W(R,s)

The R-dependence of $W(R,s=0)$ is shown in fig.1. The shape is strongly surface-peaked at all energies. This is a characteristic feature of the potential ΔV cal-culated with the Skyrme force SIII. Indeed, the density dependence of the force is such that the strength $\tilde{v}(r)$ of the particle-hole interaction in the interior region is about 10 times smaller than at the surface, thus cutting down all contri-butions to ΔV inside the nucleus. Therefore, the volume part of ΔV is probably underestimated. However, the fact that RPA calculations using the same interaction give satisfactory results for surface modes like L = 2 although they fail to describe the breathing mode indicates that the surface part of ΔV may be well described since

surface modes are mostly sensitive to the properties of the residual
interaction at the nuclear surface. Our results show that the
absorptive potential is peaked at a distance R $\simeq$ 7.3 fm for all energies considered. The width of the potential at half-maximum varies
from 0.85 fm (E = 5 MeV) to 1.10 fm (E = 30 MeV).

The non-locality of W(R,s) at the surface is shown in fig.2.
As a function of s, the shape of W(R,s) is well reproduced by a
gaussian distribution with a non-locality range parameter β of the
order of 1 fm at R = 7.2 fm. The range parameter β increases slightly
with increasing R.

From figs.1 and 2 we can also see that the absorptive potential increases with the incident energy. This is due mainly to the
larger number of inelastic channels which can contribute to the
calculated absorption when the energy goes up. However, the shape of
the potential does not change much and we do not obtain any appreciable volume absorption for the reasons explained before.

Having calculated the non local potential $V_{opt} = V_{HF} + \Delta V$
we have also computed the local equivalent potential
$V_{loc}(r) + i W_{loc}(r)$ using the Perey-Saxon expansion[16]. In fig.3 are
shown the potentials
$W_{loc}(r)$ corresponding to
E = 15 MeV (solid line)
and E = 30 MeV (dotted
line). They are both
peaked at the same dis-
tance R = 7.3 fm and
have similar shapes,
only their magnitudes
differ. The calculated
volume integral per
nucleon is only 15 MeV.
fm^3 at E = 15 MeV which
is 4-5 times smaller
than the empirical
value[17]. However, our
preliminary results con-

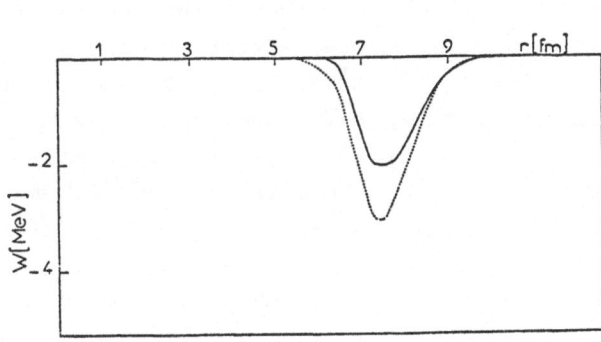

Fig.3 : Imaginary part of the local
equivalent potential

cerning reaction cross-sections calculated from the potential

$V_{loc}(r) + i W_{loc}(r)$ are only 2 times smaller than the measured values.

4. THE REAL PART OF THE OPTICAL POTENTIAL

The major contribution to the real part of V_{opt} comes from the HF potential. The local equivalent potential corresponding to the HF term is shown in figs. 4 and 5 (dotted line). The radius and surface diffuseness are in agreement with phenomenological values. The depth at the center of the nucleus is less than the values determined from optical model fits. This was already noticed in ref. 5) where a more systematic discussion of the Skyrme-HF potential can be found. The dependence of the central depth on the energy is linear, the slope α being governed by the value of the effective mass $m^*(r)$ at r = 0. The force SIII used here leads to $m^*(r = 0)/m = 0.76$ and therefore $\alpha = 1-m^*(o)/m = 0.24$.

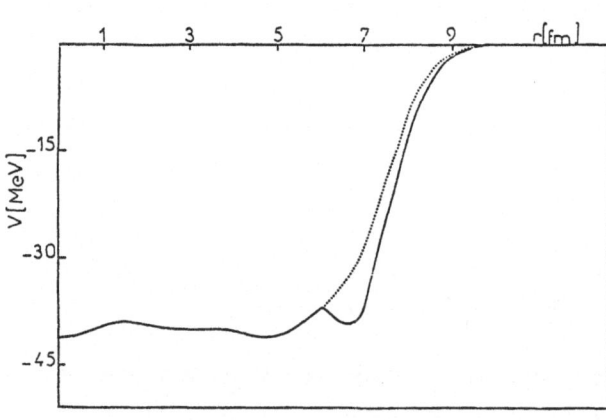

Fig. 4 : the real part $V_{loc}(r)$ at
E = 15 MeV

The real part of ΔV is localized at the nuclear surface for the same reasons as explained in the previous section. When it is added to the HF part one obtains a local equivalent potential which has a steeper

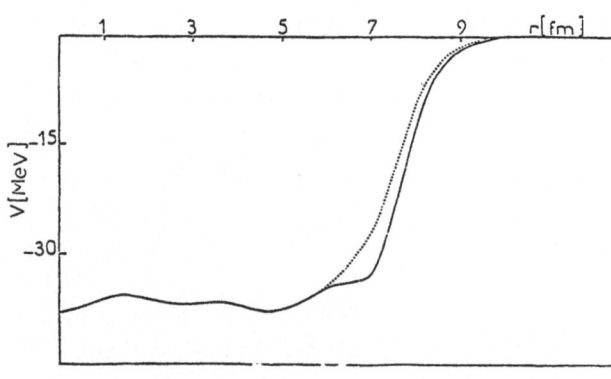

Fig. 5 : the real part $V_{loc}(r)$ at
E = 30 MeV

surface than the HF potential. The local potentials $V_{loc}(r)$ equivalent to the real part of $V_{HF} + \Delta V$ are shown in figs. 4 and 5 (solid line). One can see that the effect of the real part of ΔV is larger at E = 15 MeV than at 30 MeV.

5. SUMMARY

We have calculated the optical potential in the case of low energy neutron - ^{208}Pb scattering using an effective Skyrme interaction. In addition to the HF potential we have obtained the complex correction potential ΔV due to the excitation of the low lying collective states and the giant resonances. The potential ΔV is localized at the nuclear surface because of the particular density dependence of the force we used. A larger effect in the interior region is needed, especially for the absorptive potential which should have a volume part. Such an effect could be obtained if one used a force with a weaker density dependence than the force SIII.

To investigate further the properties of our calculated potential, scattering cross-sections must be computed. The preliminary results we have obtained recently for the elastic differential cross-section at 15 MeV without renormalizing the potential $V_{loc}(r) + iW_{loc}(r)$ are very encouraging.

1) D. Vautherin and D.M. Brink, Phys. Rev. C5 (1972) 626
2) M. Beiner, H. Flocard, Nguyen Van Giai and P. Quentin, Nucl. Phys. A238 (1975) 29
3) J.W. Negele, Phys. Rev. C1 (1970) 1260
4) X. Campi and D.W.L. Sprung, Nucl. Phys. A194 (1972) 401
5) C.B. Dover and Nguyen Van Giai, Nucl. Phys. A190 (1972) 373
6) J.P. Jeukenne, A. Lejeune and C. Mahaux, Proc. Int. Conference on Nuclear Self-Consistent Fields, Trieste (1975) p.155
7) F. Brieva and J.R. Rook, Nucl. Phys. A291 (1977) 299 ; A291 (1977) 317
8) N. Vinh Mau and A. Bouyssy, Nucl. Phys. A257 (1976) 189
9) A. Bouyssy, H. Ngô and N. Vinh Mau, Contribution to this Conference
10) V. Bernard, Thèse de 3è Cycle, Université Paris-Sud (1978)
11) V. Bernard and Nguyen van Giai, to be published
12) N. Vinh Mau, Theory of Nuclear Structure (IAEA, Vienna 1970) p.931
13) J.W. Negele and D. Vautherin, Phys. Rev. C5 (1972) 472
14) G.F. Bertsch and S.F. Tsai, Phys. Reports 18C (1975) 126
15) K.F. Liu and Nguyen van Giai, Phys. Lett. 65B (1976) 23
16) F. Perey and D.S. Saxon, Phys. Lett. 10 (1964) 107
17) S. Kailas and S.K. Gupta, Phys. Lett. 71B (1977) 271.

THE ISOSPIN DEPENDENCE OF THE NON-LOCAL OPTICAL POTENTIAL*

M.M. Giannini**, G. Ricco and A. Zucchiatti
Istituto di Scienze Fisiche dell' Università, Genova
Istituto Nazionale di Fisica Nucleare, Sezione di Genova

A non-local energy-independent potential model supplies the framework for an unified description of various data regarding the single-particle properties of nuclei, both in the bound state and positive energy range[1].

In I, the analysis has been restricted to N = Z nuclei, thus, missing the isospin dependent part of the optical potential. However, such a term is of considerable interest[2-6]. Therefore, we extend the model developed in I, in order to describe the single particle properties in groups of isotopes and test the isospin potential in the energy range from -70 MeV up to 140 MeV, by means of both proton and neutron data.

In the context of multiple scattering theory (for details see ref.[1]), the interaction potential energy of one nucleon in the presence of a core (A - 1) nucleus can be written:

$$V(\vec{r}, \vec{r}') = H(\vec{r} - \vec{r}')(U(\frac{\vec{r} + \vec{r}'}{2}) + \frac{1}{A - 1} \vec{\tau} \cdot \vec{T} \; U_t(\frac{\vec{r} + \vec{r}'}{2})) +$$
$$+ \; \vec{\sigma} \cdot \vec{L} \; H_S(\vec{r} - \vec{r}') \; U_N^S(\frac{\vec{r} + \vec{r}'}{2}) \tag{1}$$

where

$$H_{(S)}(x) = \frac{1}{(\pi \beta_{(S)}^2)^{3/2}} \; e^{-x^2/\beta_{(S)}^2} \tag{2}$$

β , β_S being non-locality ranges;

$$U_{(t)}(y) = (A - 1) \int d\vec{q} \; e^{i\vec{q} \cdot \vec{y}} \, S(q) \; t_{(T)}(q) \tag{3}$$

$$U_N^S(y) = \frac{r_\pi^2}{y} \frac{d}{dy} ((A - 1) \int d\vec{q} \; e^{i\vec{q} \cdot \vec{y}} \, S(q) \; t_S(q)) \tag{4}$$

and

$$S(q) = \int d\vec{z} \; e^{-i\vec{z} \cdot \vec{q}} \, \rho(z)$$

---

* Presented at the Workshop on Microscopic Optical Potentials,
  Hamburg, September 25-27, 1978

** Present address: Institut für Kernphysik, Universität Mainz, D-6500 Mainz

$\rho(z)$ is the density of nuclear matter; t, t_T and t_S are the isoscalar, isospin dependent and spin dependent nucleon-nucleon scattering amplitudes; T is the total isospin of the nucleus, $(\vec{\tau})\vec{\sigma}$ is twice the (iso-) spin of the extra nucleon. Eqs. (1-4) are used to guess the form of the potential and to fix some constraints on its parameters: in this way an overall consistency is achieved, possibly at the expense of less accurate fits.

From eqs. (3, 4) we see that the potential is complex and that the radial behaviour of the leading term in each well is determined by the nuclear matter density $\rho(r)$; therefore, we can write:

$$U(r) = - V_N f_N(r) - 4i W_N f_N(r) (1 - f_N(r)) \tag{5}$$

$$U_t(r) = - V_{Nt} f_N(r) - 4i W_{Nt} f_N(r) (1 - f_N(r)) \tag{6}$$

$$U_N^S(r) = V_{NS} \frac{2}{r} \frac{d}{dr} f_N(r) \tag{7}$$

with

$$f_N(r) = (1 + e^{(r-R_N)/a_N})^{-1}$$

According to the analysis performed in I, real surface and volume absorption parts are omitted and we neglect the energy dependence of the non-local potential. We are left then with nine energy independent parameters: the two non-locality ranges β, β_S, the well depths V_N, V_{Nt}, W_N, W_{Nt}, V_{NS} and the geometrical parameters R_N, a_N.

The non-local potential eq. (1) can be transformed to an equivalent local one $U_L(r)$, by extending the Fiedeldey method[7,1] to the isovector part:

$$U_L(r) = U_L^0(r) + \frac{1}{A-1} \vec{\tau} \cdot \vec{T} V_t(r) + \vec{\sigma} \cdot \vec{L} V_S(r) +$$
$$+ V_1(r) \vec{L}^2 + V_T(r) \frac{\vec{T}^2}{(A-1)^2} + V_{ST} \vec{\sigma} \cdot \vec{L} \frac{\vec{\tau} \cdot \vec{T}}{A-1} \tag{8}$$

where:

$$U_L^0(r) = U(r) e^{-\alpha(E-U_L^0(r))} ; \qquad \alpha = \frac{\beta^2 m^*}{2\hbar^2} \tag{9}$$

$$V_t(r) = \frac{U_t(r)}{1-\alpha U_L^0(r)} e^{-\alpha(E-U_L^0(r))} \tag{10}$$

$$V_S(r) = \frac{U_N^S(r)}{1-\alpha U_L^0(r)} e^{-\alpha_S(E-U_L^0(r))} ; \quad \alpha_S = \frac{\beta_S^2 m^*}{2\hbar^2} \tag{11}$$

(m* is the reduced nucleon mass) and

$$V_1(r) = \alpha_S (V_S(r))^2 \; ; \qquad V_T(r) = \alpha (V_t(r))^2$$

$$V_{ST}(r) = (\alpha + \alpha_S) \, V_S(r) \, V_t(r)$$

The three wells $U_L^0(r)$, $V_t(r)$, $V_S(r)$ can be parametrized in an analogous way to the corresponding non-local ones:

$$U_L^0(r) = - V_L \, f_V(r) - 4iW \, f_I(r) \, (1 - f_I(r)) \tag{12}$$

$$V_t(r) = - V_t \, f_t(r) - 4iW_t \, f_{It}(r) \, (1 - f_{It}(r)) \tag{13}$$

$$V_S(r) = \frac{2}{r} V_S \frac{d}{dr} f_S(r) \tag{14}$$

with

$$f_x(r) = (1 + e^{(r-R_x)/a})^{-1} \qquad x = V, I, t, It, S$$

and all a (but a_I, see below) have been taken equal to a_N.

The transformation (9-11) leads to a relationship between the local and non-local parameter sets (see table I). Owing to the non-locality, all the parameters of the equivalent local potential acquire a spurious dependence on the energy E:

$$E = E_{CM} - V_{Coul}$$

which is the center of mass nucleon energy shifted by the Coulomb term

$$V_{Coul} = -1.08 + 1.35 \, \frac{Z-1}{(A-1)^{1/3}} \quad \text{MeV}$$

Moreover the wave function calculated through the local potential (8) is a good approximation of the non-local one, provided it is divided by the Perey factor:

$$f(r) = \exp \{ - \frac{1}{2} (\alpha U_L^0(r) + \alpha_S \, \vec{L}^2 \, V_1(r) + \alpha_S \, \vec{\sigma} \cdot \vec{L} \, V_S(r) +$$

$$+ \, \alpha \, \frac{\vec{\tau} \cdot \vec{T}}{A-1} \, V_t(r) + \alpha \, \frac{\vec{T}^2}{(A-1)^2} \, V_T(r)) \}$$

We determine first the radial parameter R_N, by fitting, for one nuclide in each group of isotopes, the single particle energies of the occupied states and the charge distributions. To this end we take into account the proton and neutron

Table I: The relationship between non-local and local parameter sets.

$$V_L(E) = V_L(0) + bE + cE^2 \qquad V_L(0) = V_N e^{-\alpha V_L(0)} \qquad b = -\frac{\alpha V_L(0)}{1 + \alpha V_L(0)}$$

$$c = \frac{b^2(1 + b)}{2V_L(0)}$$

$$R_V(E) = R_N + a_N \ln(2e^{2\omega}-1) \qquad\qquad \omega = \frac{\alpha V_L(E)}{4}$$

$$W(E) = W_N e^{-(\alpha E + 2\omega)} \qquad\qquad \omega_o = \frac{\alpha V_L(0)}{4}$$

$$R_I(E) = R_N + a_I \ln \frac{1+\omega}{1-\omega}$$

$$V_S(E) = V_S(0) \frac{1+2\omega_o}{1+2\omega} \exp\left(-\alpha_S(E + \tfrac{1}{2}(V_L(E) - V_L(0)))\right)$$

$$V_S(0) = \frac{V_{NS}}{1+2\omega_o} \exp\left(-\alpha_S \tfrac{1}{2}(V_L(0))\right)$$

$$R_S(E) = R_N + a_N \ln\left(\frac{1+3\omega}{1+\omega} \frac{1+\alpha_S V_L/4}{1-\alpha_S V_L/4}\right)$$

$$V_t(E) = V_t(0) \frac{1+4\omega_o}{1+4\omega} e^{-(\alpha E + 4\omega - 4\omega_o)} \qquad\qquad V_t(0) = \frac{V_{Nt}}{1+4\omega_o} e^{-4\omega_o}$$

$$R_{Vt}(E) = R_N + a_N \ln\left((2e^{2\omega}-1)\frac{1+6\omega}{1+2\omega}\right)$$

$$W_t(E) = \frac{W_{Nt}}{1+2\omega} e^{-(\alpha E + 2\omega)} \qquad\qquad \alpha = \frac{\beta^2 m^*}{2\hbar^2}$$

$$R_{It}(E) = R_N + a_N \ln \frac{1+3\omega}{1-\omega} \qquad\qquad \alpha_S = \frac{\beta_S^2 m^*}{2\hbar^2}$$

charge form factor, the c.m. corrections and the spin-orbit and magnetic contributions[8]. In figs. 1 and 2 the results for the charge form factors and densities are reported. The fitted R_N and the corresponding r.m.s. charge radii are given in fig. 3; their behaviour is well accounted for by the two relations:

$$R_N = 1.15(A-1)^{1/3}(F) \qquad <r^2>^{1/2} = 0.84 + 0.77 A^{1/3} F \qquad (15)$$

A large set of $T_>$ and $T_<$, particle- and hole-states both for protons and neutrons[4,5] is used in order to determine the well depths of the potential. States of

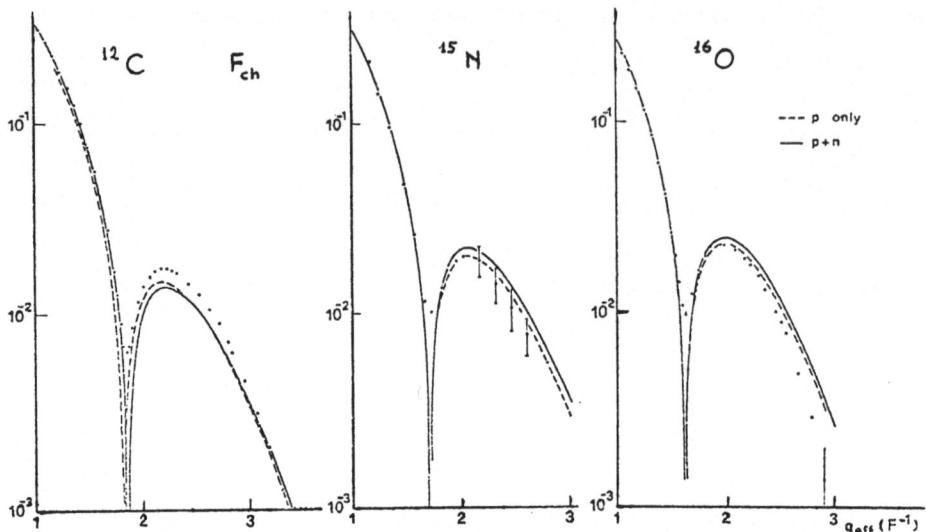

<u>Fig. 1.:</u> Elastic monopole form factors of ^{12}C, ^{15}N and ^{16}O. Data are from ref.9. Full (dashed) curves are the independent particle model fits with (without) the neutron contribution.

various isotopes but with the same quantum numbers are grouped into multiplets to which a unique centroid energy E_{CM} is assigned by taking into account the spin-orbit and isospin splittings, that is we assume:

$$E_{nJl} = E_{CM}(n_J l) + a_S <\vec{\sigma} \cdot \vec{L}> + a_T \frac{\vec{\tau} \cdot \vec{\tau}}{A-1} + a_{ST} <\vec{\sigma} \cdot \vec{L}> \frac{\vec{\tau} \cdot \vec{\tau}}{A-1}$$

Once the energy is fixed, the radial parameters of all the potential wells are determined by the real depth V_L and the non-locality ranges (see table I), which, for this purpose, have been assumed equal to those found in I.

For each multiplet, the real potential well V_L is fitted, in correspondence to various values of $V_t(0)$; plotting the results in a V_L - $V_t(0)$ diagram (see fig. 4), we get straight lines with different slopes: their intersection determines with fair accuracy the correct $V_t(0)$. The values obtained in this way are shown in fig. 5. and reported in table II: the mean value is $\sim$ 34 MeV, with a slight increase as Z increases.

The consistency of these results with the positive energy region has been checked. Here, however, the analysis must be performed in a different way, since there is a lack of proton and neutron cross sections at energies differing for the

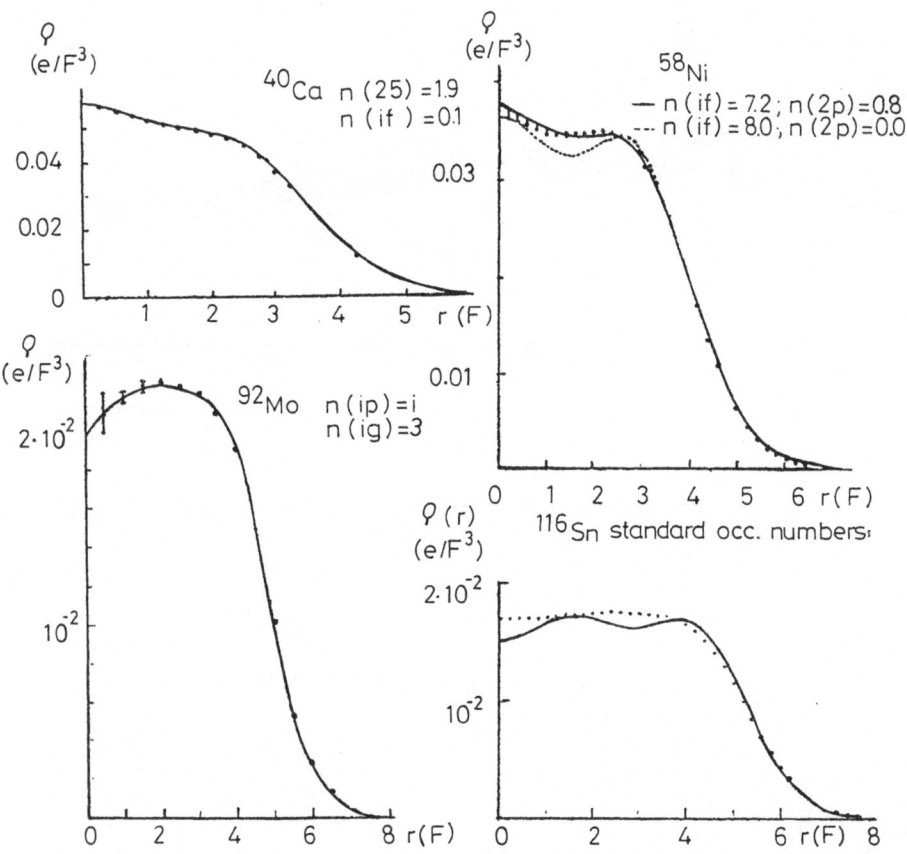

Fig. 2.: Charge distributions in ^{40}Ca, ^{58}Ni, ^{92}Mo and ^{116}Sn. Data are from refs. 10-13; the curves are the independent particle model fits. Non-standard occupation numbers are indicated; for ^{58}Ni the dependence on the $2P_{3/2}$ occupation number is shown.

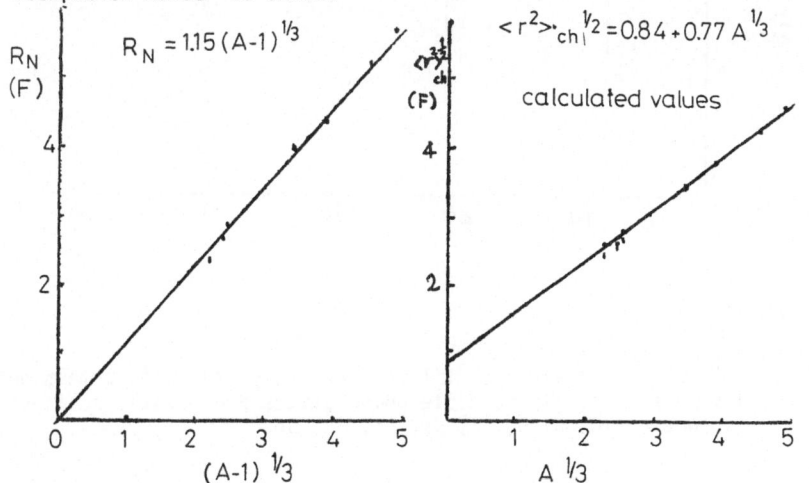

Fig. 3.: The potential radii R_N and the r.m.s. charge radii. The lines are given by eqs. (15). Data from refs. 12-14.

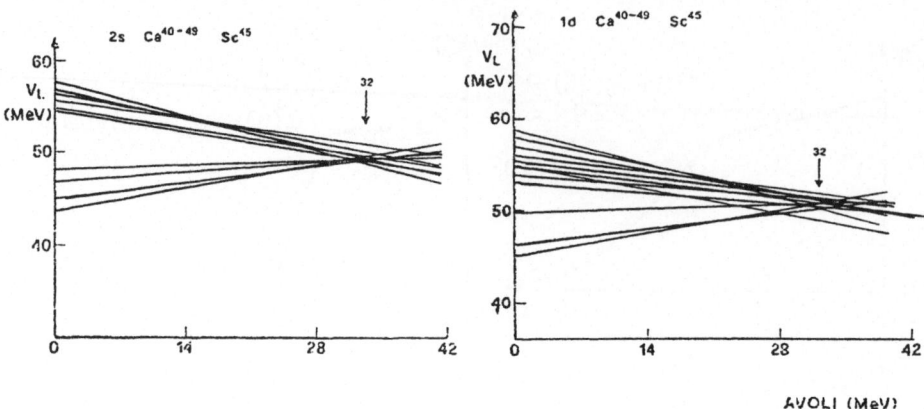

<u>Fig. 4.:</u> The fitted real central potential as a function of $V_t(0)$, (AVOLI), for the 2s, 1d-state in the Ca-isotopes.

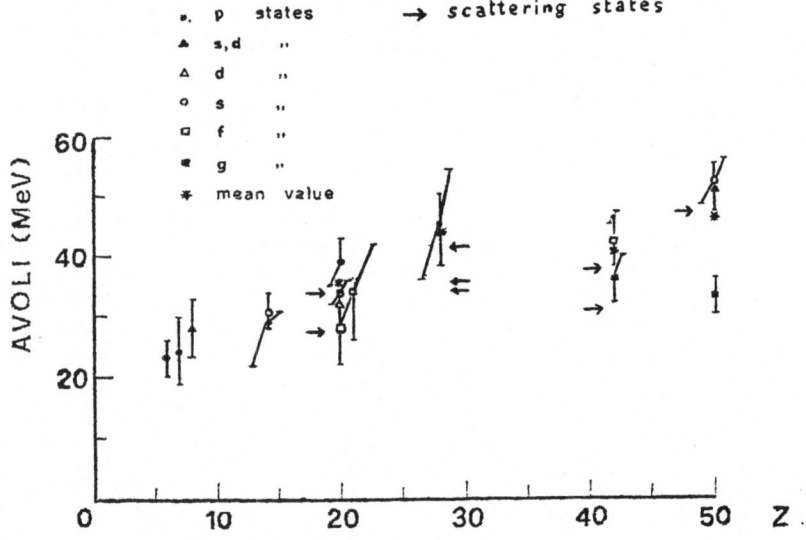

<u>Fig. 5.:</u> The isospin local potential at zero energy (AVOLI), determined by the analysis of single-particle bound states and elastic scattering in isotopes of C, N, O, Si, Ca, Ni, Mo, Sn.

Table II: The isospin local potential at zero energy determined by the analysis of single-particle bound states (a) and of proton (P) and (or) neutron (Ṅ) elastic scattering (b), in various groups of isotopes.

a)						b)			
Z	states	$V_t(0)$	Z	states	$V_t(0)$	Z	E(MeV)		
6	p	22.8±2	21	f	33.6±8	20	15.6	P	34.5±0.8
7	p	23.7±6	28,29	f	43.7±5	20	50.	P	28. ±3.4
8	s,d	28.1±5	28,29	p	45.1±9	26-27	14.7	N	42 ±1
14	s	31. ±3	42	f	42.4±4	26-27-28	14.7	N	36 ±1
14	d	30 $^{+1}_{-8}$	42	p	45.7±1	28	39.6	P	34.8±6.6
20	d	32.4±4	42	g	35.7±4	42	10.9	N	31.7±1.2
20	s	34. ±2	50	g	33 ±3	42	12.5	P	38.1±2.8
20	f	30.9±5	50	d	51 ±4	50	24.5		48.1±0.01
20	p	38.7±3.5	50	s	56 ±2				

Coulomb shift. Therefore, we analyse either proton or neutron elastic scattering data on groups of isotopes. Two parameters are fitted (V_L and W), while W_t has been fixed to the value of $\sim$ 30 MeV, which, in the mean, makes W not too strongly dependent on N-Z; in order to get reasonable fits, we let the smoothness a_I vary with A, according to the rule:

$$a_I = 0.54 + 3.26 \cdot 10^{-3}(A - 1) \qquad (F)$$

in this way the lacking of volume absorption is partially compensated in heavy nuclei. In fig. 6 we give the fitted elastic differential cross section of protons on Mo isotopes at 12.5 MeV.

For each group of isotopes, the correct value of $V_t(0)$ is determined by plotting the fitted real depth V_L as a function of $\frac{N-Z}{A}$ and interpolating the zero-slope straight line. The results, reported in table II, agree with the bound state ones.

Finally, we plot all the fitted real depths V_L as a function of energy (fig. 7). The full curve gives the energy behaviour of an equivalent local potential with (see table I)

$$V_L(0) = 48.5 \text{ MeV}; \qquad b = -0.36$$

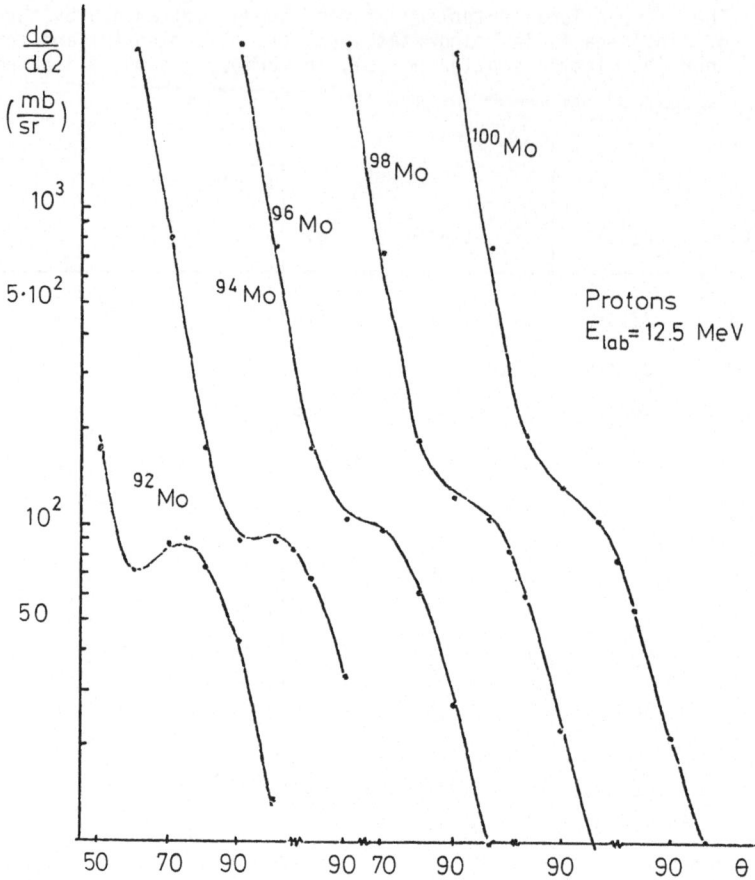

Fig. 6.: The fit to elastic scattering of protons on Mo isotopes at 12.5 MeV.
Data from ref. 15.

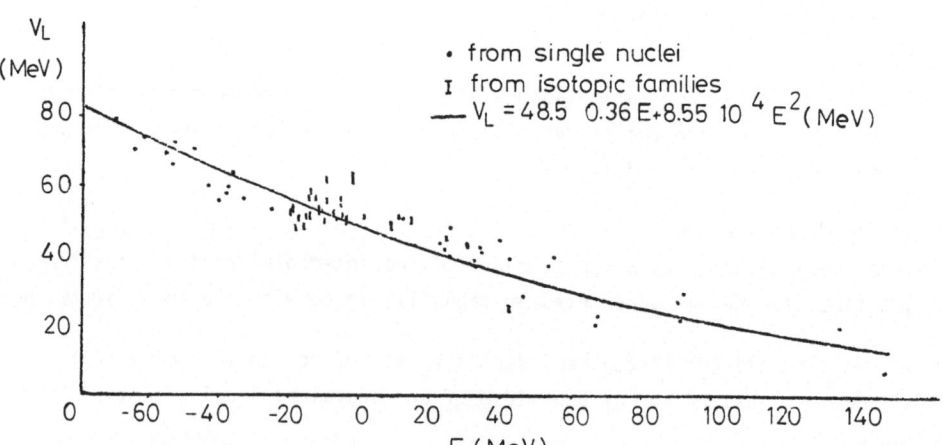

Fig. 7.: The energy dependence of the real central potential (points: fitted
values; full curve: first equation of table I with $V_L(0) = 48.5$ MeV
and $b = -0.36$).

In table III we give the final values of the fitted parameters and the corresponding non-local quantities. We note here that, according to I, a phenomenological energy dependence had to be introduced in the imaginary part.

Table III: The fitted parameters of the equivalent local potential and the related non-local ones.

Local	Non-local

$V_L(0) = 48.5$ MeV $\qquad$ $\alpha = -\dfrac{1}{V_L(0)}\,\dfrac{b}{1+b} = 1.160\ 10^{-2}(\text{MeV})^{-1}$

$b = -0.36$ $\qquad$ $V_N = V_L(0)\,e^{\alpha V_L(0)} = 85.1$ MeV

$\qquad\qquad\qquad\qquad R_N = 1.15\,(A-1)^{1/3}$ F

$V_t(0) = 34$ MeV $\qquad$ $V_{Nt} = V_t(0)\,(1+\alpha V_L(0))\,e^{\alpha V_L(0)} = 93$ MeV

$V_S(0)^a = 6$ MeV $\qquad$ $V_{NS} = V_S(0)\,(1+\alpha V_L(0)/2)\,\exp\left(\alpha_S \tfrac{1}{2}V_L(0)\right) = 9.1$ MeV

$\qquad\qquad\qquad\qquad \alpha_S^a = 0.007\ (\text{MeV})^{-1}$

$\qquad\qquad\qquad\qquad W_N^a = 29(1 - \exp\text{-}(0.0244\ E))\ \text{MeV}$

$\qquad\qquad\qquad\qquad W_{Nt} = 30$ MeV

$\qquad\qquad\qquad\qquad a_N^a = 0.57$ F; $a_I = 0.54 + 3.26 \cdot 10^{-3}(A-1)$ F

[a] Taken from ref. 1.

The preceding analysis shows that a non-local energy independent potential model can be fairly extended also to the isospin part of the interaction. In particular, the model accounts for the variation of the central depth in a wide energy interval (see fig. 7), with possibly the exception of the zero energy region[16].

The analysis of single particle energies of different types permits a reliable determination of the isospin dependent term. The mean value ($\sim$ 34 MeV) we get and the increasing trend with Z are consistent with previous bound state analyses[4] and with the description of the elastic scattering data.

A possible insight into the origin of such a behaviour can be provided by a plot of $V_t(0)$ versus $(Z-1)/(A-1)^{1/3}$ (see fig. 8), which shows a linear dependence on the Coulomb energy:

$$V_t(0) = 15.9 + 3.29\ (Z-1)/(A-1)^{1/3}\ \text{(MeV)}$$

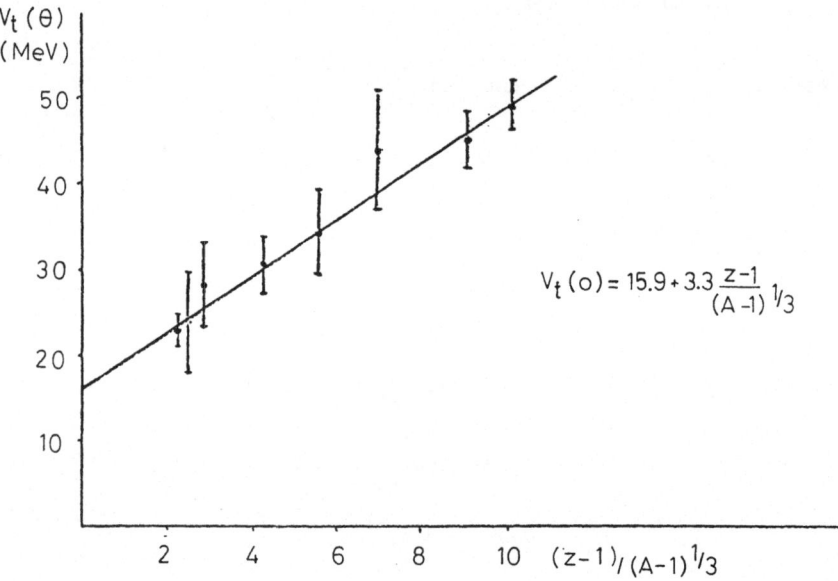

<u>Fig. 8.:</u> Relation between the isospin potential at zero energy and the Coulomb factor $(Z-1)/(A-1)^{1/3}$.

In fact, the isospin potential gives rise to coupled channel equations for the proton states at energy E and neutron states at energy $E - V_c$. The elimination of one channel affects the potential for the other one: an interference between the Coulomb energy and the isospin potential can then arise in both cases.

References:

1. M.M. Giannini and G. Ricco, Ann. of Phys. 102 (1976) 458, hereafter referred to as I.
2. G.R. Satchler, in "Isospin in Nuclear Physics" (D.H. Wilkinson ed.) p. 391, North-Holland, Amsterdam 1969
3. P.E. Hodgson, "Nuclear Reactions and Nuclear Structure", Oxford 1971
4. D.J. Millener and P.E. Hodgson, Nucl. Phys. A209 (1973) 59
   F. Malaguti and P.E. Hodgson, Nucl. Phys. A215 (1973) 243
   F. Malaguti and P.E. Hodgson, Nucl. Phys. A257 (1976) 37
5. A. Calboreanu and S. Mancas, Nucl. Phys. A266 (1976) 72
6. P.C. Sood and D.C. Agrawal, Phys. Rev. C11 (1975) 2056
   D.M. Patterson, R.R. Doering and A. Galonsky, Nucl. Phys. A263 (1976) 261
   J.D. Brandenberger and R. Schrils, Phys. Rev. C13 (1976) 2559
   J.C. Ferrer, J.D. Carlson and J. Rapaport, Phys. Lett. 62B (1976) 399
7. H. Fiedeldey, Nucl. Phys. 77 (1966) 149
8. W. Bertozzi, J. Friar, J. Heisenberg and J.W. Negele, Phys. Lett. 41B (1972) 408
9. I. Sick and J.S. McCarthy, Nucl. Phys. A150 (1970) 631
10. J.B. Bellicard et al., Phys. Rev. Lett. 19 (1967) 527
11. I. Sick et al., Phys. Rev. Lett. 35 (1975) 910
12. B. Dreher, Phys. Rev. Lett. 35 (1975) 716
13. J.R. Ficenec et al., Phys. Lett. 42B (1972) 213
14. G. Fey et al., Z. Phys. 265 (1973) 401
15. S.J. Burger and G. Heymann, Nucl. Phys. A243 (1975) 461
16. J.P. Jeukenne, A. Lejeune and C. Mahaux, Int. Conf. on the Interaction of Neutrons with Nuclei, Lowell (Mass., U.S.A.) July 1976

Shell Model Description
of the Optical Model Potential

M. Micklinghoff
Universität Hamburg, I. Institut für Experimentalphysik
Zyklotron, Luruper Chaussee 149, D-2000 Hamburg 50

For the calculation of the mass operator a Bethe-Goldstone type integral equation is to be solved for a finite system. Approximation schemes are presented in the framework of the collective model as well as using the hole line expansion. The structure of the resulting equations for the collective G-matrix and the two-body G-matrix is identical, presenting essentially a three-body problem. It is attacked by expanding the interior region in a harmonic oscillator basis and neglecting the two-body correlations outside this space.

The asymptotic region is described by modified continuum functions which are orthogonal to the harmonic oscillator space. This allows us to neglect the continuum-continuum coupling, and the bound-continuum coupling is reduced to a one-body operator.

1. Nucleon-Nucleus Optical Potential

If the target ground state is denoted by $|\phi_o\rangle$, whereas $|\phi_i\rangle$ stands for the i-th excited state, the optical potential is given by the well-known integral[1)]

$$\bar{U}_{oo} = \langle\phi_o|\bar{U}|\phi_o\rangle \qquad (1.1)$$

$$= \langle\phi_o|V|\phi_o\rangle + \langle\phi_o|VG^o\left[1-|\phi_o\rangle\langle\phi_o|\right]\ \bar{U}|\phi_o\rangle$$

This equation exhibits the typical feature that the target ground state configuration is excluded as an intermediate state. It should be noted that the quantities in eq.(1.1) represent single particle operators. For a solution of this equation, it may be written in either r-space or in the shell model representation. Although one may include exchange terms, it is difficult to take care of the blocking effect of the Pauli-principle using the r-representation. Hence, the shell model representation is used in this study. There exist several ways to solve eq.(1.1) approximately, two of which shall be discussed here.

First, the target excitations are considered as highly collective, which renders the projectile in some respects distinguishable from the target nucleons. Hence, the target excitations may be obtained in a separate calculation (RPA) or taken from experiment. In order to allow a graphical representation of such a calculation, the target ground state is represented by the vacuum and its excitations by a dashed line.

FIGURE 1

$$\langle p | \bar{U}^C_{oo} | q \rangle =$$ $+$ $+$ $+ \ldots$

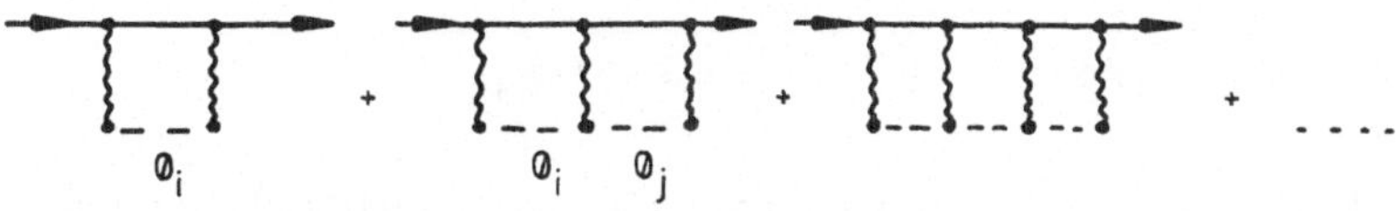

The Pauli-principle is taken care of by:
(i) prohibiting that the projectile falls below the Fermi-surface
(ii) using antisymmetrized interaction matrix elements

In a more microscopic calculation one has to antisymmetrize fully each intermediate state. In order to simplify this problem, the hole line expansion may be used. In first order only one hole is allowed in the intermediate states. This yields the following graphical representation

FIGURE 2

$$\langle p | \bar{U}_{oo}^{(1)} | q \rangle = \sum_{i}^{occ.} \cdot$$ $= \sum_{i}^{occ}$

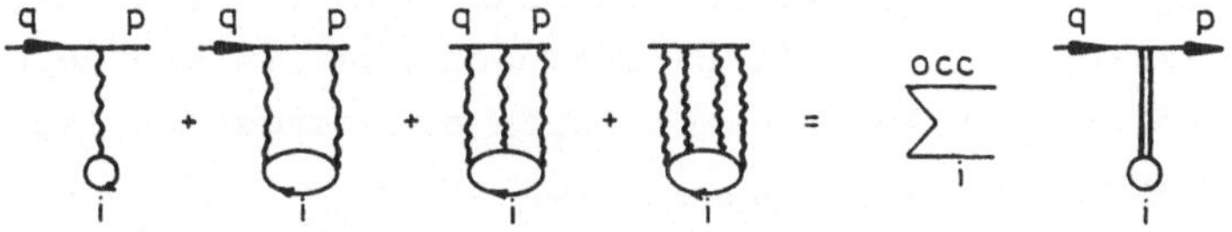

The last graph represents the two-body G-operator, which is given by the Bethe-Goldstone type integral equation:

$$\langle p | \bar{U}_{oo}^{(1)} | q \rangle = \sum_{n}^{occ.} \langle pn | t_G | qn \rangle$$

$$\text{(1.2)}$$

$$\langle pn | t^G | qn \rangle = \langle pn | v | qn \rangle + \sum_{\substack{p_1 p_2 \\ exc.}} \langle pn | v | p_1 p_2 \rangle \frac{1}{E^+ - \varepsilon_{p_1} - \varepsilon_{p_2}} \langle p_1 p_2 | t^G | qn \rangle$$

The intermediate states ($p_1 p_2$ exc.) must not be occupied in the uncorrelated target ground state. This not only reflects the presence of the other particles but makes the corresponding graphs irreducible as well.

Whereas in infinite nuclear matter the hole-line expansion is related to a series in the powers of the density, it is not clear how useful such an expansion is for finite nuclei. In infinite nuclear matter it is possible to minimize the higher order graphs by a selfconsistent choice for the single particle potential.

2. Solution of the Optical Potential Equation

In order to write the opical potential equation (1.1) in the shell model space we assume for simplicity that there exist no bound single particle states above the Fermi-surface. The shell model continuum is representated by the states $|\varepsilon\rangle$.

$$<\varepsilon_1\phi_o|\bar{U}|\varepsilon_2\phi_o> = <\varepsilon_1\phi_o|V|\varepsilon_2\phi_o>$$

$$+ \sum_{i\neq o} \int d\varepsilon' \; \frac{<\varepsilon_1\phi_o|V|\varepsilon'\phi_i><\varepsilon'\phi_i|\bar{U}|\varepsilon_2\phi_o>}{E-\varepsilon'-E_i} \tag{2.1}$$

Knowing the shell model potential and the residual interaction, the so-lution of this equation presents mainly a continuum problem. First, we have several *simple* continua: one particle in a scattering state for each channel. Secondly, there exists a *double* continuum for the target break-up. It should be mentioned that the structure of the G-matrix equation (1.2) is the same as in the above eq.(2.1). One simply must re-place the target in eq.(2.1) by one individual nucleon to obtain the Bethe-Goldstone equation (1.2). The difference may be summarized as fol-lows: In the G-matrix approach the projectile interacts only with one nucleon up to all orders. The optical potential can then be obtained by summing (folding) over all target nucleons. In the coupled channel ap-proach the projectile interacts with all target nucleons collectively up to all orders. The type of equation to be solved is the same in both cases. While in the coupled channel approach the target may be excited to a state well above the particle threshold, in the G-matrix approach both nucleons in the intermediate states may be in a scattering state. This reflects the three-body nature of the optical potential problem.

However, as we are not interested in the asymptotic behavior in the break-up channels, a bound state expansion may be possible. For the G-matrix approach a bound basis for target nucleon may be generated by diagonalization of the unperturbed shell model Hamiltonian (Hartree-Fock potential) in harmonic oscillator space of finite size (N). The states with negative energy are practically identical to the genuine bound states in the Hartree-Fock potential. The states with positive energy are either similar to the single particle resonances or present merely an expansion basis. To start with, we consider them as quasi-bound states and neglect their decay properties. Also in the coupled channel approach the target is at first assumed to always be in a bound state. Hence only the continuum of the projectile is left, for which the above mentioned bound state expansion is used as well. However, as the asymp-totic region is now also relevant, not only the bound state space $|\mu>$ is used but also the residual (modified) continuum $|\tilde{\varepsilon}>$ is included.

$$1 = \int d\varepsilon |\varepsilon><\varepsilon| = \sum_{\mu=1}^{N} |\mu><\mu| + \int d\varepsilon |\tilde{\varepsilon}><\tilde{\varepsilon}| \tag{2.2}$$

The reason for this expansion is the fact that the coupling within the residual continuum may be neglected[2]). With this approximation the optical potential becomes:

$$\langle \tilde{\varepsilon}_1 \phi_0 | \tilde{U} | \tilde{\varepsilon}_2 \phi_0 \rangle = \sum_\omega \frac{\langle \tilde{\varepsilon}_1 \phi_0 | V + \tilde{v} | \omega \rangle \langle \omega | V + \tilde{v} | \tilde{\varepsilon}_2 \phi_0 \rangle}{E - E_\omega - \Delta_\omega + \frac{i}{2} \Gamma_\omega} \qquad (2.3)$$

The states $|\omega\rangle$ are obtained by diagonalization of the full Hamiltonian only in the bound state space, excluding the target ground state

$$(E-H)|\omega\rangle = 0 \quad \text{with} \quad |\omega\rangle = \sum_{i \neq 0} \sum_\mu \alpha_{i\mu}^\omega |\mu \phi_i\rangle \ . \qquad (2.4)$$

Also the width and shift, which are assumed to be diagonal in ω, are calculated without the ground state channel (elastic channel)

$$\Delta_\omega(E) - \frac{i}{2} \Gamma_\omega(E) \cong \sum_{i \neq 0} \int d\varepsilon \frac{\langle \omega | V + \tilde{v} | \tilde{\varepsilon} \phi_i \rangle \langle \tilde{\varepsilon} \phi_i | V + \tilde{v} | \omega \rangle}{E^+ - \varepsilon - E_i} \qquad (2.5)$$

As neither the modified scattering states $|\tilde{\varepsilon}\rangle$ nor the quasi-bound states $|\mu\rangle$ are eigenstates to the unperturbed shell model Hamiltonian, we have to include a one-body residual interaction $\tilde{v}$ in addition to the two-body residual interaction

$$\langle \mu | \tilde{v} | \tilde{\varepsilon} \rangle : = \langle \mu | H^0 | \tilde{\varepsilon} \rangle \qquad (2.6)$$

This reflects the fact that the modified shell model basis states $\{|\mu\rangle, |\tilde{\varepsilon}\rangle\}$ are eigenstates of the <u>modified</u> Hamiltonian $\tilde{H}^0 = k^2 + U - \tilde{v}$, U being the Hartree-Fock potential. Hence, the total optical potential is given by

$$M = U - \tilde{v} + \langle \phi_0 | \tilde{U} | \phi_0 \rangle \qquad (2.7)$$

It is obvious that eqs. (2.3-2.5) can be transformed into the corresponding two-body G-matrix equations. One simply has to replace $|\phi_i\rangle$ by the excited shell model orbits $|\mu\rangle$ and $|\phi_0\rangle$ by the orbit $|n\rangle$ of the struck nucleon.

Finally, the target break-up is to be included approximately. The procedure is demonstrated for the case of the microscopic G-matrix. Denoting the harmonic oscillator space by Q, the "target excitations" (as considered above) are obtained by solving

$$(\varepsilon_\mu - H^0_{QQ})|\mu\rangle = 0 \qquad (2.8)$$

However, for positive energies the correct target states are scattering states $|\varepsilon\rangle$

$$(\varepsilon - H^0_{QQ} - W^0_{QQ})Q|\varepsilon\rangle = \tilde{v}|\tilde{\varepsilon}\rangle \quad \text{with} \quad \langle \mu | W^0_{QQ} | \mu \rangle = \int \frac{\langle \mu | \tilde{v} | \tilde{\varepsilon}' \rangle \langle \tilde{\varepsilon}' | \tilde{v} | \mu \rangle}{\varepsilon^+ - \varepsilon'} d\varepsilon' \qquad (2.9)$$

The approximation we are going to adopt is the following: As the asymptotic region of the target break-up is of little importance for the elastic channel, the inhomogeneous part of eq.(2.9) is neglected. However, the resulting equation is not Hermitian because of W^O_{QQ} and the eigenstates form an energy-dependent biorthogonal set $\{|\bar{\mu}>, |\bar{\bar{\mu}}>\}$ with complex energies

$$<\bar{\bar{\mu}}' (E) |H^O_{QQ} + W^O_{QQ}(E) |\bar{\mu}(E) > = (\varepsilon_\mu - \frac{i}{2}\gamma_\mu) \delta_{\mu\mu'} \qquad (2.10)$$

However, instead of solving this equation it is more practical to use the old basis $|\mu>$ (eq.2.8) and to include $W^O_{QQ}(E)$ in the Hamiltonian of eq.(2.4) to be solved for the intermediate states $|\omega>$. Now the biortho-gonal sets of ω-states with complex energies $(E_\omega - \frac{i}{2}G_\omega)$ have to be used in eq.(2.3).

A further approximation consists of including $W^O_{QQ}(E)$ only in a per-turbative way. In first order the old (real) ω-states can be retained, which acquire only a complex energy

$$E_\omega + D_\omega - \frac{i}{2}G_\omega \cong E_\omega + <\omega|W^O_{QQ}(E) |\omega> \qquad (2.11)$$

This is particularly useful for the collective approach, because now the width Γ_{ϕ_i} of the core states $|\phi_i>$ may be calculated separately. The width of the intermediate states $|\omega>$ due to the core state break-up then becomes in the diagonal approximation (cf. eq.2.4)

$$G_\omega = \sum_{i\mu} |\alpha^\omega_{i\mu}|^2 \Gamma_{\phi_i} \qquad (2.12)$$

Using the same approximation for the particle degree of freedom the total width can be written as the sum of the particle width and the core state width.

$$G_\omega + \Gamma_\omega = \sum_{\mu, i \neq o} |\alpha^\omega_{i\mu}|^2 (\Gamma_{\phi_i} + \gamma_\mu) \qquad (2.13)$$

with $\gamma_\mu = 2\pi |<\mu|\tilde{v}|\tilde{\varepsilon}>|^2$

3. Results

Calculations using the collective approach have been performed for the neutron optical potential for ^{40}Ca. Preliminary results are given in the following figures. Figure 3a shows the volume integral of the imaginary part. The volume integral of the real part (fig.3b) does not include the Hartree-Fock potential. The results demonstrate a strong energy dependence and angular momentum dependence, which could be aver-aged out. The diagonal part of the non-local potentials for 10 MeV in r-space is displayed in fig.4. The real part is peaked at the nuclear surface (R_o), because of the underlying vibrational model. The imaginary part, however, reaches further out and exhibits some shell structure.

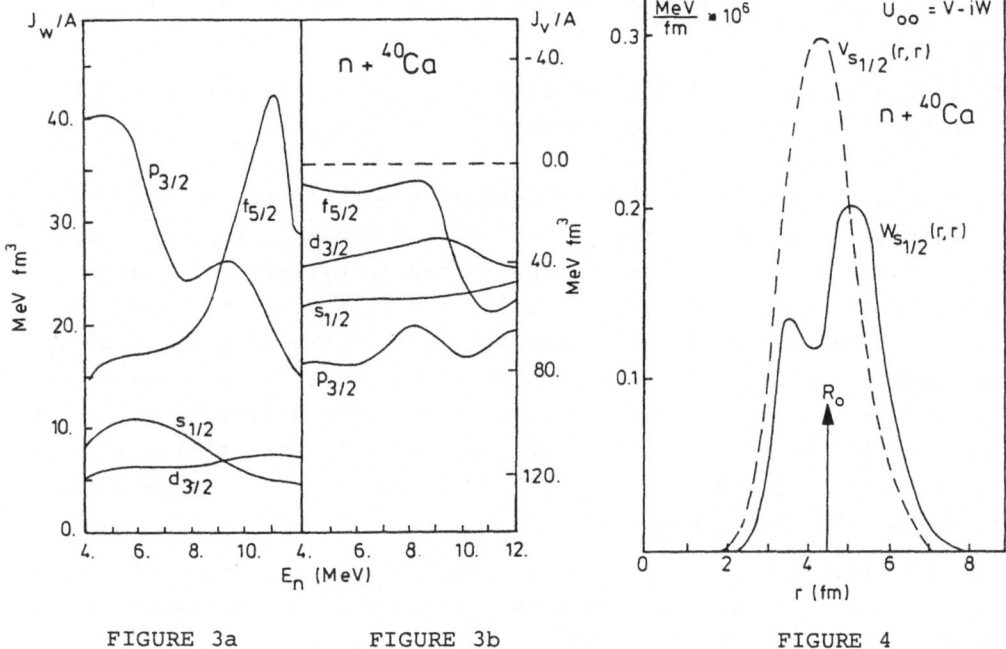

FIGURE 3a   FIGURE 3b   FIGURE 4

References

1) M. Micklinghoff, Microscopic description of the optical potential
   as a three-body problem, Internal Report, HH 1977-03, Univ.Hamburg,
   Cyclotron

2) M. Micklinghoff, Z.Phys. in press; Nucl.Phys. A295(1978)237

Three-Body Bethe-Faddeev Equations and Single-Particle Potentials
in Nuclei

John G. Zabolitzky, Institut für theoretische Physik, Ruhr-Universi-
tät Bochum, D-4630 Bochum, W.-Germany

Nucleon-nucleus optical potential models are designed to describe
the interaction between a real A+1st nucleon and an A-particle nucleus. Another problem one may be interested in is the self-energy
(single-particle energy) experienced by one of A nucleons in an A-
particle nucleus when virtually excited - without the presence of an
A+1st nucleon. This latter question is of interest if one intends to
calculate ground-state properties of the A-particle system. It is the
purpose of the present note to show that these two quantities - energy
of a real particle and of a virtually excited one - are quite different and should not be confused.

   Let me first consider the interactions of a real A+1st particle
with an A-particle nucleus. These may conveniently be expanded in the
number of particles the external, A+1st nucleon is <u>simultaneously</u>
interacting with. The first term in this expansion is the interaction
with just one particle shown schematically in fig. 1a, the second term
is the simultaneous interaction with two particles out of the A-particle nucleus, shown in fig. 1b. Lines with downgoing arrows indicate
nucleons in their occupied shell-model orbits of the A-particle system,
undirected lines indicate the external particle. The boxes stand for
the sum of <u>all</u> interaction processes possible involving the indicated
number of nucleons. The two-particle box, for example, is the sum of
all particle-particle and hole-hole ladder diagrams and all combinations of these. The three-particle box includes all three-body scattering processes, e.g. the Bethe-Faddeev sum. Let me consider the two-
body box in more detail. In lowest order approximation, it will be
equal to just a Brueckner reaction matrix (G or K matrix), i.e. involve the scattering of the two particles into unoccupied states, re-
scattering within unoccupied states, and scattering back to the original states. The second-order process of this kind is shown in fig. 2a.
Upgoing arrows indicate unoccupied states in the A-particle system.
The dashed line represents an interaction via the bare two-nucleon
interaction. Because of nuclei being rather dilute many-particle
systems the two-body processes will dominate the three-body processes,
these in turn will dominate the four-body processes, and so on. In
calculations performed so far[1) just the two-body contribution has been
included resulting in a satisfactory description of optical model po-

tentials. Looking at fig. 2a it is evident that one is truly talking about two-body processes involving the external, A+1st particle and one of the A particles of the nucleus. The other particles enter only in a statistical, mean-field sense in that they restrict the phase space available to the intermediate states. In this leading order process there is no need to talk about a third particle explicitly.

Let us forget about the preceding discussion momentarily and consider just the A-particle system by itself. This being a dynamical many-body problem, particles will undergo virtual excitation processes. Let me assume that the single-particle basis has been chosen in an optimal way, i.e., according to the condition that there are no one-particle one-hole excitations in the ground state (generalized Brueckner-Hartree-Fock or Maximum Overlap condition[2]). Consequently, the simplest process which will create a virtually excited particle is a two-particle two-hole excitation process, i.e., a two-body collision. The bottom part of fig. 2b shows a simple case of such a process. The wiggly line represents a Brueckner G-matrix producing a two-particle two-hole excitation.

We are now interested in the self-energy (single-particle energy) of the virtually excited particles. Before even starting to consider the problem in any detail we immediately notice a fundamental difference to the problem considered above, the energy of a real, A+1st nucleon added to the system. There we had to consider just this nucleon and its interactions. In the present problem, however, there are already two nucleons present in particle states. Even more severe, these two nucleons are close together and strongly correlated because they just did undergo a two-body collision. As a consequence we have to consider a third nucleon if we want to calculate particle self-energies. Any of the virtually excited nucleons will have interactions with other particles resulting in a self-energy. Again the leading order process will involve just one additional - the third - nucleon. Fig. 2b shows a simple process as an example. It is evident that in lowest order there are already three particles involved. One might be led to the assumption that this actually might not be true because one could "cut out" the right part of fig. 2b which looks so much like fig. 2a. One then could calculate the particle self-energy without reference to the process it will be used with. Unfortunately this is an illegal procedure because of various reasons.

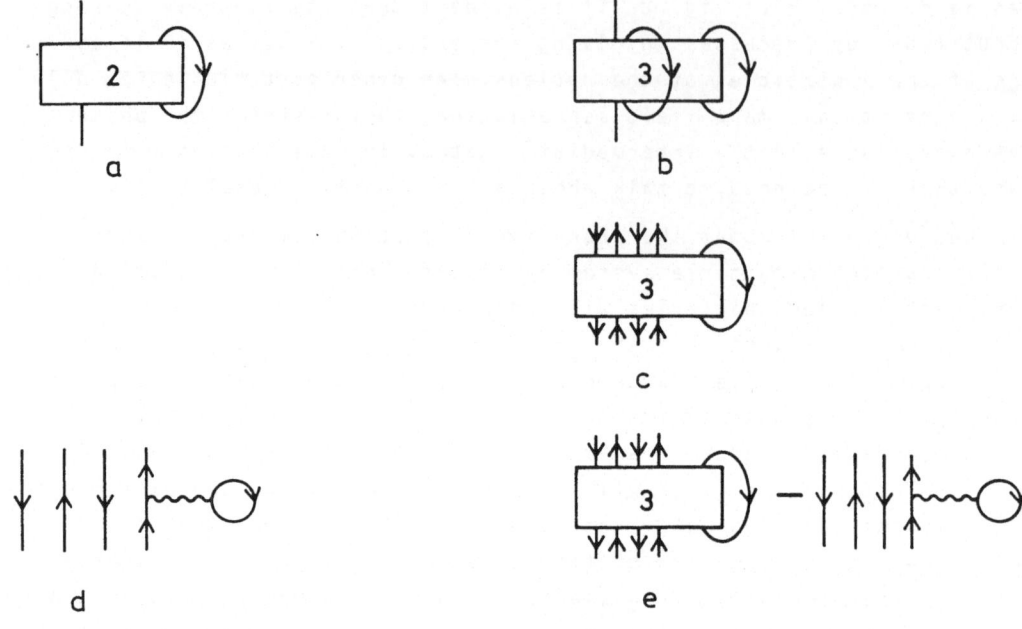

Fig. 1

Nucleon self-energy contributions. a,b-real particles, c-e virtual particles

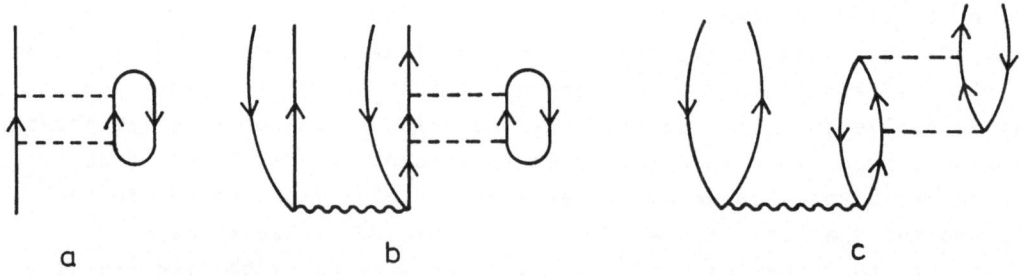

Fig. 2

Lowest order contributions. a-real particles, b,c-virtual particles.

Firstly, there is a formal reason. According to the diagram rules for Goldstone diagrams[3] one has to include in the energy denominator present between any two interactions the energies of <u>all</u> particles and holes excited at the corresponding level of the diagram under consideration. In other words, the interactions on the right side of fig. 2b are off-shell because of the presence of the extra particle and two extra holes. This fact by itself makes the "insertion" in fig. 2b already different from fig. 2a, as is very well known since many years.

Secondly, and much more important, there is a physical reason. As explained above, the two particles excited at the bottom of fig. 2b are close together and strongly correlated. As one of the particles undergoes a collision with a third particle, this cannot be far away either - in other words, we have three particles close together and are forced to take into account <u>all</u> interactions between the three particles now - we have to use a fully correlated three-body amplitude. The physical significance is that because of the strong core in the bare nucleon-nucleon interaction we must prevent the third particle in fig. 2b to overlap with the first one. This means that we cannot just take into account processes like fig. 2b only, but we must include also interactions between particles one and three. This, however, will leave the second particle uncorrelated now, so that we are forced to perform a full three-body Bethe-Faddeev summation[2,4]. Consequently we arrive at the conclusion that in order to obtain meaningful numbers the energy of a real particle hopefully may be calculated from two-body processes in lowest order, whereas for virtual particles already the lowest order involves a three-body process.

Thirdly, there is a second important physical reason why fig. 2a is not meaningful for the self-energy of a virtually excited particle. A particle self-energy is due to its interaction with holes. For a virtual particle - as opposed to a real particle - there is an additional, a second way of interaction with holes possible than fig. 2b. An example for this second kind of processes is shown in fig. 2c. It arises because of the presence of holes accompanying the virtual particles. The virtual particle is scattered back filling the hole state, and another particle-hole pair is created. At first sight it might seem strange to call fig. 2c a self-energy. However, the right part of this figure has the same general structure as fig. 2a: a particle line entering, interaction with a hole line, and a particle line emerging. As additional complication we again have three particles excited simultaneously and need to perform a full three-body Bethe-Faddeev summation.

In summary, the notion of self-energy for intermediate states (virtual particles) has not much left from the simple physical idea expressed in fig. 2a. It cannot be defined in a meaningful way without reference to the other particle and the two hole states involved in the process of exciting the virtual particle. Therefore, the only possible way to define a particle self-energy is to define it as a matrix as shown in fig. 1c. The two-particle two-hole configuration interacts with an additional hole to produce another two-particle two-hole configuration. This prescription will include the processes of fig. 2b,c plus the Bethe-Faddeev summations plus several other terms[2] and has been implemented by the author[5] several years ago. Similar ideas have been expressed by Brandow[6].

In order to demonstrate that these observations lead to considerable quantitative effects let me treat as an example a particle in the sd-shell. First I will add a physical 17th particle to a ^{16}O nucleus[7]. The first three rows of table 1 show kinetic energies, two-body potential energies, three-body contributions and resulting single-particle energies according to the first row of fig. 1, calculated from the Reid soft core potential. It is seen that the three-body term indeed is a small correction to the two-body term as anticipated.

Following the ideas expressed above, I next calculate the "self-energy matrix" pertaining to virtual excitations in the ^{16}O ground state (no 17th nucleon present) shown schematically in fig. 1c. There is no two-body term. The average matrix element divided by two (per particle) for the sd-shell is shown in the fourth row of table 1. It is seen that virtual particles experience much less attraction than real ones. This is mainly due to the Faddeev summations weakening the lowest-order contributions as well as the enlarged energy denominators (off-shell propagation). In more picturesque language, the extra particle present introduces additional repulsion by making it harder for the third particle to approach.

Is it possible to reinstate part of the simple physical picture expressed by fig. 2a for virtual particles? One could try to use as a two-body contribution an off-shell G-matrix with some average (off-shell) energy denominator, depicted schematically in fig. 1d. In order to obtain a correct result at the three-body level, one then has to subtract out this approximate two-body contribution from the known three-body term, fig. 1e. The corresponding numerical values are shown in the last line of table 1. It is seen that this approximate two-body term is much too attractive, and the three-body term is not small. The sum is correct by construction. It is evident that the three-body Bethe-Faddeev summation cannot be neglected.

	T	U_2	U_3	ϵ
d 5/2	19.9	-24.4	+1.9	-2.6
s 1/2	17.2	-19.7	+1.0	-1.5
d 3/2	17.7	-17.8	+0.9	+0.8
sd	18.7		-7.5	+11.2
sd	18.7	-12.1	+4.6	+11.2

Table 1.

Energies of particles in the sd-shell. First three rows, real particles. Last two rows, virtual particles. (in MeV)

In summary I have shown that there exist significant physical and numerical differences between potential energies for real and virtual particles. Neither of them can be used as approximation for the other. Each of them needs to be evaluated in its own right. It is amusing to note that similar confusion existed in the case of real hole and virtual hole energies some time ago[8].

References

1) C. Mahaux, this conference.

2) H. Kümmel, K.H. Lührmann, J.G. Zabolitzky, Phys. Rep. 38C (1978) 1

3) B.D. Day, Rev. Mod. Phys. 39 (1967) 719

4) R. Rajaraman, H.A. Bethe, Rev. Mod. Phys. 39 (1967) 745

5) J.G. Zabolitzky, Phys. Lett. 47B (1973) 487; Nucl. Phys. A228 (1974) 285; Phys. Rev. C14 (1976) 1207

6) B.H. Brandow, Ann. Phys. 57 (1970) 214; Ann. Phys. 74 (1972) 112

7) R. Offermann, W. Ey, H. Kümmel, Nucl. Phys. A273 (1976) 349; J.G. Zabolitzky, W. Ey, to be published

8) H. Kümmel, Phys. Rev. C8 (1973) 1134; K. Emrich, J.G. Zabolitzky, K.H. Lührmann, Phys. Rev. C16 (1977) 1650

THE IMAGINARY PART OF THE NUCLEAR OPTICAL POTENTIAL
AND INELASTIC FORM FACTOR

V.A. Madsen[+], F. Osterfeld, and J. Wambach
Institut für Kernphysik der KFA Jülich, D-5170 Jülich, W. Germany

Several authors have already discussed various approaches to the microscopic nuclear optical potential. Jeukenne, Lejeune and Mahaux[1] and Brieva and Rook[2] calculate the optical potential first for nuclear matter by making a low-density expansion of the mass operator. They then obtain the optical potential for finite nuclei by making either the local density approximation on the potential itself[1] or on the two-nucleon t-matrix which then is folded with finite nuclear wave functions to get the potential[2]. Vinh Mau and Bouyssy[3], on the other hand, calculate the optical potential for finite nuclei directly by using microscopic RPA-states for the intermediate excitations of the target and an effective interaction taken from the nuclear shell model.

Now, we suggest here a method which combines both approaches and which can be used for elastic and inelastic scattering. We divide the total Hilbert space into three parts, a model space, a near space, and a far space (see fig. 1). The first consists only of the initial and final states of the nucleus, the second of the space of known RPA doorways, and the third, everything else. The far space is formally eliminated in a procedure much like the calculation of the Brueckner K matrix except that the projection operator Q onto the far space plays the role of the Pauli operator. The boundary of the far space in projectile and nuclear energies is chosen so that the total energy is somewhat higher than the projectile energy. The formal elimination of the Q space, which takes into account primarily high-lying two-particle, one-hole states results in a Brueckner matrix

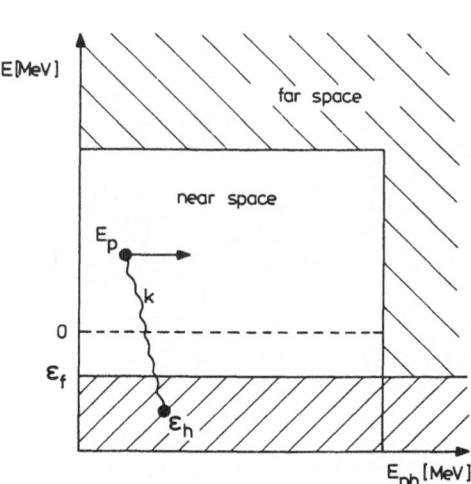

Fig. 1: Schematic representation of the near space and far space as defined by the boundary energy E_b. The shaded area specifies the occupied states in the target nucleus.

$$K = V_{PP} + V_{PQ} \frac{1}{E-H} QVP \ , \tag{1}$$

which, because of the choice of the boundary energy E_b is always real and only weakly energy dependent.

The elimination of the high-lying two-particle states should result in a K weak enough to use in perturbation theory. Thus the imaginary part of the optical potential or inelastic form factor is given by second order perturbation theory as

$$\text{Im} \sum_n <\psi_f| (K-\hat{V}_o)\mathcal{A}| \psi_n >G_n<\psi_n|\mathcal{A}(K-\hat{V}_o)| \psi_o> \tag{2}$$

where $\hat{V}_o$ is an auxiliary potential in which the intermediate projectile Greens function G_n is produced, ψ_i, ψ_n, and ψ_f are nuclear states of the target and $\mathcal{A}$ antisymmetrizes between projectile and target. The states ψ_n and ψ_f are chosen as RPA doorways, although core polarization[4] (two-particle two-hole states) is taken into account in the inelastic transition between ψ_n and ψ_f. The sum is extended over all open channels, since they all carry away flux and therefore contribute to the imaginary potential. The diagrams representing the elastic and inelastic mechanisms are shown in figs. 2 and 3. In the latter case

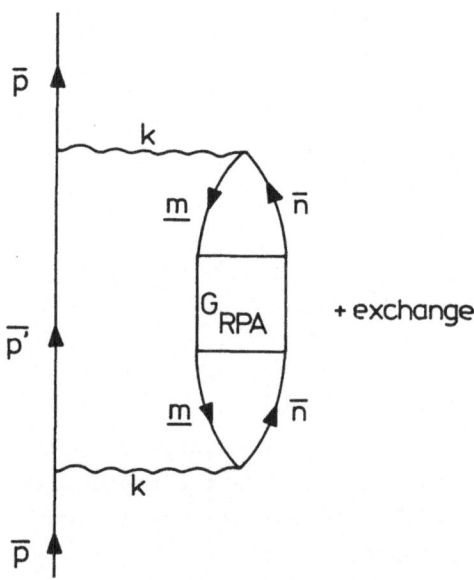

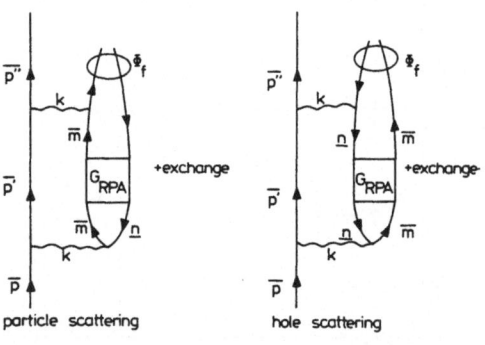

Fig. 3: Diagrams showing the particle-scattering and hole-scattering terms contributing to the inelastic form factor in second order.

Fig. 2: This diagram shows the second-order contribution to the optical potential.

we see that there are two separate kinds of terms which occur, intermediate particle scattering and intermediate hole scattering. We have
found[5] that these terms enter the form factor with opposite signs.
Since the hole states are restricted to occupied bound states this
terms subtracts in the inner region, giving rise to a reduced form
factor in the interior.

The antisymmetrization operator in each step gives rise to
several different mechanisms, direct-direct, direct-exchange, exchange
direct, and exchange-exchange. In the first the initial projectile is
also the final projectile, in the second and third a particle
different from the projectile always comes out, and in the fourth
either the initial particle or any other can come out, although the
latter is A-1 times as large in the amplitude.

We have recently developed a rapid program to compute these
exchange terms as well as the direct ones using a finite range two-
body potential to give the second order imaginary form factor or
optical potential. One can include more than 100 intermediate RPA
states without major difficulty. At this time we have only preliminary
results using an effective interaction instead of eq. (1). The first
results indicate that the inclusion of exchange gives us an imaginary
potential which extends somewhat farther out than that calculated with
direct alone. The finite range force has a similar effect on both
direct and exchange contributions. We tentatively agree with Ngô,
Bouyssy and Vinh Mau[3,6] that a combination of collective RPA states
and the exchange gives rise to peaking of the imaginary potential at
or beyond the nuclear radius as it appears in phenomenological
potentials.

In summary we make the following points: For elastic scattering
our approach is similar to that of Vinh Mau et al.[3,6] although we
can use an optical intermediate-particle Green's function; the nuclear
size and structure is treated better and more naturally than in the
nuclear matter approach; we can check the latter method, which, although elegant, has some uncontrollable approximations such as the
local density assumption. For inelastic scattering our approach has
the additional advantage that it treats intermediate hole as well as
particle scattering and that 2-particle 2-hole core polarization effects
can be included. Our method has the disadvantages that it is complicated to carry out and is limited to second order.

References

1. J.P. Jeukenne, A. Lejeune, C. Mahaux, Phys.Rep. 25, No. 2 (1976) 83-174
2. F.A. Brieva, J.R. Rook, Nucl.Phys. A291 (1977) 299-316; Nucl. Phys. A291 (1977) 317-341
3. N. Vinh Mau, A. Bouyssy, Nucl.Phys. A257 (1976) 189-220
4. P. Ring, J. Speth, Nucl.Phys. A235 (1974) 15
5. G. Baur, V.A. Madsen, F. Osterfeld, Phys.Rev. C17 (1978) 819
6. H. Ngô, A. Bouyssy, N. Vinh Mau, contribution to this conference

[+] Permanent address: Oregon State University, Department of Physics, Corvallis, Oregon 97331, USA

FOLDING DESCRIPTION OF ELASTIC AND INELASTIC SCATTERING

F. Petrovich

Department of Physics
The Florida State University
Tallahassee, Florida 32306, USA

Abstract: The essential features of the folding description of elastic and inelastic scattering reactions are reviewed. A momentum
space technique for evaluating folding integrals is emphasized and
approximate methods for treating knockout exchange are discussed.
Various "realistic" interactions which have been popular in folding
model calculations are listed and a qualitative summary of their properties is given. Several applications of the folding model are considered. These cover both elastic and inelastic scattering as well
as both nucleon-nucleus and nucleus-nucleus collisions. Specific
examples where the tensor and spin-orbit interaction components enter
are included. Some new results concerning the excitation of high spin
states in ^{208}Pb by 135 MeV are of particular interest.

1. INTRODUCTION

Work on the microscopic theory of elastic and inelastic scattering reactions dates back to the early 1950's when Watson and collaborators developed the multiple scattering formalism [1]. There was
some early success in describing proton-nucleus scattering at $E_p \approx 150$
MeV using the impulse approximation [2,3]. The experimental data at
these energies was rather limited, however, and until quite recently
the bulk of experimental nucleon-nucleus scattering studies were
restricted to the 10-60 MeV energy region. The original microscopic
approach to nucleon-nucleus scattering of these lower energies was a
phenomenological method due mainly to Satchler [4]. In this approach
suitable phenomenological forms were introduced for the effective interaction and the interaction parameters were determined by comparing
the results of theoretical calculations with experimental data. This
method enjoyed only partial success [5]. A major difficulty with the
approach is that the interaction parameters deduced depend strongly
on the nuclear wave functions used in the calculations. The latter
were often chosen rather arbitrarily.

To describe elastic and inelastic scattering reactions satisfactorily in a microscopic picture, it is essential to use a realistic
effective interaction in conjunction with nuclear wave functions or
densities that contain any collective correlations that may be present
in the transitions being considered. It is also necessary to include
amplitudes arising from antisymmetrization. A major breakthrough in
the microscopic model for nucleon-nucleus scattering in the 10-60 MeV
energy region came with the realization that the real part of the effective interaction at these energies might be closely approximated
by the bound state G matrix. The first calculations using a G matrix
interaction and satisfying the remainder of the above criteria were
made by the Michigan State group in collaboration with Madsen and
Atkinson [6]. Love and Satchler pursued this approach [7] and suggested a phenomenological method for including the contribution from
the imaginary part of the effective interaction [8]. The Saclay
group [9] and Amos and Geramb [10] carried out related work using
similar effective range interactions. More recently, the G matrix
approach has been extended to heavy-ion scattering by Love and Satchler
[11] and the Florida State group [12]. With the opening of new experimental facilities such as IUCF and LAMPF, new work on high energy
nucleon-nucleus scattering is expected in the future.

The most recent advances in the microscopic theory of elastic and
inelastic scattering concern more fundamental attempts at calculating
the effective interaction. These include estimates of the complete
effective interaction from nuclear matter [13,14] and perturbative
estimates of the imaginary potentials for scattering from finite
nuclei based on bound state G matrix interactions [15]. The nuclear
matter results support the use of the bound state G matrix to represent the real part of the effective interaction at low energies, so
the major new results from these works are the estimates of the imaginary part of the effective interaction.

Below the main features of the microscopic folding model for
elastic and inelastic scattering are reviewed. Special attention is
given to momentum space folding techniques and the properties of
knockout exchange amplitudes. Some properties of realistic effective
interactions are summarized and several applications of the folding
model with realistic interactions are discussed. Both nucleon-
nucleus and nucleus-nucleus scattering are considered and some recent

results concerning the excitation of high spin states in ^{208}Pb by 135 MeV protons are presented.

2. FOLDING MODEL

In the folding model we start with an effective interaction which is a sum of two body interactions between the nucleons of the projectile and the nucleons of the target.

$$V = \sum_{p,t} v_{pt} \tag{1}$$

The scattering potentials are obtained by taking the matrix elements of the interaction between the initial and final internal wave functions of the projectile and target.

$$U(\bar{r}) = <\psi_p^f \psi_t^f | V | \psi_p^i \psi_t^i> \tag{2}$$

In writing eq. 2 antisymmetrization between the projectile and target has been ignored so that $\psi_p \psi_t$ represents products of wave functions. The integration is over the internal coordinates $\bar{r}_p$ and $\bar{r}_t$ shown in fig. 1a and the resulting potential is a function of $\bar{r}$ - the relative coordinate between the projectile and target. The problem of antisymmetrization will be considered in the next section.

By writing the interaction in the form

$$v_{pt} = v(\bar{s}) = v(\bar{s}') \delta(\bar{r}_{p'} - \bar{r}_p) \delta(\bar{r}_{t'} - \bar{r}_t) \tag{3}$$

$$\bar{s}' = \bar{r}_{p'} + \bar{r} - \bar{r}_{t'}$$

eq. 2 is easily converted to the double folding form needed to describe heavy-ion scattering.

$$U(\bar{r}) = \int v(\bar{s}') \rho_p(\bar{r}_{p'}) \rho_t(\bar{r}_{t'}) d^3 r_{p'} d^3 r_{t'} \tag{4}$$

Here the projectile and target densities are defined by

$$\rho_x(\bar{r}_{x'}) = <\psi_x^f | \sum_x \delta(\bar{r}_{x'} - \bar{r}_x) | \psi_x^i> \tag{5}$$

with x = p or t. The single folding form needed to describe nucleon-nucleus scattering is easily obtained from eq. 4 by taking

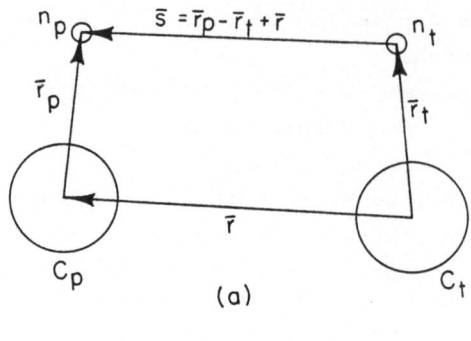

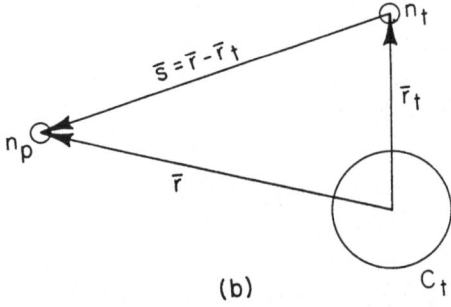

Figure 1. Coordinates used in (a) double folding calculations and (b) single folding calculations. Recoil is neglected as a matter of convenience.

$$\rho_p(\overline{r}_{p'}) = \delta(\overline{r}_{p'}).\tag{6}$$

The corresponding reduction of coordinates is shown in fig. 1b. All reference to spin, i-spin, and momentum variables has been suppressed in writing the above; however, V has been considered a function of vector $\overline{s}$ to allow for both central and non-central forces.

The folding integrals are most easily evaluated by expanding the interaction in plane waves [16,17].

$$v(\overline{s}') = (2\pi)^{-3} \int v(\overline{k}) e^{i\overline{k}\cdot\overline{s}'} d^3k\tag{7}$$

With this expansion eq. 4 becomes

$$U(\overline{r}) = (2\pi)^{-3} \int e^{i\overline{k}\cdot\overline{r}} v(\overline{k}) \rho_p^*(\overline{k}) \rho_t(\overline{k}) d^3k.\tag{8}$$

The corresponding single folding form is obtained by setting $\rho_p^*(\overline{k}) = 1$. In practice [16] it is convenient to work in a finite volume characterized by a radius R, so that

$$\int d^3k \rightarrow \sum_n k_n^2 \Delta k_n \int d\hat{k}_n \tag{9}$$

with $k_n = n\pi/R$. The convolution integrals are thus reduced to a sum of products of Fourier-Bessel transforms of the interaction and the densities.

The momentum space techniques for evaluating convolution integrals described above are now new; however, many workers still use coordinate space Legendre methods to calculate the scattering potentials. These coordinate space methods generally require the evaluation of double integrals except when special functions are used. These double integrals are difficult to avoid when heavy-ion collisions are considered. The momentum space methods avoid the problem of computing double integrals and provide a common form for the treatment of nucleon-nucleus and nucleus-nucleus scattering as well as central and non-central interactions. The effects of the effective interaction and the densities are clearly displayed in the momentum space approach and the most direct comparison of nucleon-nucleus and nucleus-nucleus collisions with electron-nucleus scattering is achieved. An interesting view of distortion effects is obtained by carrying out local distorted wave calculations completely in momentum space [16]. Additional details on the computation of scattering potentials using these methods may be found in ref. [18].

3. ANTISYMMETRIZATION

3.1 Single Nucleon Knockout Exchange

Single nucleon knockout exchange (SNKE) may be included in the folding model by making the replacement

$$v_{pt} \rightarrow v'_{pt} = v_{pt}(1-P_{pt}) \tag{10}$$

where P_{pt} is the operator which exchanges all of the coordinates of particle p with those of particle t. The operator P_{pt} is a product of space, spin, and i-spin exchange operators

$$P_{pt} = P_{pt}^x P_{pt}^\sigma P_{pt}^\tau \tag{11}$$

and the spin and i-spin exchange operators can be written in terms of the spin and i-spin projection operators

$$P_{pt}^\sigma = P_1^\sigma - P_0^\sigma \qquad P_{pt}^\tau = P_1^\tau - P_0^\tau . \tag{12}$$

The latter result follows from the symmetry (antisymmetry) of the triplet (singlet) spin and i-spin states.

For a central interaction,

$$v_{pt} = \sum_{ST} v_{TS}(s) P_S^\sigma P_T^\tau , \tag{13}$$

the idempotency of the projection operators leads immediately to

$$v'_{pt} = v_{pt} + v_{pt}^{ex} P_{pt}^x \tag{14}$$

where

$$v_{pt}^{ex} = -\sum_{ST} (-1)^{S+T} v_{TS}(s) P_S^\sigma P_T^\tau . \tag{15}$$

This result suggests a constructive interference between the direct and exchange terms for the even state interaction components (S+T=odd) and a destructive interference for the odd state interaction components (S+T=even). These constructive and destructive interferences are complete in the case of a δ-interaction for which P_{pt}^x can be set to unity. In the general case of a finite range interaction the presence of P_{pt}^x in eq. 14 gives rise to a non-locality in the scattering potentials and the constructive and destructive interferences are incomplete.

Although it is possible to deal with non-localities in the scattering potentials exactly [6,7,9,10], calculations are greatly simplified when suitable local approximates to the non-localities can be made. One approximation used often in nucleon-nucleus scattering calculations [6,19], is based on the existence of a characteristic momentum associated with the SNKE process.

To understand the approximation it is sufficient to consider the matrix elements of v'_{pt} between relative plane wave states where

$$P^x_{pt} e^{i\overline{k}_i \cdot \overline{s}} = e^{-i\overline{k}_i \cdot \overline{s}} . \tag{16}$$

The matrix elements are

$$M^C = \int e^{-i\overline{k}_f \cdot \overline{s}} v'_{pt} e^{ik_i \cdot \overline{s}} d^3s = v_{pt}(q^2) + v^{ex}_{pt}(p^2) \tag{17}$$

where $\overline{q} = \overline{k}_f - \overline{k}_i$, $\overline{p} = \overline{k}_f + \overline{k}_i$, and

$$v(x^2) = 4\pi \int j_0(xs) v(s) s^2 ds \tag{18}$$

with $x = q$ or p. The result for M^C is just the familiar expression obtained in the Born treatment of nucleon-nucleon scattering. Unlike nucleon-nucleon scattering where both q and p depend strongly on the scattering angle, in nucleon-nucleus scattering at energies high enough so that the Fermi motion of the bound nucleon is not too important, p is nearly independent of angle and is approximately given by k_{lab} - the momentum transfer required to stop the projectile in the target and eject a bound nucleon with momentum and energy close to those of the incident projectile.

The approximation then, is to take

$$v'_{pt} = v_{pt} + v^{ex}_{pt}(k^2_{lab}) \delta(\overline{s}) \tag{19}$$

in coordinate space. The energy dependence of the exchange term and its relation to the form of the interaction is directly displayed in this approximate result. Because of the δ-interaction form of the exchange term, it is also clear that the exchange term will be most important relative to the direct term when high multipoles of the interaction are considered.

Comparisons with exact calculations [6,19] have shown that the approximate form given in eq. 19 is quite reasonable even for incident energies as low as 25 MeV, but it doesn't achieve high accuracy until the incident energy is greater than 60 MeV. Part of the success of the approximation at low energies has to do with the fact that the exchange term is not very important for low multipoles where the approximation is particularly poor. By it's nature, the approximation does better for short range interactions than long range ones. When accurate estimates of scattering potentials of low multipolarity, such as the central optical potential, are needed at low energies, the

Slater exchange approximations [19,20] popular in schematic Hartree-Fock calculations [21] do much better than eq. 19.

In addition to the existence of a characteristic momentum $p \approx k_{lab}$, the approximation of eq. 19 depends on the direct and exchange terms for the central interaction having the same form in the short range limit. This may be seen from eq. 17 where $v_{pt}(q^2)$ and $v_{pt}^{ex}(p^2)$ are both constants when s may be taken much smaller than unity in the integration in eq. 18. A similar situation prevails in the case of the spin-orbit interaction where

$$M^{\ell s} = \{v_{pt}(\bar{q}) \times \bar{k}_i - v_{pt}^{ex}(\bar{p}) \times \bar{k}_i\} \cdot (\bar{\sigma}_p + \bar{\sigma}_t) . \qquad (20)$$

In the short range limit, $v_{pt}(\bar{q})$ is a constant times $\bar{q}$ and $v_{pt}^{ex}(\bar{p})$ is a constant times $\bar{p}$. Since $\bar{q} \times \bar{k}_i$ and $\bar{p} \times \bar{k}_i$ are equal, the direct and exchange terms have the same form and an approximation analogous to eq. 19 is possible. The approximation is particularly appropriate for realistic spin-orbit interactions because of their short range. It is interesting to note the difference in sign between the direct and exchange terms in eq. 20 as compared to eq. 17. This occurs because of the dependence of the spin-orbit interaction on the momentum operator $\bar{k}$ and the property

$$\bar{k} P_{pt}^{x} = -P_{pt}^{x} \bar{k} . \qquad (21)$$

This sign change leads to constructive (destructive) interference between the direct and exchange terms for the odd (even) state spin-orbit interaction components which is opposite to the result for the central interaction.

For the tensor interaction

$$M^t = \{v_{pt}(\bar{q}) + v_{pt}^{ex}(\bar{p})\} \cdot T_2(\bar{\sigma}_p, \bar{\sigma}_t), \qquad (22)$$

the relative sign between the direct and exchange terms is the same as in the case of the central interaction, but $v_{pt}(\bar{q})$ goes to a constant times $q^2 Y_2(\hat{q})$ in the short range limit and $v_{pt}^{ex}(\bar{p})$ goes to a constant times $p^2 Y_2(\hat{p})$. In this case setting $p \approx k_{lab}$ does not lead to a simple effective tensor interaction which accounts for exchange. In addition realistic tensor interactions have quite long ranges which makes the approximate prescription quite unreliable. To date, in most

calculations involving the tensor force, SNKE has been treated exactly. More work on approximate methods for including the tensor SNKE
is needed.

For heavy-ion scattering SNKE requires the computation of overlaps between the spatial configurations shown in fig. 2a and fig. 2b.

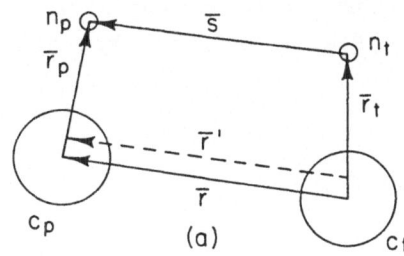

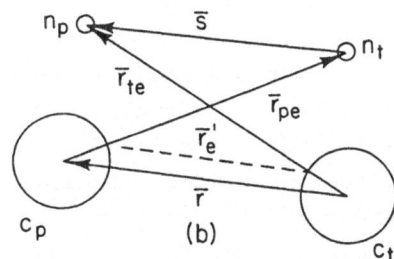

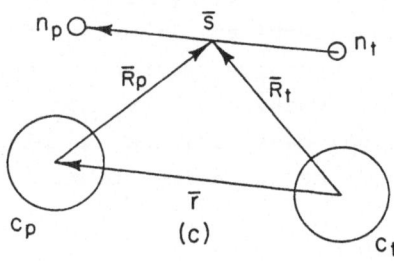

Figure 2. (a) Coordinates used in double folding model. (b) Coordinates obtained by application of P_{pt}^x. (c) Coordinates used in evaluating exchange integral. Broken arrows indicate proper relative coordinates.

Since the relative kinetic energy per nucleon is rather low in the typical heavy ion collision (3-10 MeV/nucleon), it is quite a poor approximation to neglect the Fermi motion of the bound nucleons in computing these overlaps; however, one can take advantage of the fact that the cross sections for heavy-ion collisions are primarily sensitive to the scattering potentials at the strong absorption radius $D_{1/2}$ [11]. At this distance the colliding nuclei are barely over-lapping and the range of the non-locality in the scattering potentials is rather small.

To see this note that the rela-tive coordinates $\bar{r}'$ and $\bar{r}'_e$ in fig. 2a and fig. 2b are related by $\bar{r}'_e = \bar{r}' - \alpha\bar{s}$ where $\alpha = (A_p + A_t)/A_p A_t$. For typical cases $\alpha s/D_{1/2} < 5\%$ and it is a reasonable approximation to take $\bar{r}'_e \approx \bar{r}' \approx \bar{r}$ in fig. 2a and fig. 2b. With this approximation it is possi-ble to compute the required overlaps by expanding the radial wave func-tions for the bound nucleons in os-cillator functions and making the transformation to the coordinates shown in fig. 2c. Such calculations have been carried out for the case of a central effective interaction [22]. The results of these calculations show that the SNKE contributions to the tails of the heavy-ion potentials may be included by using an effective interaction

$$v'_{pt} = v_{pt} + A\delta(\bar{s}) \tag{23}$$

where A is a constant which depends on v_{pt}. The constant A differs somewhat from the coefficient $v^{ex}_{pt}(k^2_{lab})$ introduced in eq. 19 for nucleon-nucleus scattering. More recent estimates of the SNKE contribution to heavy-ion potentials based on the Slater exchange approximation [19] give results similar to those of ref. [22].

3.2 Other Exchange Terms

SNKE is but one of the exchange terms which arise when the condition of complete antisymmetry between the projectile and target wave functions is imposed. Unlike the SNKE term, which can be of the order of the direct folding term in the exterior region of the scattering potentials, the other exchange terms are related to the non-orthoganalities between the projectile and target wave functions and tend to be most important in the interior regions of the scattering potentials. These other exchange terms are generally neglected in the folding model. This is justified when absorption leads to an insensitivity to the interior of the potentials or when the non-orthoganalities are sufficiently small to make these terms negligible. The first situation is realized in many heavy-ion collisions. An example of the second situation would be a high energy proton scattering off a nucleus. In situations where these other exchange terms are important it is, of course, necessary to include them. Low energy nucleon-nucleus scattering collisions where the absorption is weak, and heavy-ion collisions involving particles of nearly equal mass [23] are examples of cases where these other exchange terms can be quite important.

4. REALISTIC INTERACTIONS

The initial microscopic calculations employing realistic interactions to describe nucleon-nucleus scattering in the 10-60 MeV energy region [6] emphasized the use of the long range part of the Kallio-Kolltveit potential [24] for the effective interaction. This is a G matrix interaction obtained from an s-state nucleon-nucleon potential by means of the Scott-Moskowski separation method [25]. Subsequent work [7,26-28] stressed the long range part of the Hamada-Johnston

potential popularized by Kuo and Brown [29]. This is an even state G
matrix interaction which is similar to, but somewhat more realistic
than the long range part of the Kallio-Kolltveit potential. The odd
state components of the central G-matrix interaction were generally
neglected because they were unavailable in representations useful for
folding model calculations. This seemed justified in view of the cancellation of the odd state contribution to the scattering potentials
due to SNKE.

A closer look at this cancellation shows that it is fairly complete in the interior, not so complete in the surface, and rather incomplete in the far tail of the nucleon-nucleus potential. The lack
of cancellation in the far tail of the potential is not too important,
since nucleon-nucleus scattering is rather insensitive to potential
tails; however, it is important in the calculation of heavy-ion potentials where the far tail of the potential is all important. Double
folding calculations [30] using these even state G matrix interactions
produced nucleus-nucleus potentials which were typically 2-3 times too
deep near $D_{1/2}$. This problem was resolved when a local representation
of the complete G matrix was introduced [31]. The odd state central
components of this interaction do produce substantial contributions to
the nucleus-nucleus potentials near $D_{1/2}$ [11,12]. These odd state
contributions cancel against the even state contributions producing
potentials 2-3 times weaker than when the even state central interaction is used alone.

Further calculations [32] of nucleon-nucleus potentials using
the complete G matrix of ref. [31] have shown that the odd state interaction components are somewhat more effective in the surface region
than originally supposed - leading to nucleon-nucleus potentials with
radii which are somewhat too small. This problem can be traced to
the fact that the representations of the G matrix interaction introduced in ref. [31] do not account for the saturation properties of
these interactions. Consistent results for nucleon-nucleus and
nucleus-nucleus potentials have been obtained [12] in calculations
using a modified density dependent version of the complete G matrix
interaction of ref. [31]. Specifically, the following form was assumed for the effective interaction

$$v_{pt} = \alpha g + \beta \rho(r) \delta(\overline{s}) \qquad (24)$$

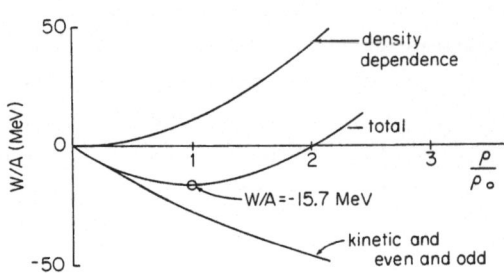

Figure 3. Saturation curve for symmetric nuclear matter obtained by introducing a linear density dependent term in the G matrix interaction of ref. [31].

where g denotes the G matrix from ref. [31] and α and β are constants determined from nuclear matter calculations. The resulting saturation curve for symmetric nuclear matter is shown in fig. 3. In this case α = 1.1. Some of the results of ref. [12] will be discussed in the next section of this paper.

The trend throughout the above has been to examine a widening variety of collisions by means of the folding model. At certain times this process has led to modification in the representations of the G matrix which have been used in the calculations. This has been a progressive process in that the successive representations of the G matrix have become more realistic. The density dependent form introduced in ref. [12] seems the best to date. This will undoubtedly be superceded by the more realistic complex G matrix interactions of ref. [13,14] in future applications of the folding model.

Two local representations of the free two nucleon t matrix have appeared in the literature recently [33,34]. These were motivated by interest in the impulse approximation description of nucleon-nucleus scattering at E_p = 100-200 MeV which is currently being studied experimentally at IUCF. The modulus of the Fourier transforms of the components of the t matrix of ref. [34] are shown in fig. 4. The main purpose of this figure is to provide some feeling for the relative importance of the various interaction components. Analogous figures obtained by considering the G matrix interactions discussed above are qualitatively similar.

In elastic scattering and normal parity inelastic transitions ($\Delta\pi = (-1)^J$), the spin independent central components and the spin-orbit components are most important. These are shown on the left in fig. 4. The tensor force enters only through a spin-transfer amplitude which is inhibited by nuclear collective effects [28] and is negligible in most cases. From the figure it is clear that the

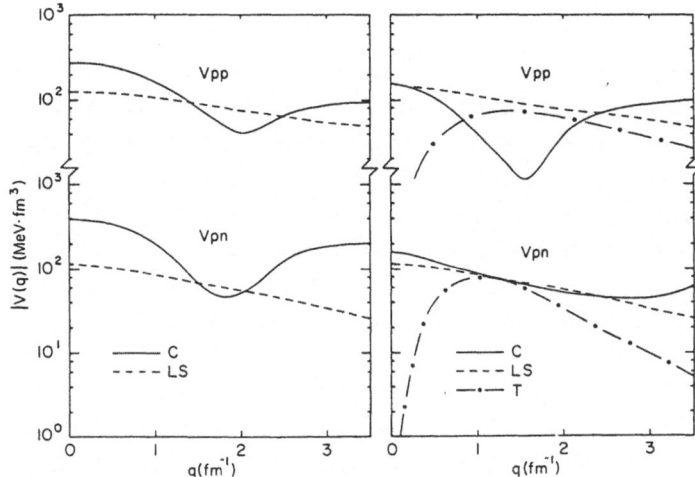

central proton-neu-
tron component (v_{pn})
is stronger than the
central proton-pro-
ton component (V_{pp})
of the interaction,
the spin-orbit in-
teraction is rela-
tively more impor-
tant in V_{pp} than in
V_{pn}, and the spin-
orbit component
falls off more slow-
ly than the central
components with in-
creasing q. The
first of the above
indicates that pro-
ton-nucleus scatter-
ing is complimentary
to electron-nucleus
scattering [35],
i.e. the former
"sees" the neutrons

Figure 4. The modulus of the Fourier transforms of free t-matrix interaction of ref.
[35]. The interaction components which enter
for normal parity and abnormal parity transitions are shown on the left and right, respectively. Central (C), spin-orbit (LS), and
tensor (T) components are shown separately.
SNKE contributions corresponding to E_{lab}=135.2
MeV have been included in the C and LS interaction components via the approximation of
eq. 19.

in the target while the latter "sees" the protons. The last of the
above shows in part why the spin-orbit component of the interaction
is most important in inelastic transitions requiring larger angular
momentum transfer [36]. For abnormal parity inelastic transitions
($\Delta\pi = (-1)^{J+1}$) only the weak spin dependent components of the central
interaction enter. These are comparable in strength to the spin-orbit
and tensor components and all must be considered in treating these
transitions. The interaction components for abnormal parity transitions are shown on the right in fig. 4.

5. APPLICATIONS

5.1 Real Part of Central Optical Potential

The density dependent version of the G matrix interaction of ref.
[31] described in the preceding section has been used in folding calculations of the real part of the optical potential for a variety of

nucleon-nucleus and nucleus-nucleus systems [12]. The energy ranges considered were 10-60 MeV for the former and 3-10 MeV/nucleon for the latter. In the calculations proton densities were taken from electron scattering and several models for neutron densities were considered. In all of these models neutron radii and proton radii were nearly the same. The Slater exchange approximation [19-21] was used to include SNKE in the nucleon-nucleus calculations and the approximation of ref. [22] was used to include SNKE in the nucleus-nucleus calculations. The necessary rearrangement contributions to the potentials, which are associated with the density dependence of the interaction [37] were included.

Elastic differential cross sections have been generated by adding phenomenological imaginary potentials to the folded potentials [12]. In the nucleon-nucleus scattering calculations the imaginary potentials were assumed to be of Woods-Saxon shape with parameters varied to fit experimental cross sections. The theoretical real potentials have volume integrals and mean square radii within 10% of those deduced from emperical studies and no adjustments to the real folded potentials were made in this case. In the nucleus-nucleus scattering calculations it was most often assumed that the real and imaginary potentials were of the same shape. Here both the real and imaginary potential depths were varied to fit experimental cross sections. For many systems the required adjustment of the real potential depth is less than 10%. Notable exceptions are ^{6,7}Li scattering [11] where the required adjustment is nearly a factor of two. Some heavy-ion systems show a definite preference for differences in shape between the real and imaginary potentials. ^{6,7}Li scattering stands out here also – requiring long range absorption.

Examples of potentials for p + ^{40}Ca and ^{16}O + ^{60}Ni at E = 10 MeV/ nucleon are shown in fig. 5 and fig. 6. The density dependence results in a significant decrease in the depth of the p + ^{40}Ca potential in the nuclear interior and a definite increase in the potential radius. More quantitatively the density dependence produces a 14% decrease in the volume integral and a 10% increase in $<r^2>$ for the potential in this case. The density dependence also produces a decrease in the central depth of ^{16}O + ^{60}Ni potential and effects the slope of the potential for larger radii; however, it makes no contribution at $D_{1/2} \approx 9.75$ fm - the part of the potential which is important in elastic scattering. The effect of the density dependence on

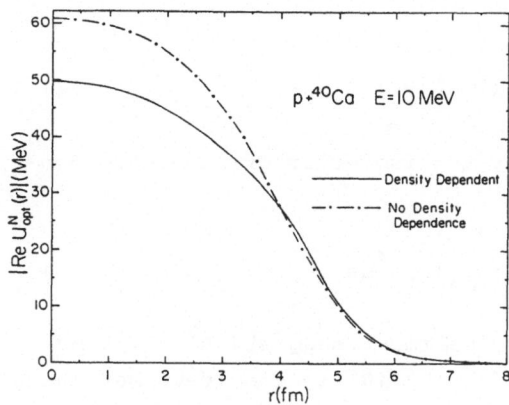

Figure 5. The effect of density dependence on the real part of the optical potential for the p + ^{40}Ca system at E_p = 10 MeV.

slope of the nucleus-nucleus potential as one moves from $D_{1/2}$ to smaller radii might be important in relating fusion and elastic scattering processes [38].

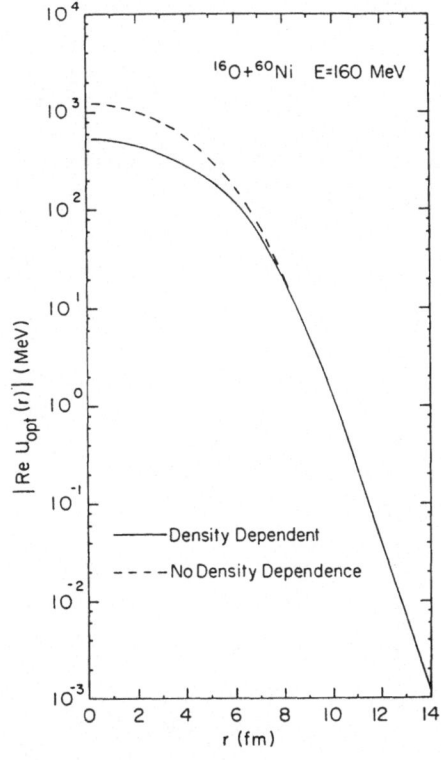

Figure 6. Same as fig. 5 for the ^{16}O + ^{60}Ni system at E_{16_O} = 160 MeV.

Typical cross sections obtained with the folded potentials are shown in fig. 7 and fig. 8. The experimental data for = p + ^{60}Ni at E_p = 8.56 MeV shows a definite preference for the real potential obtained with the density dependent interaction. The effect is more dramatic in cases of larger E_p. For ^{16}O + ^{58}Ni at E_p = 42 and 48 MeV the effect of the density dependence in the interaction does not show in the cross sections. The adjustment of the real potential depths required to produce the fits of fig. 8 is less than 10% and essentially energy independent.

5.2 Non-Central Terms in the Optical Potential

Whenever the spin of the projectile or target is non-zero, the optical potential can contain non-central terms. The best known of these is the spin-orbit term. A useful approximate expression for the proton-nucleus spin-orbit potential is obtained by treating the two body spin-orbit potential in the short range limit [39]. This result is

$$V_{\ell s} = \frac{1}{r} \frac{d}{dr} [\sum_{T} K^T \rho^T(r)] \; \overline{\ell} \cdot \overline{s} \tag{24}$$

where ρ^0 and ρ^1 denote the isoscalar and isovector ground state densities and

$$K^T = - \frac{4\pi}{3} \int g^T_{odd}(s) s^4 ds \; . \tag{25}$$

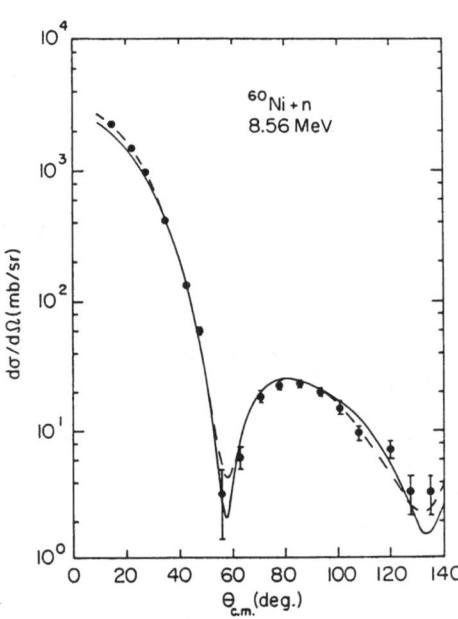

^{60}Ni + n
8.56 MeV

Experiment suggests $K^0 \approx 168$ MeV fm^5 [39,40] in the 10-60 MeV energy which correlates nicely with the value $K^0 \approx 160$ MeV fm^5 obtained from the G matrix interaction of ref. [31]. This exercise also yields $K^0 \approx 3K^1$ which corresponds to a negative symmetry term in the spin-orbit part of the nucleon-nucleus optical potential. The symmetry term in the central part is positive. Available experimental data on the analyzing power in quasi-elastic (p,n) reactions [41] and single particle levels in heavy nuclei [42] is consistent with a nucleon-nucleus spin-orbit potential with a negative symmetry term. This feature of the spin-orbit potential has important bearing on stability estimates for super-heavy elements [43].

Figure 7. Elastic differential cross sections for p + ^{60}Ni at E_p = 8.56 MeV. Dashed curve is result with density dependent interaction and solid curve is result with density independent interaction.

Recent experiments with polarized ^{6}Li beams performed by the Heidelberg group [44] provide information about the spin-orbit term in the heavy-ion optical potential. The potentials obtained in a double folding calculation [45] using the G matrix interaction of ref [31] provide a reasonable description of the experimental vector analyzing powers for the elastic scattering of ^{6}Li from ^{16}O, ^{28}Si, and ^{58}Ni. These results have already been shown in fig. 15 of ref. [11] and will not be repeated here.

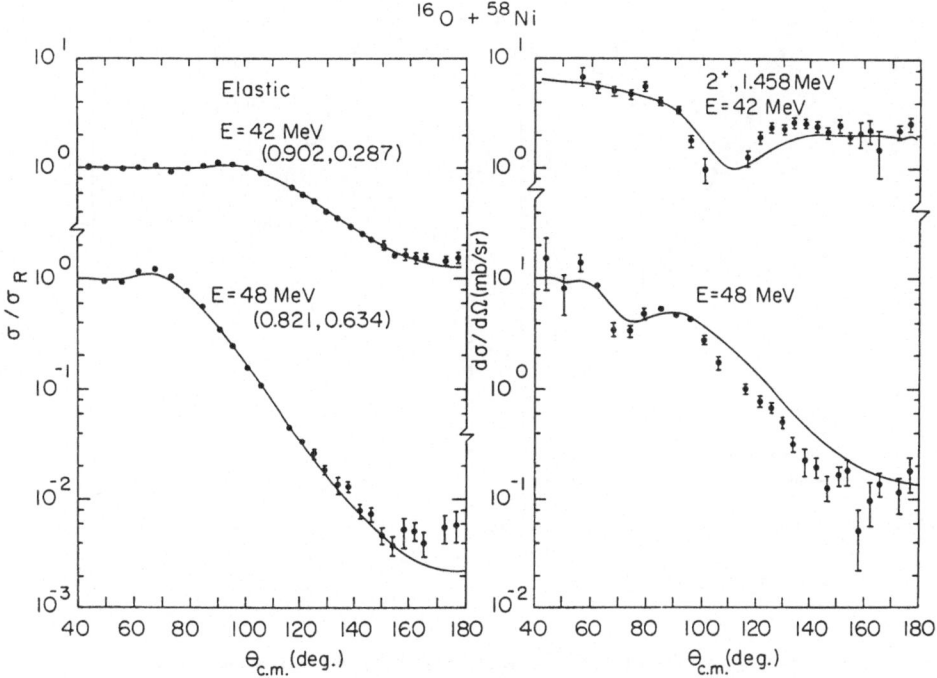

Figure 8. Elastic differential cross sections for $^{16}O + ^{58}Ni$ at $E_{16_O} = 42$ and 48 MeV are shown on the left. Corresponding inelastic differential cross sections for the excitation of the first 2^+ level in ^{58}Ni are shown at right.

When the spin of the projectile or target is greater than or equal to one the optical potential may contain terms with quadrupole (and possibly more complicated) shapes. Possible effects of such terms in heavy-ion scattering are illustrated in fig. 9 which shows the results of a double folding calculation [46], again based on the G matrix interaction of ref. [31], for the elastic scattering of $^{10,11}B$ from ^{27}Al. More direct evidence on the quadrupole terms in the potentials for ^7Li scattering has been obtained in recent experimental studies using tensor polarized ^7Li projectiles [47]. This data has yet to be investigated in terms of the G matrix folding model.

The results discussed above indicate that the small terms in the optical potentials are qualitatively well understood in terms of the folding model. Careful studies of the deficiencies in folding model estimates for the small terms in the optical potential may provide important information about dynamical polarization effects in the

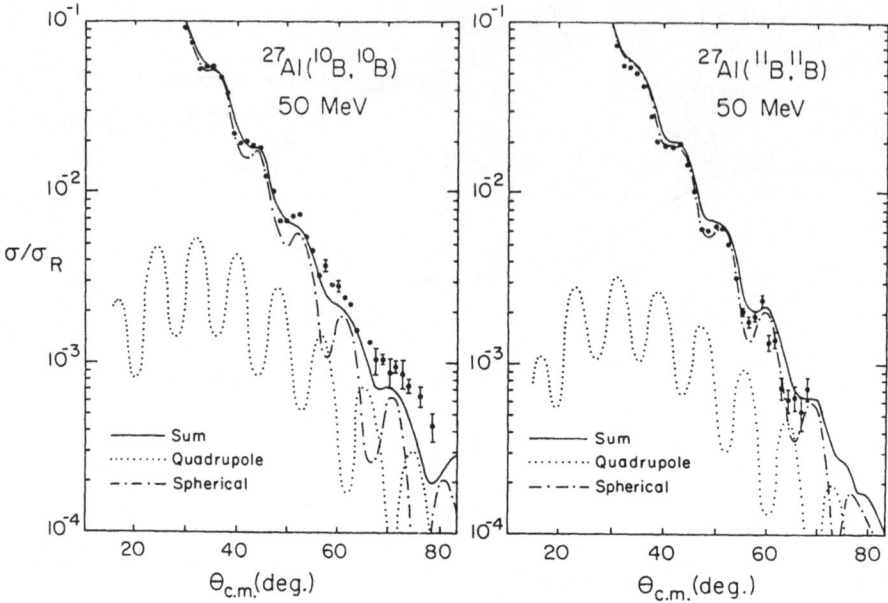

Figure 9. The filling of the diffraction minima in elastic differential cross sections due to quadrupole terms in the elastic scattering potentials.

collisions. This is particularly true in the case of heavy-ion collisions.

5.3 Inelastic Scattering

Inelastic scattering is the primary source of information about the higher momentum components of the effective interaction. This is evident from the k-space expression for the scattering potentials, eq. 8 , if it is noted that the Fourier transforms of the non-spherical transition densities entering in inelastic scattering peak at larger momentum values than the Fourier transforms of the spherical ground state densities which enter in the description of elastic scattering. To deduce information about the high momentum components of the interaction one must consider nuclear transitions which are "known", e.g. transitions in N=Z nuclei where proton transition densities are known from electron scattering and it is reasonable to assume $\rho_p \approx \rho_n$. If the effective interaction is known with some confidence, inelastic scattering can be used to gain information on effects associated with nuclear structure, e.g. differences between proton and neutron transition densities in nuclei with N≠Z [27,35] and core polarization effects in nuclei [28].

The results of folding model calculations [27] for the excitation of low lying normal parity levels in ^{208}Pb by 61.2 and 35 MeV protons are shown in fig. 10 and fig. 11, respectively. In these calculations

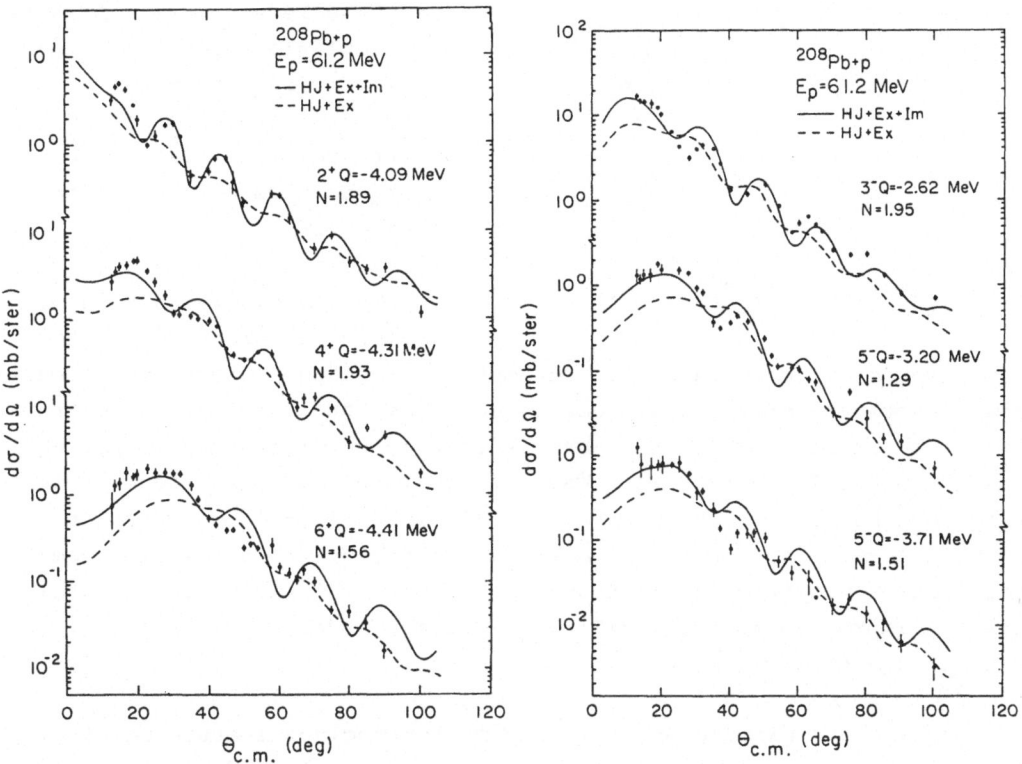

Figure 10. Results of folding model calculations for excitation of low lying normal parity levels in ^{208}Pb by 61.2 MeV protons. Dashed curves are results with real interaction and solid curves are results with complex interaction. The theoretical results have been divided by the factors of N shown.

the long range part of the HJ was used for the real part of the effective interaction, the imaginary term was introduced phenomenologically [8], and it was assumed that $\rho_n = N/Z \, \rho_p$ with ρ_p taken from electron scattering [48]. The normalization factors required to fit the experimental data measure the deficiency in the assumption $\rho_n = N/Z \, \rho_p$ and do not reflect inadequacies in the reaction model. A normalization

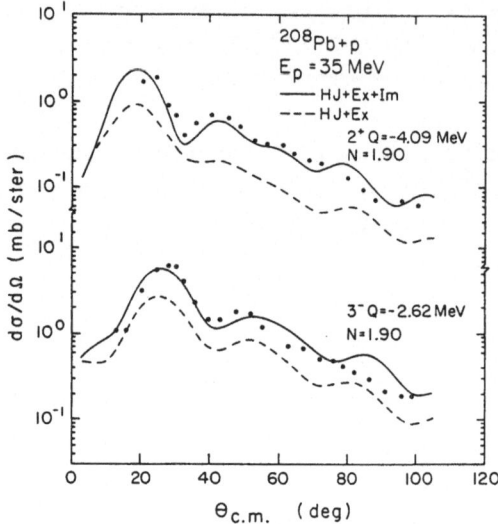

^{208}Pb+p
$E_p = 35$ MeV
—— HJ+Ex+Im
---- HJ+Ex

2^+ Q=-4.09 MeV
N=1.90

3^- Q=-2.62 MeV
N=1.90

Figure 11. Same as fig. 10 for E_p = 35 MeV.

of approximately two implies $\rho_n \approx \rho_p$ instead of the assumed $\rho_n \approx N/Z \, \rho_p$. The results suggest $\rho_n \gtrsim \rho_p$ for the transitions in Pb. The deviations from $\rho_n \approx \rho_p$ are roughly consistent with differences suggested by R.P.A. wave functions for these levels [49]. Results of calculations for the 2^+ and 3^- excitations where the transition densities have been constructed from R.P.A. wave functions have been obtained in ref. [26].

The imaginary term in the effective interaction is the major source of uncertainty in attempting to deduce information about small proton-neutron differences in inelastic nucleon-nucleus scattering. The importance of this term is seen quite clearly by comparing the 35 and 61.2 MeV results. The effect of the imaginary term on the shape of the cross sections is quite different at the two energies. If this term is not included, misleading results are obtained. The new complex G matrix interactions of ref. [13,14] should prove valuable in improving the reliability of folding model studies of inelastic scattering.

As another example of inelastic proton-nucleus scattering, fig. 12 contains the results of folding model calculations for the excitation of the first 2^+, 4^+, 6^+, and 8^+ levels in ^{40}Zr by 61.2 MeV protons [28]. In the simplest picture these levels are thought to be the excited states of two protons in the $(1g_{9/2})^2$ configuration outside an inert ^{88}Sr core. The results obtained using these wave functions are shown as small dashes in fig. 12. They underestimate the experimental data by more than an order of magnitude. The remaining curves are the results obtained when 3p-1h configurations, associated with the polarization of the core by the valence protons, are included in the wave functions. These core polarization effects are dominant and persist even in transitions of high multipolarity. The theoretical foundations of these core polarization effects is still uncertain [28]

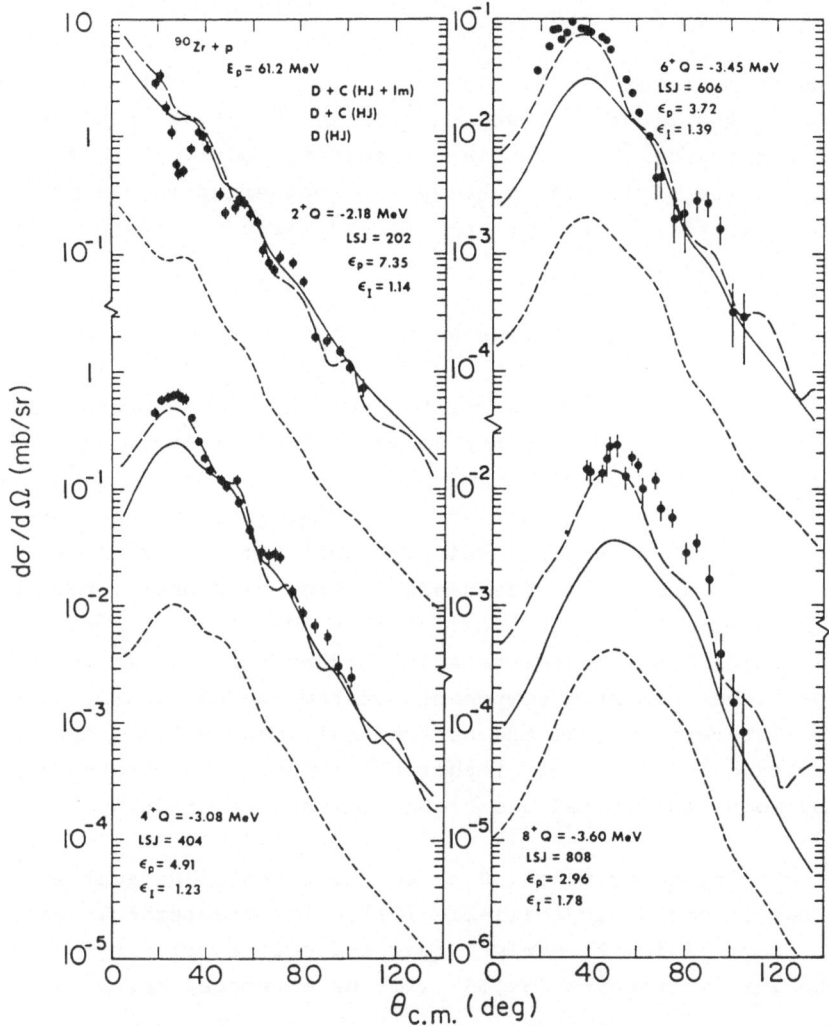

Figure 12. Theoretical cross sections for the excitation of first 2^+, 4^+, 6^+, and 8^+ levels in ^{90}Zr by 61.2 MeV protons. D and D+C denote the results with and without core polarization included.

and continued study of these effects promises to be interesting.

The folding model description of inelastic excitations in heavy-ion scattering has yet to receive as much attention as nucleon-nucleus inelastic scattering. Here the transitions observed tend to be of low multipolarity because the nucleus producing the excitation behaves as a long range interaction with respect to the nucleus being excited, i.e. there is a short range momentum distribution for either ρ_p or ρ_t in eq.8 which results in a strong L dependence in the inelastic scattering form factors [50]. The inelastic cross sections for the excitation of the first 2^+ level in ^{58}Ni by ^{16}O at 42 and 48 MeV shown in fig. 8 are typical of the G matrix folding model [12]. Another interesting feature of inelastic heavy-ion scattering is the possibility of mutual excitation of the projectile and target. This has been discussed in ref. [17,18].

5.4 Inelastic Proton Scattering at E_p = 135 MeV

There are several factors which have made proton-nucleus scattering in the 100-200 MeV energy attractive since the early days of the microscopic model. First, the impulse approximation provides a reasonable first guess at the complete effective interaction at high energies. Second, the proton-nucleus reaction mechanisms are expected to be simpler at the higher energies. Third, high energy protons can transfer more momentum to the nucleus than lower energy protons. This makes it possible to observe states of higher spin and at the same time probe higher momentum components of the nuclear distributions. It is also interesting that the non-central components of the effective interaction will be more important at these higher energies because they more readily mediate higher momentum transfers.

High quality experimental data for proton-nucleus scattering at E_p = 135 MeV is presently available [51]. The theoretical analysis of this data is still in its early stages and only a brief synopsis of two of the initial reports [52,53] will be attempted here.

The first report [52] concerns impulse approximation estimates of the cross sections for the excitation of low lying normal parity levels in ^{208}Pb. Specifically, the 3^-_1, 5^-_1, 5^-_2, 2^+_1, 4^+_1, 6^+_1, 8^+_1, 10^+_1, and 10^+_2 levels in ^{208}Pb have been considered. The same levels have been observed in (e,e') studies [48,54,55]. In the calculations the proton transition densities were taken from the electron scattering studies. Neutron transition densities were assumed to have the same shape as the proton transition densities and their strengths were adjusted to fit the (p,p') data.

The ratios of neutron to proton transition densities obtained in the calculations were on the order of unity and consistent with results

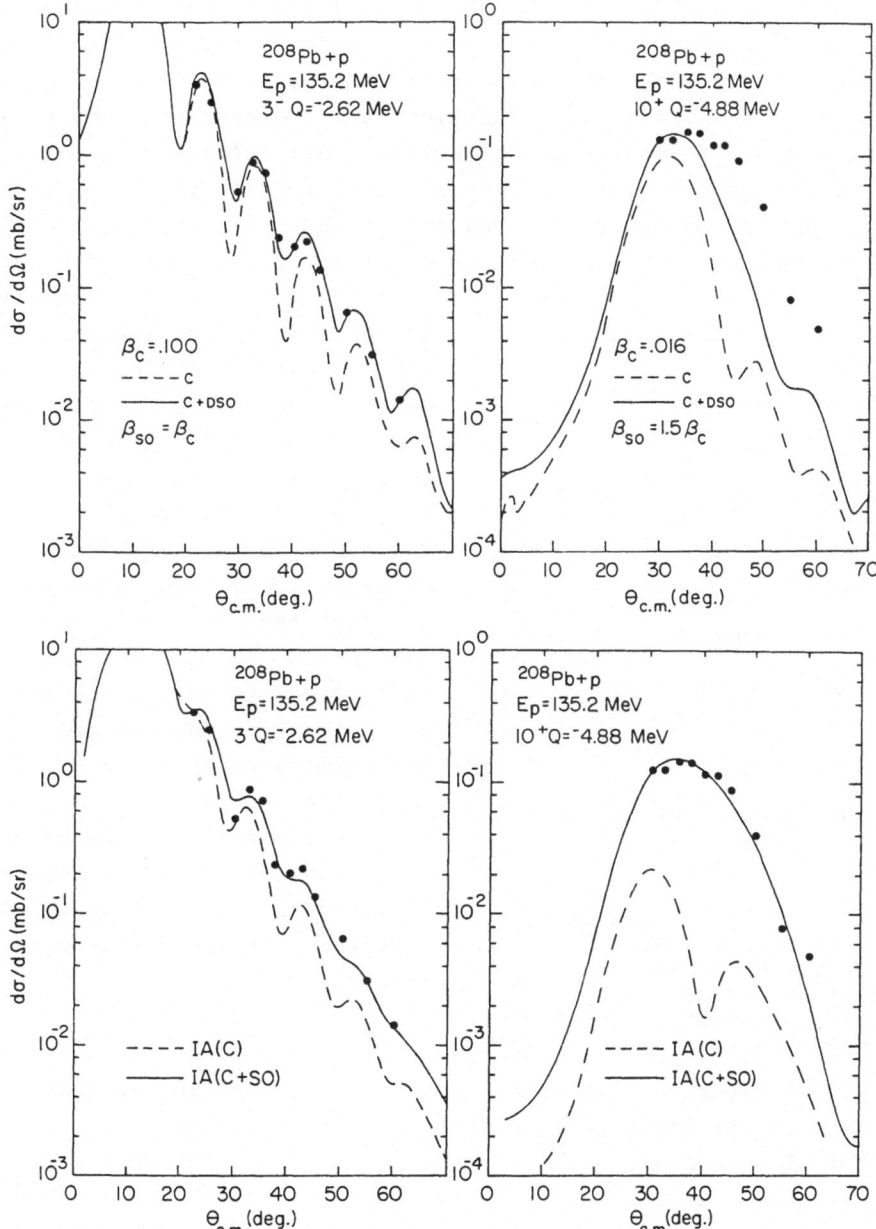

Figure 13. Collective model (above) and microscopic distorted wave impulse approximation (below) results for the excitation of the first 3⁻ and 10⁺ levels in ²⁰⁸Pb by 135.2 MeV protons. Solid and dashed curves are results with and without contributions from the spin orbit potentials, respectively.

obtained in earlier low energy proton scattering studies discussed in the previous sections. The magnitude of the free t matrix interactions thus appears reasonable for use at these energies. Results of the calculations for the 3_1^- and 10_1^+ excitations are compared with the experimental data and the results of collective model calculations in fig. 13. One of the most striking features of the results is the important contributions to the cross sections due to the spin-orbit interaction. For the 3_1^- excitation the collective model calculation provides a some-what better fit to the experimental data than does the impulse approximation calculation. The central interaction components are most important here and the deficiencies in the impulse approximation results are quite possibly due to the neglect of density dependent Pauli corrections which are important for these interaction components [13,14]. For the 10_1^+ excitation the collective model is inadequate, but the impulse approximation result is quite good. The contribution to the cross section due to the spin-orbit force is dominant here, and it is concluded that the impulse approximation gives this more correctly.

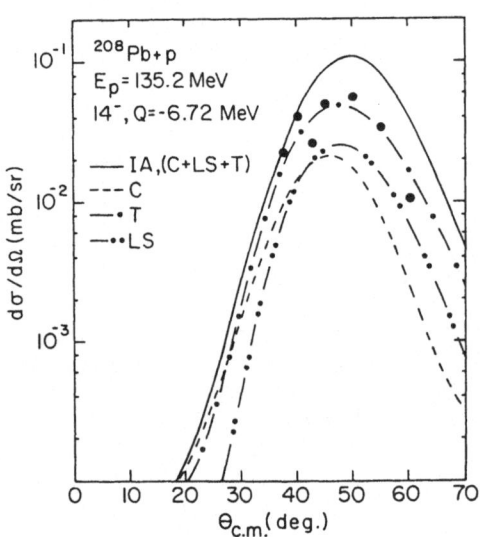

Figure 14. Impulse approximation results for the 14⁻ level in ^{208}Pb. C, T, and LS denote results with central, tensor, and spin-orbit interaction components separately.

The second report [53] concerns impulse approximation estimates of the cross sections for high spin abnormal parity excitations in ^{208}Pb. The states of interest are the 12⁻ and 14⁻ levels which were first reported in the electron scattering work of ref. [55]. The electron scattering data confirms the expected $1i_{13/2}-1h_{11/2}^{-1}$ proton-proton hole and $1j_{15/2}-1i_{13/2}^{-1}$ neutron-neutron hole structure of these levels [56]. The impulse approximation calculation give a reasonable description of the cross sections for the 12⁻ and 14⁻ neutron excitations, but significantly overestimates the cross section for the 12⁻ proton excitation. The tensor force gives the largest contributions to these cross sections and the results suggest some difficulty with the high momentum components of the tensor force given by the impulse approximation. The results for the 14$^-$ level are shown in fig. 14.

6. CONCLUSION

In conclusion the folding model calculations based on <u>realistic</u> interactions give a reasonably consistent description of a large variety of elastic and inelastic scattering data. The model is becoming progressively more realistic and comprehensive with time. Future applications of the model to high energy proton nucleus collisions and heavy-ion collisions are sure to provide interesting new information on the structure of nuclear levels and dynamic polarization processes in nuclear collisions.

ACKNOWLEDGEMENTS

Special thanks are due to W.G. Love who has been either a correspondent, collaborator, or competitor on most of the material presented here, to D. Stanley who has carried out most of the heavy-ion calculations shown within, and to Janice Finney for typing the manuscript.

REFERENCES

1. K.M. Watson, Phys. Rev. <u>89</u>, 575 (1953); N. Francis and K.M. Watson, Phys. Rev. <u>92</u>. 291 (1953); G. Takeda and K.M. Watson, Phys. Rev. <u>97</u>, 1336 (1955); K.M. Watson, Phys. Rev. <u>105</u>, 1388 (1957).
2. A.K. Kerman, H. McManus, and R.M. Thaler, Ann. Phys. (N.Y.) <u>8</u>, 551, (1959).
3. R.M. Haybron and H. McManus, Phys. Rev. <u>136</u>, B1730 (1964); <u>140</u>, B638 (1965); R.M. Haybron, M.B. Johnson, and R.J. Metzger, Phys. Rev. <u>156</u>, 1136 (1967); H.K. Lee and H. McManus, Phys. Rev. <u>161</u>, 1087 (1967).
4. G.R. Satchler, Nucl. Phys. <u>A95</u>, 1 (1967).
5. S.M. Austin, in <u>The Two-Body Force in Nuclei</u>, ed. S.M. Austin and G.M. Crawley (Plenum, New York, 1972).
6. D. Slanina and H. McManus, Nucl. Phys. <u>A116</u>, 271 (1968); F. Petrovich <u>et al.</u>, Phys. Rev. Lett. <u>22</u>, 895 (1969).
7. W.G. Love <u>et al.</u>, Phys. Lett. <u>29B</u>, 478 (1969); W.G. Love and G.R. Satchler, Nucl. Phys. <u>A159</u>, 1 (1970); L.W. Owen and G.R. Satchler, Phys. Rev. Lett. <u>25</u>, 1720 (1970).
8. G.R. Satchler, Phys. Lett. <u>35B</u>, 279 (1971); G.R. Satchler, Comm. Nucl. Particle Phys. <u>5</u>, 379 (1972).
9. D.A. Agassi and R. Schaeffer, Phys. Lett. <u>26B</u>, 703 (1968); R. Schaeffer, Nucl. Phys. <u>A132</u>, 186 (1969).
10. H.V. Geramb and K. Amos, Nucl. Phys. <u>A163</u>, 337 (1971); H.V. Geramb, R. Sprickmann, and G. Strobel, Nucl. Phys. <u>A199</u>, 545 (1973).
11. G.R. Satchler and W.G. Love, Phys. Lett. <u>65B</u>, 415 (1976); W.G. Love, paper at this conference.
12. H. Wojciechowski <u>et al.</u>, Phys. Rev. <u>C17</u>, 2126 (1978); D. Stanley and F. Petrovich, to be published.
13. C. Mahaux, paper at this conference.
14. F.A. Brieva, paper at this conference.
15. N. Vinh Mau, paper at this conference.

16. F. Petrovich, Nucl. Phys. A251, 143 (1975).
17. P.J. Moffa, C.B. Dover, and J.P. Vary, Phys. Rev. C16, 1857 (1977).
18. F. Petrovich and D. Stanley, Nucl. Phys. A275, 487 (1977); M.E. Williams-Norton et al., Nucl. Phys. A275, 509 (1977)
19. W.G. Love, Part. and Nuclei 3, 318 (1972); W.G. Love and L.W. Owen, Nucl. Phys. A239, 74 (1975); W.G. Love, to be published.
20. B. Sinha, Phys. Rep. 20C, 1 (1975).
21. J.W. Negele and D. Vautherin, Phys. Rev. C5, 1472 (1972).
22. M. Golin, F. Petrovich, and D. Robson, Phys. Lett. 64B, 253 (1976).
23. Y.C. Tang, paper at this conference.
24. A. Kallio and K. Kolltveit, Nucl. Phys. 53, 87 (1964).
25. S. Moskowski and B. Scott, Ann. Phys. (N.Y.) 11, 65 (1960).
26. E.C. Halbert and G.R. Satchler, Nucl. Phys. A233, 265 (1974).
27. A. Scott, N.P. Mathur, and F. Petrovich, Nucl. Phys. A285, 222 (1977)
28. F. Petrovich et al., Phys. Rev. C16, 839 (1977).
29. G.E. Brown, Unified Theory of Nuclear Models and Nucleon-Nucleon Forces, (North-Holland, Amsterdam, 1967) 2nd. ed.
30. G.R. Satchler, Phys. Lett. 59B, 121 (1975).
31. G. Bertsch et al., Nucl. Phys. A284, 399 (1977).
32. F. Petrovich, D. Stanley, and J.J. Bevelacqua, Phys. Lett. 71B, 259 (1977).
33. A. Picklesimer and G. Walker, Phys. Rev. C17, 237 (1978).
34. W.G. Love et al., Phys. Lett. 73B, 277 (1978).
35. G.R. Hammerstein, R.H. Howell, and F. Petrovich, Nucl. Phys. A213, 45 (1973).
36. W.G. Love, Nucl. Phys. A192, 49 (1972).
37. S.A. Moskowski, Phys. Rev. C2, 402 (1970).
38. W.G. Love, Phys. Lett. 72B, 4 (1977).
39. A. Bohr and B. Mottelson, Nucl. Structure (Benjamin, New York, 1969), Vol. I, p. 259.
40. G.W. Greenlees, G.J. Pyle and Y.C. Tang, Phys. Rev. 171, 1115 (1968).
41. J.M. Moss et al., Phys. Rev. C6, 1698 (1972).
42. E. Rost, Phys. Lett. 26B, 184 (1968).
43. F. Petrovich, in Proc. Int. Symp. on Superheavy Elements, Lubbock, Texas, 1978, to be published; J.M. Moss, Phys. Rev. C17, 813 (1978).
44. W. Weiss et al., Phys. Lett. 61B, 237 (1967).
45. F. Petrovich et al., Phys. Rev. C17, 1642 (1978).
46. L.A. Parks et al., Phys. Lett. 70B, 27 (1977).
47. G. Tungate, contribution to this conference.
48. M. Nagao and Y. Torizuka, Phys. Lett. 37B, 383 (1971)
49. V. Gillet, A. Green, and E. Sanderson, Nucl. Phys. 88, 321 (1966).
50. P.J. Moffa et al., Phys. Rev. Lett. 35, 992 (1975).
51. G.S. Adams, thesis, Indiana University, 1977 (unpublished), available as IUCF Internal Report 77-3.
52. F. Petrovich and W.G. Love, Bull. Am Phys. Soc. 23, 963 (1978) and to be published.
53. A.D. Bacher et al., Bull. Am. Phys. Soc. 23, 945 (1978) and to be published.
54. J. Friedrich, K. Voegler, and H. Entenaur, Phys. Lett. 64B, 269 (1976).
55. J. Lichtenstadt, J. Heisenberg, C.N. Papanicolas, C.P. Sargent, A.N. Courtemanche, and J.S. McCarthy, submitted to Phys. Rev. Lett.
56. W.W. True, C.W. Ma, and W.T. Pinkston, Phys. Rev. C3, 2421 (1971).

A NEW TYPE OF PARAMETER SYSTEMATICS FOR PROTON-NUCLEUS SCATTERING[+]

H.LEEB and G.EDER

Atominstitut der Österreichischen Universitäten, A-1020 Vienna, Austria

1.Introduction

The phenomenological analysis of scattering data in the framework of the optical potential is not unique. Therefore it is very hard to extract the physics out of these potentials. Particularly the ambiguity between the potential depth and the geometry is evident. Much work on this problem has been done by Greenlees and collaborators[1],[2]. Their investigations show clearly that the volume integral pro nucleon and the mean square radius of the potential are the well determined quantities by the elastic scattering data. This result justifies the assumption of a certain geometry parameter set in order to perform a consistent analysis. The so constructed potentials reproduce the scattering data quite well. Nevertheless these parameter systematics are unsatisfactory if you want information about the physics of the potential. The present work concerns with this problem. For this purpose a new parameter systematics is established which includes the valley ambiguity between the potential depth and the geometry. The main interest does not lie in the extremly good reproduction of scattering data but on a maximum of information about the physics of the potential.

2. Method of Analysis

The potential is constructed from individual fits of scattering data with chosen geometry parameter sets. This method is not so straightforward like a simultaneous fitting procedure but gives the true functional form for the potential. The analysis of the scattering data - differential cross section, polarisation and total nonelastic cross section - is made phenomenologically based on the potential

$$U_{opt}(r) = -Vf(x_0) - i\left(W - 4W_s \frac{d}{dx_I}\right)f(x_I) + 2.0 \ \text{fm}^2 \ V_{so} \ \vec{\sigma}\vec{l}\frac{1}{r}\frac{d}{dr}f(x_{so}) + V_c(r) \quad (1)$$

with $\quad f(x_i) = (e^{x_i} + 1)^{-1}\quad$ and $\quad x_i = (r - r_i A^{1/3})/a_i$. For V_c the Coulomb potential of a homogeneous charged sphere with the radius[3]

$$R_c = (1.089 \ A^{1/3} + 0.317) fm \tag{2}$$

is used. The parameter are analysed and fixed in the sequence (V, W_s, V_{so}) according to their influence on the scattering data. The experimental data are taken from the familiar literature of nuclear physics. For the construction of the potential 75 sets of the differential cross section, 60 sets of the polarisation and 77 values of total nonelastic cross section are used. The analysis is performed with the computer code SPFWL[4].

3. Theoretical Assumptions

The aim of the analysis requires to minimize the number of assumptions. In spite of the extensive study of the valley ambiguity assumptions are necessary to guarantee the uniqueness of the potential. In order to get pure information about the physics of the potential the required assumptions are based on a realistic theoretical evaluation.

3.1 The Volume Absorption

Since a phenomenological analysis does not say anything about the radial shape of the potential an assumption is necessary to separate volume and surface absorption consistently. Unfortunatly there does not exist any satisfactory microscopic calculation which can give a guideline for the separation. Therefore we made the following consideration.
For high energies the impulse approximation is a good and handy tool to calculate the imaginary part of the optical potential from the total nucleon-nucleon cross sections. For energies lower than 100 MeV this method can not be used. Microscopic calculations in the nuclear matter approach predicts for the imaginary part of the optical potential a nearly linear dependence on energy[5],[6],[7]. Since the condition of nuclear matter is performed only in the inner region of a realistic nucleus the imaginary part obtained can be interpreted as a volume absorption. Now the idea is to connect these predictions continuosly at the pion production threshold. Thus we fixed for the volume absorption

$$W = \frac{23.55441}{r_I^3 (1 + (a_I \pi / (r_I A^{1/3}))^2)} \ \frac{E^2 + 5170E}{E^2 + 5170E + 938.8^2} (4.35 - 1.75(e^{(E-425)/70} + 1)^{-1}). \tag{3}$$

3.2 The Coulomb Correction Term in the Real Central Potential

The energy dependence of the local optical potential reduces the effect of the Coulomb potential. In order to take into account this effect one usually adds a term of $0.4Z/A^{1/3}$ in the potential depth V as proposed by Lane[8]. We have recalculated this term.

In the low energy range the energy dependence is mainly caused by the nonlocality of the potential[9]. Assuming an energy independent nonlocal potential of the form of Perey and Buck[10] we can calculate an equivalent local potential by the Perey-Saxon approximation[11]. Neglecting higher order terms we obtain for the real central potential

$$U_L^c(r) = (1 + \frac{1}{4} \beta^2 \frac{2m}{h^2} \frac{\alpha_f hc}{r_c} \bar{g} \frac{Z}{A^{1/3}}) V_L(r) \tag{4}$$

where $V_L(r)$ is the equivalent local potential and β is the nonlocality parameter. $\bar{g}$ is an averaged value for the formfactor which describes the shape difference between the Coulomb correction term and $V_L(r)$. $\bar{g}$ is defined by

$$\bar{g} = \frac{3r_c A^{1/3}}{\alpha_f hcR^3} \int_0^R \frac{V_c(r) \ r^2}{1 - \frac{1}{4} \beta^2 \frac{2m}{h^2} V_L(r)} dr \tag{5}$$

with $R = (r_o A^{1/3} + 2a \ln3)$. Using the value $\beta = 0.85$ fm^{-1} of Perey and Buck[10] and an averaged form for the potential depth of the equivalent local potential $V_L = 50 - 0.3E + 24(N-Z)/A$ we have calculated $\bar{g}$. From this we can establish a parametrisation for V of the phenomenological analysis

$$V = (1 + (0.0085 + 1.2 \ 10^{-5}E)\frac{Z}{A^{1/3}}) V_L . \tag{6}$$

4. Phenomenological Parametrisation

If the geometry parameter are fixed, the assumption of W guarantees a unique solution of the phenomenological analysis. So we are able to study the dependence of V, W_s and V_{so}.

4.1 The Real Central Potential Depth V

The real central potential is the dominant part of the optical potential and can be considered independently from other terms. To study the valley ambiguity several analyses with various geometry parameters are made. Knowing the result of Greenlees et al.[1] we assume that the volume integral pro nucleon is in first order a good representation also for the arguments r_o and a. Therefore we make the ansatz

$$V_L = \left(V_{NN} - \alpha_r (r_o - 1.25) - \alpha_a (a - 0.6)\right) \left(\frac{1.25}{r_o}\right)^3 \left(1 + \left(\frac{\pi \, a}{r_o A^{1/3}}\right)^2\right)^{-1} \tag{7}$$

which is referred to the average values $r_o = 1.25$ fm and $a = 0.6$ fm. The analyses show that (7) is a good description. It yields the coefficients

$$\alpha_r = E - 200(r_I - r_o) \quad \text{and} \quad \alpha_a = 0 \tag{8}$$

The dependence of α_r on r_I and E is also confirmed by the study of the parameter sets summarized by Perey-Perey[12].

V_{NN} contains the whole information about the proton-nucleus force. The analyses show that the energy dependence of V_{NN} can be described by a linear relation for the energy range 20MeV < E < 100 MeV given by

$$V_{NN} = V_{NN}^0 - 0.36 \, E \tag{9}$$

V_{NN}^0 depends only on the values of the target nucleus. According to Lane[8] we know that this part contains an isospin dependence. Normally optical potential analyses can not give information about the value of the symmetry term because the separation between size and isospin dependence causes some troubles[13]. The quantitative description of the valley ambiguity allows a separation. Since U_{symm} must be independent of the size of the nucleus but a dependence of the geometry must still remain we make the ansatz

$$V_{NN}^0 = V^0(A) + V_{symm}(1 + 2.274 \, A^{-2/3})\frac{N-Z}{A} \quad . \tag{10}$$

The analyses give for the parameters

$$V^0 = 45 + 45A^{-1/3} , \qquad V_{symm} = 22 \pm 3 \text{ MeV}. \tag{11}$$

4.2 The Surface Absorption W_s

Since the imaginary part is the dominant term in the calculation of the nonelastic cross section σ_{ne} we have determined W_s directly from σ_{ne}-data. We study first the valley ambiguity between r_I, a_I and W_s. Since W_s is a second order term we must also consider a dependence on r_o. The analyses show a very strong depndence on r_I, a_I and r_o. Neither the assumption that the volume integral is well determined nor the ansatz of a polinomial of low order gives a good description. The reason for this is a very strong coupling of the energy dependence with the geometry. The energy dependence can be parametrised by

$$W_s = (\delta_1 + \delta_2 E) \, E \, e^{-\delta_3 E} \quad . \tag{12}$$

An expansion of low order for δ_1 and δ_2 in r_o, r_I and a_I then gives a satisfactory description. The direct analysis of σ_{ne} demonstrates a strong dependence on the mass number which can not be obtained by over all fitting procedures.

4.3 The Spin-Orbit Term V_{so}

The determination of V_{so} is performed by fitting polarisation data. The study of the energy dependence does not give a clear result caused by great fluctuations. In the energy range from 20MeV to 50MeV an average value of about 6MeV gives a good description. This is in good agreement with the study of Hodgson[14] besides the fact that we have evidence for a weak energy dependence. This is confirmed by analysis at high energies where the values are consistently smaller than 5MeV. From the averages at each energy for $r_o = r_I = r_{so} = 1.25$ fm and $a = a_I = a_{so} = 0.6$ fm we get

$$V_{so} = 7.5 \ e^{-0.008E} \tag{13}$$

The study of the valley ambiguity is limited to the dependence on r_o, r_{so} and a_{so}. Due to the great uncertainties only a linear expansion in these parameters is performed.

5. Discussion and Conclusion

Summarising the results we obtain the following parameter systematics for proton-nucleus scattering: (All values in MeV or fm)

$$V = \{45 + 45A^{-1/3} + 22(1 + 2.274A^{-2/3})\frac{N-Z}{A} - (E - 200(r_I - r_o))(r_o - 1.25) - 0.36 \ E\}$$

$$\left(\frac{1.25}{r_o}\right)^3 \{1 + (0.0085 + 1.2 \ 10^{-5}E)\frac{Z}{A^{1/3}}\}\left[1 + \left(\frac{\pi a}{r_o A^{1/3}}\right)^2\right]^{-1},$$

$$W_s = \{0.26 + 0.68(r_o - 1.25 + 2.38(r_I - 1.25) + 1.3(a_I - 0.6) - 0.006A + 0.015E +$$

$$(0.014 - 0.014(r_o - 1.25) - 0.039(r_I - 1.25) - 0.046(a_I - 0.6) - 0.25(r_I - 1.25)^2)$$

$$e^{0.034A}E\} \ E \ \exp\left(-(0.065 + 0.00011A^{4/3})E\right),$$

$$V_{so} = 7.5\left(1 - 2.0(r_o - 1.25) - 2.3(r_{so} - 1.25) - 1.4(a_{so} - 0.6)\right)e^{-0.008E} \tag{14}$$

which reproduces the scattering data quite well, if the conditions

$$r_o, r_I, r_{so}: \quad 1.05 < r_i < 1.40 \text{ fm} \qquad 0 < r_I - r_o < 0.3 \text{ fm}$$
$$a, a_I, a_{so}: \quad 0.45 < a_i < 0.75 \text{ fm} \qquad 20 < E < 140 \text{ MeV} \qquad A > 12 \tag{15}$$

are satisfied. In spite of the extended range the reproduction of the scattering data is very good and comparable with special potentials.

Iso-tope	Yukawa formf. Ref. 2	Gauß-formf. Ref. 2	present work	Ref. 15
^{40}Ca			-0.19	0.03
^{57}Fe	0.22	-0.04	-0.05	
^{59}Co	0.20	-0.03	-0.03	
^{58}Ni	0.24	0.01	0.03	0.14
^{120}Sn	0.27	0.15	0.19	0.44
^{208}Pb	0.19	0.13	0.34	0.62

Tab. 1 Comparison of δ with other estimates. "form" refers to the used formfactor in the folding integral of Ref. 2.

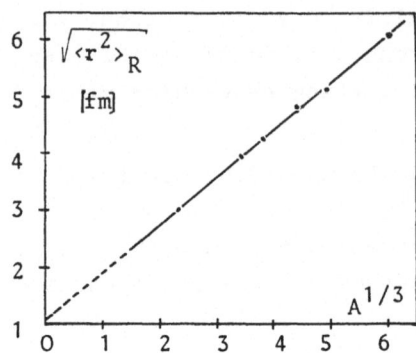

Fig. 1 Mass dependence of the mean square radius of the real central part.

The quantitative description of the valley ambiguity allows a consistent calculation of the optimal geometry. We obtain for the mean square radius of the real part the results illustrated in fig.1 which can be described by

$$\sqrt{<r^2>_R} = 0.84\ A^{1/3} + 1.05\quad\text{fm}\ . \tag{16}$$

For the imaginary part the analysis gives

$$\sqrt{<r^2>_I} = 0.90\ A^{1/3} + 1.85\quad\text{fm}\ . \tag{17}$$

In agreement with individual analyses we obtain

$$r_I > r_o > r_{so}\ . \tag{18}$$

In order to compare the potential with microscopic calculations we have calculated

$$\delta = \sqrt{<r^2>_n} - \sqrt{<r^2>_p} \tag{19}$$

where $<r^2>_n$ is the mean square radius of the neutron distribution and $< r^2>_p$ is the same for the protons. Using the expression of Eder and Oberhummer[3] for $<r^2>_p$ and assuming the mean square radius for the nucleon-nucleon interaction to be $<r^2>_{nn} = 4.27$ fm^2 we obtain the results summarised in Tab. 1. The agreement with the results of Greenlees et al.[2] is very good, especially for the Gauß form. From this it can be concluded that the use of a Gauß formfactor in the folding model gives better agreement with results of phenomenological analyses than a Yukawa form. Besides this conclusion the mass dependence of δ indicates a better agreement with realistic microscopic calculations, e.g.[15].

Thus we can state that the presented potential contains the pure physics and can be used for comparison of over all dependences with microscopic calculations. The observed isospin potential is determined quite well

and in full agreement with results of (p,n)-reactions. There is evidence
of an energy dependence of the spin-orbit potential as predicted by
microscopic calculations.

The general formulation of the presented potential allows a direct comparison with individual analyses as well as the recalculation of other
parameter systematics. Finally it should be mentioned that the potential
can be extended easily to high energies.

References

+   Work aupported by Fonds zur Förderung der wissenschaftlichen Forschung
    (Österreich).
1. G.W.Greenlees, G.J.Pyle and Y.C.Tang, Phys. Rev. 171,1115(1968)
2. G.W.Greenlees, W.Makofske and G.J.Pyle, Phys. Rev. C1,1145(1970).
3. G.Eder and H.Oberhummer, Lett. Nuovo Cim. 15,609(1976).
4. H.Leeb, not published.
5. A.S.Reiner, Phys. Rev. 133,B1105(1964).
6. J.P.Jeukenne, A.Lejeune and C.Mahaux, Phys. Rev. C10,1391(1974).
7. F.A.Brieva and J.R.Rook, Nucl. Phys. 291,299(1977).
8. A.M.Lane, Nucl. Phys. 35,676(1962).
9. G.Passatore, Nucl. Phys. A95,694(1967)
10. F.G.Perey and B.Buck, Nucl. Phys. 32,353(1962).
11. F.G.Perey and D.S.Saxon, Phys. Lett. 10,107(1964).
12. C.M.Perey and F.G.Perey, Atomic Data and Nuclear Data Tables
    17,1(1976).
13. G.W.Greenlees, G.J.Pyle and Y.C.Tang, Phys. Lett. B26,658(1968).
    P.E.Hodgson, Nucl. Phys. A150,1(1970).
14. P.E.Hodgson, Nuclear Reactions and Nuclear Structure, Clarendon Press
    Oxford,1971.
15. C.J.Batty and G.W.Greenlees, Nucl.Phys. A133,673(1969).

L-dependent Optical Potentials: What Experiment Tells us about Local Density Models

R.S. Mackintosh, Daresbury Laboratory, England

and

A.M. Kobos, Institute of Nuclear Physics, Cracow, Poland

1. Introduction

In this paper we shall attempt to show how the full information content of proton elastic scattering data for light target nuclei (which have not previously been fitted by a simple potential model) can be exploited to illuminate various fundamental features of the proton-nucleus interaction. We emphasize that our approach is complementary to and not in competition with various first principles calculations and global fitting studies. We achieve χ^2 values often one or two orders of magnitude lower, with the largest improvements for ^{16}O which has previously proven very hard to fit. Among the features of the nucleon-nucleus interaction upon which we now appear to be able to throw light are:

(a) the range of validity of the single particle model and the direct interaction picture

(b) the appropriateness of "local density" models, i.e. models which take the finite size (not to mention collectivity etc. of a particular nucleus) of the nucleus into account only by doing a folding model over $\rho(r)$ with the folding interaction calculated from some local density. In somewhat metaphorical terms, we are studying "shell corrections" to the "liquid drop" calculations based on a g-matrix.

(c) the specific nature and contribution of coupled reaction channel (CRC) effects such as the coupling to pickup channels, and

(d) various "resonance-like" phenomena.

There are certain by-products of this work.

(a) New spin-orbit terms [1]: an imaginary term which changes sign at 60 MeV and new real spin-orbit parameters which may be important for nuclear structure.

(b) New optical potentials for application to DWBA calculations such as (p, p'). This is important since p + ^{16}O ℓ-dependent potentials which actually fit the data accurately give backward angle peaks

in (p, p') and this should be considered when giant resonance
information is extracted by fitting backward angle (p, p') data.

2. The Potential

Our potential [1 - 4] has the form $U(r) = V^{LI} + V^{LD}$ where V^{LI} is
the standard form of Perey and Perey [5] (their equation 1) and V^{LD} is
an ℓ-dependent term

$$V^{LD} = -f(\ell^2, L^2, \Delta^2)\{U^L g(r, R^R, a^R) + i\, W^L\, g(r, R^I, a^I)\} \tag{1}$$

where f is the standard Fermi form and $g(r,R,a) = -4a\frac{d}{dr}f(r,R,a)$.
Derivative terms were chosen as being appropriate to an ℓ-dependent
term. The appearance of squares in the fermi-form overall ℓ-dependence
factor is historical and inessential; for values of Δ actually found
this factor has the effect of introducing a new potential for $\ell < L$,
which is near zero for $\ell > L$ with fairly rapid transition near L (which
need not be integral). In some early calculations [2] we allowed
separate L and Δ for the imaginary term, but this proved unnecessary.

3. The Results of Fitting

We have mostly restricted the analysis to energies where there
are angular distributions and analysing powers out to backward angles.
We have fitted $p + {}^{16}O$ at about 18 energies between 23 and 65 MeV,
$p + {}^{40}Ca$ at 10 energies between 17.3 and 48.0 MeV, $p + {}^{56}Fe$ at 30 MeV,
$p + {}^{58}Ni$ at 40 and 100 MeV and a range of nuclei from ${}^{12}C$ to ${}^{18}O$ at
35 MeV. We have also studied some cases of helion scattering. In
figure 1 we show quite a typical fit. For the angular distribution,
χ^2 was reduced by two orders of magnitude over the best ℓ-independent
Woods-Saxon fit. Similar fits were obtained in every case we studied
although the χ^2 improvement factor was not so large for heavier nuclei
where moderately good fits were previously obtained. Nevertheless, the
ℓ-dependent terms had the same qualitative features in every case and
were not small for heavier nuclei at moderate energies. Thus ℓ-dependence
appears to be a universal phenomenon even where, as for heavier nuclei,
the lack of deep diffraction minima allow reasonable fits to be obtained
with ℓ-independent potentials. We find that the ℓ-dependent potentials
have the following properties in all cases:

(1) There is a rapid transition region defined by parameter L such that
the potential for $\ell < L$ is substantially different than for $\ell > L$.
We have explicitly shown that the change in the potential is not
confined to one or two partial waves.

(2) The new terms which come in for low partial waves are always
repulsive ($U^L < 0$) and absorptive ($W^L > 0$). We have found no

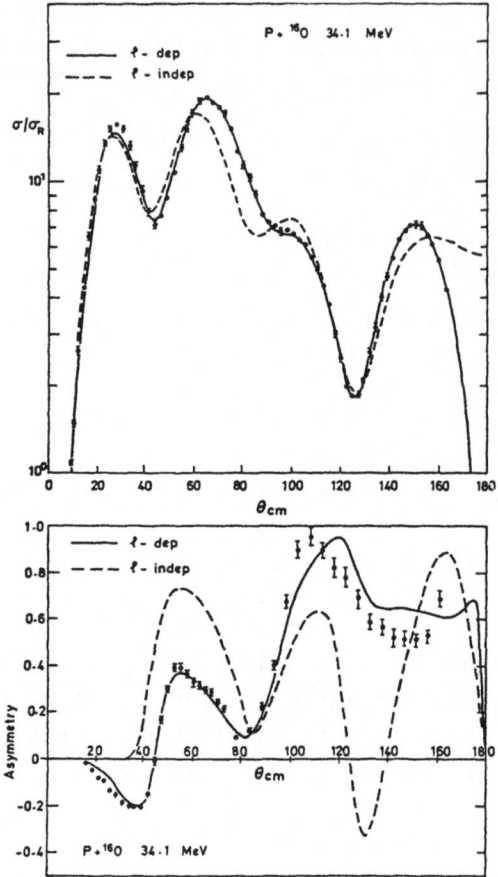

Figure 1
Proton scattering from ^{16}O at 34.1 MeV
showing improvement to fit when ℓ-dependent
terms included

exceptions to this pattern and there appear to be no alternative
solutions with different combinations of signs.

(3)   The repulsive real term is almost always peaked within the nuclear
surface and is generally narrow.

(4)   The absorptive term is always peaked at the far surface of the
nuclei and is also generally narrow. (That is for protons; for
helions the two terms peak at nearly the same radii).

(5)   The ℓ-dependent terms are particularly strong from 20-40 MeV
becoming less at higher energies. They appear to vary smoothly
through "resonances" which appear in the ℓ-independent parts of

the ℓ-dependent potentials.

(6) These "resonances" were a very well defined feature of the energy
     dependence of the quantities J_R, J_I and $<r^2>$ calculated from the
     ℓ-independent parts V^{LI} of the ℓ-dependent potentials. These
     quantities varied <u>much</u> more smoothly with energy than for
     ℓ-independent potentials and show both "resonance" regions and
     non-resonance regions over which the ℓ-dependent parts did not
     show rapid variations, see figure 2.

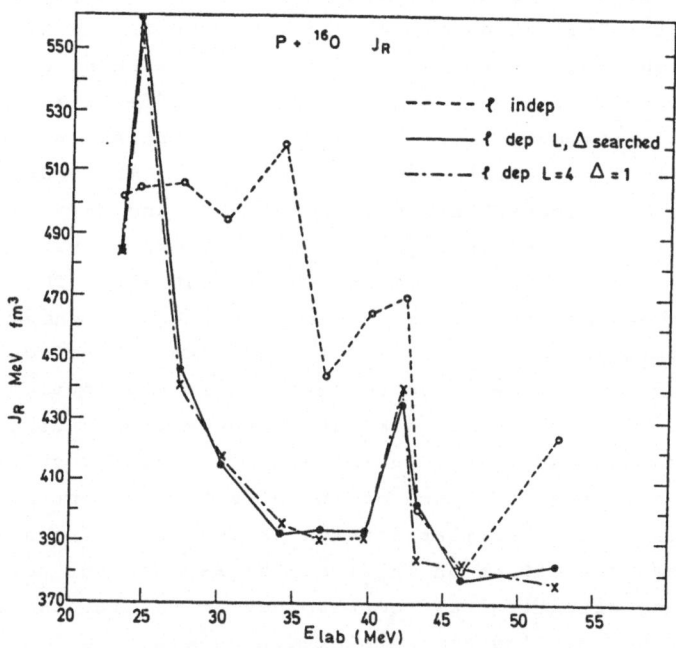

Figure 2
Energy dependence of the volume integral of the real part of
the ℓ-independent potential compared with that of the
ℓ-independent part of the ℓ-dependent potential. More recently
we have more cases that fit perfectly onto the smooth line.
We have clear "resonance" and non-resonance regions but the
ℓ-dependent term is much the same at 25 or 35 MeV

(7) The ℓ-dependent parameters show a discontinuity in magnitude but
     not in general character at the ^{16}O shell closure, as found in
     fitting proton elastic scattering at 35 MeV for the nuclei ^{12}C
     through ^{18}O.

In interpreting the narrowness of the ℓ-dependent derivative terms,
note that what we are finding are local equivalents to very complicated
non-local terms whose range of non-locality is not very small compared

to the sizes of light nuclei. They evidently have to generate reflection.

4.   Connection with Pickup (CRC) Processes

Calculations in which deuteron channels are coupled to the proton elastic scattering chennels suggest [6,7] that pickup channels make an important contribution to the proton optical potential. Indeed, for the case of proton scattering from ^{40}Ca at 30 MeV, these [7] CRC calculations readily fitted features of the backward angle differential cross section which could not be fitted with standard potentials. There is now a mounting body of evidence that deuteron intermediate states play an important role in nucleon scattering. We have now found a remarkable correspondence between pickup (CRC) effects and ℓ-dependent potentials which strongly implies that the latter is, in large part, a phenomenological representation of the former. The ℓ-dependent components in the potential and the pickup coupling have very similar effects on arg $(\eta_{\ell j})$ and on $|\eta_{\ell j}|$ as can be shown by switching off the respective terms without changing the potentials. The effects upon arg $(\eta_{\ell j})$ are compared in figures 3 and 4. It can be shown [4] that for proton scattering arg $(\eta_{\ell j})$ largely reflects the effect of the real potential. We find that when deuteron channels are coupled to the proton channels through a CRC calculation the corresponding perturbation in arg $(\eta_{\ell j})$ changes sign as a function of ℓ. This is sufficient to establish the ℓ-dependent nature of the effective potential generated by this coupling. In a similar fashion there is, [4], a close corres-pondence between the effect on $|\eta_{\ell j}|$ of the ℓ-dependent terms and the channel coupling effects. This corresponds to the imaginary ℓ-dependent term generated by coupling to the pickup channels. We thus conclude that ℓ-dependence is a phenomenological representation of pickup terms in the optical potential.

4.   Connection with Model Independent Fits

We can also make a connection between our ℓ-dependent potentials and model independent (spline interpolation) fits to proton elastic scattering. The results [8] of these spline interpolation fits are preliminary and do not yet correspond to fits which are as good as the ℓ-dependent potentials. This is particularly the case for p + ^{16}O. Nevertheless the fits are better than those for Woods-Saxon potentials. There is a strong tendency for spline potentials to have quite sharp dips (real part) and peaks (imaginary part) at the nuclear centre. These appear to be giving extra repulsion and absorption for low partial waves, just the properties of our ℓ-dependent terms. We are

193

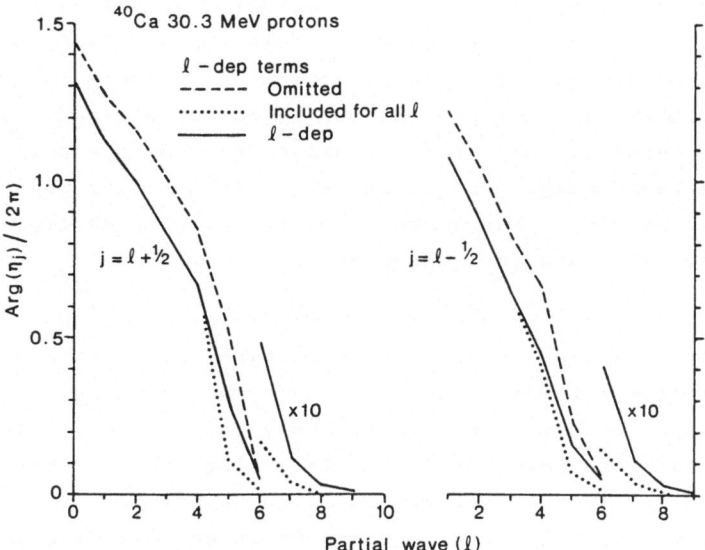

Figure 3
The effect on arg (η_ℓ) of turning off the ℓ-dependent terms
(keeping the rest of the potential fixed). For explanation of
dotted line see ref. [4].

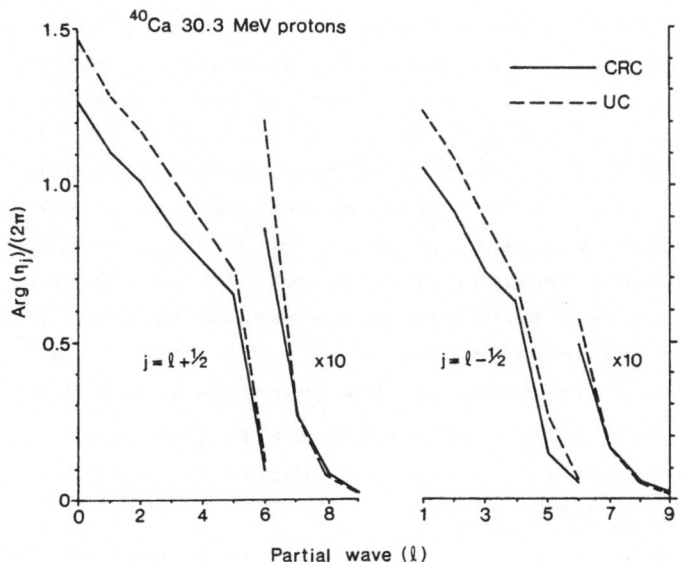

Figure 4
The effect on arg (η_ℓ) of turning off the coupling
to the deuteron channels (leaving the potentials fixed).

currently exploring the way in which scattering generated by ℓ-dependent
potentials may be fitted by spline-interpolation ℓ-independent potentials.
It appears to be the case that there exists an ℓ-independent potential
to fit any scattering. If this scattering is actually due to an
ℓ-dependent potential, then it is important to know how this is
reflected in the ℓ-independent potential.   This latter may be somewhat
pathological. We would like to know how to relate pathologies in our
spline fits to real data to ℓ-dependence.

5. General Discussion

We mentioned in the introduction two spin-offs of this work resulting
from the precision fits which we get. More specifically, precision
fits to analysing powers consistently imply surprisingly small values
of the spin-orbit radius parameters. This also appears necessary if
the level ordering in the lead region is to be correct, and, as Nix
has pointed out [9], this is of importance in establishing shell closures
in the superheavy region. There is also some evidence that anomalously
small spin-orbit radii are required in the oxygen region (we find this
small radius in almost all cases examined). This result and the
imaginary spin orbit term [1] appear to be closely connected with
pickup effects.

It is possible to construct physical models of p$\leftrightarrow$d processes which
suggest ℓ-dependence effects; they are, of course, dependent on the
finite spatial extension of the nucleus and the particular filled
orbitals. It is just the properties of this kind which cannot be
derived from the usual kind of local density folding model. We therefore regard our approach as a means of tapping more of the information
present but often unexploited in proton elastic scattering data in
order to explore the applicability of local density models. It is now
very desirable to have precision proton scattering data, going out to
backward angles and including analysing powers, at a wide range of
energies and for nuclei which are not necessarily closed shell.

We remark that some of our results fit in very well with the work
of Coulter and Satchler [10] who have studied inelastic contributions
to scattering in a different way.

Much of this work was done at the Nuclear Physics Laboratory,
Oxford.

References

[1]   R.S. Mackintosh and A.M. Kobos, J. Phys. G. $\underline{4}$ (1978) L135

[2]   R.S. Mackintosh and L.A. Cordero, Phys. Lett. $\underline{68B}$ (1977) 213

[3]   A.M. Kobos and R.S. Mackintosh, J. Phys. G., in press

[4]   R.S. Mackintosh and A.M. Kobos "The Physical Interpretation of Angular Momentum Dependent Potentials", Daresbury preprint

[5]   C.M. Perey and F.G. Perey, Atomic Data and Nuclear Data Tables $\underline{17}$ (1976) 1

[6]   R.S. Mackintosh, Phys. Lett. $\underline{44B}$ (1973) 437; Nucl. Phys. $\underline{A230}$ (1974) 195

[7]   R.S. Mackintosh and A.M. Kobos, Phys. Lett. $\underline{62B}$ (1976) 127

[8]   A.M. Kobos, R.S. Mackintosh and R. Dymarz, work in progress

[9]   J.R. Nix, "Superheavy Elements", Proc. Int. Conf. on Dynamical Properties of Heavy-Ion Reactions", Johannesburg, 1978. LA-UR-78-1689

[10] P.W. Coulter and G.R. Satchler, Nucl. Phys. $\underline{A293}$ (1977) 269

MICROSCOPIC ANALYSIS OF p-^{40}Ca ELASTIC SCATTERING AT 30.3 MeV

A. Bouyssy[*], H. Ngô, N. Vinh Mau
Division de Physique Théorique[+], Institut de Physique Nucléaire -
91406 ORSAY Cedex, France

The main properties of the nucleon-nucleus optical potential can be obtained in a microscopic theoretical approach. On the
other side an optical model analysis, assuming for the real and imaginary parts of the potential, given analytic forms, fits parameters
to experimental data. The set of such parameters, obtained by a best
fit of cross sections is not unique and the comparison between theoretical and phenomenological potentials is not sufficient to test the
reliability of the microscopic model. That is why we have performed,
with our potential, a direct analysis of p-^{40}Ca elastic scattering
experimental data. Even though a large energy range (10-60 MeV) has
been studied, we restrict ourselves here to one energy. We have
chosen the incident energy E_O = 30.3 MeV which lies above most of the
target RPA collective states so that the most important inelastic
channels are included. On the other hand, this energy is not large
enough for the assumptions to break down.

In many body theory framework, N. Vinh Mau[1] has chosen a
Green's function formalism to derive a fully antisymmetrized potential. The whole derivation is based on the RPA. The first results[2] obtained for the absorptive part of the optical potential with a zero
range effective nucleon-nucleon interaction were satisfactory but
only qualitative. It was interesting to go further and to calculate
the whole potential with a more realistic finite range effective
interaction. Then we have calculated[3] the non local energy dependent
optical potential and tested its sensitivity to the choice of the
effective interaction. The results are presented here for the two
particular cases :

[1] : for a Gillet Sanderson[4] (GS) effective interaction : it
has been adjusted to reproduce energies and RPA wave functions of the
lowest collective ^{40}Ca excited states. We have used it to calculate
both the real and the imaginary potentials.

---

[*]Institut des Sciences Nucléaires, B.P. 257, Centre de Tri ,
38044 Grenoble-Cedex

[+]Laboratoire associé au C.N.R.S.

[2] for a Reichstein-Tang[5] (RT) one : less specific to ^{40}Ca than the above one, it gives better inelastic DWBA[6] form factors. It has been adjusted on nucleon-nucleon scattering length and effective range at low energy.

To show our main results as simply as possible, we have determined, besides a non local potential, its local equivalent form according to Perey Saxon[7] approximation . Many tests[9] performed on this method have proved its validity when the range of the non locality is smaller than the range of the potential, what is true for p-^{40}Ca potential.

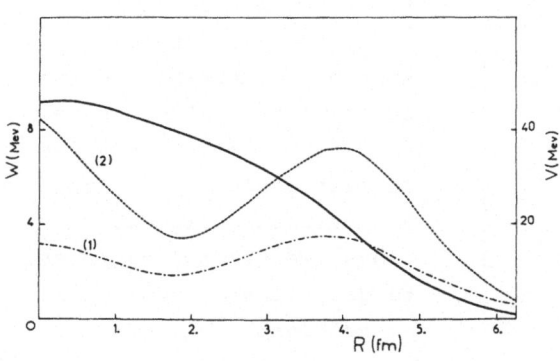

Fig.1: local equivalent optical
        potential :
        ——Real part ; [1]——·—— Imaginary
part with GS effective interaction ; [2]---- Imaginary part
with RT effective interaction.

On fig.1 are plotted the absorptive potentials obtained with both inter-actions, and the real poten-tial evaluated with the GS one. The absorptive part of the potential varies strongly with the effective interaction. This differen-ce between the potentials leads to a factor 2 between the volume integrals, while the root mean square radius, in good agreement with phe-nomenological results, is almost independent of the interaction :

$$\langle r_w^2 \rangle^{1/2} = 4.60 \text{ fm} \qquad (J_w)_{RT} = 103 \text{ MeV fm}^3$$

while the experimental values are about 4.7 fm for the RMS and 120 MeV fm^3 for the volume integral.

For the real potential, we reproduce well the root mean square radius while the volume integral is about 10 % smaller than phenomenological results[8].

Our imaginary potential has not the same shape as the phe-nomenological one, even though it is possible to distinguish a

surface term and a volume term. For instance the surface contribution
to the absorptive potential is nearly twice as broad as the surface
term of Becchetti-Greenlees[8]. So the best critical discussion on our
potential comes from a direct analyses of experimental data. The
differential elastic and total reaction cross sections have been calculated with the local potential equivalent to our non local one. The
spin orbit potential plays a non negligible role in the calculation
of the cross section. As we have not calculated this term we have
added to our potential a phenomenological[8] spin orbit component.
The differential cross sections are plotted in fig.2a. They show an
overall good qualitative agreement with experimental data, quite
satisfactory since the calculation include no adjustable parameter.

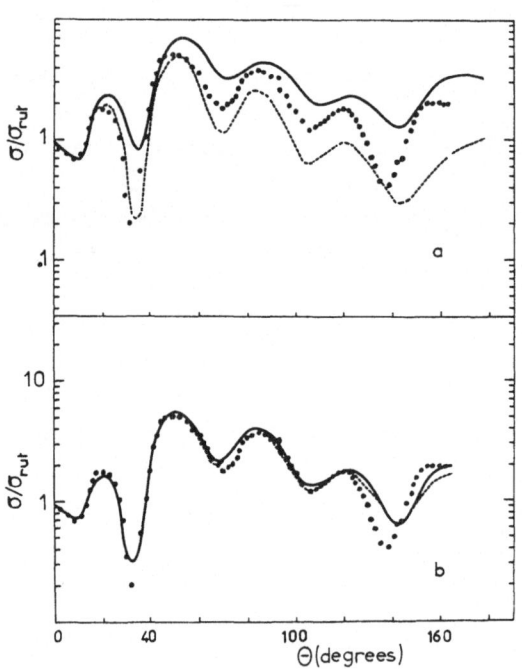

<u>Fig.2</u> : Differential elastic scatte-
ring cross sections :
—— Calculation made in case
[1] ; ---- calculation made
in case [2] ; a) without
renormalization of the potential, b) best fit after
renormalization.

In the two first columns of
the table we report the
total reaction cross
section compared with the
experimental value. As we
take into account only the
inelastic channels given by
RPA particle hole excita-
tions, we are not surprised
to get slightly too small
absorptions. It is then
interesting to evaluate to
what extent the potential
must be modified to remove
this discrepancy : we have
simply renormalized separately the real and absorptive
parts of the potential in
order to obtain the best
fits both for the angular
distribution and for the
total reaction cross section.
The renormalization factors,
α and β for the real and
imaginary potentials respectively, are given in the
third and fourth columns of
the table while the corresponding cross sections are

plotted in fig.2b.

		σ_R(mb) initial	σ_R exp (mb)	Real: α	Imaginary: β	σ_R(mb) after renormalization
[1]	GS	632	915 ± 38	1.12	1.80	850
[2]	RT	890		"	.95	889

The value 1.12 of α is quite large and tells that our real potential
is not deep enough. An explanation for this may be that the harmonic
oscillator density we have used, although it gives the right nuclear
radius, does not reproduce very accurately the nuclear density : as
the main contribution to the real potential is the Hartree-Fock term,
which is strongly connected to the density, the potential has too
large a diffuseness. Then the introduction of a renormalization factor, leading to an increase of the potential at the nuclear surface,
removes at least partly this discrepancy. It shifts the curves
upwards and shows that, with very little renormalization, the absorption obtained in the case [2] fits quite accurately the data. In
the case [1] of the GS effective interaction, we can see that a
quite large renormalization factor is needed to get a correct
absorptive potential. Nevertheless the reaction cross section,
although strongly connected to the absorption, seems to be less sensitive to the choice of the effective force since a rate 2 between
the absorptive potentials leads to about 25 % difference for σ_R.

However, even after renormalization, it remains a small
discrepancy at backward angles for both effective interactions. This
could come from the approximation used to derive a local equivalent
potential. Furthermore a simple renormalization is, perhaps, not the
kind of modification which gives the most accurate results.

1) N. Vinh Mau, Theory of nuclear structure (IAEA, Vienna 1970) p.931
2) N. Vinh Mau and A. Bouyssy, Nucl. Phys. A229 (1974) 1
3) A. Bouyssy, H. Ngô and N. Vinh Mau, to be published
4) V. Gillet and E.A. Sanderson, Nucl. Phys. A91 (1967) 292
5) I. Reichstein and Y.C. Tang, Nucl. Phys. A139 (1969) 144
6) G.R. Satchler, Z. Physik 260 (1973) 209
7) F.G. Perey and D.S. Saxon, Phys. Lett. 10 (1964) 107
8) F.D. Becchetti and G.W. Greenlees, Phys. Rev. 182 (1969) 1190
9) R.E. Peierls and N. Vinh Mau, to be published.

A SEMI-PHENOMENOLOGICAL ANALYSIS OF PROTON ELASTIC SCATTERING

A. Tarrats and J.L. Escudié
Département de Physique Nucléaire
CEN Saclay, BP 2, 91190 Gif-sur-Yvette, France

The analysis of nuclear reactions usually involves, through the distorted waves, a description of the elastic scattering. A part of the difficulties of these analyses has its origin in the inadequacy of this description. For example, ambiguities in the phenomenological optical models and lack of consistency in the description of the elastic and inelastic scatterings often plague the extraction of the nuclear structure information from the reaction data. Thus a careful analysis of the elastic scattering using optical potentials other than purely phenomenological ones is very important from the view point of general nuclear reaction studies. As a matter of fact, it were indeed the problems we encountered in the analysis of other nuclear reactions that urged us to undertake the present semi-phenomenological analysis of the elastic scattering. Besides, it is legitimate to hope that such analyses could provide some information or at least enable one to place some constraint on the effective bound-free N-N interaction in some situations.

In recent years, the field of optical model has seen a considerable renewal of theoretical interest and is presently going from a qualitative theoretical understanding to some convincing quantitative description. Taking advantage of the great amount of data that we and other experimentalists have collected, often as a by product, we chose a much more limited point of view : between the standard phenomenological optical potential and these fundamental theoretical works, based on realistic N-N interactions, in the following we adopt a semi-phenomenological approach which is an improvement over the early folding models. We limit ourselves to the first order, so that the description of the experimental data depend on the nuclear density and some effective interaction, which we take as density dependent and a slowly varying function of energy.

I. THE EQUIVALENT LOCAL OPTICAL POTENTIAL

Given a two-body interaction $t(|\vec{r}-\vec{r}'|)$, the first order optical potential acting on the incident particle described by the wave function $\psi(\vec{r})$ may be written as :

$$\mathcal{V}(\vec{r})\ \psi(\vec{r}) = V_D(\vec{r})\ \psi(\vec{r}) + \int V_E(\vec{r},\vec{r}')\ \psi(\vec{r}')\ d\vec{r}', \qquad (1)$$

where

$$V_D(\vec{r}) = \int \rho(\vec{r}') \; t(|\vec{r}-\vec{r}'|) \; d\vec{r}' \qquad (2)$$

with $\rho(\vec{r})$ being the nuclear density, and

$$V_E(\vec{r}) = -\rho(\vec{r},\vec{r}') \; P_6 \; P_\tau \; t(\vec{r}-\vec{r}') \qquad (3)$$

with P_6 and P_τ being the spin and isospin exchange operators, and $\rho(\vec{r},\vec{r}')$ the Dirac mixed density.

The analysis of elastic scattering can be performed using the above non-local potential and, as a reference for testing the following approximation, we indeed made some calculations this way. But the calculations are lengthy and expensive. Moreover the resulting potentials are difficult to handle, particularly for making a comparison with the phenomenological ones. In the following we replace the effective interaction in the exchange term by some simple approximation that leads to a local potential. Replacing the force by its zero-range limit yields a trivial local term, but this crude approximation loses the characteristic features of the localisation of a non local potential, namely the Perey effect.

Proceeding one step further, that is using an approximation of the force quadratic in the momentum space

$$\mathcal{W}(q) = a - b \; q^2 \qquad (4)$$

we used, following Perey [1] for example, a general method of localisation in which the term of the Schrödinger equation which includes the derivative of the wave function is eliminated by an ad-hoc change of function. This leads an effective mass

$$\frac{m^*}{m} = \frac{1}{1 - b \; \rho(r)} \qquad (5)$$

that accounts for the Perey effect, and a local equivalent potential

$$\mathcal{V}_{LE}(r) = \frac{1}{1 - b \; \rho(r)} \left\{ V_D(r) - \right.$$
$$\left. - b \; \rho(r) \; E - \frac{1}{4} b^2 \frac{\rho'(r)^2}{1 - b \; \rho(r)} + a \; \rho(r) - b(\tau(r) + \frac{1}{2} \nabla^2 \rho(r)) \right\} \qquad (6)$$

where $\tau(r)$ is the local kinetic energy density.

The energy dependence of this potential is contained for one part in the term where the energy E appears explicity, for another part by the implicit energy dependence of a and b.

As for the determination of the parameters a and b, that is the fitting of the force in the relevant momentum region by a quadratic form, we use a semic-classical recipe, empirically tuned to give the best possible fit to the exact non local calculation. This gives satisfactory results for the elastic observables and even for the distorted waves inside the nucleus.

The spin orbit term also gives an exchange contribution, the approximation of which provides a Thomas shape potential.

II. THE NUCLEAR DENSITY DISTRIBUTIONS

Different types of density were used, with equal or different proton and neutron densities, mainly based either on the experimental charge form factors, or on the droplet model derived by Myers [2]. Though this type of calculation in this energy region is not the best way to provide information about proton-neutron radii differences, the agreement with experimental data was sizeably improved by using different proton and neutron densities, and we generally used either the droplet model for both proton and neutron densities, or the proton densities from experimental charge form factors, together with the neutron-proton differences predicted by the droplet model. Some calculations have also been performed with all the densities calculated from nuclear wave functions computed in a Woods-Saxon well.

III. THE EFFECTIVE INTERACTION

Our aim is, starting from a reasonable guess of the interaction, to derive, by empirical adjustment, an ad-hoc interaction. We claim provided that the starting point reproduces the general properties of the transition matrix, the adjustment will give an interaction which it will make sense to compare with a realistic t-matrix.

Namely, we used as a starting point of the real part of the interaction, a two Yukawa force close to the Y3P derived by the MSU group [3] which is an approximate reaction matrix in a low energy density region, and which proved useful for the description of the inelastic scattering. Moreover, from its success in structure calculations, its need in realistic transition matrix calculation, and its phenomenological need evidenced by Myers [4] to get realistic potential shapes, we found it necessary to introduce some density dependence in the real force. We choose to multiply the entire finite range force by a factor :

$$F_{DD} = 1 - C \, \rho(r)^{2/3} \qquad (7)$$

as introduced by many authors [5].

Concerning the imaginary part, the starting point is a 0 - range force determined by the optical theorem :

$$\text{Im}(t(0)) = \frac{4\pi}{k}\,\bar{\sigma}$$

where $\bar{\sigma}$ is an energy averaged reaction cross-section obtained from the parametrisation A/(B+E) taking into account the fact that the Pauli principle forbids final momentum of the scattered particle lower than the Fermi momentum, defined in the local density approximation. With the chosen parametrization, this average is analytical [6] and we introduced some simple corrections due to the presence of nuclear matter (other than Pauli principle effects) [7]. This simple model already accounts for the transition of the absorbing potential from surface to volume shape with the increase of the incoming energy. Besides, as it is clearly necessary on fundamental as well as phenomenological grounds, a finite range has been introduced in the imaginary force, and is considered as a free parameter in the subsequent fitting procedure.

Calculating the elastic observables from the optical potential defined in I, using the interaction just mentioned, we explored the sensitivity of the observables to the different parameters, and retained in the fitting procedure those for which the sensitivity is sizeable, namely :

- for the real potential a global weight for the central part P_c
  an isospin dependence (V_{SE}/V_{TE}) R_{ST}
  the root mean square range μ
  the density dependence parameter C
  a weight for the LS part P_{LS}
- for the imaginary part a global weight P_i
  the rms range λ.

An automatic search of these parameters has been performed to reproduce simultaneously a dozen of experimental sets. Such an adjustment was made separately around 20, 25, 30, 35 and 50 MeV and it shows a smooth variation of the parameters.

IV. RESULTS

Though the described fitting procedure is not yet completed we obtained some significant results concerning the effect of different parameters on the optical potential and the elastic observables, and the variation with the incident energy.

The effects of exchange and of the density dependence of the interaction are shown in Figs. 1 and 2. It appears that they have similar effects on the geometry of the potential, but opposite effects on

its volume, and that, including both of them, we obtain potentials similar to the phenomenological ones and a good agreement with the experimental data.

Likewise, Fig. 3 features the effect of the finite range correction for the imaginary part and the sensitivity of the cross-section
to the parameters A and B demonstrating the need of the finite range
correction, and the variation of surface to volume form of the imaginary potential.

Figs. 4 and 6 show the results for different energies and nuclei, obtained with the same set of parameters slowly varying with the
incident energy. In particular, the density dependence parameter C and
rms μ are slowly decreasing ; so does the global weight for the real
part, with a slope about - 0.003. The experimental data used were collected for one part at Saclay, for another part from the Eindhoven
and Milan groups, whom we thank for sending it prior to publication.

So far, we have neither used the presently available realistic
t-matrix nor we have compared them with our fitted t-matrix. We intend
to do it in the near future.

REFERENCES

[1] F. Perey and B. Buck, Nucl. Phys. 32 (1962) 353.

[2] W.D. Myers, Nucl. Phys. A145 (1970) 387.

[3] G. Bertsh et al., Nucl. Phys. A284 (1977) 399.

[4] W.D. Myers, Nucl. Phys. A204 (1973) 465.

[5] A.M. Green, Phys. Lett. 24B (1967) 384.

[6] E. Clementel and C. Villi, Il Nuovo Cimento II (1955) 176.

[7] C. Gomes, Phys. Rev. 116 (1959) 1226.

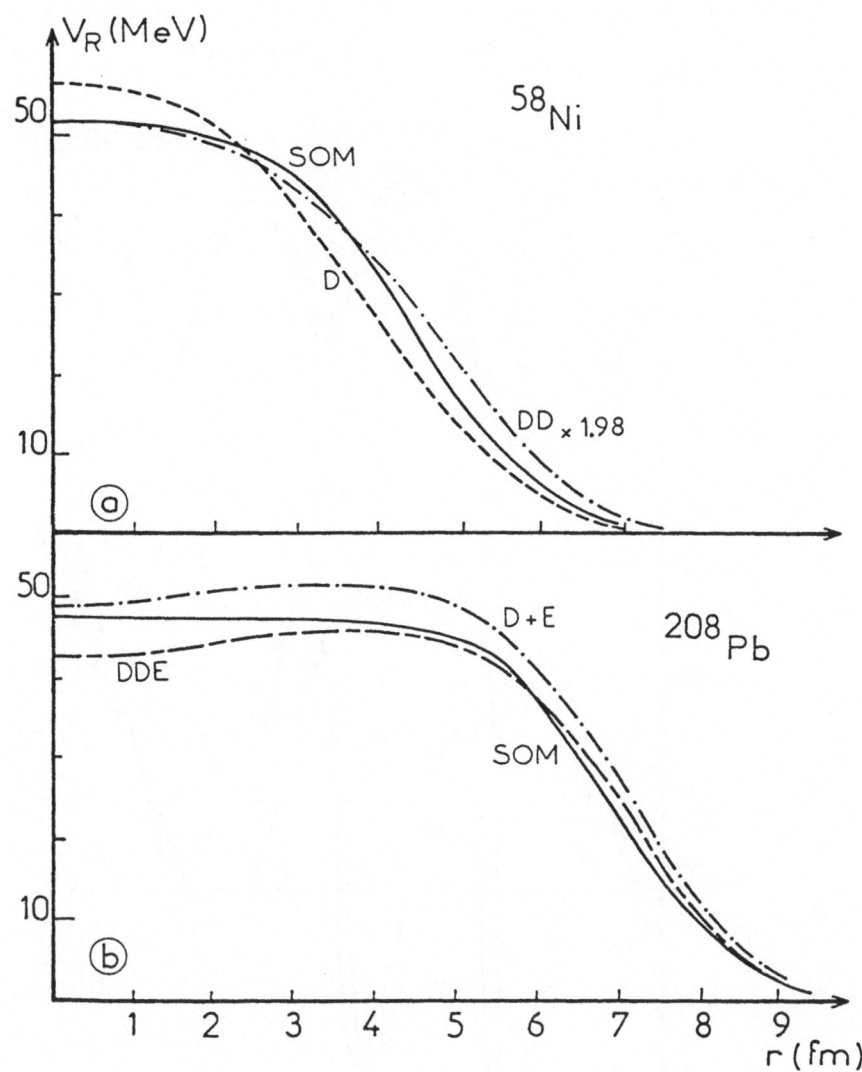

Fig. 1 - The real part of the optical potential at 24.5 MeV, as obtained :

- from a phenomenological calculation (SOM)

- without exchange or density dependence effect (D)

- without exchange with the same density dependence term as
  Myers (C = 2) with a normalization factor 1.98 (DD)

- with the exchange contribution (D+E)

- with the exchange and density dependence contributions (DDE)

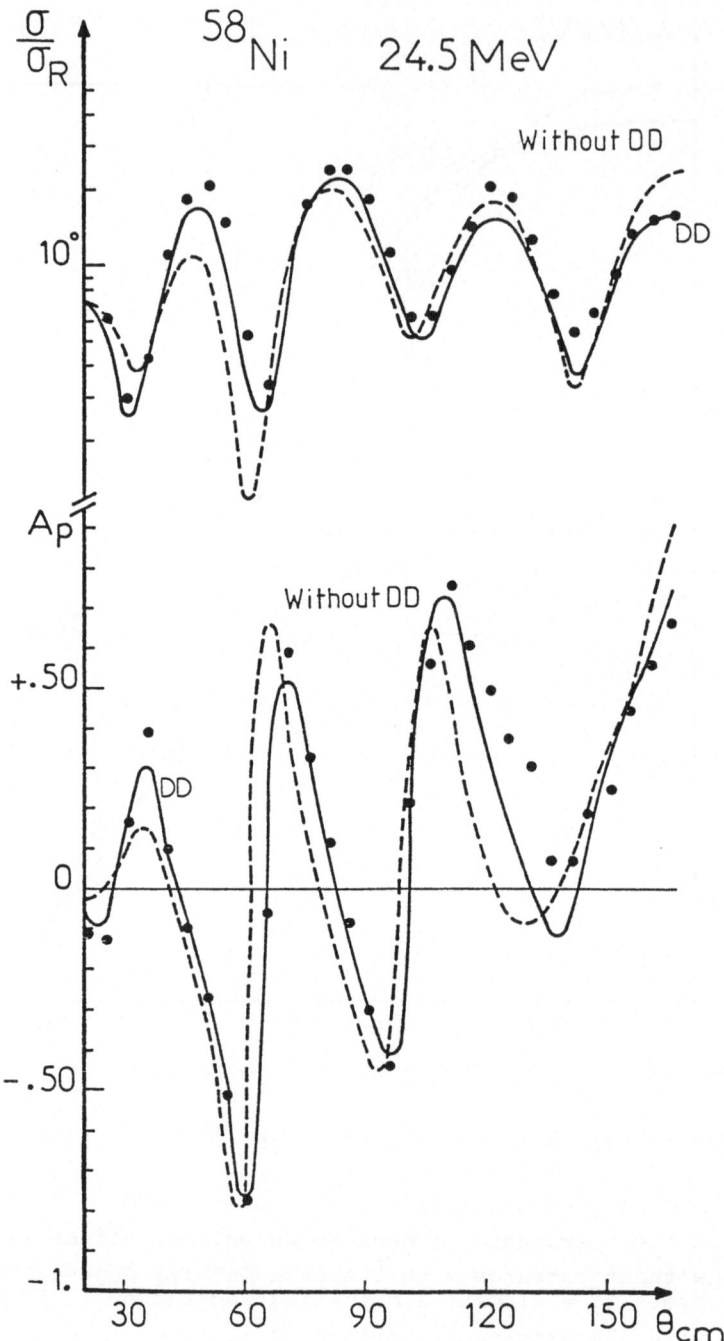

Fig. 2 - The results for ^{58}Ni at 24.5 MeV from calculations without (without DD) and with (DD) the density dependence factor

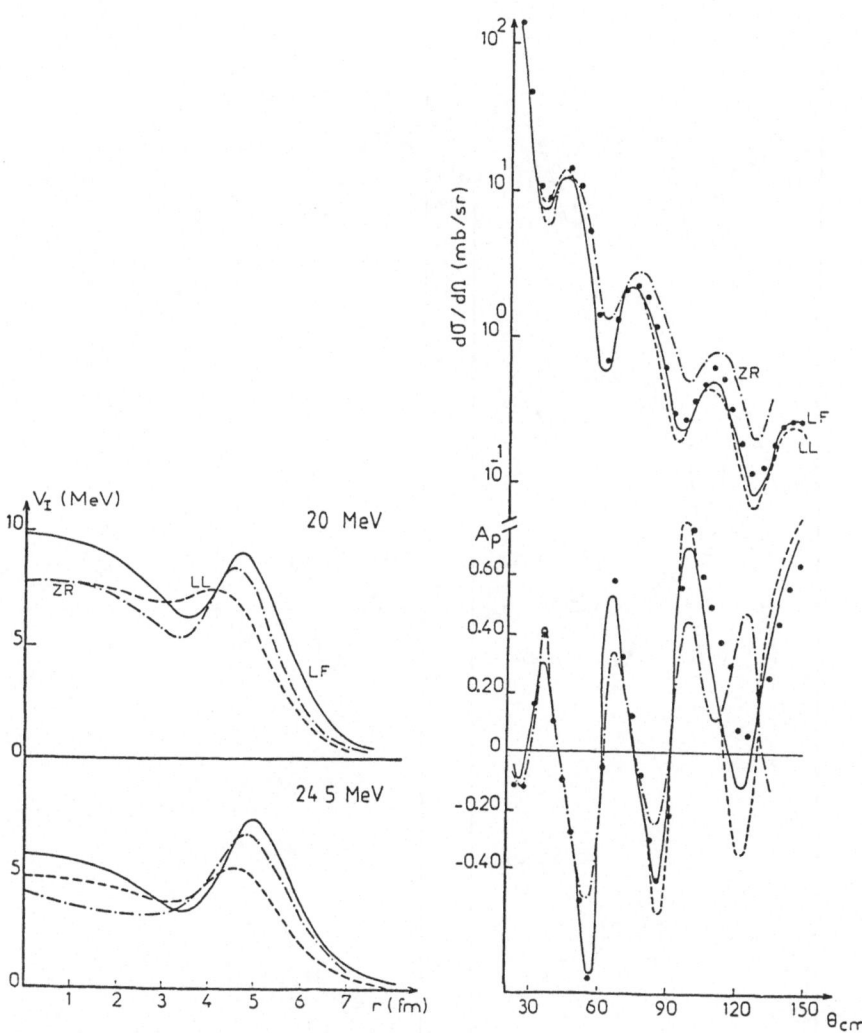

Fig. 3 - The imaginary part of the optical potential for ^{58}Ni at 24.5 and 40 MeV and the observables for ^{58}Ni at 24.5 MeV using either zero-range (ZR) or a finite range ($<r^2> = 1.5$ fm^2) ; imaginary part with A and B parameters taken from free N-N scattering (LL) or density dependent A and B parameters (LF).

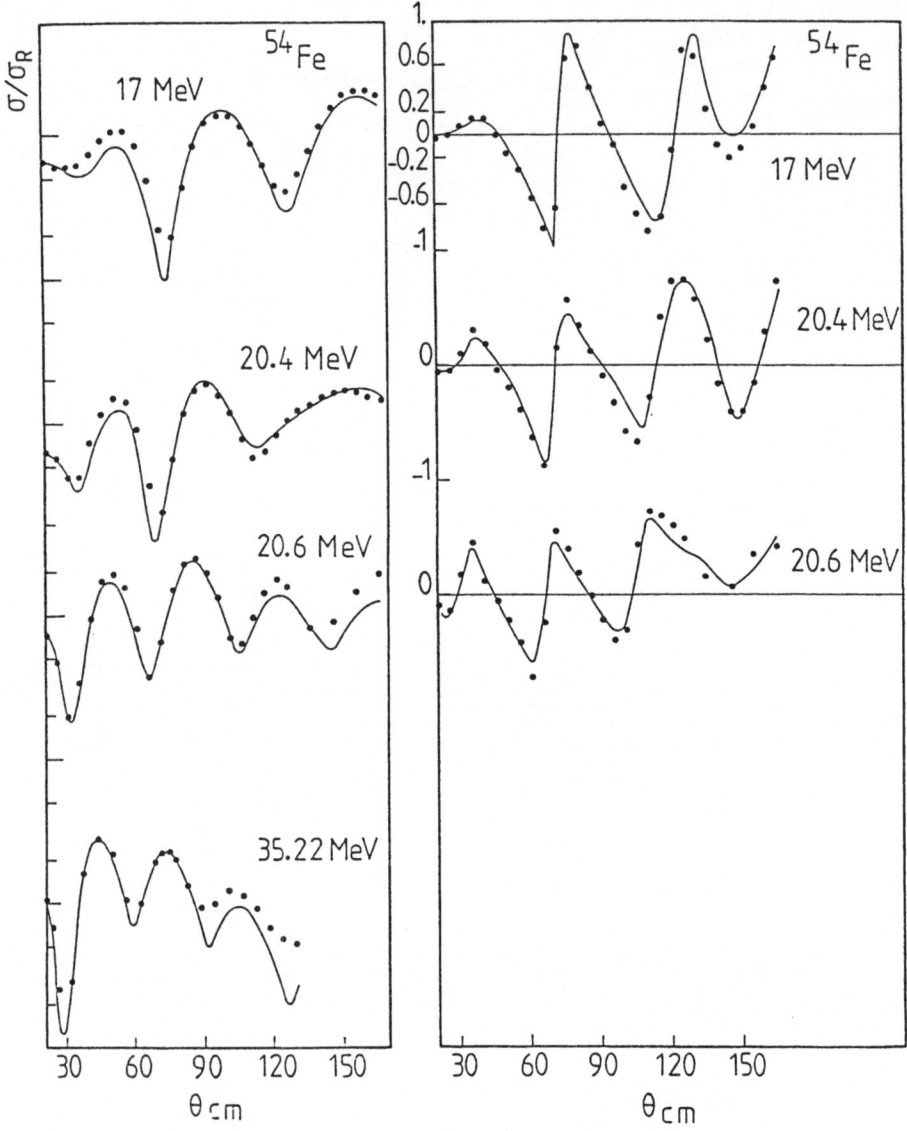

Fig. 4 - Differential cross-sections and asymetries for ^{54}Fe in the 17-35 MeV energy range, using parameters slowly varying with the in-coming energy.

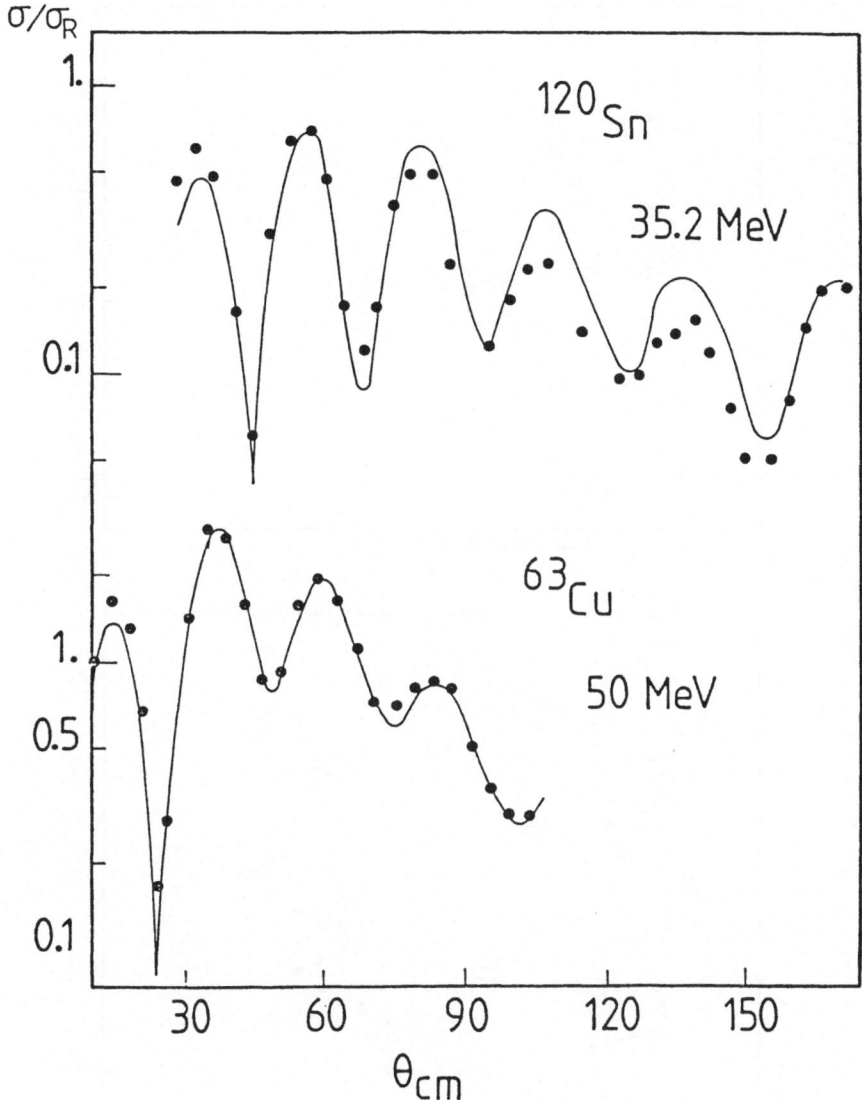

Fig. 5 - Differential cross-sections for ^{120}Sn at 35.2 MeV and ^{63}Cu at 50 MeV.

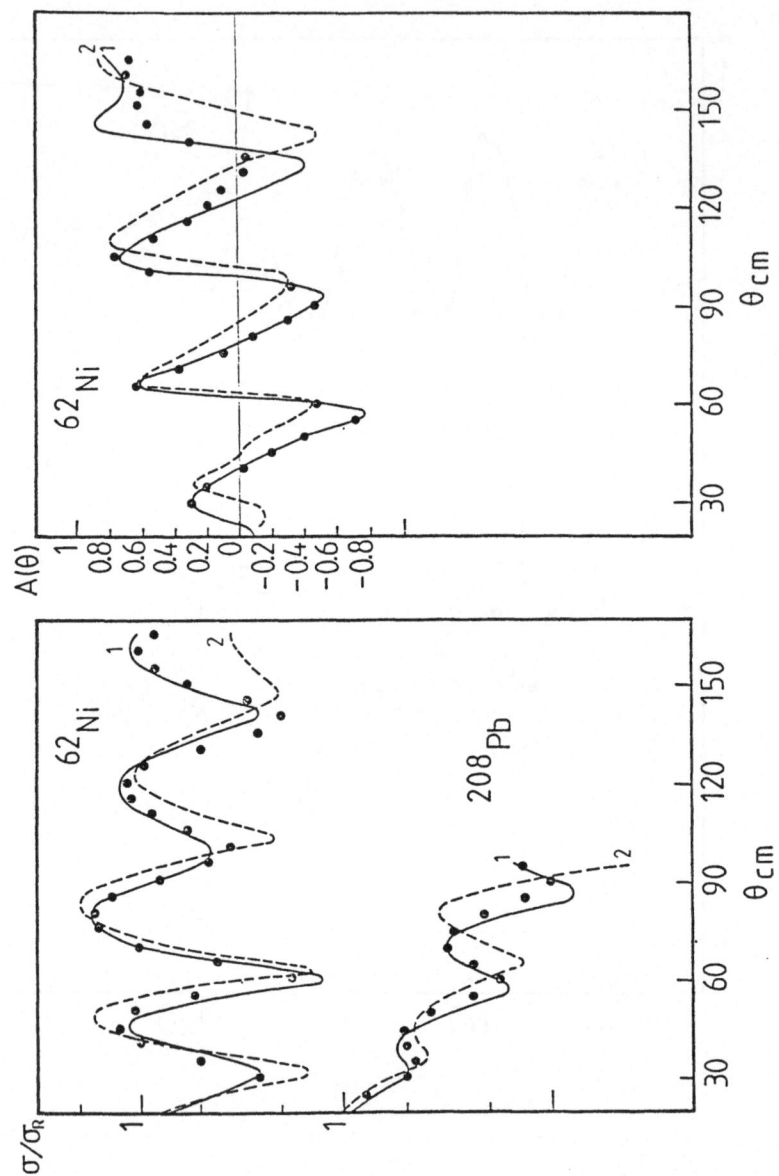

Fig. 6 - Differential cross-sections and asymetries for 24.6 MeV proton scattering obtained with (1) or without (2) the exchange contribution.

PROMINENT FEATURES OF PROTON ELASTIC SCATTERING ON NUCLEI BELOW A=70 AT INCIDENT ENERGIES BETWEEN 10 AND 50 MeV.

E. Fabrici, S. Micheletti, M. Pignanelli and F. Resmini
Istituto di Fisica dell'Università and INFN, Milano, Italy.

R. De Leo, G. D'Erasmo and A. Pantaleo
Istituto di Fisica dell'Università and INFN, Bari, Italy.

1 - INTRODUCTION

Recent progress in folding model calculations[1,2,3] has brought renewed interest to nucleon scattering on light nuclei. These calculations are particularly justified in the case of proton scattering on light magic nuclei, because of the poor fits to the differential cross sections obtained by a standard optical model calculation.[4,5] However when a new calculation is tested, as recently done,[2,6,7] comparing its results with a very limited set of data, actually with proton scattering on light doubly magic nuclei at 30 MeV, misleading conclusions could be obtained.

For this reason we have collected a larger sample of data covering nearly every isotope in the light mass region.[8-11] In the two following sections the data are reviewed and a classification on the basis of their relevant features is attempted.

2 - ENERGY DEPENDENCE OF PROTON ELASTIC SCATTERING

Proton elastic scattering experiments at incident energies between 10 and 50 MeV on nuclei with mass numbers $A \leqslant 40$ are here shortly described. Only fairly complete measurements will be considered, namely: full angular range differential cross sections; data collected at several incident energies in the same laboratory, consistent data when obtained in different laboratories, analyzing power at least at some incident energies. Light nuclei which fulfil these requirements are: ^{9}Be (UCLA, Nucl. Phys. A199(1973)433; ibidem A157(1970)145), ^{12}C (Jülich, Milano, Z. Physik, A274(1975)339; Ref. 8), ^{14}N (UCLA, Nucl. Phys. A198(1972)257; ibidem 58(1964)32; A166(1971)378, Ref. 11), ^{15}N (Milano, 1978, Ref. 11), ^{16}O (UCLA, Ratherford Lab., Phys. Rev. 167(1968)908; Nucl. Phys. A132

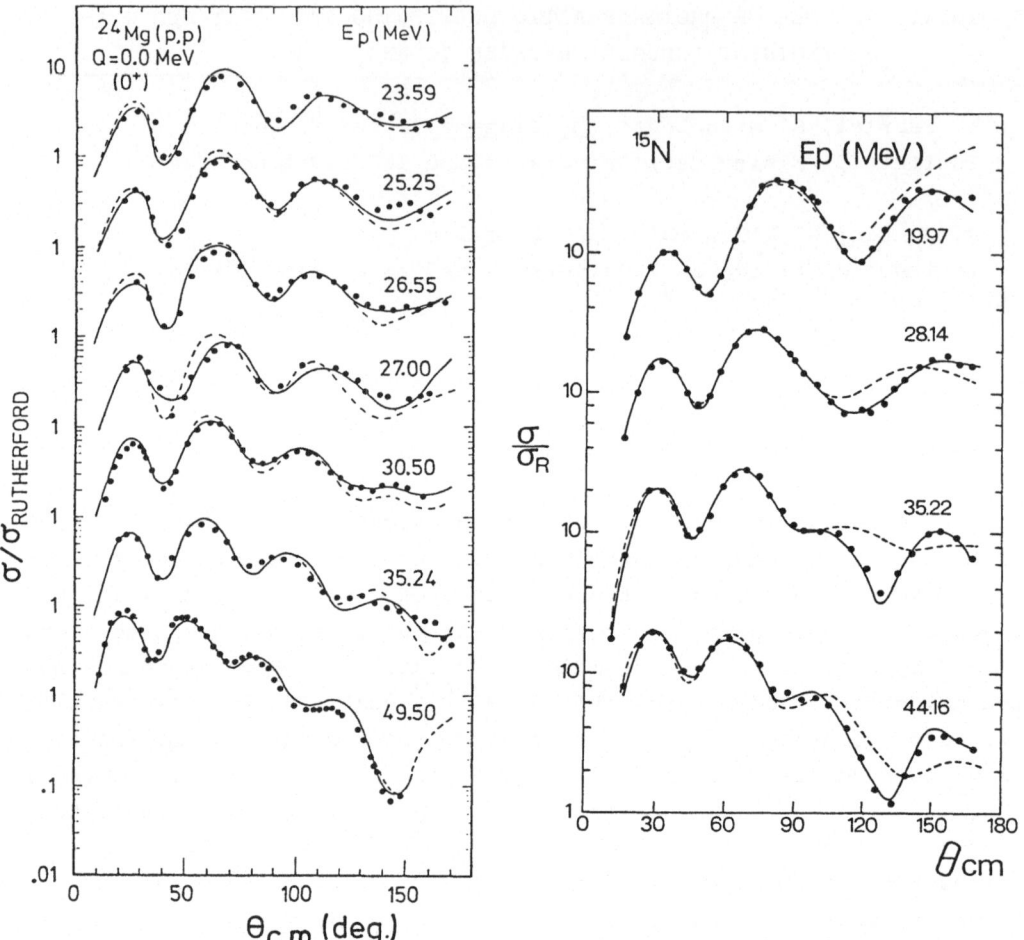

Figure 1 - Proton elastic scattering cross sections at different
incident energies and optical model
fits obtained with standard geometrical parameters. The full lines
are the results obtained by adjusting the well depths at each incident energy, the dashed lines by
using instead average values with
a linear energy dependence. The
agreement is typical of a deformed
nucleus.

Figure 2 - ^{15}N differential cross
sections and optical model fits
given as an example of proton
elastic scattering on a spherical
light nucleus. The dashed lines
give the fits obtained by using
standard parameters, while the
continuous lines give the results
of non-conventional potentials
(R_w=1.45A$^{1/3}$ fm, a_w=0.2-0.4 fm,
R_{so}=1.62A$^{1/3}$fm).

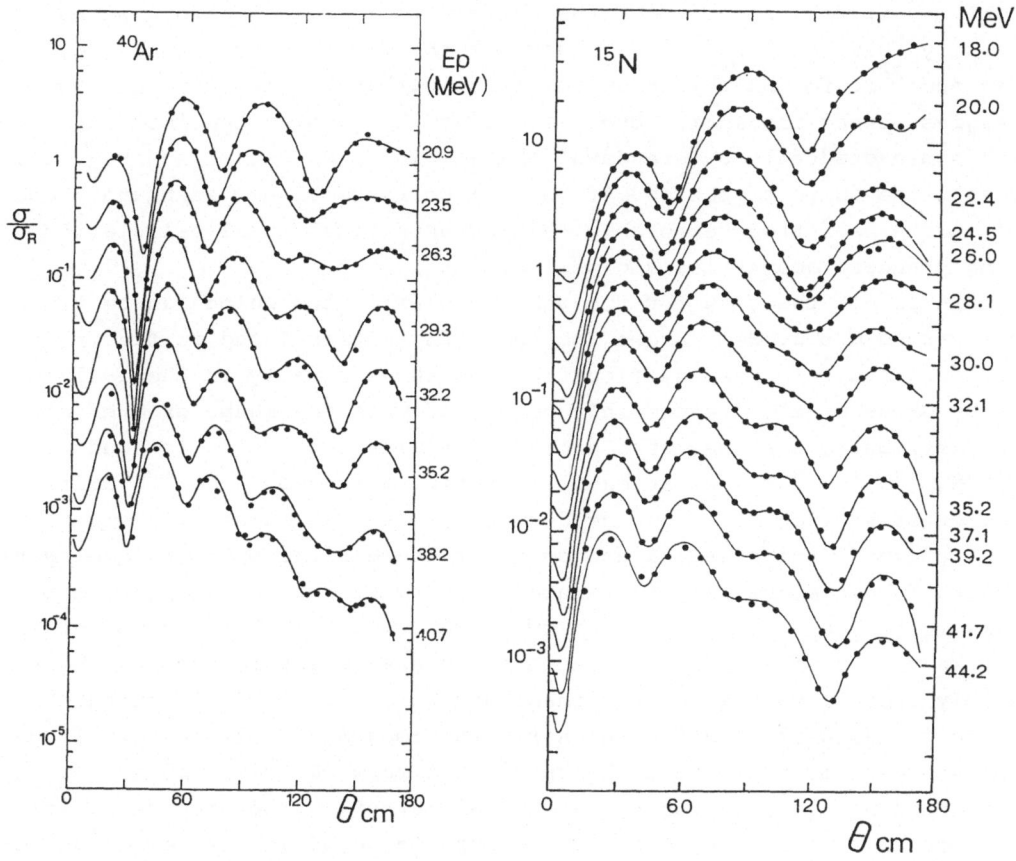

Fig. 3 - Elastic scattering differential cross sections (points)
together with the results of a phase shift analysis (full lines).
The angular distributions are drawn displaced vertically proportionally to incident energy increments. True cross section values
are obtained multipling by $\exp(0.37(E_p-20.9))$ and by $\exp(0.3(E_p-18))$
for ^{40}Ar and ^{15}N respectively.

(1969)348), ^{18}O (Milano, 1978, Ref. 11), ^{24}Mg (Jülich, Milano, Nucl.
Phys. A268(1977)12, Ref. 10), ^{28}Si (Milano, 1978, Ref. 9), ^{40}Ar
(Milano, 1978, Ref. 11) and ^{40}Ca (Oak Ridge, Manitoba, Ref. 4; Nucl.
Phys. A167(1971)57).

From a first inspection of the differential cross sections we can classify these nuclei in two groups. In the first we can place ^{12}C, ^{24}Mg and ^{28}Si for which, in spite of noticeable fluctuations at low energies, the differential cross sections, increasing the incident energy, assume gradually a "standard" shape (Fig. 2 of Ref. 8 and Fig. 1). We call the shape of these differential cross sections "standard" because it can be fitted reasonably well by an optical model calculation using standard geometries and well depths.

This is not the case for the other nuclei listed which can be placed in a second group. Well known are the difficulties found in fitting ^{16}O and ^{40}Ca data.[4,5] These difficulties, as shown clearly in the early paper by Gross et al.,[4] are mainly connected with the shape of the angular distributions at backward angles. A similar situation is found for ^{14,15}N, ^{18}O and ^{40}Ar. As an example in Fig. 2 is shown the ^{15}N differential cross section at a few incident energies. The dashed lines give the fits obtained with standard and constant geometrical optical model parameters. The data are well reproduced at forward angles, but not in the backward region. Improved fits (continuous lines) can however be obtained also keeping Saxon-Woods form factors. Several prescriptions can be profitably used; a radius for the imaginary well much larger than that for the real one, a small diffuseness for the imaginary well, a very large (and unphysical) radius for the spin-orbit term (R_{so}=1.4-1.6 $A^{1/3}$). The third prescription is probably a trivial way of introducing an additional angular momentum dependence. The importance of angular momentum dependent terms in optical potentials has been stressed by Mackintosh and collaborators.[7]

As noted above, the disagreement between experiments and optical model predictions is more evident at backward angles. In Fig. 3 the differential cross sections for proton elastic scattering on ^{15}N and ^{40}Ar are given. The curves reported in the same figures are the results of a phase shift analysis. The "anomalous" backward peak appears above 25 MeV, moves at first toward larger angles and then stays essentially fixed as shown in Fig. 4. It is still present at 44 MeV for ^{15}N and at 50 MeV for ^{14}N and ^{16}O, while it clearly disappears above 40 MeV for ^{40}Ar (Fig. 3) and ^{40}Ca (Ref. 4). It must be remarked that the effect appears for all nuclei at 26-27 MeV.

Finally one can observe that the nuclei showing an anomalous elastic angular distribution are all spherical (^{14,15}N, ^{16}O, ^{40}Ar and ^{40}Ca) or slightly deformed (^{18}O), while those of the first group (^{12}C, ^{24}Mg and ^{28}Si) are known to have large permanent deformations.

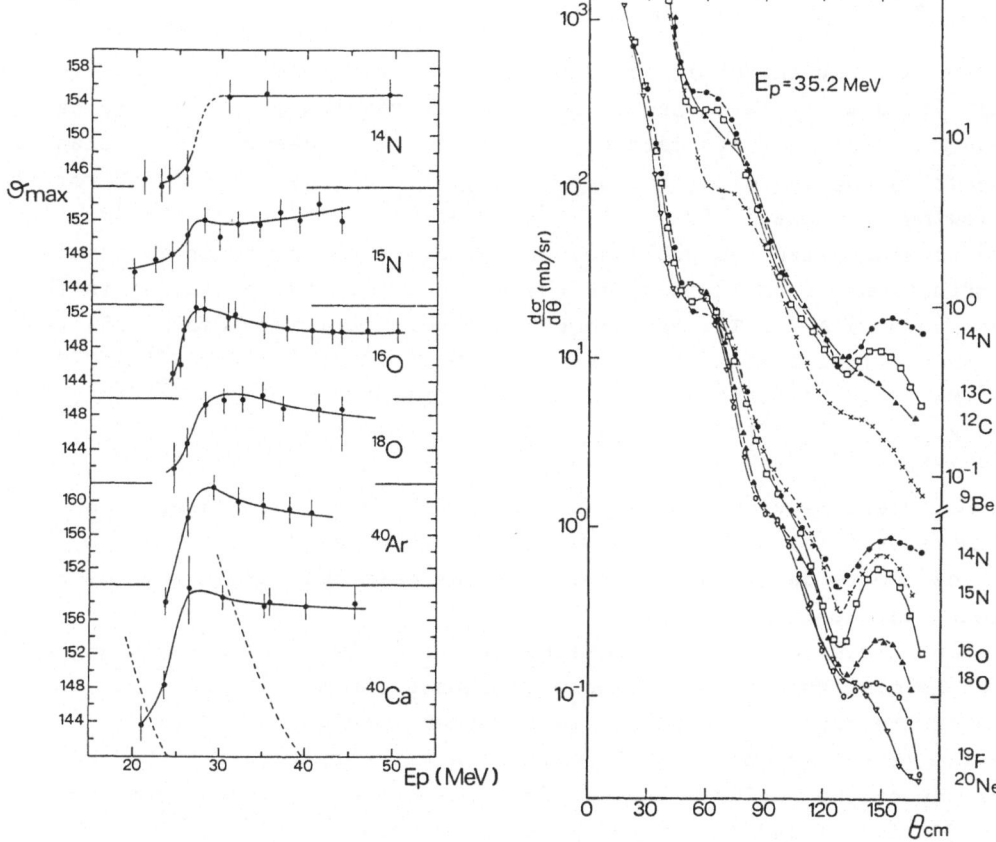

Fig. 4 - Angular position of the backward maximum in the angular distribution plotted against the incident energy. The error bars give the estimated uncertainty, while the continuous lines are only for eye guide. The dashed lines, given for ^{40}Ca, show the energy dependence predicted by an optical model calculation.

Fig. 5 - Differential cross sections for proton elastic scattering from different nuclei in the ^{16}O region. Data for ^{17}O have been omitted, being very similar to those for ^{16}O. The lines drawn through the experimental points are only for eye guide. The strong mass dependence of the backward maximum is shown.

3 - MASS DEPENDENCE OF PROTON ELASTIC SCATTERING ON LIGHT NUCLEI

From a first inspection of the data reported in sect. 2, the "anomalous" yield at backward angles seems to be maximum in the energy region 30-36 MeV. To study its mass dependence we have measured[11] the elastic scattering angular distributions at a proton energy of 29.8 MeV for the following 10 nuclei: ^{23}Na, ^{31}P, ^{39}K, ^{45}Sc, ^{50,53,54}Cr and ^{62,64}Ni. These data together with the published ones, form a set of about 40 angular distributions at an incident energy close to 30 MeV for nuclei in the mass region 9<A<70. The measurement has been performed also at 35.2 MeV for the following 61 nuclei: ^{9}Be, ^{10,11}B, ^{12,13}C, ^{14,15}N, ^{16,17,18}O, ^{19}F, ^{20,22}Ne, ^{23}Na, ^{24,25,26}Mg, ^{27}Al, ^{28,29,30}Si, ^{31}P, ^{32,34}S, ^{35,37}Cl, ^{40}Ar, ^{39,41}K, ^{40,42,43,44,48}Ca, ^{45}Sc, ^{46,47,48,50}Ti, ^{51}V, ^{50,52,53,54}Cr, ^{55}Mn, ^{54,56,58}Fe, ^{59}Co, ^{58,60,61,62,64}Ni, ^{63,65}Cu, and ^{64,66,67,68,70}Zn. The general trend is very similar at the two energies and therefore only the more extended set at 35.2 MeV will be discussed.

From a more detailed inspection of the data emerges a number of significant features:

1 - Many nuclei exhibit an anomalously large yield at backward angles, which is most evident in two mass regions close to ^{16}O and ^{40}Ca. Examples of the resulting angular distributions are given in Fig. 5.

2 - The effect is more likely related to the structure than to the dimensions of the nucleus. Support for this statement comes from the large difference found for the couples of isobars ^{40}Ar-^{40}Ca, ^{48}Ca-^{48}Ti, ^{54}Cr-^{54}Fe, ^{58}Fe-^{58}Ni and ^{64}Ni-^{64}Zn, in which the closed shell nuclei (^{40,48}Ca, ^{54}Fe and ^{58,64}Ni) show a larger backward yield. Besides, the effect is completely absent in strongly deformed nuclei like ^{12}C, ^{20,22}Ne, and ^{24}Mg as it can be appreciated for ^{12}C and ^{20}Ne in Fig. 5.

3 - It also turns out that the anomaly is not limited to the aforementioned backward region. In this respect we have so far performed a carefull study for nuclei up to A=22, where the effect is larger and where it is possible, owing the use of gaseous targets, to obtain data whose relative values are more precisely determined. The cross sections integrated over the forward angles (20°-40°) are shown in Fig. 6. Comparison of Fig.s 5 and 6 shows at a glance that those nuclei for which the backward effect is most enhanced, e.g. ^{13}C, ^{14,15}N, ^{16,17}O, are also characterized by a larger forward cross section. As a help in judging the extent of this "anomaly" we have used as a reference the results of a folding model calculation (Ref. 3 and contribution to the present Conference by A. Tarrats). The folding model gives predictions with a clear resemblance with the experiment, however the experimental cross sections

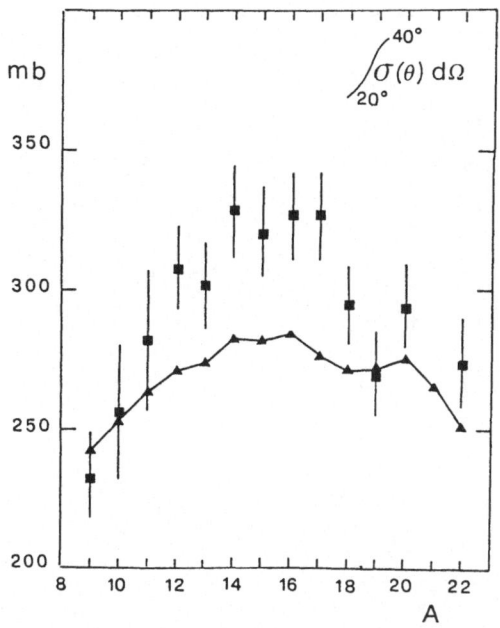

Fig. 6 - Comparison of experimental cross sections integrated over
forward angles (squares) with values calculated by using a folding
model potential (triangles and full line). The bars give an upper
limit to the errors, which takes into account also uncertainties
in the absolute normalizations.

show a more marked mass dependence. The forward effect reaches sizeable absolute values (judging from Fig. 6, of the order of 50 mb)
however it is relatively less important than that at backward angles
and can be, for this reason, easily reproduced by standard optical
model calculations with only minor adjustments in the parameters.

A straightforward way to parametrize the effect here discussed is
to consider the peak value (σ_{peak}) at the backward cross section maximum. The σ_{peak} values for the studied nuclei at 35.2 MeV are plotted
in Fig. 7 against the mass number. Spurred by the evidence outlined
in point 2) above, we have made several attempts to find, if any,
quantitative correlations with the collective properties of the target
nucleus. In this connection we report a striking result obtained in
the comparison between the elastic cross sections at backward angles
(σ_{peak}) and β_2, the quadrupole deformation parameters. The latters
have been deduced, for even-even nuclei, from electromagnetic rates[12].
The values of $0.1/\beta_2$ give, except for the nuclei ^{12}C, $^{16,18}O$ and ^{48}Ca,
almost exactly the numerical values of the observed σ_{peak} expressed
in mb/sr (Fig. 7, part b), at least within the estimated errors. These
are generally of the order of 10% both for β_2 and σ_{peak}. A better

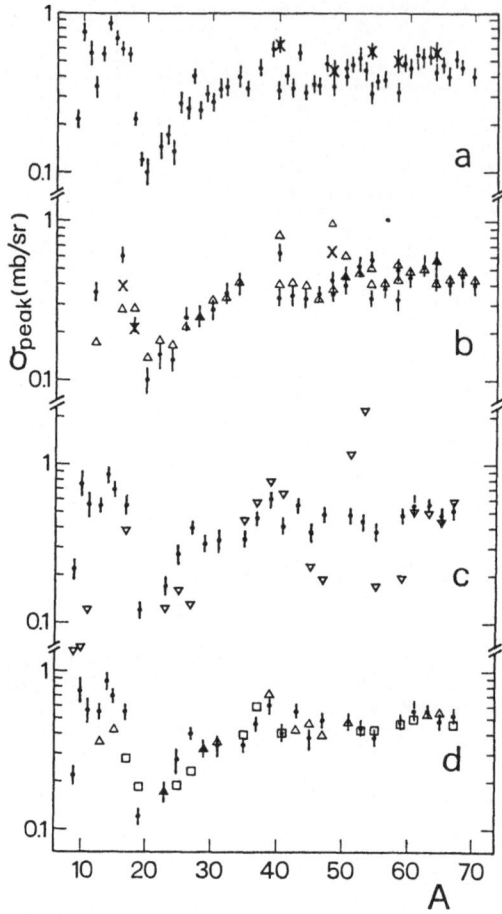

Fig. 7 - Mass dependence of peak value of the backward maximum
(σ_{peak}) and comparison with β_2, the quadrupole deformation parameters.
Part a): experimental σ_{peak} values; vertical bars give estimated uncertainties. The crosses mark magic number nuclei (^{40}Ca, ^{48}Ca, ^{54}Fe, ^{58}Ni
and ^{64}Ni) for couples of isobars. Part b): comparison for even-even
nuclei with numerical values of $0.1/\beta_2$ (triangles). The latter values
are deduced from electromagnetic transition rates. For ^{16,18}O and ^{48}Ca
the values obtained from inelastic scattering experiments are also
given (crosses). Part c): comparison for odd-A nuclei. The triangles
represent values of $0.018/\beta_2$ where the latter are deduced from g.s.
quadrupole moments. Part d): the same data as in c) are compared with
$0.1/\beta_2$ where the β_2 are deduced from core excitation strengths (triangles) or from transition rates of neighbouring even-even nuclei
(squares).

agreement for the nuclei ^{16,18}O and ^{48}Ca is obtained if one uses the
β_2 deduced from inelastic scattering experiments. For these nuclei, in
fact, contrary to what is obtained generally, the β_2 from inelastic
scattering differ from those deduced from electromagnetic transition
rates. - The β_2 values for odd-A and odd-odd nuclei (^{10}B, ^{14}N) have been
deduced, as usually, from g.s. electric quadrupole moments[13] and are there-

fore related to the g.s. static deformations. The gross structure of the
A-dependence of σ_{peak} is still reproduced (Fig. 7, part c), but less satisfactorily and with a different normalization $(0.018/\beta_2)$. A better
correlation with electromagnetic transition strengths can be established
also for odd-A nuclei (Fig. 7, part d). For several odd nuclei, here
considered, it has been shown[14] that the low-lying levels can be described by coupling the extra particle or hole to collective states of an
even core and that the collective excitation of the low-lying multiplets
has a strength which corresponds to that of the parent 2_1^+ state in the
neighbouring nucleus. If we use the β_2 derived from these core excitations or if, for want of data, we use the β_2 of the neighbouring even
nucleus, the agreement found for even-even nuclei is obtained also for
odd-A nuclei. This correlation, which could hardly be fortuitous, indicates that at 30-40 MeV the proton elastic scattering is strongly affected by collective couplings. This finding suggests to attempt a coupled
channel calculation. It can be readily found that the widely studied
couplings to the relevant quadrupole transitions, leading to low-lying
states, do not give the cross section enhancement needed for spherical
nuclei, at least if one uses conventional optical model potentials. This
point will be discussed in detail in sect. 6.

The effect outlined above shows some analogy with the "anomalous large angle scattering", ALAS, found for α-particles.[15] Other features, like the presence of the effect also at forward angles and the above correlation with collective properties, seem however typical of proton scattering.

4 - INELASTIC SCATTERING CROSS SECTIONS

Further data that can be considered concern inelastic scattering. Systematic measurements over a wide range of nuclear masses and incident
energies have been performed for the transition to the first excited
state $(J^\pi = 2^+)$ in even-even nuclei. In the same experiment above cited,[11]
the angular distributions for this transition have been measured on the
following 30 nuclei: ^{12}C, $^{16,18}O$, $^{20,22}Ne$, $^{24,26}Mg$, $^{28,30}Si$, ^{32}S, ^{40}Ar,
$^{42,44}Ca$, $^{46,48,50}Ti$, $^{50,52,54}Cr$, $^{54,56,58}Fe$, $^{58,60,62,64}Ni$ and
$^{64,66,68,70}Zn$.

The measured angular distributions show sizeable differences, for
the different nuclei, both in absolute values and in shape. The absolute values can be parametrized in terms of $R\beta_2$, where now β_2 is defined as
the normalization of the collective form factor. The β_2 values, obtained from the above data by a coupled-channels calculation, are, except
for the nuclei $^{16,18}O$ and ^{48}Ca, in good agreement with the values de-

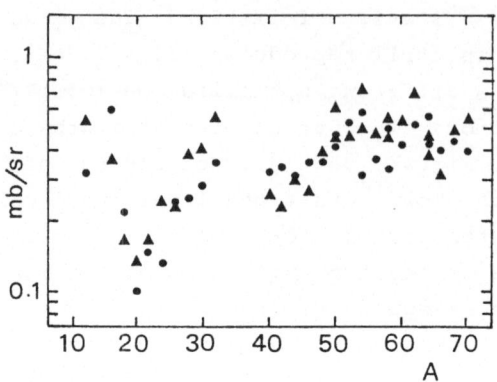

Fig. 8 - Mass dependence of peak values (triangles) of the backward maximum in the angular distribution for inelastic scattering (to 2^+_1 states) divided by $2(R\,\beta_2)^2$, where R is the nuclear radius. These ratios are proportional to $\sigma(\text{backward})/\langle\sigma\rangle$. The points give, for comparison, the values of σ_{peak} for elastic cross sections already given in Fig. 7.

duced from B(E2) values. If one takes the cross section at the backward maximum in the angular distribution (σ_{peak}) and calculates the ratios $\sigma_{\text{peak}}/R\,\beta_2)^2$, which correspond approximately to $\sigma(160^\circ)/\langle\sigma\rangle$, one obtains an indication of the mass dependence of the shape of the differential cross sections. In Fig. 8 these ratios are plotted against the mass number and compared with σ_{peak} for the elastic scattering. The mass dependence of the two kind of data shows a rather good similarity.

The "anomalous" backward yield seems therefore to affect also inelastic transitions, at least strong collective transitions as those to the low-lying 2^+ states.

5 – PHASE SHIFT ANALYSIS AND ANGULAR MOMENTUM LOCALIZATION

This analysis, which is model independent, has been performed using the search program SNOOPY to ascertain if some particular wave could be held responsible for the effect observed.

The data analyzed are those at 35.2 MeV in the oxygen and calcium regions and those for ^{15}N, ^{18}O, ^{40}Ar and ^{40}Ca at different incident energies. The starting set of phase shifts, used in the search, was obtained from an average optical model potential. To minimize ambiguities, which could be furtherly reduced if polarization data were at disposal at each incident energy, several optical potentials have been used to obtain other starting sets of phase shifts. Moreover the partial waves, for each L-value up to 9, were first singularly searched on, in order to find for every nucleus and incident energy those waves which have the largest effect in decreasing the initial x^2. Curves like those of Fig. 9 are obtained. Then every partial wave was let to vary in succession starting, for each angular distribution, with the one with the largest

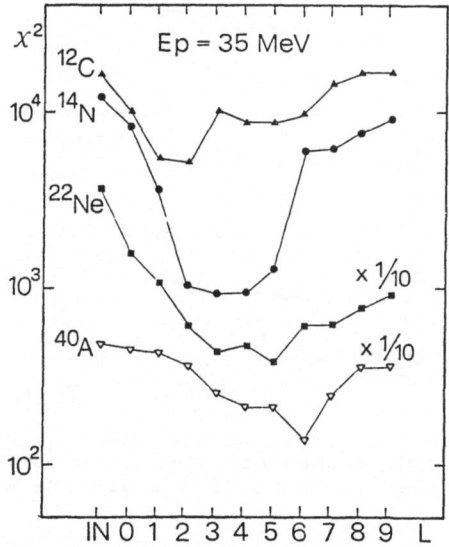

Fig. 9 – Results of a best fit procedure to the experimental elastic cross sections, for the nuclei indicated in the figure, at 35.2 MeV proton energy. The partial waves, for each L-value up to 9, have been singularly searched on to find out those waves which have the largest effect on the initial x^2 value (IN in the figure).

effect in x^2. To a very large degree of confidence, in spite of the ambiguities which cannot be avoided completely in such kind of analysis, we can draw the following conclusions: a) the partial waves responsible for the backward enhancement completely account also for the forward effect; b) the fits to the full angular distributions, which are indeed excellent (Fig. 3) are accomplished by small variations of the real and imaginary parts of the phase shifts; c) the waves involved start being L=2 and 3 for nuclei below oxygen at 35.2 MeV and gradually shift to L=3 and 4 with increasing A. The L=5 plays some role only after A=18, while L=5 and 6 are the dominant waves in the calcium region. In this connection we recall that at 35.2 MeV the angular momentum of the grazing wave goes from L=2 or 3 up to L=5 or 6 in going from ^{9}Be to ^{40}Ca. The role of the grazing waves is confirmed also by the energy dependence of the data.

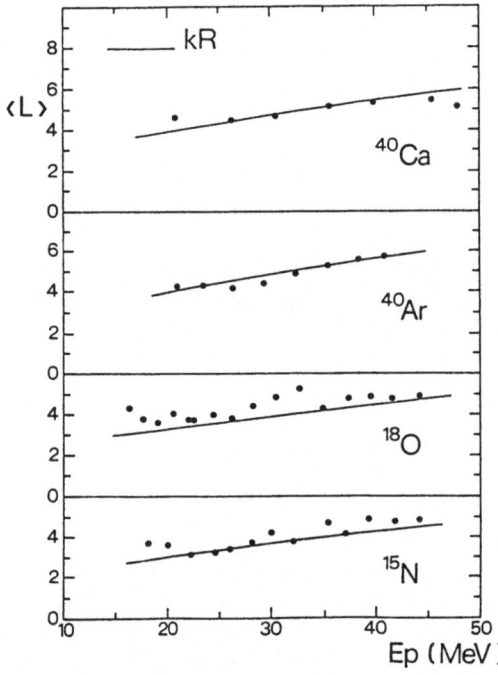

Fig. 10 - Average (see text) angular momentum <L> of the partial waves involved in the enhanced "anomalous" yield. The average values are compared with the angular momentum of the grazing wave L=kR.

If one plots the average angular momentum, defined as:
<L> = $(\Sigma_i L_i/\chi_i^2)/(\Sigma_i 1/\chi_i^2)$, against the incident energy, the graphs of Fig. 10 are obtained. Generally <L> is well reproduced by kR, where k is the wave number of the incident proton and R is the r.m.s. radius, as derived from the real part of the optical model potential used in the search to give the starting set of phase shifts.

One can then conclude that the effect does not involve a specific L value, but two or more, close to the momentum of the grazing wave.

6 - SUMMARY AND COMMENTS

The above phenomenology gives at least some clear indications:

1) The proton elastic scattering on light nuclei shows a strong systematic mass dependence, with features that must be considered "anomalous" in comparison to a standard optical model calculation.

2) The "anomaly" is systematically related to the collective properties of the target nucleus and can be parametrized in terms of β_2, the quadrupole deformation parameter, being very evident for spherical nuclei and completely absent for strongly deformed nuclei.

3) The "anomaly" is energy dependent. It appears at 26-27 MeV, reaches its maximum in the energy region between 30 and 40 MeV and seems to disappear (at least in the calcium region) above 45 MeV.

As already mentioned in the introduction, the data more widely used to test recent theories for optical potentials pertain to proton scattering, at an incident energy of about 30 MeV, on magic nuclei. It is shown in the present study that cross sections at incident energies between 28 and 45 MeV on spherical light nuclei, and therefore on ^{16}O and ^{40}Ca, must be considered as a "particular" class of experimental data. The more recent and successful calculations, derived from complex G-matrix evaluations, show that optical potentials should have a radial dependence different from that of a conventional optical model. Such non standard form factors produce differential cross sections with an enhanced yield which are in better agreement with the experimental data for magic nuclei. A relevant example, for proton scattering on ^{40}Ca at 30-40 MeV, is reported in a recent paper by Brieva and Rook.[2] The authors claim that a t-matrix calculation can reproduce satisfactorily the experiment without extra terms or processes.[6,7] However, to asses the general validity of the model, comparison with the experiments should be performed both for spherical and deformed nuclei. In other words, it would be, in our opinion, very interesting to test the new microscopic theories on the correlations found between proton scattering and collective properties. A complex t-matrix, obtained for a finite nucleus, would automatically include any specific effect of the excitation spectrum for that particular nucleus. However the use of t-matrix folded on nuclear matter is in fact an approximation which includes excitation effects, but only in some averaged way. It is therefore at the best of our knowledge, an open question if folding models, also including a more detailed microscopic description of the nuclear structure, will adequatel describe the mass dependent features of proton scattering.

Another approach, adopted by some authors,[6,7] is to explicitly take into account the non-elastic channels. The most important "door ways" for the absorption of protons are expected to be direct inelastic scattering and direct rearrangements. Recently the coupling with high-lying inelastic channels and with (p,d) reaction have been considered.[6,7] Both couplings give an enhancement in the elastic channel yield, which results in a backward maximum. The angular position of this maximum shows an energy dependence in acceptable agreement with the experiment. The effect on the elastic channel is maximum at an energy of about 10 MeV for the intermediate particle (proton or deuteron). Considering the energy position of giant resonances, which act as a clustering of intermediate states for the (p,p',p) process, and the values of the (p,d) tresholds, one can understand why the "anomalous" backward yield is maximum between 30 and 40 MeV. Less straightforward is the A-dependence. In fact the differences found between magic and collective nuclei cannot be explained on the basis of simple considerations on tresholds, single-particle energies or on the energy distribution of quadrupole strength. More explicitly the excitation of the intermediate states should be rather similar for magic and collective nuclei. The structure of the involved nucleus might determine the depletion process of the intermediate states and therefore their effect on the elastic channel. A detailed calculation would require however a better knowledge of spectroscopic factors for transitions between excited states.

In spite of some success in fitting specific data, present theoretical interpretations leave the problem of explaining the entire proton scattering phenomenology still open. This phenomenology, on the other hand, has reached a satisfactory degree of definiteness; its main features are now better known and should allow exhaustive tests of theoretical models.

REFERENCES

1 - J.P.Jeukenne, A.Lejenne and C.Mahaux, Phys. Rev., C15(1977)10.
2 - F.A.Brieva and J.R.Rook, Nucl. Phys., A291(1977)317.
3 - J.L.Escudié and A.Tarrats, Compte Rendu d'Activité, Report CEA-N-1861 Saclay (1975)187.
4 - E.E.Gross, R.H.Bassel, L.N.Blumberg, B.J.Morton, A.Van der Woude and A.Zuker, Nucl. Phys., A132(1967)673.
5 - W.T.H. van Oers and J.M.Cameron, Phys. Rev., 184(1969)1061.
6 - P.W.Coulter and G.R.Satchler, Nucl. Phys., A293(1977)269.
7 - R.S.Mackintosh and A.M.Kobos, Phys. Lett., 62B(1977)127.
8 - R.De Leo, G.D'Erasmo, F.Ferrero, A.Pantaleo, M.Pignanelli, Nucl. Phys., A254(1975)156.
9 - R.De Leo, G.D'Erasmo, A.Pantaleo, G.Pasquariello, G.Viesti, M.Pignanelli and H.V. von Geramb, to be published.
10- R.De Leo, G.D'Erasmo, A.Pantaleo and M.Pignanelli, to be published.

11- E.Colombo, R.De Leo, J.L.Escudié, E.Fabrici, S.Micheletti, M.Pigna-
    nelli and F.Resmini, J. Phys. Soc. Jap. $\underline{44}$(1978)543 and to be pu-
    blished.
12- J.Raman et al., Proc. of Int. Conf. on Nuclear Structure, Tokyo
    (1977)79 and to be published on Nucl. Data Tab.
13- G.H.Fuller and V.W.Cohen, Nucl. Data Tab., $\underline{A5}$(1969)433.
14- R.J.Peterson, Phys. Rev., $\underline{172}$(1968)1098; W.Scholz and F.B.Malik,
    ibidem $\underline{153}$(1966)1071.
15- H.P.Gubler, U.Kiebele, H.O.Meyer, G.B.Plattner and I.Sick, Phys.
    Lett., $\underline{74B}$(1978)202 and the literature cited therein.

Optical Model Proton Parameters at Subcoulomb Energies [*])

W. Drenckhahn, A. Feigel, E. Finckh, G. Gademann, K. Rüskamp, M. Wangler

Physikalisches Institut, Universität Erlangen-Nürnberg, D 8520 Erlangen

Proton parameters of the optical model are well determined at energies above the Coulomb barrier[1] (column 1 in table 1). An extrapolation of these parameters to lower energies completely fails to reproduce the measured absorption cross section of tin isotopes[2]. The cross section, measured via the neutron yield of the (p,n) reaction, shows a broad size resonance due to the 3p-wave. To reproduce the measured cross section, the depth of the real and depth and diffuseness of the imaginary potential had to be changed. The real potential, which mainly influences the position of the resonance, shows an energy dependence of b = 0.9/MeV instead of 0.3/MeV. The imaginary potential has a depth of W_D = 9 MeV and a diffuseness of a_I = 0.4 fm [3-5] (column 2 in table 1), which is much smaller than the usual value of a_I = 0.65 fm. Without the reduction of the diffuseness it is not possible to fit the absorption data.

These changes of the optical model potential have negligible influence on the elastic scattering because at the low energies the nuclear potential is small compared to the Coulomb potential. However, the analyzing power of elastic scattering shows significant differences by changing the potential. We therefore investigated the analyzing power at three energies, E_p = 6.8, 7.8, and 8.8 MeV. The data at E_p = 9.8 MeV are taken from Greenlees et al.[6]

The measurements were made using the Erlangen Lamb Shift ion source and the large scattering chamber equipped with 12 detectors. The proton spectra were recorded with an on-line-PDP 11/40 computer. The beam intensity was about 20 nA on the target. The polarization of the beam (P_Z=0.75) was switched on - off with a frequency of 1 Hz and was monitored by a He-4-polarimeter.

The optical model analysis was started from a parameter set which fitted the absorption data (set D ref.1). A small change in the depth of the real and imaginary potential (column 3 in table 1) gives good agreement for the analyzing power (dotted line in fig.1). But this change already destroys the agreement with the absorption data (dotted line in fig.2).

This example shows again, that the investigation of analyzing power alone or of absorption alone does not give correct results.

For the final analysis we used the parameters of column 2 in table 1, which describe the absorption data very well with a constant value of the diffuseness of the imaginary potential. We varied in several steps the depth of the real potential, all parameters of the imaginary potential and of the spin-orbit potential. To obtain a good fit to all data (fig. 1 and fig.2, full curve) the radius and diffuseness of the spin-orbit potential had to be reduced (column 4 in table 1).

Similar experiments for the Zirkonium isotopes confirm that correct results at subcoulomb energies cannot be obtained by extrapolation of optical model parameters to lower energies nor by analyzing absorption or polarization data alone. Only the combination of both data seem to determine the parameters uniquely.

References:

1) F.D. Becchetti, G.W. Greenlees, Phys. Rev. 182, 1190 (1969)
2) C.H. Johnson, R.L. Kernell, Phys. Rev. C2, 639 (1970)
3) W. Drenckhahn, A. Feigel, E. Finckh, R. Kempf, M.Koenner, P. Krämmer, K.-H. Uebel, 4th Int. Symp. on Polarization Phenomena in Nucl. React. Zürich, Aug. 1975, pg. 613
4) C.H. Johnson, J.K. Bair, C.M. Jones, S.K. Penny, D.W. Smith; Phys. Rev. C15, 196 (1977)
5) C.H. Johnson, A. Galonsky, R.L. Kernell, Phys. Rev. Lett 39, 1604 (1977)
6) G.W. Greenlees, C.H. Poppe, J.A. Sievers, and D.L. Watson, Phys. Rev. C3, 1231 (1971)
7) W. Kretschmer, G. Böhner, E. Finckh, 4th Int. Symp. on Polar. Phenomena in Nucl. React., Zürich, Aug. 1975, pg. 611
8) A. Feigel, W. Küfner, M. Pflüger, W. Drenckhahn, E. Finckh, Proc. Int. Conf. on Nucl. Struct., Tokyo, Sept. 1977, pg. 498

*) Work supported by Deutsche Forschungsgemeinschaft

Table 1: Optical model parameters.
Potential depth in MeV, length in fm.
Symbols are used as in ref. 1.

	a	b	c	d
U	$54.0-0.32 \cdot E+$ $+0.4 \cdot Z \cdot A^{-1/3}+$ $+24.0(N-Z)/A$	$59.8-0.9 \cdot E+$ $+0.4 \cdot Z \cdot A^{-1/3}+$ $+24.0(N-Z)/A$	$59.75-0.84 \cdot E+$ $+0.4 \cdot Z \cdot A^{-1/3}+$ $+24.0(N-Z)/A$	$57.7-0.84 \cdot E+$ $+0.4 \cdot Z \cdot A^{-1/3}+$ $+24.0(N-Z)/A$
r_R	1.17	1.17	1.20	1.20
a_R	0.75	0.75	0.7	0.7
W_D	$11.8-0.25 \cdot E+$ $+12.0(N-Z)/A$	9.00	9.91	9.91
r_I	1.32	1.32	1.31	1.31
a_I	$0.51+0.7(N-Z)/A$	0.4	$0.093+0.052 \cdot E$	0.45
V_{so}	6.2	6.20	6.0	6.0
r_{so}	1.01	1.01	1.1	1.03
a_{so}	0.75	0.75	0.7	0.3
r_c	1.21	1.21	1.21	1.21

a) ref.1;
b) ref.2;
c) and d) see text

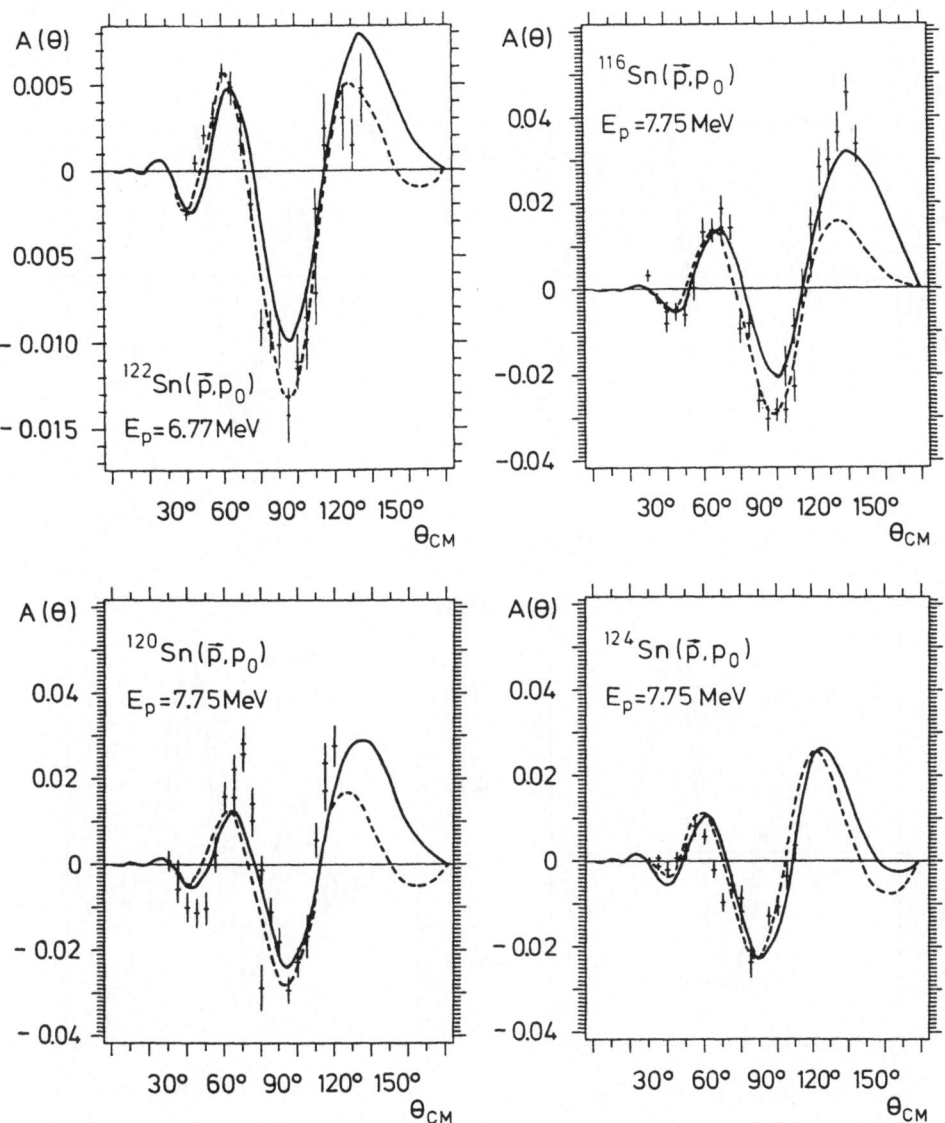

Fig.1a: Analyzing power of elastic proton scattering on tin isotopes
at various energies. Dotted and full curve are calculated
with the parameters of column 3 and 4 in table 1.

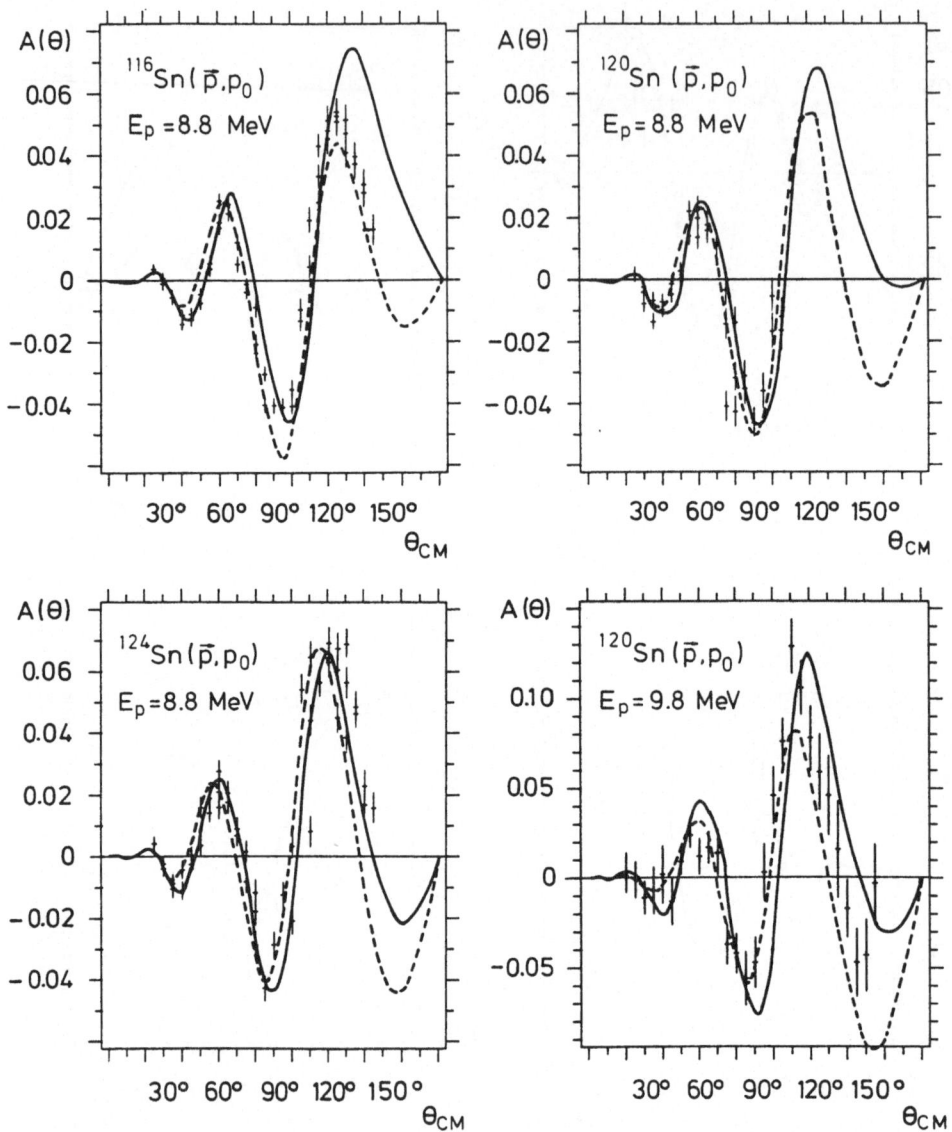

Fig.1b: Analyzing power of elastic proton scattering on tin isotopes at various energies. Dotted and full curve are calculated with the parameters of column 3 and 4 in table 1.

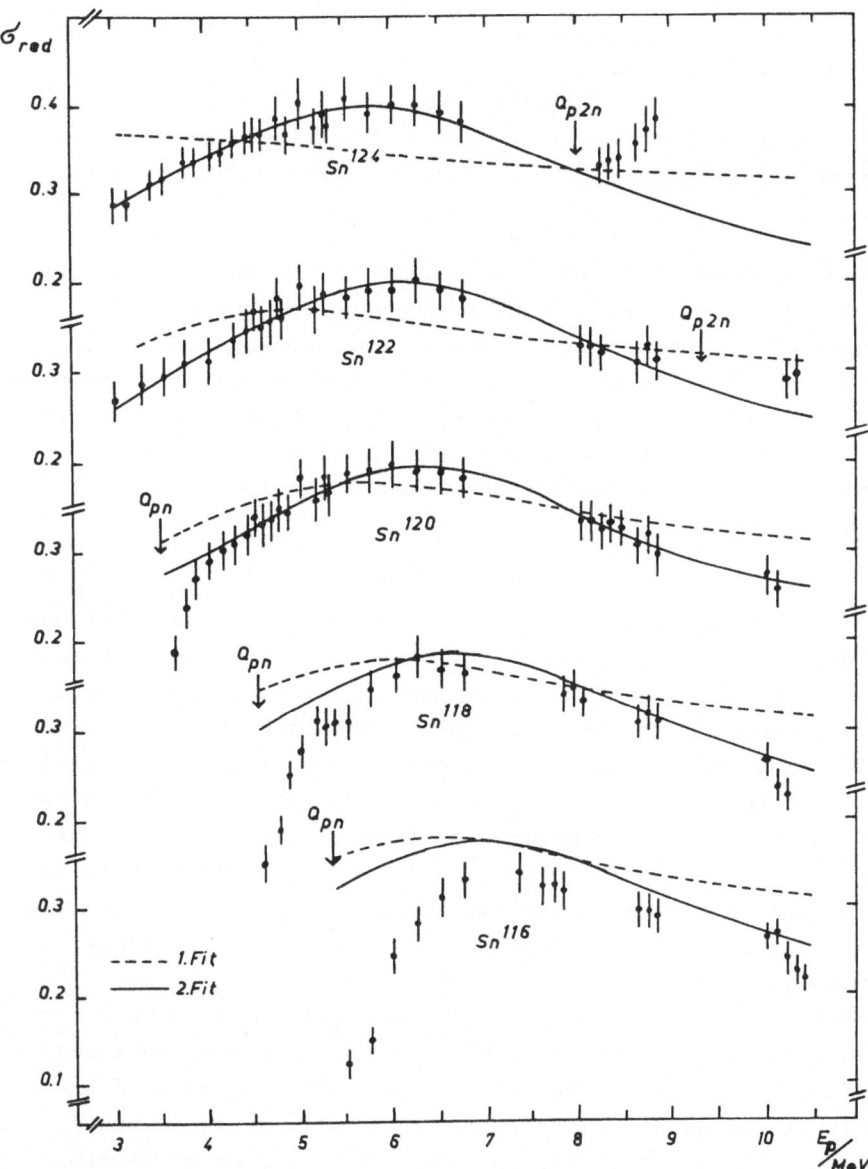

Fig.2: Absorption cross section of tin isotopes. Data and calculations are plotted as reduced cross section σ_{red} which has taken out the exponential rise due to Coulomb barrier transmissions according to ref.2. Dotted and full curve are calculated with the parameters of column 3 and 4 in table 1.

A NEW TECHNIQUE FOR MEASURING RATIOS OF ELASTIC SCATTERING CROSS SECTIONS: AN APPLICATION TO THE CALCIUM ISOTOPES*

Sam M. Austin

Cyclotron Laboratory and Physics Department

Michigan State University, East Lansing, Michigan 48824

Abstract

A technique has been developed which allows one to measure ratios of elastic scattering cross sections for nearby nuclei and which should be free of most systematic uncertainties. Protons scattered from a mixed target are momentum-analyzed in a high resolution spectrograph and the scatterer is identified by the kinematic shift. The result of an application to scattering from the calcium isotopes ^{40,44,48}Ca is discussed.

I. Introduction

The recent rebirth of interest in the optical model (OM) has been sparked by several developments. Perhaps the most important of these are the theoretical advances, discussed in detail at this conference, which have made it possible to calculate OM potentials from first principles, but phenomenlogical analyses have also led to significant results. Information about neutron density distributions has been obtained from OM analyses of elastic scattering data. And it has been realized[1,2] that phenomenological fits can yield substantial information about the details and complexities of the elastic scattering process if one requires fits which are precise ($\chi^2_\nu \approx 1$) rather than just "reasonably good."

Progress on the experimental side has perhaps been less rapid, and elastic scattering data of truly high quality are available only in limited ranges of mass and energy. Measurements of elastic scattering are relatively straightforward compared to most measurements in nuclear science, but it is not trival to obtain accurate cross sections free of systematic uncertainties. The problems are particularly acute when production of uniform, well characterized targets is difficult, as is often the case for rare or reactive nuclides; when angular variation is rapid, as is the case for forward angles, especially at higher energies; and when current integration is difficult as is the case for heavy ions. I describe here a new technique in which protons scattered from

*This lecture describes research performed by S.M. Austin, E. Kashy, C.H. King, R.G. Markham, I. Redmount and R.M. Ronningen. Research was supported by the U.S. National Science Foundation Grant 78-01684.

a mixed target containing nuclides of different mass are momentum-
analyzed in a high resolution spectrograph and the scattering nuclide
is identified by the kinematic shift (see Fig. 1). From a measurement
of the relative yields one obtains the <u>ratio</u> of cross sections for the
various nuclides, given the relative composition of the target. This
is determined from a similar ratio measurement carried out, perhaps with
a different beam, in the Rutherford scattering region. Measurements of
beam currents, target thicknesses, spectrograph apertures, etc. are not
required and the measured ratios should be free of most systematic uncertainties.

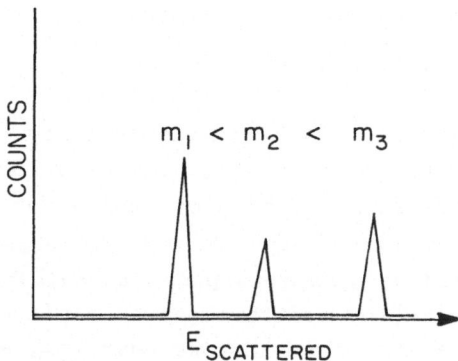

Fig. 1. Schematic spectrum of particles elastically scattered from
a mixed target consisting of nuclides of masses m_1, m_2, and m_3.

It is obvious that this procedure is not universally applicable:
scattering from the heaviest nuclei and near $0°$ yields kinematic separations which are not resolable. Since the range of accessable nuclides
depends on the attainable resolution we next review briefly what can be
achieved in a practical experiment. Figure 2 shows the best that has
been done, a resolution $\Delta E(FWHM)/E$ of about $1/23,000$ for 35 MeV protons,
and also illustrates the fact that with targets thicker than 100 $\mu g/cm^2$
one is inevitably limited to resolutions worse than $1/10,000$. It is
also worth noting that these resolutions were attained with nuclear
emulsions for particle detection; similar resolutions have been attained
with on-line detectors in test situations, but not yet in practical
experiments. This distinction is important since track counting of
emulsions does not easily yield results of high accuracy. To summarize,
a practical limit on resolution at present is $\Delta E/E \simeq 1/4000$ with on-line
detectors, with factor of two or three improvement, to perhaps $1/10,000$
expected in the near future.

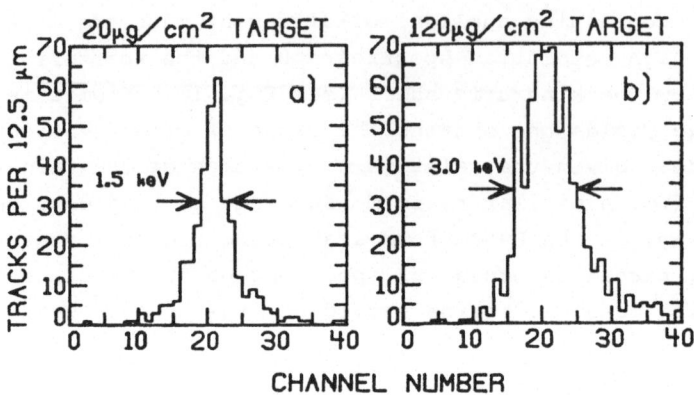

Fig. 2. Spectra of 35 MeV protons elastically scattered from thin Ni targets[3]. The spectra were recorded on nuclear track plates in the focal plane of an Enge split-pole spectrograph.

Given the resolution one can determine from Fig. 3 what nuclei one can study. Plotted for proton scattering is the value of the kinematic energy separation $E´/\Delta E´$ as a function of the center of mass angle θ c.m. Here $E´/\Delta E´$ is calculated for a target mass difference $\Delta M = 1$ amu. If ΔM is larger, then of course $E´/\Delta E´$ is proportionately smaller. As an example, if one is studying scattering from a mixed target of ^{40,44}Ca, at 30° the value of $\frac{E´}{\Delta E´} = 6800$. Since $\Delta M = 4$ we have $\frac{E´}{\Delta M \Delta E´} = 1700$. Attainable resolutions are better than this so the measurement can be done.

Whether any particular measurement is possible is a detailed question, but several general comments can be made. First of all, the technique will mainly be useful at low energies, since the bulk of the cross section moves to small angles as the energy increases and separation at small angles is difficult regardless of energy. Secondly, the method may be more useful for projectiles heavier than protons, since the kinematic shifts increase with projectile mass. For example, the kinematic shifts are approximately four times larger for alpha particles than for protons.

I now will turn to the measurements we have made, combining a description of the experiment with a detailed discussion of the problems and limitations encountered when they seem of general importance. We chose to study proton scattering from the calcium isotopes ^{40,44,48}Ca at 30.3 MeV for both technical and "physics" reasons. High resolution techniques are most highly developed for protons, at least at MSU, and the necessary resolution is easy to achieve. Furthermore, high quality data from the Rutherford laboratory[4,5] is available for ^{40}Ca at this energy. The major reason for the choice, however, is that the proton-neutron interaction V^{pn} is much stronger than the proton-proton inter-action V^{pp}. This follows from the fact that the triplet-even and sin-glet-even interactions are of comparable magnitude and are both attract-ive.

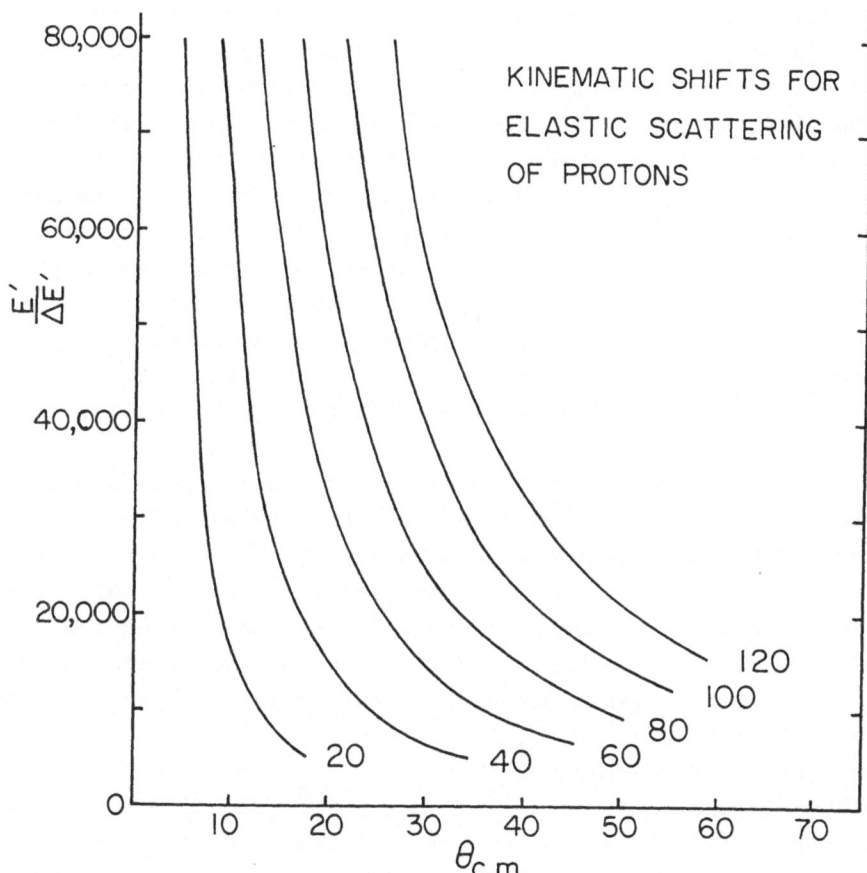

Fig. 3.   Values of the kinematic energy shift $E'/\Delta E'$ for proton scat-tering as a function of center-of-mass scattering angle. The curves are drawn for the scatterer masses noted on the figure. For details see the text.

Thus the like-nucleon interaction, which can occur only in the singlet state, is weaker than that for unlike nucleons. Theoretical estimates and empirical determinations yield values in the range V^{pn}/V^{pp} = 2 to 3.5. The optical potential for protons in this energy range should therefore mainly reflect the neutron density distribution. In contrast, α particles and high energy protons are about equally sensitive to protons and neutrons. It is not yet clear whether the increased sensitivity is sufficient to compensate for the greater complexity of the proton scattering process in the 20-50 MeV range, but the possibility clearly justifies a substantial investigatory effort.

II. Application to the Calcium Isotopes

A mixed target, 100 μg/cm^2 thick, of roughly equal parts of ^{40,44,48}Ca, was prepared and its relative isotopic composition was determined by elastic scattering of 18 MeV ^{12}C ions. At this energy, OM calculations indicate that the scattering is within 1% of the Rutherford value at all angles. Another consideration strongly influenced the choice of bombarding energy. When the scattered particles are not fully stripped, (as in the present case where the dominant charge state is ^{12}C^{5+}), the charge state fractions will depend on the energy of the scattered particles, and this is different for particles scattered from different nuclides. One must therefore choose the bombarding energy to minimize the energy dependence of the charge state fraction. In the present case this effect contributes at most 1% to the uncertainty in the target composition.

The ^{12}C ions were momentum-analyzed in an Enge split-pole spectrograph and detected at the focal plane in a position-sensitive proportional counter (PSPC). Resolution was dominated by the effects of small angle scattering in the target. It was the need to clearly resolve the three peaks in the heavy ion experiment that limited the target thickness to 100 μg/cm^2, too thin for optimum statistics in the proton scattering experiments. These measurements yielded the target ratios with an overall uncertainty of ± (2-3)%.

The proton scattering experiment was also performed in the Enge spectrograph, but a special slanted-electrode PSPC[6] was used to detect the scattered particles. Resolutions were 8-10 keV FWHM (E/ΔE≈3500). Spectra were taken for the mixed target and for ^{40}Ca and ^{44}Ca targets of the same thickness, the latter to provide lineshapes for use in peak fitting. Mechanical limitations of the spectrograph system limited the maximum scattering angle to 125°. The spectrum at 35° is shown in Fig. 4, along with a fit to the data using a line shape derived from the ^{44}Ca spectrum. The fit is quite good even for this forward angle, and similar fits at other angles yielded reliable peak areas into at least 30°. The greatest uncertainty in the analysis is tracable to the long tails on the peaks which lead, at forward angles, to peak area uncertainties substantially greater than the statistical error. These tails arise when delta rays produced by the detected proton travel roughly parallel to the counter wires and slightly displace the centroid of the ionization region.

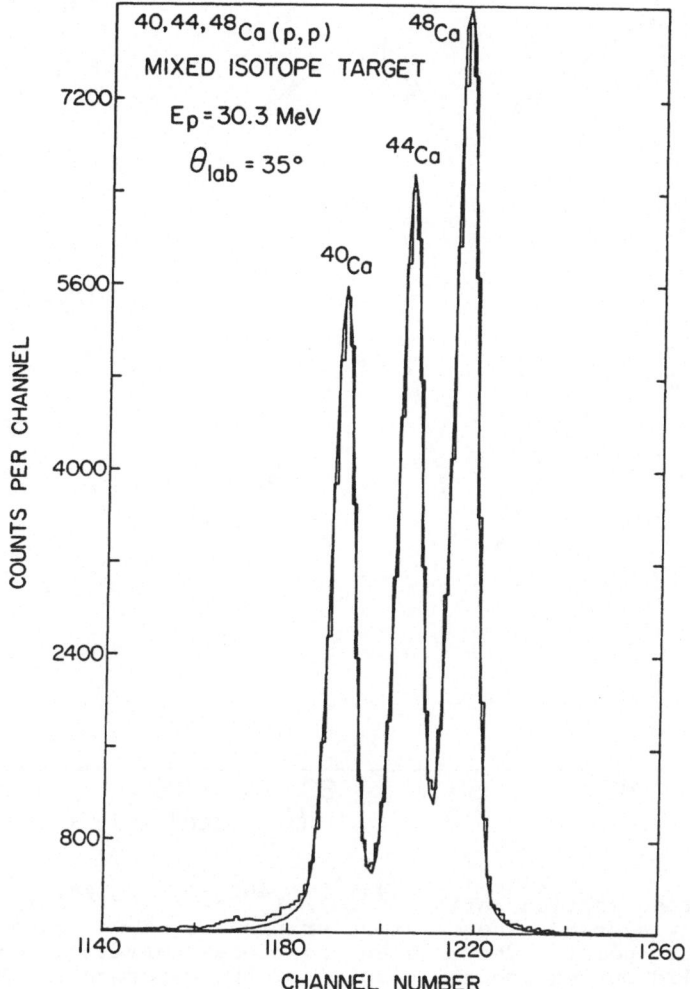

Fig. 4. Spectrum of 30.3 MeV protons scattered from a mixed isotope ^{40,44,48}Ca target at θ_{lab} = 35°. The solid curve is the result of a fit using line shapes derived from a spectrum taken with a ^{44}Ca target at the same angle.

The cross section ratios are shown in Fig. 5. Typical relative errors are in the 1-5% range, and are dominated by fitting uncertainties at the forward angles and by statistical uncertainties at the backward angles.

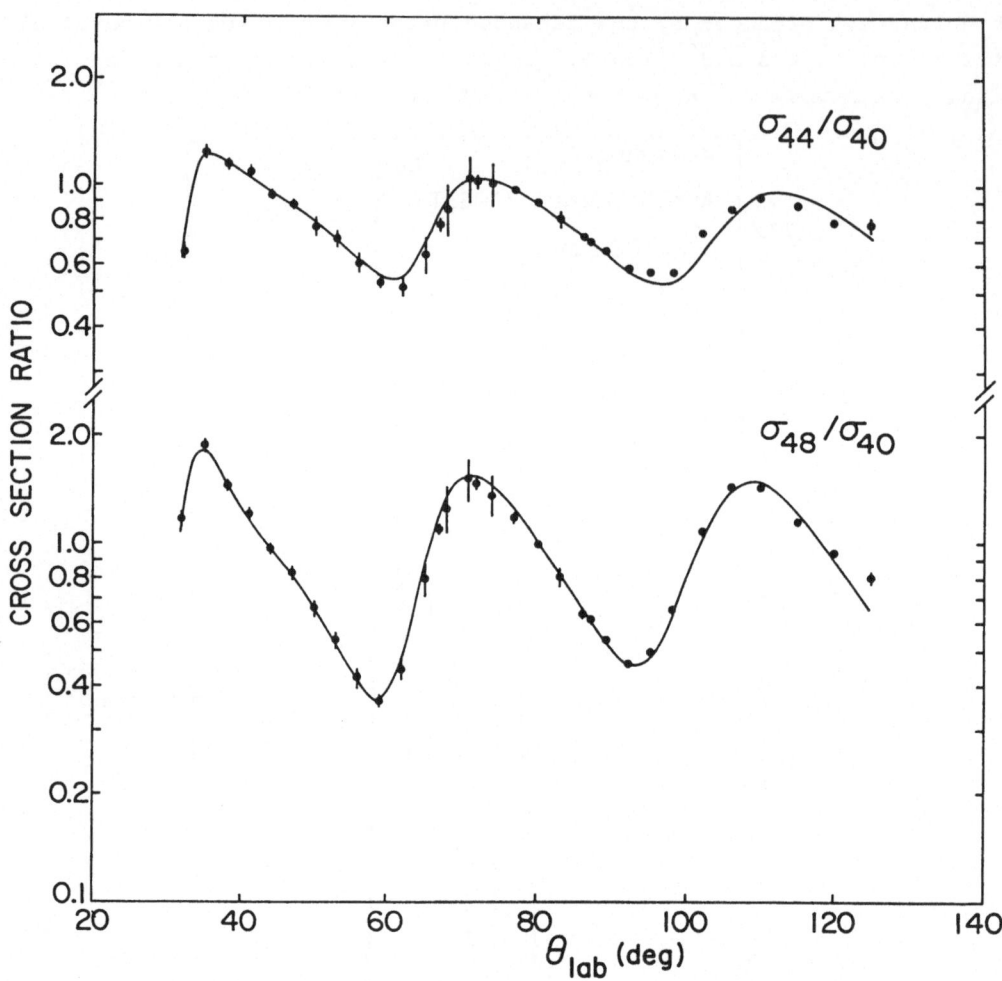

Fig. 5. Cross section ratio $\sigma(^{44}Ca)/\sigma(^{40}Ca)$ and $\sigma(^{48}Ca)/\sigma(^{40}Ca)$ in the laboratory system. The relative errors are shown whenever they are larger than the point. There is an additional normalization error of ±3%. The solid curves are the results of the 6-parameter OM searches.

It is not clear what approach is best for the analysis of the ratio data. If a reliable OM potential were available for one of the nuclides one could fit the ratios directly in turns of differences of OM potentials from nuclide to nuclide. This approach has a great intrinsic appeal, but reliable OM potentials have not been available, even for the thoroughly studied case of ^{40}Ca. It is clear that complex reaction mechanisms, e.g. effects of deuteron channels, contribute to the scattering process for ^{40}Ca and that these are poorly reproduced by standard

OM potentials.[1] For the initial analysis we have therefore chosen an entirely conventional approach. Cross sections were obtained from the measured ratios of the present experiment and the ^{40}Ca cross sections of Ridley and Turner[4]. These were fitted with standard OM potentials, including an imaginary spin-orbit term[7].

Searches with all parameters free (11 parameters) gave very good fits to the data ($\chi^2_{\nu\nu} \lesssim 1$) as is shown in Fig. 6. However, it was clear that the data were not sufficent to fix the parameters of the spin-orbit potential, its unambigious specification apparently requiring asymmetry data or cross section data at more backward angles. We therefore constrained the spin-orbit parameters in two different ways: 1) Fixing the spin-orbit diffuseness and radius at the values obtained by Mackintosh and Kobos in their fits to ^{40}Ca data,[7] yielding 9 free parameters and 2) Fixing the spin orbit potential at the values of Becchetti and Greenlees[8] (and also the volume absorbtion term at 0.4 MeV) yielding 6 free parameters. The results of these searches are quite good, also yielding values of $\chi^2_{\nu\nu} \lesssim 1$. Results of the 6-and 9-parameter searches are shown in Fig. 5 and Table I, along with the volume integrals J_R and r.m.s. radii ($<r^2>^{\frac{1}{2}}_R$) of the real potentials. It is not certain that the more constrained results yield reliable values of the r.m.s. radii, but taken in the most naive sense, neglecting any contribution from protons and all higher order effects, they imply that the increase in the r.m.s. neutron radius from ^{40}Ca to ^{44}Ca is substantially greater than from ^{44}Ca to ^{48}Ca. These results are consistent with analyses of the 800 MeV proton scattering data from LAMPF.

III. Discussion

The ratio technique appears to have the promise of providing accurate information for a reasonable number of isotopes - we discuss briefly here what might be the most likely use of such information. Clearly, cross sections for use in phenomenological analyses can be obtained whenever reliable absolute cross section data are available for any one of the target nuclides. But it seems more useful to take advantage of the special character of the measurements and to cast the analysis directly in terms of the cross section ratios. One might hope that theoretical models could predict differences in scattering among nearby nuclei more reliably than absolute cross sections for a given nucleus. As noted previously, phenomenological analyses might usefully be made in terms of differences in parameters from nucleus to nucleus.

Another phenomenological approach that may have some advantages is use of the reformulated optical model (ROM)[9]. In this model, one obtains the optical potential by folding the two-body interaction V_{ip}

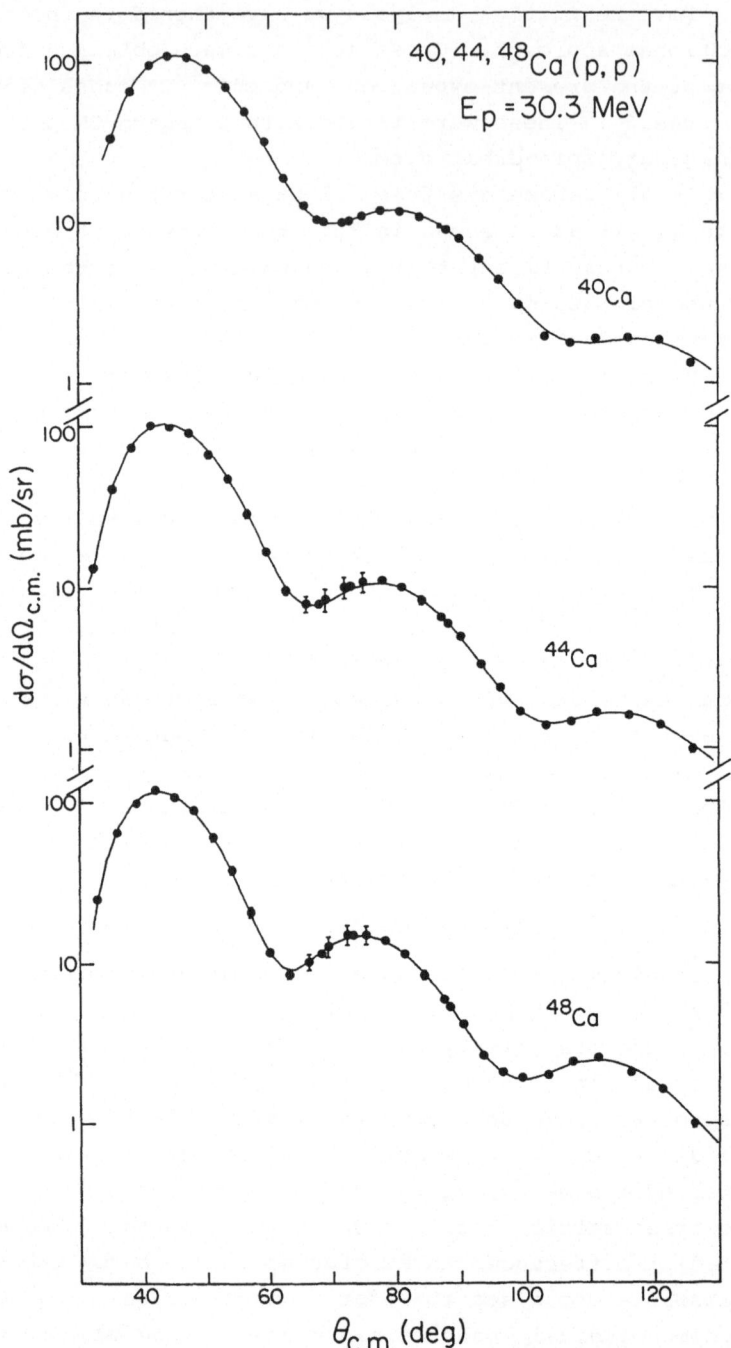

Fig. 6. Differential cross sections for ^{40,44,48}Ca. The ^{40}Ca data are from Ridley and Turner (Ref. 3) and the cross sections for ^{44,48}Ca were obtained from the ^{40}Ca cross sections and the ratios measured in this experiment. The curves are the results of the 11-parameter optical model searches.

Table I. Parameters[a] resulting from optical model searches

Nuclide	V (MeV)	r_R (fm)	a_r (fm)	W_D (MeV)	r_I (fm)	a_I (fm)	$\frac{J_R}{A}$ (MeV fm^3)	$<r^2>^{\frac{1}{2}}_R$ (fm)	χ^2_ν
			9 parameter	searches					
^{40}Ca	45.6	1.255	0.624	9.8	1.114	0.512	456	4.05	0.8
^{44}Ca	46.0	1.202	0.687	6.4	1.072	0.829	421	4.16	0.3
^{48}Ca	50.1	1.191	0.682	9.0	1.057	0.756	441	4.20	1.1
			6 parameter	searches					
^{40}Ca	48.4	1.161	0.686	5.7	1.259	0.667	411	4.00	2.1
^{44}Ca	45.7	1.223	0.663	9.0	1.159	0.609	432	4.16	0.6
^{48}Ca	46.5	1.274	0.586	12.7	1.161	0.473	467	4.20	0.6

a) Notation is that of ref. 10.
b) Values of (W, V_{so}, W_{so}) were (0.05, 3.3, 2.4), (1.2, 0.4, 0.4) and
     (0.1, 4.5, 0.1) all in MeV, for ^{40}Ca, ^{44}Ca and ^{48}Ca, respectively.
     r_{so} = 0.759 fm, a_{so} = 0.517 fm.
c) W = 0.4 MeV, V_{so} = 6.2 MeV, W_{so} = 0.0, r_{so} = 1.010 fm, a_{so} = 0.750 fm.

Table II. Results of estimates using the reformulated optical model (ROM)

Nuclide	$<r^2>^{\frac{1}{2}}_R$ (fm)		J_R (MeV-fm^3)	
	Expt[a]	ROM	Expt[a]	ROM[b]
^{40}Ca	4.00	3.98	411	411
^{44}Ca	4.16	4.06	432	427
^{48}Ca	4.20	4.08	467	441

a) From 6-parameter fit of Table I.
b) Normalized to ^{40}Ca. Original values were 9% higher.

between the projectile (p) and the target nucleons (i) with the neutron
and proton density distributions. This is essentially a first order model,
but it does appear to provide a reasonable description of at least the
real part of the OM potential, provided exchange and density dependence

effects are included. Two points of view are possible in fitting the data. One might assume that V_{ip} is well known, take proton densities from electron scattering data and then adjust the parameters of the neutron density distribution to fit the data. Requiring that the neutron and proton parameters are essentially identical for ^{40}Ca could provide a useful renormalization of the two-body force to account for inadequacies in V_{ip} or neglected effects. One could then obtain information about the neutron distributions of ^{44,48}Ca from directly fitting the cross section ratios. Whether the special sensitivity to neutrons, of proton scattering in this energy region is sufficient to overcome apparent complexities in the reaction mechanism remains to be seen.

Alternatively, for an N>Z case when the proton and neutron densities are well known, one might search on the parameters of V_{ip} to fix its isospin dependence.

As an indication of what one might expect from a ROM analysis, we have used the proton and neutron densities for ^{40,44,48}Ca from the preliminary analysis[11] of the LAMPF 800 MeV scattering data and the V_{ip} of Greenlees et al.[9] to estimate the variation of J_R and $<r^2>_R^{\frac{1}{2}}$ among the calcium isotopes. The results of this analysis are shown in Table II and are in reasonable agreement with the results of the constrained fits, particularly when one considers the approximations of the estimate (which, for example, entirely neglects density dependence).[12] It appears that the data do not allow an increase of more than 0.1 fm in the r.m.s. radius for neutrons between ^{44}Ca and ^{48}Ca. This is in good agreement with the results from the analysis of the 800 MeV data.

Turning to the technical aspects of the experiment, it appears that the precision obtained in the present experiment can be substantially improved. Two main effects dominated the uncertainties. First, the target thickness was somewhat too large to permit completely clean separation of peaks in the ^{12}C scattering experiment performed to determine the target composition. At the same time the target was of marginal thickness to give adequate statistics for the proton scattering measurements at backward angles. This problem could be attacked in two ways. The factor-of-ten larger solid angle of Q3D spectrographs would provide an immediate solution allowing use of thinner targets. Alternatively one could measure the isotopic ratios for a thin (20-30 $\mu g/cm^2$) target and use it as a secondary standard to determine the composition of a thicker target for measurements at backward angles. Either of these approaches should make possible measurements of the target composition with 1-2% precision and yield 1% statistics (in reasonable times) for proton scattering at all angles.

The other limitation involves fitting uncertainties at small angles due to the long tails on the peaks and uncertainties in the line shapes. The tails could be somewhat reduced in the present counter by setting tighter gates on the energy-loss signal, but great care would be necessary to avoid biasing the measured ratios. A new counter is presently under construction and should essentially eliminate these tails. Mixed targets containing only two nuclides could also substantially reduce fitting errors by allowing the line shape to be derived by combining the high energy side of the high energy peak and the low energy side of the low energy peak. It appears that fitting errors could be reduced by these two means to 1-2% (when the energy shift is equal to the energy resolution) and that cross sections could be obtained for substantially smaller angles.

IV. Summary

We have developed a technique employing a mixed target and kinematic shift identification by which ratios of cross sections for nearby nuclides can be determined essentially free of systematic uncertainties. The technique has been applied to a study of ^{40,44,48}Ca, yielding ratios with a relative uncertainty of 1-5%. In addition there is a overall normalization uncertainty of 2-3%. Straight-forward improvements in technique discussed in section III could reduce both these uncertainties to about 1-2%.

The present data are well-described by a standard optical model, and in a reformulated optical model estimate, yield differences among isotopes for r.m.s. radii of the neutron density distributions which are in agreement with results of analysis of 800 MeV proton scattering data.

V. References

1. R.S. Mackintosh and L.A. Cordero, Phys. Lett. 68B, 213 (1977).
2. A.M. Kobos and R.S. Mackintosh, J. Phys. G, to be published.
3. J.A. Nolen and P.S. Miller, Proceedings of the 7th Int'l Conf. on Cyclotrons and their Applications. (Birkhäuser, Basel, 1975) p.249.
4. B.W. Ridley and J.F. Turner, Nucl. Phys. 58, 497 (1964).
5. V. Hnizdo, O. Karban, J. Lowe, G.W. Greenlees, and W. Makofske, Phys. Rev. C 3, 1560 (1971).
6. R.G. Markham and R.G.H. Robertson, Nucl. Instr. and Meth. 129, 131 (1975).
7. R.S. Mackintosh and A.M. Kobos, J. Phys. G4, L135 (1978).
8. F.D. Becchetti, Jr. and G.W. Greenlees, Phys. Rev. 182, 1190 (1969).
9. G.W. Greenlees, W. Makofske, and G.J. Pyle, Phys. Rev. C 1, 1145 (1970) and references therein.
10. C.M. Perey and F.G. Perey, Atomic Data and Nucl. Data Tables 17, 1 (1976).
11. L. Ray, private communication.
12. B. Sinha, Phys. Rev. 20, 1 (1975).

Propagation of a deuteron in nuclear matter and the spin dependence
of the deuteron optical potential

A.A. Ioannides and R.C. Johnson
Department of Physics, University of Surrey,
Guildford, Surrey, GU2 5XH, England

1.   INTRODUCTION

It has been increasingly realised recently that the Pauli exclusion principle
plays a very important role in nuclear reactions involving multi-nucleon bound
projectiles and targets.   The simplest such example is the scattering of deuterons
from heavy nuclei.   Even in this case however effects due to antisymmetrization
cannot be clearly separated from other equally important effects unless drastic
approximations and assumptions are used.   The major part of this contribution is
devoted to the study of a much simpler problem, namely the propagation of a deuteron
through infinite nuclear matter.   This problem was studied in detail by Gambhir and
Griffin[1], as part of their deuteron nucleus studies, for the case that V_{np}, the
neutron-proton interaction, is the rank-1 S-wave potential of
Yamaguchi[2].   In this work we extend their method to the case when V_{np} contains a
tensor force component of the type of Yamaguchi and Yamaguchi[3].

The emphasis here will be on a clear and thorough description of the
underlying physical mechanism and the important role played by the tensor force
component in V_{np}, at the expense of a more formal and mathematical presentation.
These later aspects of the problem are treated in detail in a recent publication[4].

An outline of the extension of the nuclear matter calculations to the
realistic case of a high energy deuteron scattered by a heavy nucleus will be
presented at the end.   A brief account of this work can be found in reference 5
and a detailed description in reference 4.   The relevant literature on deuteron
nucleus scattering has been reviewed recently by Pong and Austern[6] who also
proposed an alternative method for calculating the spin independent part of the
effect under study here.   An alternative method for the spin-dependent part was
also proposed by Austern[7].

2.   BASIC FORMALISM

The wave function describing the internal motion of a bound n-p pair, which we
simply call deuteron hereafter, propagating with a centre of mass momentum $\vec{K}$,
through nuclear matter of Fermi momentum k_F, satisfies[4]

$$|\phi_{\vec{K}, \varepsilon_M}> = \frac{1}{\varepsilon_M - t_r} Q_{\vec{K}}(k_F) V_{np} |\phi_{\vec{K}, \varepsilon_M}> \qquad (2.1)$$

where the quantum number M is the eigenvalue of the operator

$$J_M = \vec{J}.\vec{K}/K, \quad \vec{J} = \text{deuteron total spin} \qquad (2.2)$$

and $B_M = -\varepsilon_M$ is the binding energy of the state with the corresponding M value.
The operator $Q_{\vec{K}}(k_F)$, which expresses the Pauli exclusion principle requirements,
acts on the spins and relative coordinates of the neutron and proton and is given by

$$Q_{\vec{K}}(k_F) = \sum_{\sigma_p, \sigma_n} \int d\vec{k} |\vec{k}, \sigma_p, \sigma_n><\vec{k}, \sigma_p, \sigma_n| \Theta_{k_F}(\vec{K}, \vec{k}) \qquad (2.3)$$

In nuclear matter $\Theta_{k_F}(\vec{K}, \vec{k})$ is a simple step function given by

$$\Theta_{k_F}(\vec{K},\vec{k}) = 1 \text{ if } k_p \text{ and } k_n > k_F$$

$$\Theta_{k_F}(\vec{K},\vec{k}) = 0 \text{ if } k_p \text{ or } k_n \leq k_F \qquad (2.4)$$

where k_p and k_n are the proton and neutron total momenta,

$$k_p = \frac{\vec{K}}{2} + \vec{k} \; ; \qquad k_n = \frac{\vec{K}}{2} - \vec{k} \qquad (2.5)$$

The angular dependence of $\Theta_{k_F}(\vec{K},\vec{k})$ involves only the relative angle, θ, between $\vec{K}$ and $\vec{k}$; it can be expressed in the form

$$\Theta_{k_F}(\vec{K},\vec{k}) = \frac{1}{2} \sum_{\ell \text{ even}} \Theta_\ell(K,k,k_F) \sqrt{\frac{4\pi}{2\ell+1}} \; Y_{\ell 0}(\theta) \qquad (2.6)$$

The projection operator $Q_{\vec{K}}(k_F)$ is spherically symmetric in $\vec{k}$-space if only the first term, with $\ell=0$, survives in the last equation, as for example in free space ($k_F=0$)

$$\Theta_{k_{F=0}}(\vec{K},\vec{k}) = \frac{1}{2} \Theta_0(k_F=0) = 1 \qquad (2.7)$$

The method used for the calculation of various properties of the deuteron in nuclear matter and the corresponding results are described in detail in reference 4. The most important feature of these results is the behaviour of the deuteron binding energy in nuclear matter which we outline below.

The total spin of the deuteron J is not a good quantum number in nuclear matter but its projection along $\vec{K}$ is. A deuteron which in free space is triply degenerate (M=±1,0), with binding energy B_f, splits in nuclear matter into two states of different binding energy, B_M: a non-degenerate state with M=0, $B_0 = -\varepsilon_0$, which has a vanishing angular momentum projection along $\vec{K}$ and is termed "ortho-state", and a doubly degenerate state with M=±1, $B_{\pm1} = -\varepsilon_{\pm1}$, which has a non-vanishing angular momentum projection along $\vec{K}$ and is termed "para-state".

The binding energy in nuclear matter is always less or equal to the binding energy in free space,

$$B_f \geq B_{\pm1}, \; B_0 \qquad (2.8)$$

The binding energy of the ortho-state, B_0, is very sensitive, in nuclear matter, to details of the tensor interaction in V_{np} but $B_{\pm1}$ is not.

The quantity, $B_f - B_M$, can be interpreted as an effective potential for a deuteron, and it can be written as

$$B_f - B_M = \Delta\varepsilon_c + \frac{\Delta\varepsilon}{K^2} [(\vec{J}.\vec{K})^2 - \frac{2}{3} K^2] \qquad (2.9)$$

The spin independent part, $\Delta\varepsilon_c$, is largest when both K and k_F are small, and it becomes negligible when K becomes large. The second term in equation (2.9) has the form of a second rank tensor potential, of the T_p type, in the original classification by Satchler[8]. Its magnitude, $\Delta\varepsilon = B_{\pm1} - B_0$, is small when both K and k_F are small and has a maximum when K is of the order of 5 fm^{-1} for $k_F \sim 1.36$ fm^{-1}.

A simple picture for the underlying physical mechanism, which explains these results is presented in the next section.

3. A simple picture

In figure 1 we depict pictorially the space of relative momentum-$\vec{k}$-space, for a deuteron propagating with a center of mass momentum $\vec{K}$, through nuclear matter of Fermi momentum k_F. Figure 1A and 1B represent the cases $K>2k_F$ and $K<2k_F$ respectively. In both cases the line $\overrightarrow{PN}$ represents the center of mass momentum $\vec{K}$ in both magnitude and direction. With the middle point of PN,O, as origin, the part of $\vec{k}$-space forbidden by the Pauli exclusion principle lies within the two Fermi spheres of radii k_F, with centers at P and N. For a given relative momentum, $\vec{k}$, represented on the figure by $\overrightarrow{OR}$ the proton and neutron total momenta are respectively $\overrightarrow{PR}$ and $\overrightarrow{RN}$. Thus $\vec{k}$ ($\overrightarrow{OR}$) is allowed by the Pauli exclusion principle if the point R lies outside the two Fermi spheres and forbidden otherwise.

The $\vec{k}$-space is naturally divided into three spherically symmetric regions, as shown in figure 1, and defined as follows:

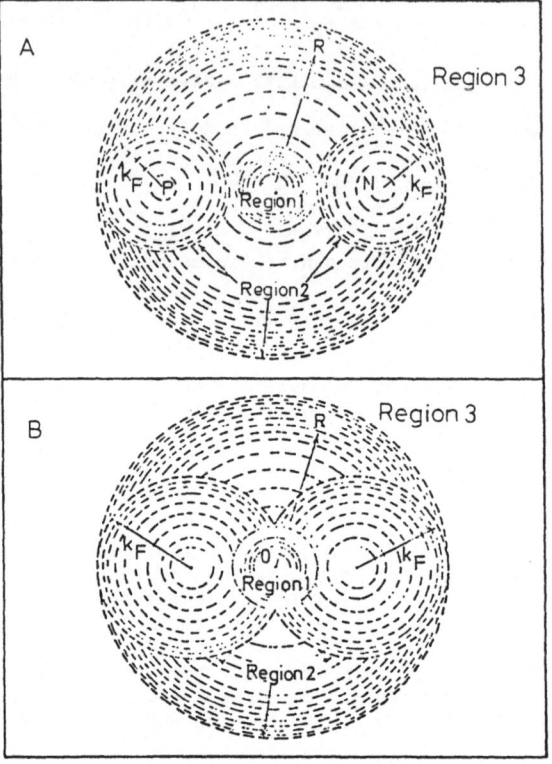

Fig.1 - A representation of the space of relative momenta ($\vec{k}$-space) for a deuteron in nuclear matter of Fermi momentum k_f. The deuteron center of mass momentum is $\vec{K}=\overrightarrow{PN}$ and the origin of $\vec{k}$-space is located at O. In the top part (a) $K > 2k_F$ and in the lower part (b) $K < 2k_F$.

Region 1: $0<k<\mu$;
$$\mu=\frac{K}{2}-k_F \quad \text{if} \quad k > 2k_F \qquad \text{(Fig.1A)}$$
$$\mu=\sqrt{k_F^2-\frac{K^2}{4}} \quad \text{if} \quad k < 2k_F \qquad \text{(Fig.1B)}$$
$$\tag{3.1}$$

Region 2: $\mu<k<\nu$; $\nu=\frac{K}{2}+k_F$ $\qquad\qquad\qquad\qquad$ (3.2)

Region 3: $\nu<k<\infty$ $\qquad\qquad\qquad\qquad\qquad\qquad\qquad$ (3.3)

In region 3 all orientations for the relative momentum are allowed. In region 1 all orientations for $\vec{k}$ are allowed if $K>2k_F$ (Fig.1A), and forbidden if $K<2k_F$ (Fig.1B). In both regions 1 and 3 the allowed part of $\vec{k}$-space is spherically symmetrical and the expansion of the function $\theta_{k_F}(\vec{K},\vec{k})$ in terms of $Y_{\ell 0}(\theta)$ in equation (2.6) contains just the first term with $\ell=0$. In these two regions the Pauli requirements preserve the free space symmetry of the deuteron wave function.

The allowed part of region 2, the spherically symmetric space lying between regions 1 and 3, is not spherically symmetric; in this region some angular orientations for a given magnitude, k, for the relative momentum are allowed while others are forbidden. The expansion of $\Theta_{k_F}(\vec{K},\vec{k})$ in this region contains terms with $\ell > 0$ and the deuteron wave function cannot be represented by an admixture of S-and-D-state components only, as in free space but states with even ℓ values and $\ell > 2$ must be introduced.

In figure 2 the S-state, $P_0(k)$, and D-state, $P_2(k)$, distributions of relative momenta are plotted for the Yamaguchi[3] wavefunction. These distributions correspond to a free space D-state probability, $P_D=4\%$, and they are normalized so that their respective maxima located at $k=k_0$ and $k=k_2$ are one. The maximum of the S-state distribution is mainly determined by the binding energy, B_f, and is $k_0 \approx 0.2$ fm^{-1}. The maximum of the D-state distribution is very sensitive to details of the tensor interaction assumed and it is roughly $k_2 \approx 1$ fm^{-1}.

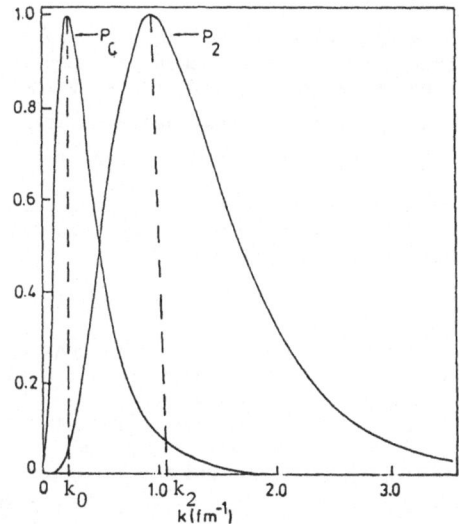

Fig.2 - Distribution of relative momenta in the S state (P_0) and the D state (P_2) corresponding to a D-state probability of 4%. P_0 and P_2 are normalized in such a way that their respective maxima are equal to one.

The severity of the Pauli effects, discussed here, is determined by how large a portion of the momentum space probability distributions $P_0(k)$ and $P_2(k)$ lies in the forbidden space. The overlap of $P_2(k)$ with the forbidden space is of great importance because of the strong binding effect of the tensor interaction. For a given V_{np} ($P_0(k)$ and $P_2(k)$ given) effects due to the Pauli exclusion principle are characterized by the magnitudes K, k_F, k_0 and k_2. For $2k_F > K > 2k_0$ the deuteron tends to be unbound.

For small deuteron energies E_K, $K \lesssim 2k_0$, and $k_F \lesssim k_0$, the S-state distribution has a strong overlap with the forbidden space. The D-state distribution lies mainly in region 3, which has no forbidden part. For this case Pauli effects are spin independent, e.g. $\Delta\varepsilon = B_{\pm 1} - B_0$ is small but $\Delta\varepsilon_c$ is large.

For large deuteron energies, $K \gtrsim 2k_2$, a bound deuteron usually exists when $\frac{K}{2} - k_0 > k_F$ (Fig.1A). In this case the S-state distribution lies mainly in region 1, which contains no forbidden part and the D-state distribution in region 2, provided k_F is not too small. The tensor force component in V_{np} couples the plane of the orbit of the neutron and proton about their center of mass to the deuteron total spin J. The net effect of this coupling is the preferential alignment of the D-state relative momenta around a plane perpendicular to $\vec{J}$. Thus in the ortho-state configuration ($\vec{J}.\vec{K}=M=0$) this plane, around which the D-state relative momenta are concentrated, contains the center of mass momentum $\vec{PN}$, in Fig.1A, and it therefore has the maximum overlap with the forbidden space. In contrast the corresponding plane of preferential alignment for $\vec{k}$, in the para-state configuration is the plane perpendicular to $\vec{PN}$ through 0. This plane lies in between the two Fermi spheres and it therefore has the least overlap with the forbidden space. Pauli effects are therefore stronger, and more sensitive to details of the tensor interaction in V_{np}, in the ortho-state configuration than corresponding effects in the para-state configuration. The $\vec{J}.\vec{K}$ dependence of the deuteron binding energy, thus produced, was found to be appreciable over a very wide energy range ($E_K < 1 BeV$) for k_F values corresponding to densities similar to the ones at the center of real heavy nuclei ($k_F \sim 1.36$ fm^{-1}). This is due to the large range of k values in the D-state distribution $P_2(k)$.

It is worth noting that a mechanism, similar to the one described above, is believed to play an important role in the understanding of the saturation properties of nuclear forces[9].

4. Finite size target nucleus

The nuclear matter calculations can be extended to the realistic case of a deuteron scattered by a heavy target nucleus[4] by assuming that each value, R, of the radial distance of the deuteron center of mass from the center of the target, the deuteron has a definite center of mass momentum, K(R), determined by the local value of the Coulomb energy and the nucleon nucleus optical potentials. The binding energies $B_M(R)$ and the binding energy difference, $\Delta\varepsilon(R) = B_{\pm 1}(R) - B_0(R)$, are determined in this model by solving a Bethe-Goldstone equation like equation (2.1) with k_F, the Fermi momentum corresponding to the local density (obtained from the point charge distribution[10]). For incident deuteron energies, E_d, $E_d \gtrsim 150$ MeV, where this model is expected to be valid[4,11], the T_p force thus produced, has a strength, $\Delta\varepsilon(R)$, which is very sensitive to details of the high momentum components in the deuteron wavefunction, and in particular to high momentum details in the tensor interaction in V_{np}.

In figure 3 we plot $\Delta\varepsilon(R)$ for a deuteron scattered of ^{92}Zr with E_d=250 MeV, and with the Yamaguchi[3] neutron proton interaction. The sensitivity on the tensor force part of V_{np} is reflected by the strong dependence of $\Delta\varepsilon(R)$ on the free space D-state probability, P_D, assumed.

For any deuteron wavefunction of the standard form

$$\phi_M(\vec{k}) = [u_s(k) + S_{12}(\hat{R}) u_D(k)] \chi_{1M} \qquad (4.1)$$

$\Delta\varepsilon$ can be estimated from the perturbation theory result;

$$\Delta\varepsilon \underset{K \gg k_F}{\approx} <\phi_1|QV_{np}|\phi_1> - <\phi_0|QV_{np}|\phi_0> \qquad (4.2)$$

which yields

$$\Delta\varepsilon(R,K) \underset{K \gg K_F}{\approx} 48\pi^3\rho(R)(E_d+B_f)u_D(\tfrac{K}{2}) \times$$

$$[u_D(\tfrac{K}{2}) - u_s(\tfrac{K}{2})] \qquad (4.3)$$

where $\rho(R)$ is the value of the target density at R.

This formula is accurate to within 15% for all nuclear regions for $E_d \gtrsim 300$ MeV and for regions

Fig.3 - Binding energy difference between para-state and ortho-state in a finite nucleus for the Yamaguchi potential and E_d=290 MeV.

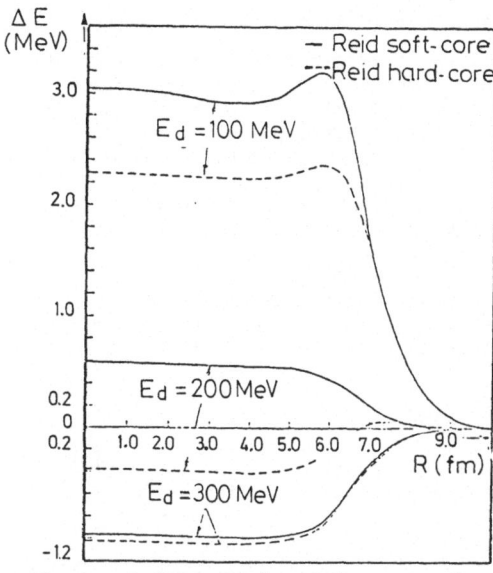

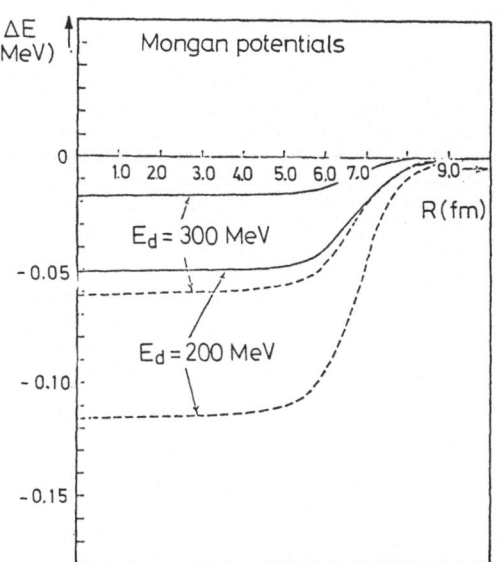

Fig.4 – $\Delta\varepsilon(R)$ calculated from equation (4.3) for ^{208}Pb(d,d) with the Reid soft core and hard core potentials.

Fig.5 – As in Fig.4, but for the Mongan potentials. The parameters of the V_{np} potential are taken from Table V (dashed line) and Table VI (solid line) in reference 13(b).

beyond the half-density radius for lower energies. The last equation exhibits clearly the dependence of $\Delta\varepsilon$ on the momentum distribution in the deuteron.

In Figure 4 $\Delta\varepsilon(R,K)$, calculated for the Reid[12] soft and hard core potentials, from equation (4.3), is plotted for deuteron ^{208}Pb elastic scattering at 100, 200 and 300 MeV. The unphysically large value for $\Delta\varepsilon$ for the interior region of the target at 100 MeV($\Delta\varepsilon > B_f$) is due to the fact that equation (4.3) is not accurate in this range. Differences between the soft and hard core results, for the higher energies, arise from differences in the high momentum components of the corresponding deuteron wave functions (the low momentum components are almost identical for the soft and hard core cases).

Another example of the sensitivity of $\Delta\varepsilon$ on the tensor interaction in V_{np} is shown in Figure 5, where $\Delta\varepsilon(R,K)$, calculated from equation (4.3), is plotted for some of the Mongan potentials[13]. The much smaller values for $\Delta\varepsilon$ in this case, compared to the corresponding values in Figures 3 and 4, are due to the small D-state probabilities of the Mongan potentials ($P_D < 1.1\%$).

5. Discussion

The study of Pauli effects in heavy ion collisions is nowadays a fashionable field of research. Approximations, made to reduce the great complexity and length of these calculations, are often judged by the agreement between theoretical predictions and experiment. The observables in heavy ion reactions depend little on spin because of the 1/A dependence of the spin-dependent potentials, and for this reason spin-dependent observables are not often measured.

In deuteron-nucleus elastic scattering, phenomenological potentials, resembling the ones derived from theory, provide reasonable fits for the differential cross-section, $d\sigma \over d\Omega$. For deuteron energies, however, above the Coulomb barrier energy, no theory proposed so far produced predictions for both

$\frac{d\sigma}{d\Omega}$ and the spin-dependent observables in reasonable agreement with experiment. Theoretical calculations, for the deuteron-nucleus effective interaction, have not treated, in the past, simultaneously the D-state component of the deuteron wave function and the Pauli exclusion principle. The calculations described in earlier sections indicate that the Pauli mechanism, under study here, plays a very important role in low energy deuteron-nucleus scattering. The model outlined in section 4 is not however expected to be valid in this energy region[11]. Calculations with an improved version of this model[14], which also accounts for the D-state component and, at least approximately the Pauli exclusion principle, based on the perturbed stationary state method[11,15,16], are in progress.

It seems to us that an extensive study of antisymmetrisation effects in deuteron-nucleus scattering, the simplest example of multi-nucleon bound projectile and target collision, can be a profitable prelude to investigations of the Fermi dynamics in the more complicated heavy ion case.

ACKNOWLEDGEMENTS

The calculations shown in Figures 4 and 5 were performed by J. Butcher[17]. The authors are grateful to W.S. Pong for useful and stimulating discussions.

References

1. B.L. Gambhir and J.J. Griffin, Phys.Rev.C7,590 (1973)
2. Y. Yamaguchi, Phys.Rev.95,1628 (1954)
3. Y. Yamaguchi and Y. Yamaguchi, Phys.Rev.95, 1634 (1954)
4. A.A. Ioannides and R.C. Johnson, Phys.Rev.C17, 1331 (1978)
5. R.C. Johnson and A.A. Ioannides, in Proc.Second. International Conf. on clustering phenomena in nuclei, Vol.II, Univ. of Maryland, 1975
    A.A. Ioannides and R.C. Johnson, Phys.Lett.61B, 4 (1976)
6. W.S. Pong and N. Austern, Ann.Phys.(N.Y.)93, 369 (1975)
7. N. Austern, Phys.Lett.61B, 7 (1976)
8. G.R. Satchler, Nucl.Phys.21, 116 (1960)
9. H.A. Bethe, Ann.Rev.Nucl.Sci.21, 95 (1971)
10. R. Engfer et al, At.Data Nucl.Data Tables 14, 509 (1974)
11. A.A. Ioannides, W.S. Pong and R.C. Johnson (unpublished)
12. R.V. Reid, Ann.Phys.(N.Y.) 50, 411 (1968)
13. (a) T.R. Mongan, Phys.Rev.175, 1260 (1968)
    (b) T.R. Mongan, Phys.Rev.178, 1597 (1969)
14. A.A. Ioannides (unpublished)
15. D.R. Bates et al, Proc. of the Royal Soc. A216, 437 (1953)
    T.Y. Wu and T. Ohmura, The Quantum theory of Scattering, (1963) pp228-231
16. W.S. Pong, private communication
17. J. Butcher, University of Surrey report (1977), (unpublished)

Description of Low Energy Deuteron Scattering
Using Multishell Form Factors

J. Stumm and A. Hofmann

Physikalisches Institut der Universität Erlangen-Nürnberg

8520 Erlangen, Erwin-Rommel-Straße 1, Germany

Abstract: Using multishell wave functions obtained by a fit to spectroscopic data of the (p,f)-shell we calculated the elastic and inelastic scattering data for deuteron scattering on ^{58}Ni at E_d = 10 MeV. The experimental data are described very well by these calculations.

Usually data of inelastic hadron scattering are described in the framework of the DWBA or of the coupled channels method (CC) on the basis of a "macroscopic" collective model description of the target nucleus. In these models the structure of the target nucleus is represented by the optical model potential.

During the last years intensive shell model calculations have been performed in the region of the (s,d) and (p,f) shell in order to reproduce mainly static properties of the nuclei [1]. There are only a few attempts to include inelastic scattering data in these shell model calculations.

After first encouraging results in a shell model description of low energy proton scattering on ^{58}Ni at different energies [2], we studied the low energy deuteron scattering on ^{58}Ni at E_d = 10 MeV using multishell model wave functions.

Thereby we used the same MSDI-multishell model wave functions which have been successfully used in nuclear structure calculations in the region A = 57-68 by Koops et al.[3]. The form factor for the inelastic deuteron scattering has been derived from a gaussian interaction between the projectile and the target nucleons with a realistic range of 2.5 fm [4]. The coupled channels calculations have been performed using the CC-Code INCH [5]. For the optical model potential we used a "standard" parameter set, given in table 1, which is similar to that used by Perey et al. |6|.

Table 1: Parameters of the optical model potential for the inelastic deuteron scattering on ^{58}Ni at E_d = 10 MeV (strength in Mev, length in fm).

V_r	r_r	a_r	W_d	r_d	a_d	V_{ls}	r_{ls}	a_{ls}
97.8	1.2	.81	15.5	1.34	.68	6.8	.95	.50

The results are shown in fig.1 together with the experimental values. In the elastic as well as in the inelastic channel the MSDI-multishell calculations show a rather good agreement with the experimental data.

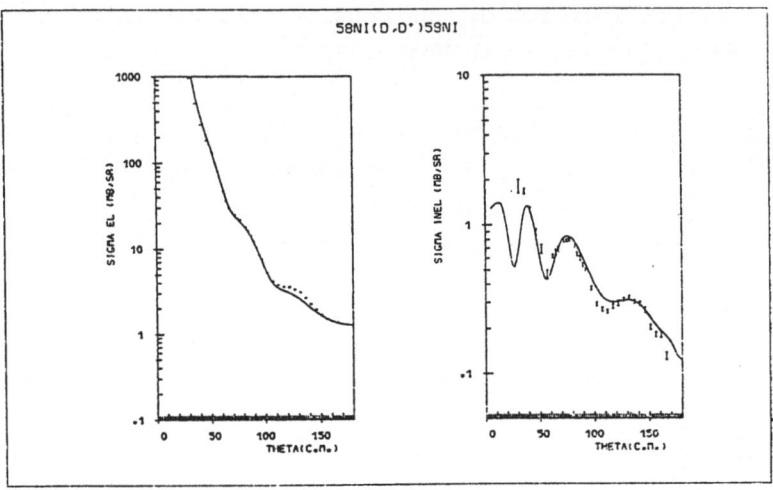

Fig.1: Calculated (MSDI) and experimental cross sections for elastic and inelastic deuteron scattering on ^{58}Ni at E_d= 10 MeV.

From these fairly good results we conclude that the analysis of low energy scattering data using multishell model wave functions for the description of the inelastic form factor may be able to connect the spectroscopic and the scattering data of the nucleus.

The next important step, however, must be a folding model description of the nuclear optical model potential in the framework of a multishell model description of the nuclear structure.

References

[1] F. Meuders, P.W.M. Glaudemans, J.F.A. van Hienen, G.A. Timmer, Z.Phys. A276(1976)113
[2] J. Stumm, A. Hofmann, J.Phys.G: Nucl.Phys.Vol.4, No.9,L229 (1978)
[3] J.E. Koops, P.W.M. Glaudemans, Z.Phys. A280(1977)181
[4] J. Stumm, Thesis, to be published
[5] A. Hill, computer code INCH (Oxford, unpublished)
[6] C.M. Perey, F.G. Perey, Phys.Review Vol. 132, No.2, 755 (1963)

Deuteron and ^{3}He Scattering:

Discussion on the Uniqueness of their Optical Potentials

A. Djaloeis

Institut für Kernphysik der Kernforschungsanlage Jülich GmbH, 5170 Jülich, W.Germany

Angular distributions of hadron elastic scatterings on nuclei have been widely analyzed in terms of the phenomenological optical model. Here the effective inter-action between the two colliding nuclei is represented by a complex potential with parameters essentially determined from the best fit to the experimental data. The theoretical cross section is obtained from a set of phase shifts δ_ℓ extracted after solving the corresponding Schrödinger equations. Complexities of the physical pro-blem, e.g., the internal nuclear structure and the possible virtual excitations etc. during the collision time, are hidden in the values of the fit parameters.

The fact that in this model the δ_ℓ's are calculated at a radius far larger than the nuclear radius leads to the cross section values which are rather insensitive to the details of the optical model wave functions in the nuclear interior. Indeed, certain variations of the optical model parameters leave the theoretical cross sec-tion invariant. Hence, knowledge of the elastic cross section alone usually leads to ambiguous information on the optical potential. In particular it is well-known that generally there exist several discrete potential families, each characterized by a certain depth V of the real part, giving equivalent fits to the experimental angular distribution.

Although the optical model can be regarded as a way of parametrizing the experi-mental data, it would be desirable to relate the extracted phenomenological fit parameters to other fundamental physical quantities, e.g., those of the matter dis-tribution and of the N-N potential. Indeed, the physically meaningful optical model parameters should be derived from these more basic quantities and not extracted from the best-fit criterion alone.

For the nucleon optical potential, the relation between the parameters of the real part to those of the nuclear matter and the V_{NN} potential seems to have been established. The real well is taken to have the Woods-Saxon form, following approxi-mately the matter distribution. The best-fit parameters were generally found to be $V \sim 50$ MeV, $r_V \sim 1.15 - 1.25$ fm and $a_V \sim 0.65 - 0.75$ fm. This potential has been widely used in the structure calculations for bound-state nuclear wave functions and found to give satisfactory agreement with the experimental data. Using the Greenlees pre-scription[1] it also gives consistent results to the mean square radius $\langle r^2_{nn} \rangle$ of the N-N potential.

While, in view of the above discussion, the real part of the proton phenomenological optical potential can be regarded as physically meaningful, the perscription to
select a corresponding one for the composite projectiles, such as deuteron (d) and
^{3}He, is not clear. Microscopic derivation of the parameters is more complicated and
has so far not been widely explored. In fact, here it is even questionable whether
the optical model description is still meaningful when the composite projectile is
at a short distance to the nucleus. Despite these complications, for loosely bound
projectiles, e.g. deuterons, it has been argued that the depth V of the real well
should be $V \sim nV_N$ where V_N is the depth of the nucleon well and n is number of nucleons constituting the projectiles. Recently based on semi-classical ideas, criteria
to select the unique potential family have been proposed and applied to α-scattering[2]. Here the elastic cross section at all angles is assumed to be solely due to
the potential scattering. According to these criteria, the experimental data have to
be taken at an incident energy such that the classical rainbow angle $\theta_r < \pi$ and span
an angular range extending beyond θ_r. It is worth noting at this point that the cross
sections at $\theta > \theta_r$ are generally very small and therefore it is questionable whether a
standard optical model analysis can still be applied here since in this region contributions from, e.g., two step processes may be significant.

To investigate this optical model problem for the deuteron and ^{3}He projectiles,
angular distributions of the elastic scattering of deuterons on ^{27}Al, ^{89}Y, ^{120}Sn and
^{208}Pb at E_d = 59 and 85 MeV and of ^{3}He on ^{24}Mg, ^{90}Zr, ^{120}Sn and ^{208}Pb at E_{3He}=130 MeV
have been measured in angular ranges extending beyond θ_r. The experiments were per-
formed at the Jülich isochronous cyclotron JULIC. The reaction products were detected
by means of two ΔE(Si)-E(Ge-Li) telescopes. The Li-drifted Ge-detectors were produced by the detector group[3] of the laboratory.

The experimental data have been analyzed in terms of the phenomenological optical
potential, with the real well having the Woods-Saxon (WS) shape. For the deuteron
case the imaginary potential was assumed to have the Woods-Saxon-Derivative (WSD)
shape; both WS and WSD shapes were investigated in the ^{3}He analysis. Despite the
spin 1 of the deuteron, only a spin-orbit term was considered. Preliminary calculations of the ^{3}He angular distributions indicated that the influence of the spin-orbit
potential could be compensated satisfactorily by slight changes in the values of
other parameters. Since no polarization data were available at this incident energy,
it was decided to neglect the ^{3}He spin-orbit term in the subsequent calculations.
The searches for the best-fit parameters were performed using the program MAGALI of
Raynal[4].

Fig. 1 shows the result of the V-grid searches for ^{120}Sn with the data extending
only up to $\theta_{c.m.} \sim 50°$, just before the exponential fall-off (refraction region)
starts. It can be seen that for both types of absorption two potential families,

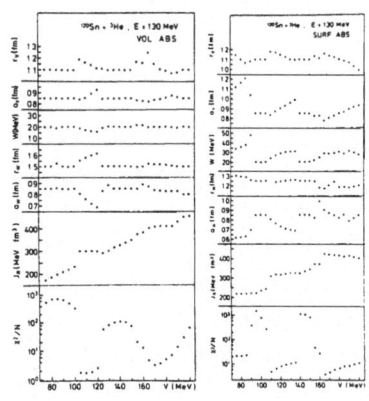

shallow and deep, characterized by $J_R \sim 330$ and 450 MeV fm^3, respectively, give equivalent fits. The same results were obtained for other nuclei studied here. Using the full set of data points extensive searches were performed to obtain the best fits and, in particular, to see whether the deep potential could be eliminated. The results are shown in fig. 2. In all four cases the shallow family gives consistently good fits. Here the surface and volume absorptions are found to give equivalent results. Furthermore, although the data extend considerably into the refraction region, it is generally still possible to obtain fits with the deep family; difficulty, however, seems to appear for ^{24}Mg and ^{120}Sn where the data extend furthest backward. This may imply that, by including data points at larger angles ($\theta > \theta_r$) in

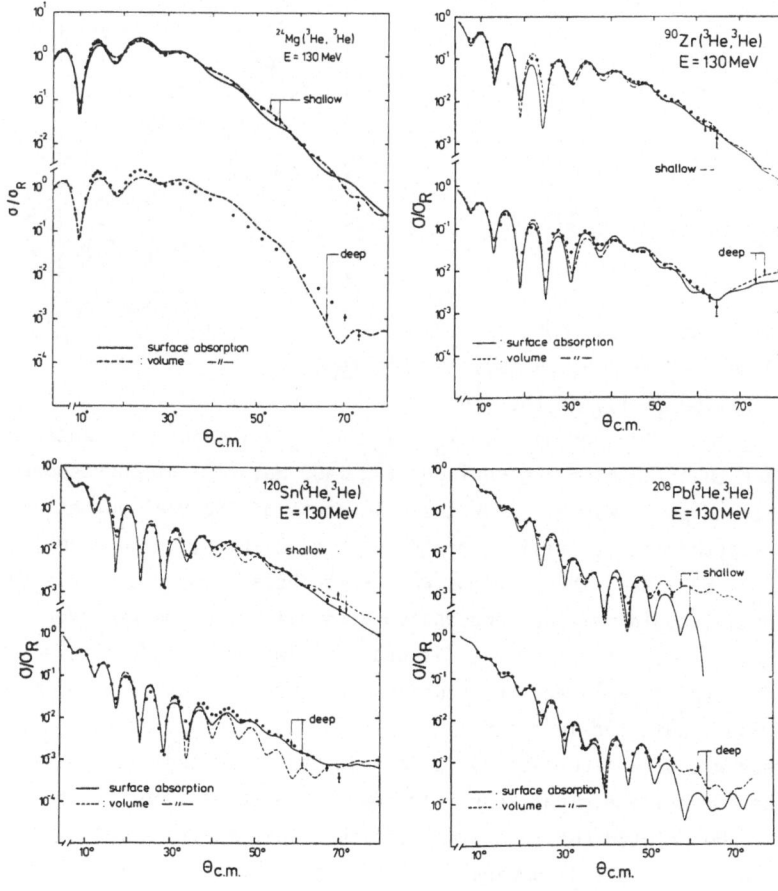

the search, the deep solution based on the standard optical model analysis alone might disappear. However, the physical meaning of such a procedure is questionable since beyond the rainbow region the cross section drops rapidly to very small values so that contributions, e.g., from two-step processes may be significant. Indeed, this has been found to be the case in an analysis of the ^{3}He elastic scattering on ^{58}Ni at 83 MeV incident

Figure 2

energy[5]. Explicit inclusion of the $(^3He,\alpha)(\alpha,^3He)$ reaction in the calculation resulted in a better fit by the deep family relative to the shallow one[5].

Fig. 3 shows the σ/σ_R angular distributions of the deuteron elastic scattering on ^{27}Al, ^{89}Y, ^{120}Sn and ^{208}Pb at E_d = 85 MeV. The data clearly exhibit the three well-known features, namely, a strong oscillation in the diffraction region (forward angles), a large enhancement in the rainbow region and an exponential fall-off at larger angles (refraction region). The optical model analysis was concentrated on two main points. The first was obtain the best fits using the deuteron real potential depth $V \lesssim 100$ MeV (approximately the sum of those of the constituent nucleons, i.e. $V_n + V_p$) as well as to study the dependence of the optical model parameters on the incident energy and the target mass. The results for 85 MeV are shown in fig. 3.

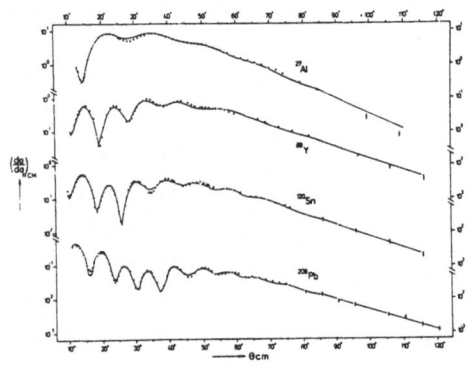

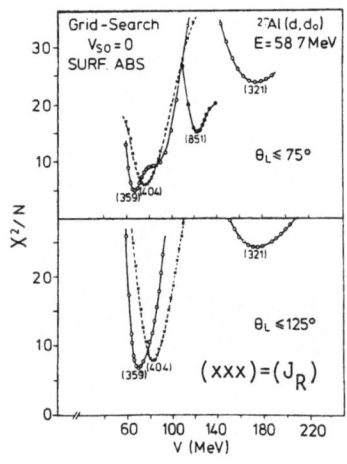

Figure 3

For all four nuclei the J_R-value was found to be $J_R \sim 300$ MeV·fm^3 ($V \sim 80$ MeV, $r_V \sim 1.10-1.15$ fm), close to that for the shallow family of the ^3He

Figure 4

case. It is worth noting that although here $V \sim V_n + V_p$ the J_R-value for the deuteron is significantly smaller than those of the nucleons ($J_R \sim 400$ MeV·fm^3) at corresponding energies[6]. The second point was to study the uniqueness of the real potential family. This has been started and is still in progress. Fig. 4 shows, as an example, the case for ^{27}Al at E_d = 58.7 MeV. This nucleus was chosen because here the absence of the spin-orbit potential could well be compensated by small variation of other parameters without significantly affecting the fit quality. Two sets of data were analyzed, namely the full ($\theta_L \leq 125^0$) and the truncated set ($\theta_L \leq 75^0$). In both cases only one family ($J_R \leq 359$, indicated in bracket in the figure) seems to exist. An indication of the appearance of another family is seen in the truncated set (curve labelled (851)). Further truncation might result in lower χ^2/N values, but has not been attempted. It is to be pointed out that ,in addition, there exists a set of potentials (labelled (404)) that gives comparable fits to the data. However its physical meaning is questionable since r_W seems too small ($r_W \sim 0.9$ fm) and the imaginary potential depth is large (~ 20 MeV). Similar features are observed also for ^{89}Y.

References:

1. G.W. Greenlees, G.J. Pyle and Y.C. Tang, Phys. Rev. 171 (1968) 1115

2. D.A. Goldberg and S.M. Smith, Phys. Rev. Lett. 29 (1972) 500

3. G. Riepe and D. Protič, Nucl. Instr. 101 (1972) 77

4. J. Raynal, Optical Model Program MAGALI, CEN, Saclay

5. J.R. Shepard, P.D. Kunz and J.J. Kraushaar, Phys. Lett. 56 (1975) 135

6. P.E. Hodgson, Nuclear Reactions and Nuclear Structure, Clarendon Press, Oxford, 1971

The form of the spin-orbit potential for spin-$\frac{1}{2}$ particles

S. Roman

Department of Physics, The University of Birmingham, England

Abstract

The spin-orbit part of the ^{3}He optical model potential determined
from the elastic scattering measurements including ^{3}He polarization
data for a range of nuclei, is characterized by a small diffuseness
parameter within the range $0.2 \leqslant a_{so} \leqslant 0.4$ fm for most nuclei. This
recent finding for ^{3}He is compared with the available triton (Los
Alamos) and nucleon elastic scattering data, which are re-examined in
terms of the conventional parametrization with derivative Woods-Saxon
form of the spin-orbit potential. It is found that the neutron elastic
scattering data are consistent with a spin-orbit potential with sharp
surface localization, in contrast to the established proton and triton
potentials, confirmed in the present analysis. The suggestion that
the difference between the triton and ^{3}He spin-orbit potential reflects
a similar difference in the behaviour of protons and neutrons in the
asymptotic region is admittedly based on the very scarce fast neutron
elastic scattering polarization data available. It is shown how
additional information may be obtained from study of (d,n) reactions,
which are sensitive to the neutron spin-orbit potential geometry.

Since the first measurements of the polarization in ^{3}He elastic scattering have become available with the work on ^{12}C[1], the sharp surface localization of the spin-orbit potential for ^{3}He has been confirmed in every case subsequently studied. The following nuclei have been investigated to date: a) energy dependence for the scattering by ^{12}C[2]; b) ^{16}O[3]; c) ^{26}Mg[4]; d) ^{27}Al[5]; e) ^{32}S[6]; f) ^{40}Ca[3]; g) ^{58}Ni[7]. The range of the ^{3}He results and their accuracy are not of a standard comparable to the precision of the available proton or deuteron polarization data. Nevertheless, the consistency with which the 'small geometry' of the ^{3}He spin-orbit potential is determined, i.e. the diffuseness parameter within the range $0.2 \leqslant a_s \leqslant 0.35$ fm or smaller, suggest that this effect is genuine within the conventional parametrization using derivative Woods-Saxon form of the spin-orbit potential.

These findings for the ^{3}He spin-orbit potential are in contrast with the 15 MeV Los Alamos triton elastic scattering polarization data[8] which yielded a conventional spin-orbit duffueseness within the range $0.63 \leqslant a_s \leqslant 0.92$ fm for the scattering of tritons by ^{52}Cr, ^{60}Ni, ^{90}Zr, ^{116}Sn and ^{208}Pb, which is similar to the accepted geometry for protons. Very recently the triton results have been augmented by further measurements on light targets, ^{9}Be and ^{12}C, which have been described by a similarly 'large' diffuseness.

To rule out any possibility of having reached wrong conclusions, by e.g. the optical model program search converging to a subsidiary χ^2 minimum or (and) parameters ambiguities, the Los Alamos ^{52}Cr and ^{90}Zr data[8] have been fitted using the code 'RAROMP' as used with ^{3}He. The parameters obtained in multiparameter searches confirmed the triton potentials of ref. 8 which have been obtained with the code 'SNOOPY' using a different search routine. Clearly then, the triton spin-orbit geometry is similar to the well established spin-orbit geometry for the case of the proton-nucleus elastic scattering, as determined by Becchetti and Greenlees[10] in their 'global' analysis.

The similarity of the triton and proton behaviour suggests that the polarization effects in the ^{3}He scattering ought to be instead compared with the neutron case since neutron gives the residual spin of the ^{3}He. For this purpose it was decided to carry out an analysis of all suitable neutron elastic scattering polarization data. Unfortunately very few polarization measurements of the elastic neutron scattering have been made at energies above a few MeV and the existing data are of a poor quality. The low energy data are not suitable since the compound nucleus contribution is hard to allow for.

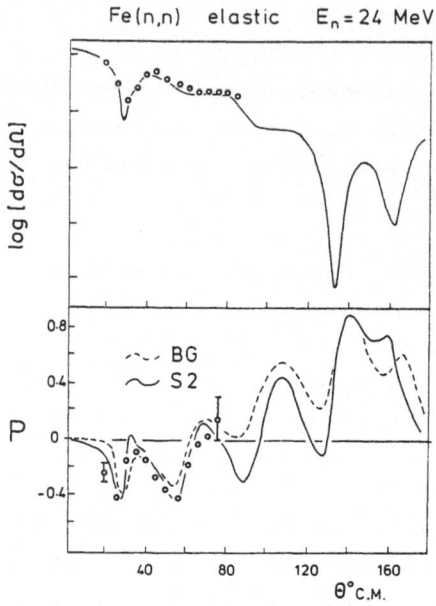

Fig. 1. The elastic scattering of 24 MeV polarized neutrons by Iron and optical model predictions obtained with parameters given in Table 1. The data fre from ref. 11.

Only three sets of fast neutron polarization measurements have been found in the literature which could be compared with the proton scattering: a) the 24 MeV data for C, Al, Fe and Pb[11] which have been used in the analysis of Becchetti and Greenlees[10]; b) the 14.1 MeV data for Oxygen[12]; c) the 10.4 MeV measurements for Pb and Bi[13]. The analysis of these neutron polarization data together with the associated cross sections (except for Pb at 24 MeV, where cross section data are not available) has been carried out starting with the average parameters of Becchetti and Greenlees (BG), searching in the spin-orbit potential (three parameters) and the central potential strength V_R (parameter sets label S); in addition W_D required adjustment for ^{16}O in order to be able to fit the cross section (set labelled R). The parameters obtained are shown in Table 1. It appears that the 'small' geometry is required except for ^{16}O, whereas the a_s values for Pb at both energies are somewhat larger. As an example, some of the fits obtained for Fe are shown in fig. 1. Clearly more and better neutron polarization data are needed to confirm this trend.

It is also interesting to note that for all four spin-½ particles the spin-orbit radius parameters r_s tend to have values very near to the radii of real central potential r_R, which is consistent with the respective values of 1.17 and 1.01 fm of Greenlees and Becchetti[10] for both protons and neutrons. Since the optical model predictions are not very sensitive to changes of the radius parameter r_s, more definite conclusions must be postponed until higher precision neutron data become available.

Table 1

Optical Model Parameters for the Elastic Scattering of Neutrons by Nuclei

	E_n	V_R	W_D	W_v	V_s	r_s	a_s	$\chi^2\sigma$	χ^2p	label
^{12}C	24.0	48.62	7.0	3.72	6.20	1.010	0.750	10.0	10.0	BG
		51.11	7.0	3.72	1.12	1.010	0.220	9.1	10.2	S1
		50.89	7.0	3.72	2.82	0.659	0.196	9.2	8.3	S2
^{16}O	14.1	51.79	9.48	1.54	6.20	1.010	0.750	10.0	10.0	BG
		51.80	2.72	1.54	5.87	1.010	0.859	1.9	1.1	R1
		51.66	2.72	1.54	4.88	1.162	0.673	1.7	1.4	R2
^{27}Al	24.0	47.73	6.56	3.72	6.20	1.010	0.750	10.0	10.0	BG
		47.57	6.56	3.72	1.29	1.110	0.123	8.7	16.3	S1
		45.13	5.19	3.72	1.59	0.985	0.452	3.7	15.6	R1
^{56}Fe	24.0	46.91	6.14	3.72	6.20	1.010	0.750	10.0	10.0	BG
		47.47	6.14	3.72	2.72	1.040	0.128	9.6	2.2	S1
		47.25	6.14	3.72	3.20	1.020	0.157	9.7	1.2	S2
^{208}Pb	10.4	47.89	7.86	0.73	6.20	1.010	0.750	10.0	10.0	BG
		48.40	7.86	0.73	7.42	0.926	0.380	7.9	11.4	S1
		48.44	7.86	0.73	5.28	1.226	0.438	9.4	8.0	S2
^{208}Pb	24.0	43.54	4.46	3.72	6.20	1.010	0.750	no data	10.0	BG
		46.32	4.46	3.72	2.79	0.989	0.456		2.2	S1
		46.25	2.60	3.72	3.05	1.010	0.540		2.0	R1
^{209}Bi	10.4	48.03	7.93	0.73	6.20	1.010	0.750	10.0	10.0	BG
		48.00	7.93	0.73	6.50	1.073	0.487	9.5	10.0	S1
		47.88	7.93	0.73	5.35	1.015	0.424	9.8	8.5	S2

r_R = 1.17, a_R = 0.75; r_w = 1.26, a_w = 0.58 throughout; all depths in MeV, lengths in fermi.

In search for additional information concerning the neutron spin-orbit potential geometry, the sensitivity of the analysing power of (d,n) reactions to the outgoing neutron distorting potential was investigated. Vector analysing power measurements for (d,n) reactions on ^{89}Y and Ni isotopes have been reported by the Wisconsin group[14,15] and the data extend to very small reaction angles down to 2.5°. In the present work DWBA calculations have shown that for some transitions the analysing power predictions in the small angles region are very sensitive to the choice of the outgoing neutron spin-orbit geometry. In particular for the $^{89}Y(d,n)^{90}Zr$ reaction DWBA calculations using various r_{so}, a_{so} combinations have shown a considerable sensitivity of the predictions to the choice of the geometry parameters at small reaction angles, fig. 2, confirming the need of a small spin-orbit potential diffuseness parameter and a radius of somewhat larger than the usually accepted value[10].

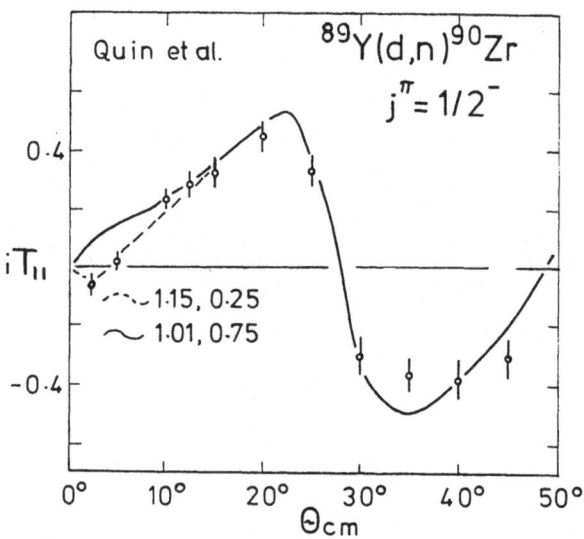

Fig. 2. DWBA calculations for the $^{89}Y(d,n)^{90}Zr$ reaction compared with the data of ref. 14 obtained at E_d = 11 MeV. The prediction shown by the continuous line was obtained using outgoing neutron parameters of ref. 10: the prediction shown by the broken line was obtained using neutron spin-orbit potential geometry parameters adjusted to the values indicated.

In conclusion, the different spin-orbit potential geometry of the triton and ^{3}He elastic scattering is confirmed in the present analysis. Furthermore it is suggested, admittedly on the basis of scarce neutron elastic scattering polarization data available, that the difference between tritons and ^{3}He reflects a similar difference in the behaviour of protons and neutrons in the asymptotic region.

1. W.E. Burcham et al., Nucl. Phys. A246 (1975) 269.
2. O. Karban et al., Nucl. Phys. A292 (1977) 1.
3. Y.-W. Lui et al., to be published.
4. M.D. Cohler et al., J. Phys. G, 2 (1976) L151.
5. N.M. Clarke et al., unpublished; S. Roman, Proc. 4th Polarization Symposium, Zurich, 1975, p.259.
6. J.F. Barnwell et al., to be published.
7. S. Roman et al., Nucl. Phys. A284 (1977) 365.
8. R.A. Hardekopf et al., Phys. Rev. Lett. 35 (1975) 1623.
9. P.A. Schmelzbach et al., Phys. Rev. C17 (1978) 16.
10. F.D. Becchetti Jr. and G.W. Greenlees, Phys. Rev. 182 (1969) 1190.
11. C. Wong et al. Phys. Rev. 238 (1962) 2339 (polarization), T.P. Stuart et al., Phys. Rev. 125 (1962) (cross sections).
12. R. Sene et al., Proc. 3rd Polarization Symposium, Madison, 1971, P.611, P.L. Beach et al., Phys. Rev. 156 (1967) 1201 (cross sections).
13. A.H. Hussein et al., Phys. Rev. C15 (1977) 233.
14. P.A. Quin et al., Nucl. Phys. A183 (1972) 173.
15. B.P. Hichwa et al., Nucl. Phys. A258 (1976) 21.

Polarization effects in elastic scattering of ^{3}He

O. Karban

Department of Physics, University of Birmingham, England

The elastic scattering of complex particles and the search for a model description has been of basic interest since the first application of the optical model to the nucleon-nucleus scattering. A comparison of the phenomenological potential with that derived from folding models provides a test of the wave function of the composite particle and various microscopic theories of the interaction. In the case of deuteron scattering the optical potential is now well established (with a possible exception of the tensor part) and the same can be said about the central part of the ^{3}He and triton potential. However, information on the spin-dependent interaction of the A = 3 particles with nuclei was until recently very limited. This requires measurements of elastic scattering polarizations which were made practically possibly only with the development of polarized ^{3}He and triton sources.

The data shown below were obtained at Birmingham University using the Lamb-shift type of polarized ^{3}He source installed at the Radial Ridge Cyclotron producing a 33 MeV ^{3}He beam. The on-target intensity was $\sim$ 500 pA with the beam polarization $P_b \sim \pm 0.60$, continuously monitored in a down-stream polarimeter based on the d-^{3}He scattering and absolutely calibrated within 10%. The left-right asymmetries ε for both beam polarization states were measured using six telescopes at three scattering angles simultaneously. The analysing power was calculated as $A = \varepsilon P_b^{-1}$, where $\varepsilon = (R-1)/(R+1)$, $R = \left[N_L^+ N_R^- / (N_L^- N_R^+) \right]^{\frac{1}{2}}$, to eliminate spurious asymmetries. In order to maximise the yield the target thickness was typically 1-2 MeV. Except for the very light nuclei the angular region covered was limited by the running time, typically three weeks per nucleus.

To-date the analysing powers in elastic scattering of ^{3}He were measured for the following nuclei: ^{58}Ni, ^{40}Ca, ^{32}S, ^{27}Al, ^{26}Mg, ^{16}O, ^{12}C, ^{9}Be, ^{6,7}Li, ^{3,4}He and ^{1,2,3}H. In all cases the magnitude of the analysing power is comparable to that observed in the proton elastic scattering. This fact already contradicts the folding model which predicts[1] that the ^{3}He polarization should be about nine times smaller than the proton polarization.

In order to obtain the phenomenological ^{3}He optical potentials the polarization data together with the cross-section measurements were analysed using the standard optical model. The search code[2] RAROMP was employed to minimise the χ^2-function varying nine parameters of the

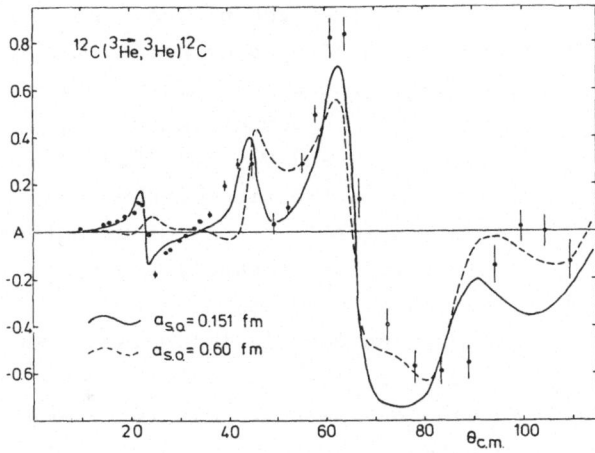

$^{12}C(^3\overrightarrow{He},^3He)^{12}C$

— $a_{so} = 0.151$ fm

---- $a_{so} = 0.60$ fm

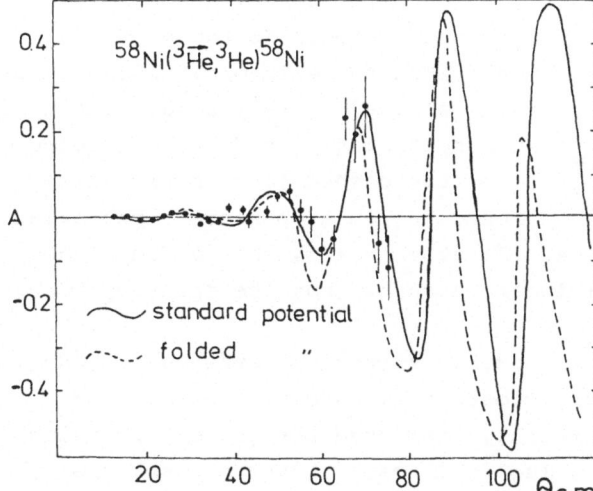

$^{58}Ni(^3\overrightarrow{He},^3He)^{58}Ni$

∿ standard potential

····· folded "

Fig. 1. Polarization of ^{3}He in elastic scattering by ^{12}C and ^{58}Ni.

usual optical potential with r_C = 1.3 fm. The results can be summarised as follows:
1) The search procedure leads to a set of optical-model parameters which can well reproduce both the cross section and polarization data. Typical polarization fits are shown in fig. 1 for ^{12}C and ^{58}Ni (solid lines). The spin-orbit interaction is described by parameters given for some nuclei in the table. The effect of the spin-orbit potential on the polarization cannot be separated since it is correlated with other parameters, particularly W.
2) The striking feature of the spin-orbit potential is its small diffuseness value, typically a_{so} = 0.2 fm. This means strong localisation of the spin-orbit potential at the nuclear surface (see fig. 2 where the real, imaginary and spin-orbit potentials for ^{58}Ni are plotted as a function of the radial distance) Alternatively, only 2-3 partial waves with L = kr give a significant contribution to the polarization (top scale in fig. 2).

The ^{3}He spin-orbit potential

	^{4}He	^{7}Li	^{9}Be	^{12}C	^{16}O	^{27}Al	^{40}Ca	^{58}Ni
V_{so}	0.61	2.44	3.55	2.29	2.99	1.89	3.14	4.69
r_{so}	1.10	1.409	1.391	1.212	1.255	1.37	1.227	1.276
a_{so}	0.16	0.122	0.178	0.151	0.217	0.189	0.232	0.293

3) There is no strong preference for a particular family of the ^{3}He

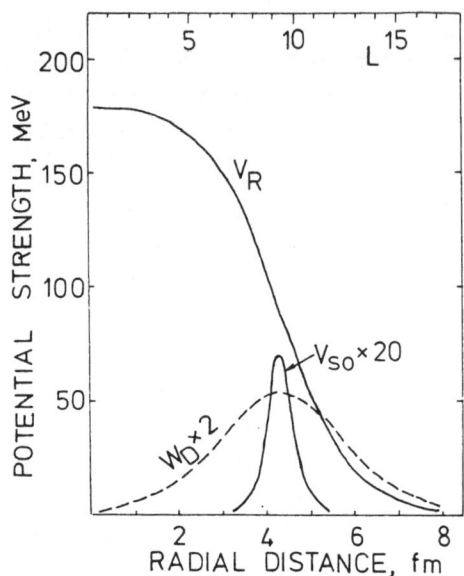

Fig. 2. Radial dependence of the real imaginary and spin-orbit potentials for ^{58}Ni.

potential, i.e. the presence of polarization data does not eliminate the well known discrete ambiguity. This is demonstrated in fig. 3 where V, W and V_{so} for ^{40}Ca are plotted as a function of the real volume integral J_R. All four potentials give approximately the same χ^2-value. There is, however, an unambiguous correlation between the three strengths.

4) The ^{58}Ni experimental data were compared with predictions using a folding-model potential obtained by the procedure of Keaton et al.[1]. Only when the spin-orbit parameters were allowed to vary a good fit to the polarization data was restored (see the dashed line in fig. 1), resulting in a spin-orbit potential close to the phenomenological one. In other words, the folding model describes correctly the central interaction but not the spin-dependent part.

With the aim to investigate possible structural effects the ^{3}He polarization was measured for two closed-shell nuclei ^{16}O and ^{40}Ca. The optical-model analysis essentially confirmed the above conclusions and the fits are shown in fig. 4 (solid lines). Nevertheless, the results indicate that in both cases the choice of the spin-orbit diffuseness is less critical than for other nuclei investigated. This is best demonstrated in fig. 5 where the χ^2-function is plotted as a function of the diffuseness parameter a_{so}. Not surprisingly, the χ^2 for the cross section data depends weakly on a_{so}. However, comparing ^{12}C with ^{16}O and ^{27}Al with ^{40}Ca, the deterioration in the polarization fit with increasing a_{so} is far less pronounced for the spherical nuclei. The dashed lines in figs. 1 and 4 show the predictions with a_{so} = 0.6 fm. It seems that the increased diffuseness affects fits mostly at the small scattering angles.

It has been recently suggested[3]) by the King's College group that the microscopic spin-orbit potential can be split into two components corresponding to the core and valence nucleons in the target. The above data, which apparently indicate some differences in the ^{3}He

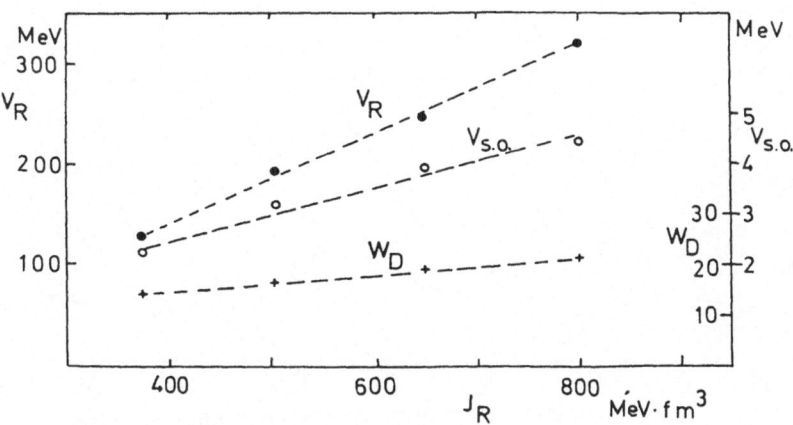

Fig. 3. Discrete ambiguity of the ^{3}He potential strengths for ^{40}Ca.

polarization effects for spherical and deformed nuclei, could provide
an interesting test of these ideas.

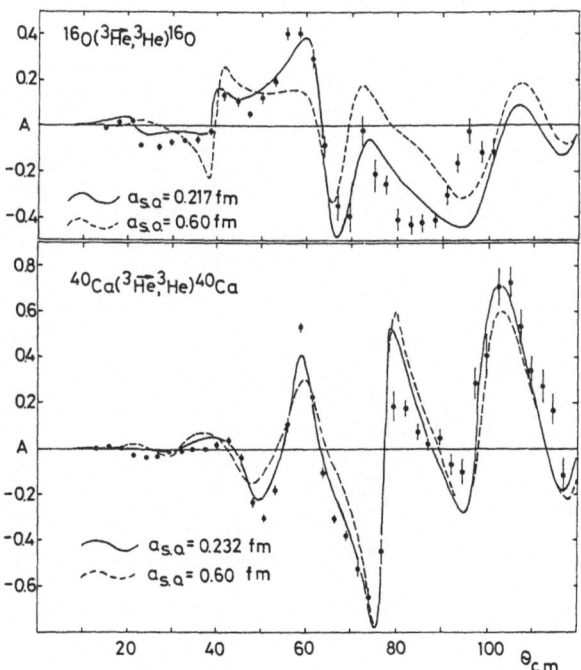

Fig. 4. Polarization of ^{3}He in elastic
scattering by ^{16}O and ^{40}Ca.

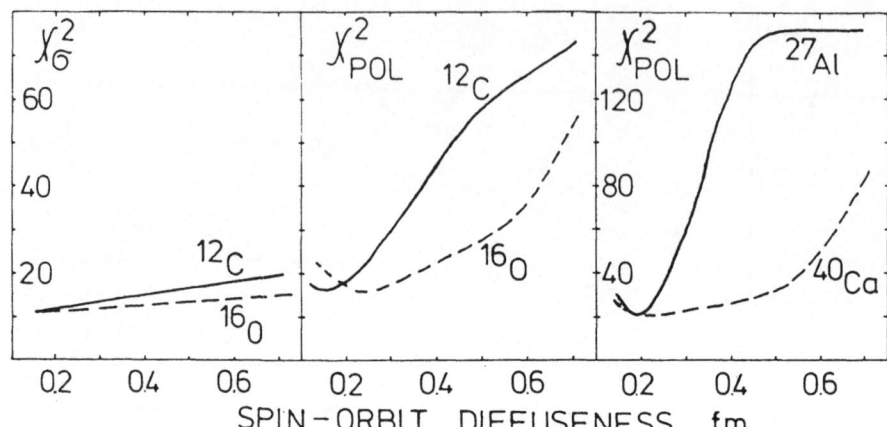

Fig. 5. The χ^2-values as functions of the spin-orbit diffuseness.

1) P.W. Keaton et al., Los Alamos Report LA-4379-MS (1970).
2) G.J. Pyle, University of Minnesota Report COO-1264-64 (1964).
3) M.D. Cohler et al., to be published.

Sensitivity of Alpha-Decay to the Real Alpha-Nucleus Potential

Daphne F. Jackson

Department of Physics, University of Surrey, Guildford, U.K.

Abstract

The information which can be obtained from studies of low energy alpha-particle scattering from heavy nuclei and from alpha-decay is discussed. The sensitivity of calculated widths and lifetimes for alpha-decay to the real nuclear potential is examined in detail using a formalism based on the unified theory of nuclear reactions. It is shown that a combined study of alpha-decay and alpha-particle scattering at energies near the Coulomb barrier should give a very precise determination of the barrier height and radius, although there is a more uniquely defined separation distance some way beyond the barrier.

Introduction

The elastic scattering of alpha-particles at energies well above the Coulomb barrier has been studied extensively in order to determine appropriate size parameters and to investigate microscopic properties of the target nuclei[1]. The original investigation[2] of the strong interaction radius was, however, based on the combined interpretation of data on low energy alpha-particle scattering and on alpha-decay. In this paper, we return to the question of what information can be obtained from these low energy studies in the light of present microscopic models of the alpha-nucleus interaction.

Information from Low Energy Elastic Scattering

Elastic scattering of alpha-particles from heavy nuclei at energies comparable with the height of the Coulomb barrier has been shown to be rather insensitive to the detailed behaviour of the alpha-nucleus optical potential, particularly in the interior region[3-8]. Imaginary potentials of surface and volume shape give equally good fits to the data[5] and for a volume imaginary potential the quality of fit changes very little[7] when the central depth is varied over the range 20-60 MeV. However, several studies[3-6] have indicated that the position r_b of the peak of the barrier formed by the Coulomb potential plus the real nuclear potential and the maximum height $U(r_b)$ of this barrier are relatively well-determined. The sensitivity of r_b and $U(r_b)$ to optical potential parameters has since been examined in greater detail[7,9] and values obtained from sets of potential parameters *all of which yield fits to the elastic scattering data* are given in Table 1.

It has also been shown[4,5,7,9] that a single-folding model for the real potential can yield a behaviour which gives very satisfactory agreement with the data, although it is necessary to adjust the imaginary potential in order to obtain equally good fits for different choices of the alpha-nucleon potential and the nuclear density distribution[7,9]. Values of the barrier radius and height are given in Table 2 for three different alpha-nucleon forces $V_{\alpha n}$ which are defined as

Table 1. Values of the barrier radius r_b and barrier height $U(r_b)$ for ^{208}Pb calculated from phenomenological optical potentials which fit the elastic scattering data

Potential	r_b (fm)	$U(r_b)$ (MeV)
Ref 3	10.94	20.42
Ref 4, set A(BL)	10.9	20.49
Ref 7, a=0.52	11.14	20.21
Ref 7, a=0.58	10.96	20.42
Ref 7, a=0.62	10.83	20.57

Table 2. Values of the barrier radius r_b and barrier height $U(r_b)$ for ^{208}Pb calculated from microscopic optical potentials which fit the elastic scattering data

Potential	r_b (fm)	$U(r_b)$ (MeV)
H	10.8	20.60
H'	11.0	20.33
M	11.0	20.56

Table 3. Penetrability calculated with the WKB approximation for the decay $^{212}Po(gs) \rightarrow {}^{208}Pb(gs)$

Potential	Inner turning point (fm)	Penetrability
H	8.90	2.16×10^{-14}
H', BL	9.11	2.72×10^{-14}
M	9.20	5.13×10^{-14}

$$\left. \begin{array}{l} \text{H: Gaussian form, } V_o = 43 \text{ MeV, } K = 0.526 \text{fm}^{-1} \\ \text{H':Gaussian form, } V_o = 53.75 \text{ MeV, } K = 0.526 \text{ fm}^{-1} \\ \text{M: Saxon-Woods form, } V_o = 42.5 \text{ MeV, } a = 0.34 \text{ fm} \\ \qquad\qquad R = 1.43 - 0.0009E \text{ fm} \end{array} \right\} \qquad (1)$$

where E is the lab energy for nucleon scattering from the alpha-particle.

Penetrabilities in Alpha-Decay

The calculation of absolute alpha-decay rates has two aspects. The nuclear structure part of the problem is concerned with the extent to which the alpha-particle is preformed in the nucleus; this is measured by the spectroscopic factor S_α which can be expressed as the ratio of the experimentally observed width Γ to the one-body width Γ_{ob}, i.e.

$$\cdot\Gamma = S_\alpha \, \Gamma_{ob} \quad , \quad \Gamma = 0.693 \, h/T_{\frac{1}{2}} \tag{2}$$

where $T_{\frac{1}{2}}$ is the half-life. The nuclear reaction part of the problem is concerned with the penetration of the alpha-particle through the barrier. This is frequently described in terms of the R-matrix theory of low energy nuclear reactions in which it is assumed that the configuration space is separated into two regions; in the external region defined by $R > R_c$ the total wavefunction can be expressed as a sum of channel functions of simple product form while the wavefunction in the internal region is replaced by a complete set of eigenfunctions X_λ with real eigenvalues E_λ. In a single channel approximation, the width of a narrow resonance is given by

$$\Gamma = 2 \, P \, \gamma_\lambda^2 \tag{3}$$

where γ_λ^2 is the reduced width determined by the overlap of X_λ with the channel function at the channel surface and P is the penetrability

$$P = k \, R_c \, / [\bar{F}^2(R_c) + \bar{G}^2(R_c)]. \tag{4}$$

The functions $\bar{F}$ and $\bar{G}$ are those solutions of the one-body Schrödinger equation containing the Coulomb potential and the nuclear optical potential whose asymptotic behaviour corresponds to the usual Coulomb functions F and G. In Gamow's theory the penetrability P is calculated in a WKB approximation assuming that the α-particle is in an s-state relative to the residual nucleus, and it is then given by

$$P = \exp \left\{ \frac{-2\sqrt{2\mu}}{\hbar} \int_{r_i}^{r_t} [U(R) - T_\alpha]^{\frac{1}{2}} \, dR \right\} \tag{5}$$

where μ is the reduced mass of the alpha-particle, T_α is the measured kinetic energy of the emitted alpha-particle, $U(R)$ is the total real potential and r_i, r_t are the inner and outer turning points respectively.

The R-matrix theory does not define the channel radius R_c beyond the rather general requirement that it should be sufficiently large to exclude coupling between various exit channels. It is well known[10,11,12] that a small change in R_c can change the penetrability by an order of magnitude. Table 3 shows values of the penetrability calculated[9] using equation (5) for some of the potentials included in Tables 1 and 2. Penetrabilities calculated[10] using equation (4) also show variations by a factor of $\sim$ 2.5 due to ambiguities in the real optical potential.

Unified Reaction Theory for Alpha-Decay

The R-matrix theory clearly yields too much sensitivity to the alpha-nucleus potential which arises from ambiguities in the potential and uncertainty in the choice of channel radius. For this reason we have developed a formalism[11,12,13] which is based on the unified reaction theory of Feshbach[15] and MacDonald[16]. A similar approach has been adopted by Tobocman[17]. This method avoids the use of arbitrary radii and ambiguous phenomenological alpha-nucleus potentials and is based on a cluster representation of the decaying nucleus. It gives a first-order expression for the width of the form

$$\Gamma = 2\pi \int d\rho(E) \ |<\Psi|H-K|\psi^+>|^2 \tag{6}$$

where $\rho(E)$ is the density of final states and

$$H = H_\alpha + H_A + T_R + V_c + V_{\alpha A} \tag{7}$$

$$K = H_\alpha + H_A + T_R + U_R \tag{8}$$

where H_α, H_A are the Hamiltonians for the internal behaviour of the alpha-particle and the residual nucleus A, T_R is the kinetic energy operator for the motion of the alpha-particle relative to the residual nucleus, V_c is the Coulomb potential, $V_{\alpha A}$ is a many-body operator representing the total nuclear interaction between the nucleons in the alpha-particle and those in the residual nucleus, and U_R is a real potential chosen so that the initial state is a bound state and not a decaying state.

Initially, we neglected antisymmetrization between the α-particle and the residual nucleus A. The initial wavefunction can then be written as

$$\Psi = \sum C_n(n_\alpha, n_A) \ \phi_\alpha^{n_\alpha} \ \phi_A^{n_A} \ \bar{\chi}_n \tag{9}$$

where the motion of the alpha-particle relative to the residual nucleus is described by the bound state function $\bar{\chi}$ which corresponds to angular momentum L and a specified principal quantum number N. This function is generated in the potential

$$
\begin{aligned}
U_R(R) &= g[U_N(R) + V_c(R)] + (1-g)U_R(r_b) \quad && R \leq r_b \\
&= U_R(r_b) && R > r_b
\end{aligned}
\left.\vphantom{\begin{aligned}U_R(R)\\U_R\end{aligned}}\right\} \tag{10}
$$

where r_b is the barrier radius, U_N is the folding model potential

$$U_N(R) = <\phi_\alpha \ \phi_A|V_{\alpha A}|\phi_\alpha \ \phi_A> \ = \ <\phi_A|V_{\alpha n}|\phi_A> \tag{11}$$

and g is a parameter which is varied to give a bound state with the required number of nodes. The binding energy of the alpha-particle in this potential is

$$\varepsilon_\alpha = T_\alpha - U_R(r_b) - 2\mu \ L(L+1)/\hbar^2 \ r_b^2 \ . \tag{12}$$

The parameter g can be regarded as renormalizing the total interaction to take account of the use of an equivalent local potential. It is not needed when antisymmetrization is taken into account in a more exact way[18].

The wavefunction for the final state is taken to be

$$\psi^+ = \phi_\alpha \ \phi_A^{n_A} \ \chi^+ \tag{13}$$

where χ^+ is a non-resonant scattering state generated in the potential

$$
\begin{aligned}
U(R) &= U_R(R) \quad && R \leq r_b \\
&= U_N(R) + V_c(R) && R > r_b
\end{aligned}
\left.\vphantom{\begin{aligned}U(R)\\U\end{aligned}}\right\} \tag{14}
$$

which has the correct behaviour beyond r_b. Thus we are using what Tobocman[17] calls the "alternative" version of the Feshbach-MacDonald formalism with an off-resonance continuum wavefunction.

The quantum numbers NL of the bound state function $\bar{\chi}_\alpha$ are deduced in the un-antisymmetrized theory from the oscillator rule

$$2(N-1) + L = \sum_{i=1}^{4} [2(n_i - 1) + \ell_i] \tag{15}$$

where $n_i \, \ell_i$ are the quantum numbers for the two protons and two neutrons in the least bound single-particle states in the initial nucleus. The principal quantum number N deduced for the lowest relative s-state in a range of heavy and superheavy nuclei is given in Table 4. It changes by one unit as the neutron shell is closed

Table 4. Application of the oscillator rule for the decay of heavy and superheavy nuclei

Initial nucleus	Lowest four-nucleon configuration	2(N-1)+L	Lowest s-state
^{210}Po	$(1h_{9/2})^2 (3p_{1/2})^2$	20	11
^{212}Po	$(1h_{9/2})^2 (2g_{9/2})^2$	22	12
^{294}110	$(2f_{7/2})^2 (4s_{1/2})^2$	22	12
^{298}114	$(2f_{7/2})^2 (4s_{1/2})^2$	22	12
^{302}118	$(2f_{5/2})^2 (4s_{1/2})^2$	22	12
^{342}114	$(2f_{7/2})^2 (2h_{11/2})^2$	24	13

at a neutron number of 126 and does not change again until a neutron number of 228 is reached. This is due to the high degeneracy of the oscillator model and raises doubts[18-20] about the adequacy of the oscillator rule as an approximate means of taking the exclusion principle into account. In consequence we have reformulated[18] the theory to take antisymmetrization between the alpha-particle and the residual nucleus into account.

Sensitivity of Alpha-Decay Widths to the Optical Potential

We have made calculations[12] with the unantisymmetrized theory for the ground state to ground state transitions from ^{210,212,214}Po leading to isotopes of lead. These are all $0^+ \to 0^+$ transitions. In most of these calculations we constructed the microscopic alpha-nucleus potential U_N by folding the matter distribution of the residual nucleus as determined by other data with the forms of the alpha-nucleon interaction $V_{\alpha n}$ given in equation (1). We also used two phenomenological potentials of Saxon-Woods form, denoted by BL and DLF, which had been obtained[5,21] by fitting the low energy alpha-particle scattering from ^{208}Pb. Results obtained for the one-body width for the decay of ^{212}Po are given in Table 5. It can be seen that our microscopic potentials yield very similar results but that a small variation remains

Table 5. Calculated values of the one-body width for the decay
$^{212}Po(gs) \rightarrow {}^{208}Pb(gs)$

Potential	Γ_{ob} (MeV)	
	Correct barrier	Barrier fixed at 20.4 MeV
Microscopic H	1.02×10^{-14}	1.27×10^{-14}
H'	1.46×10^{-14}	1.37×10^{-14}
M	1.12×10^{-14}	1.32×10^{-14}
Phenomenological BL	1.45×10^{-13}	-
DLF	3.84×10^{-14}	-

which is due to the slightly different values of the barrier radius and height
(see Table 2).   The difference between the results for the phenomenological
potentials and the microscopic potentials can easily be understood because the
phenomenological potentials have smaller depths and larger half-way radii than the
microscopic potentials (see Figure 3 of ref.12) and hence the alpha-particle
wavefunctions generated in the phenomenological potentials extend farther out and
yield a larger overlap with the perturbing interaction and the scattering function.

A comparison of the results for the one-body width obtained using the microscopic potential H with and without exchange is given in Table 6, from which it can
be seen that antisymmetrization of the folding model brings the result into line
with the results for phenomenological potentials given in Table 5.   Table 6 also

Table 6. Results obtained for the decay $^{212}Po(gs) \rightarrow {}^{208}Pb(gs)$ using the
microscopic potential H

	Γ_{ob} (MeV)	$<r^2_\alpha>^{\frac{1}{2}}$ (fm)	R_{max} (fm)
No exchange	1.0×10^{-14}	6.99	8.0
With exchange	7.0×10^{-14}	7.80	8.5

gives the rms radius $<r^2_\alpha>^{\frac{1}{2}}$ of the bound state function and the position R_{max} of
maximum probability of finding the alpha-particle.   It should be noted that the rms
radius of the nuclear matter distribution of ^{208}Pb and the barrier radius of the
$\alpha + {}^{208}Pb$ interaction are the same in both calculations.   Thus the large values for
$<r^2_\alpha>^{\frac{1}{2}}$ and R_{max} are not incompatible with an acceptable rms radius for the residual
nucleus[19].   The effect of antisymmetrization is to include many semi-redundant
states which combine to suppress the interior region[18] and shift R_{max} out by 0.5 fm.
A similar result has been observed[22] for the $^{16}O + {}^{16}O$ system.

Calculations[14,23] have also been carried out for superheavy nuclei including
the N=184 isotones.   Since rather little is known about these nuclei, the nuclear

density distributions of the residual nuclei have been generated in a selection of single-particle potentials all of which can be partially justified in relation to properties of nuclei near the lead closed shell or in the actinide region.   Typical results obtained[23] in this way are given in Table 7 for the decay $^{294}110 \rightarrow {}^{290}108$, from which it can be seen that a change in the barrier height of 1.2 MeV changes the predicted lifetime by more than one order of magnitude.

Table 7.   Results obtained for the decay $^{294}110 \rightarrow {}^{290}108$ using a selection of single-particle potentials to generate the density distribution of the residual nucleus

Single-particle potential	r_b (fm)	$U(r_b)$ (MeV)	R_{max} (fm)	$T_{\frac{1}{2}}$ (s)
A	11.3	26.0	8.3	5.48×10^{18}
B	11.3	26.2	8.3	6.07×10^{18}
C	11.2	26.3	8.3	6.48×10^{18}
R	11.9	24.8	8.8	5.30×10^{17}

Determination of Size Parameters

It is evident that our method eliminates the major uncertainty in the calculation of alpha-decay rates in heavy nuclei by avoiding arbitrary radius parameters and removing potential ambiguities.   The barrier height and radius play a very important role in this method and hence it should be possible to combine the analysis of alpha-decay with alpha-particle scattering at energies near the Coulomb barrier in order to determine these quantities very precisely.   We consider that these are important quantities because of their clear physical meaning.   Also, for Saxon-Woods potentials we have shown[20] that plot of log S_α against r_b gives a straight line.

There is evidence[7], however, that distance at which the nuclear potential has fallen to -0.2 MeV is more uniquely determined from elastic alpha-particle scattering near the Coulomb barrier.   For the lead isotopes this distance is 1.02 fm beyond r_b. We have studied[20] the radial part of the integrand defined in equation (6) and find that for all cases the integrand for the decay $^{212}Po(gs) \rightarrow {}^{208}Pb(gs)$ shows a broad peak about a separation distance of r_b + 3.7 fm.   For the superheavy nuclei we have studied the integrand always peaks at a distance of 3.65-3.95 fm beyond r_b.

1.   R.C. Barrett and D.F. Jackson, *Nuclear Sizes and Structure* (Oxford University Press, 1977) Chapter 7 and references given there

2.   E. Rutherford, Proc.Roy.Soc.A123, 323 (1929)

3.   G. Goldring, M. Samuel, B.A. Watson, M.C. Bertin and S.L. Tabor, Phys.Lett.32B, 465 (1970)

4. C.J. Batty and E. Friedman, Phys.Lett.34B, 1 (1971)

5. A.R. Barnett and J.S. Lilley, Phys.Rev.C9, 2010 (1974)

6. Y. Eisen, E. Abramson, G. Engler, M. Samuel, U. Smilanksy and Z. Vager, Nucl. Phys.A236, 327 (1974)

7. I. Badawy, B. Berthier, P. Charles, M. Dost, B. Fernandez, J. Gastebois and S.M. Lee, preprint

8. D. Werdecker, Thesis, University of Cologne (1973), unpublished.   Results quoted in ref.7

9. D.F. Jackson and M. Rhoades-Brown, Nucl.Phys.A266, 61 (1976)

10. L. Scherk and E.W. Vogt,Canad.J. Phys.46, 1119 (1968)

11. E.A. Rauscher, J.O. Rasmussen and K. Harada, Nucl.Phys.A94, 33 (1967)

12. D.F. Jackson and M. Rhoades-Brown, Ann.Phys.105, 151 (1977)

13. D.F. Jackson and M. Rhoades-Brown, Nature 267, 593 (1977)

14. D.F. Jackson, *International Symposium on Superheavy Elements* (Lubbock, 1978) to be published by Pergamon Press

15. H. Feshbach, *Reaction Dynamics* (Gordon and Breach, New York, 1973)

16. W.M. MacDonald, Nucl.Phys.54, 393 (1964)

17. W. Tobocman, Phys.Rev.C13, 790(1976);   Phys.Rev.C17, 2205 (1978)

18. D.F. Jackson and M. Rhoades-Brown, J. Phys.G: Nucl.Phys, in the press

19. D.F. Jackson, J. Phys.G: Nucl.Phys 4, 1287 (1978)

20. D.F. Jackson and E.J. Wolstenholme, *Third International Conference on Clustering Aspects of Nuclei* (Winnipeg, 1978) to be published

21. R.M. DeVries, J.S. Lilley and M.A. Franey, Phys.Rev.Lett.37 481 (1976)

22. A. Tohsaki-Suzuki, K. Naito, T. Ando and K. Ikeda, *International Pre symposium on Clustering in Nuclei* (Tokyo, 1977)

23. D.F. Jackson, E.J. Wolstenholme, L.S. Julien and C.J. Batty, preprint

The Effective Surface Potential for α Particles and its OCM Justifications

K.A. Gridnev, V.M. Semjonov, V.B. Subbotin
Leningrad State University, Leningrad, USSR
E.F. Hefter
Technical University Hannover, 3 Hannover, Fed. Rep. of Germany

The development of heavy ion beams made it possible to pursue by aid of α transfer reactions systematic and intensive investigations of light nuclei. The studies showed that in the region of high excitation energies (10-30 MeV) there are many pure α cluster states in light nuclei. The reduced α widths they exhibit in their decay into the ground-state are of the order of the Wigner limit [1]. These states can be interpreted in terms of the Pauli principle. In theoretical studies with the phenomenological effective surface potential (ESP) this model has been shown to be a useful concept for further investigations [2].

The ESP interaction is made up of an attractive part, which is usually taken to have a Woods-Saxon geometry, and an ℓ-dependent repulsive hard or soft core localized in the region of the nuclear surface. Such a potential not only reproduces extremely well positions and widths of the α cluster levels but also the angular distributions of α particle scattering by light nuclei. In previous computations we considered mainly bombarding energies up to 12 MeV; for such low energies the role of open channels, but for the elastic one, is not important, implying that the imaginary part of the potential can be neglected.

Recently we applied the ESP successfully to transfer reactions, i.e. to $^{12}C(^{6}Li,d)^{16}O$. The experimental points displayed in fig. 3 stem from the work of Baccani et al. [3]. The broken line corresponds to calculations with the finite range code LOLA. The full line is based on the zero range DWBA with an ESP potential for the alpha particle; obviously the agreement of this curve to experiment is better than the one of LOLA. To save space we did not include figures related to the 4^{+}(10.35 MeV) and to the 6^{+}(16.3 MeV). The fit to the latter is of the same order as the one shown, the only difference being that experiment is underestimated for large angles. The angular distribution for the 4^{+} is almost exactly reproduced. But these data are to be discussed elsewhere.

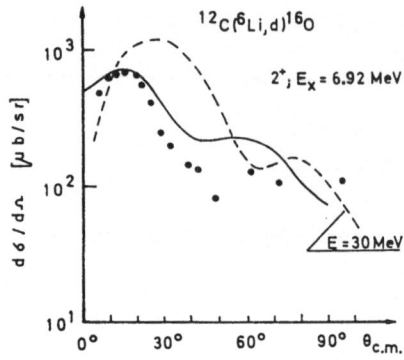

$^{12}C(^6Li,d)^{16}O$

$2^+; E_x = 6.92$ MeV

$E = 30$ MeV

Figure 1:

The angular distributions for the α transfer reaction are shown. Full curve – zero range DWBA with ESP potential for the transferred α-cluster; broken line – finite range code LOLA with a Woods-Saxon well for the α particle; the potential parameters are taken from ref.[4].

In spite of the good results obtained by aid of the ESP we would like to compare it to the microscopic OCM and RGM hoping to obtain further support for our model. Since the ESP reproduces experimental energy levels and widths to a high precision [2] the data labelled below 'experiment' can also be viewed as representing the ESP. The nucleus ^{20}Ne serves as the example to be considered in the following. In our calculations we solved the OCM equation [8]

$$\lambda_\ell P_\ell \; \lambda_\ell u_\ell = 0$$

where the projection operator λ eliminates the redundant solutions $(u_{N\ell}(r); M = 2N + \ell < 8)$, which are prohibited by the Pauli principle;

$$\lambda_\ell(r,r') = \delta_\ell(r,r') - \sum_{N=0}^{M} u_{N\ell}(r) \, u_{N\ell}(r') \; .$$

The harmonic oscillator wavefunction is represented by $u_{N\ell}$ and the symbol P_ℓ stands for

$$P_\ell = -\frac{\hbar^2}{2\mu}\frac{d^2}{dr^2} + \frac{\hbar^2}{2\mu r^2}\ell(\ell+1) + V_D^n(r) + V_D^c(r) - E \; .$$

In this relation $V_D^n(r)$ and $V_D^c(r)$ denote the direct nuclear and Coulomb potentials, respectively.

Elimination of the forbidden states gives rise to energy independent oscillations of the wavefunction of relative motion in the nuclear interior. The resulting behaviour of the OCM wavefunctions suggests that standing waves are formed in the region of the nuclear surface: Considering e.g. the 6_1^+ and 6_2^+ wavefunctions it is seen that they have one and two zeros, resp., in the region of the nuclear surface. The one of the 6_1^+ is almost exactly in the middle between the ones for the 6_2^+. All three zeros remain at the same positions no matter which values are taken on by the energy. This feature can be described

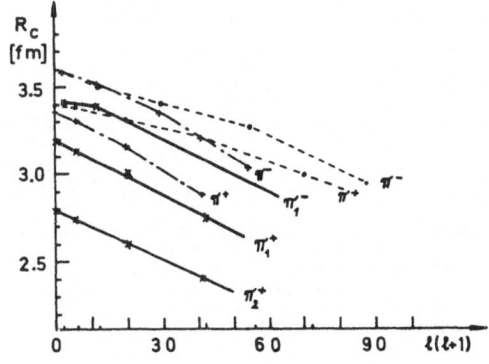

For the nucleus ^{20}Ne the radius R_c as obtained within the ESP (broken lines), by aid of Laguerre polynomials (dashed-dotted lines) and via the OCM is plotted versus $\ell(\ell+1)$.

by introducing a repulsive core into the interaction [2,9] . In this picture the position of the hard core is determined by the last zeros of the wavefunctions $(X_\ell^i (r) ; \lambda_i = 1)$ for the allowed states.

Now we would like to study the zeros as functions of $\ell(\ell+1)$. To this end we remember that the Pauli principle is known to lead to solutions $X_\ell^i (r)$, which are eigenfunctions of the exchange operator $K(r,r')$ with $\lambda_i = 1$. If the oscillator shell-model with the same oscillator parameters for ^{16}O and the α particle is used then the harmonic oscillator wavefunctions are proportional to the Laguerre polynomials,

$$L_{N-1}^{\ell + 1/2} (a^2 r^2) = L_{N-1}^{\ell + 1/2} (\xi^2)$$

with $a^2 = \mu \nu$. The frequency ν is determined by the relation $\nu^{-1} = 1.01 A^{1/3} fm^2$. N is the principal quantum number of the shell model. In order to obtain the radius of the repulsive core, R_c, one has to determine this quantum number.

In the case of ^{20}Ne the even rotational band $\pi_1^+ (0^+,2^+,4^+,6^+,8^+)$ consists of 2s-1d nucleons and hence the number of the oscillator quanta is N = 8. According to Thalmi's rule $(2(N-1)+\ell = 8)$, the first allowed states for $J^\pi = 0^+,2^+,4^+,6^+,8^+$ are characterized by the quantum numbers N = 5,4,3,2,1. The states with N = 1,2,3,4 for $J^\pi = 0^+$, with N = 1,2,3 for $J^\pi = 2^+$, with N = 1,2 for $J^\pi = 4^+$ and with N = 1 for $J^\pi = 6^+$ are forbidden by the Pauli principle.

For the π_1^+ the zeros of the Laguerre polynomials with $(N, \ell) = (5,0)$, (4,2), (3,4) (2,6) are calculated to study the dependence of the radius of the core as a function of $\ell(\ell + 1)$. Similarly the zeros for the π_1^- band were obtained. The dotted lines in Fig. 2 demonstrate that the bands are splitted according to their parity. For a comparison, earlier results of the ESP are included in the figure (dashed lines). Finally, the OCM was used to compute the wavefunction of the relative motion; its zeros are represented by the upper two full curves in fig. 2.

No matter which method we take, all curves clearly illustrate the splitting of even and odd rotational bands and show that the radius of the repulsive core decreases with increasing $\ell(\ell+1)$. Since we are only interested in a qualitative understanding of these features we did not attempt to modify any of these parameters, which would have allowed us to produce identical results for all methods used. The lowest full curve in the same figure corresponds to the zeros of the 6_2^+ band being formed from nucleons out of the $(sd)^2(fp)^2$ shells. It is strongly forbidden in comparison to the even band 6_1^+ $(N=10)$.

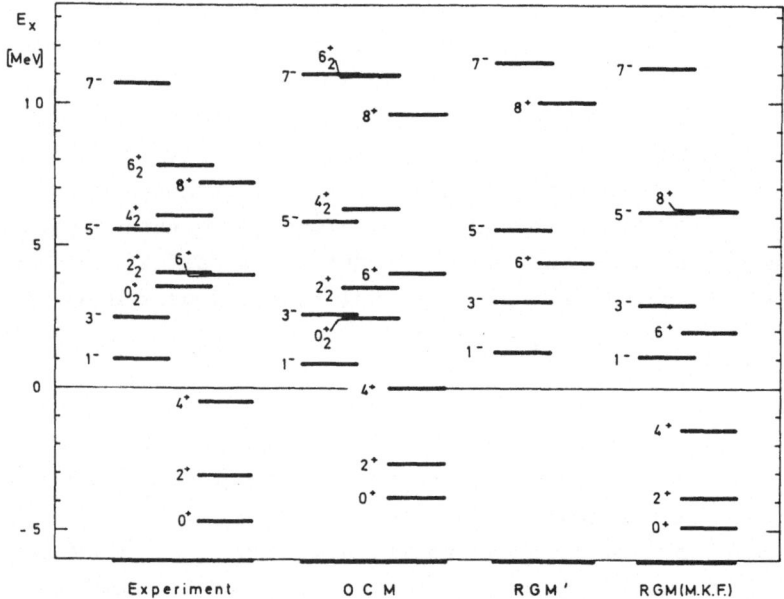

Figure 3: For the system $^{16}O+\alpha$ experimental excitation energies are compared to calculations performed by aid of the OCM, RGM' and the RGM(MKF).

In the level scheme presented in fig. 3, the positions of the experimentally observed energy levels of $^{16}O+\alpha=^{20}Ne$ are put in relation to computed numbers. The ones labelled RGM(MKF) are taken from the work of Matsuse et al. [6]. Our computations within the OCM are based on the nucleon-nucleon potentials of Hasegawa et al. [10]. We use the same oscillator parameter $\nu=0.32$ fm^{-2} for ^{16}O and the α particle. Our preliminary results denoted by RGM' rely on a simplified

Table 1

J^π	E_{ex} [MeV]	E_{th} [MeV]	Γ_{ex} [keV]	Γ_{th} [keV]	θ^2_{ex}	θ^2_{th}	R_c [fm]
0^+_1	-4.73	-3.65					3.20
2^+_1	-3.10	-2.64					3.15
4^+_1	-0.48	-0.34					2.95
6^+_1	4.05	3.10	0.11 ± 0.025	0.06	0.01	0.07	2.70
8^+_1	7.22	7.36	0.035 ± 0.01	0.40	0.001	0.01	
10^+_1		26.30		2130.00		0.40	
1^-	1.06	1.07	0.01	0.03	0.14	0.30	3.42
3^-	2.44	2.86	8.00	37.00	0.26	0.37	3.40
5^-	5.53	6.15	141.00	280.00	0.31	0.36	3.20
7^-	10.70	11.05	380.00	480.00	0.25	0.28	2.95
9^-	18.00	17.05	500.00	250.00	0.17	0.10	
11^-		35.73		6200.00		0.80	
0^+_2	3.57	2.84	800.00	630.00	0.42	0.87	2.79
2^+_2	4.07	4.00	800.00	1470.00	0.61	1.24	2.74
4^+_2	6.07	7.15	350.00	3800.00	0.23	1.55	2.61
6^+_2	7.83	12.01	100.00	5170.00		1.44	2.40

Table 1: For the alpha cluster system $^{16}O + \alpha$ experimental and theoretical excitation energies, widths and reduced widths are compared to each other. The interaction employed is a direct Woods-Saxon potential with $v_0 = 85$ MeV, $r_0 = 1.428$ fm, $\alpha = 1/a = 1.7$ fm^{-1}, $\nu = 0.36$ fm^{-2}. R_c stands for the radius of the repulsive core.

version of the RGM, which is somewhat intermediate to the original RGM and the OCM. The equations were solved approximating the non-local kernel by an expansion in terms of local kernels. To achieve the values shown in the figure the RGM' required an increase in the radius R_c by 0.2 fm. The level scheme illustrates the varying degrees of success in reproducing the experimental data which are almost identical to the ESP.

Coming to the end we would like to compare in table 1 experimental
and OCM results for the positions of the energy levels, their widths
and reduced widths. The last column gives the core radii. Our results
do not correspond to the ones of Matsuse et al. [6] but rather to the
ones of Sünkel and Wildermuth [5] and of Lemere et al. [7] ; i.e. the
energies of the last members of the even rotational bands are larger
than the ones experimentally observed.

To summarize it should be recalled that the ESP reproduces quite
nicely the experimental energies, widths and the angular distributions
of α-cluster systems of the type $\alpha + \alpha$, ^{12}C, ^{16}O and ^{24}Mg [2] allowing as well to describe transfer reactions [11] . In the present con-
tribution it is found that the microscopic OCM and RGM display similar
feature as the ESP although their agreement to experiment is less good
than the one of the ESP. Its simplicity and its applicability to structure calculations, elastic scattering and to transfer reactions makes
the ESP an attractive tool since it does not require tedious computations like the RGM. Another attractive feature of the model is seen in
its ability to predict new cluster states.

References

[1] K.P. Artemov, V.Z. Goldberg, I.P. Petrov, V.P. Rudakov, I.N.
     Serikov and V.A. Timofeev, J. de Physique 32 (1971) C6-125;
     Phys. Letters 37B (1971) 61
[2] A.I. Baz, V.Z. Goldberg, K.A. Gridnev, V.M. Semjonov and E.F.
     Hefter, Z. Physik A280 (1977) 171
     A.I. Baz, V.Z. Goldberg, N.Z. Darwisch, K.A. Gridnev, V.M. Sem-
     jonov and E.F. Hefter, Lett. al Nuovo Cim. 18 (1977) 227
     N.Z. Darwisch, K.A. Gridnev, E.F. Hefter and V.M. Semjonov,
     Nuovo Cim. 42A (1977) 303
[3] G. Baccani et al., Annual Report of Saclay, 1973
[4] C.M. Perey and F.G. Perey, Atomic Data and Nucl. Data Tables,
     17, 1 (1976)
[5] W. Sünkel and K. Wildermuth, Phys. Letters 41B (1972) 439
[6] T. Matsuse, M. Kamimura and Y. Fukushima, Prog. Theor. Phys.
     53 (1975) 706
[7] M. Lemere, Y.C. Tang and D.R. Thomson, Phys. Rev. 14C (1976) 23
[8] S. Saito, Prog. Theor. Phys. 41 (1969) 709
[9] Supplement of Prog. of Theor. Phys. 52 (1972)
[10] A. Hasegawa and S. Nagata, Prog. Theor. Phys. 45 (1971) 786
     Y. Samamoto, Prog. Theor. Phys. 52 (1974) 471
[11] K.A. Gridnev, N.Z. Darwisch, V.M. Semjonov and E.F. Hefter,
     Izv. akad. nauk. SSSR, ser. fis. (in press)

FOURIER-BESSEL-ANALYSIS OF ALPHA-PARTICLE SCATTERING OPTICAL POTENTIALS AND   NUCLEAR MATTER DENSITIES

H.J. Gils, E. Friedman[+], H. Rebel, Z. Majka[++]
Kernforschungszentrum Karlsruhe GmbH
Institut für Angewandte Kernphysik
Postfach 3640, D-7500 Karlsruhe
Federal Republic of Germany

Abstract

Elastic scattering cross sections of 104 MeV alpha-particles from ^{40,48}Ca were analyzed using a Fourier-Bessel-series description of the real optical potential and also using density dependent folding models with Fourier-Bessel expansions of the nuclear matter densities. The uncertainties in the derived parameters are discussed for the real potentials and nuclear densities, respectively, as a function of radius and additionally for integral quantities such as volume integral, root-mean-square radius or higher moments. It is shown that the errors of the radial form factors are more realistic as compared to model descriptions using e.g. squared Saxon-Woods form-factors. The analyses yield for the nuclear matter rms radii $r_m(48) - r_m(40) = 0.12 \pm 0.07$ fm.

[+] Racah Institute of Physics, The Hebrew University, Jerusalem, Isreal
[++] On leave from Institute of Physics, Jagellonian University, Cracow, Poland

1. INTRODUCTION

In optical model descriptions of elastic scattering there is usually some residual dependence of the results on the particular choice of the parametrized form for the potential. The Saxon-Woods (SW) form has been widely and very successfully used to describe optical potentials for nucleons as well as for composite projectiles. However, this form gives an implicit coupling between the surface region and the interior of the potential and this could introduce undesirable constraints in the analysis. Some attempts have been made to improve the SW form, e.g. by using the square of the SW form factor [1] or by adding an extra term to the SW potential which is centered at the nuclear surface. Both procedures yield a distinctly better representation of elastic scattering cross sections (e.g. for 104 MeV α-particles), in particular, when large scattering angles beyond the nuclear rainbow are included in the data.

In the first part of this contribution we present the Fourier-Bessel method, the main aim of which is to remove the coupling between the surface and the interior part of the potential and to provide realistic estimates of the errors of the potentials at each radial point. A similar coupling between surface and interior region as for the potentials occurs also for the nucleon densities when performing folding model analyses with common parametrizations such as two or three parameter Fermi distributions. Thus, we applied similar procedures to the nucleon densities in folding model calculations. In these analyses, density dependent α-particle-bound nucleon interactions have been introduced, which are necessary in order to account for saturation effects occurring for the large scattering angle region.

2. THE FOURIER-BESSEL DESCRIPTION OF THE OPTICAL POTENTIAL

Following descriptions of nuclear charge distributions [3] we choose a Fourier-Bessel (FB) expansion of the real potential [2], namely

$$V(r) = - V_o/(1-e^x) - \sum_{n=1}^{N} b_n \, j_o \, (\frac{n\pi r}{R_c}) \tag{2.1}$$

where $x = (r-r_o \cdot A^{1/3})/a$ and V_o, r_o and a are the parameters of the best-fit SW potential, $j_o (\frac{n\pi r}{R_c})$ are spherical Bessel functions and R_c is a suitably chosen cut-off radius beyond which the extra term of the potential vanishes.

The refined parametrization (2.1) was applied to the analysis of elastic scattering of 104 MeV α-particles from ^{40,48}Ca measured at the Karlsruhe

Isochronous Cyclotron [4]. The most important experimental prerequisites of the analyses are characterized as follows:

- angular range of cross sections $3 \leq \theta_{cm} \leq 116^{O}$
- angular acceptance 0.15^{O}
- angular steps 0.5^{O}
- absolute angular accuracy $< 0.1^{O}$

Figure 1 shows the experimental results together with fits obtained using the FB parametrization of the real part of the optical potential. The cut-off radius was in the range of 9-14 fm, the coefficients b_n as well as the parameters of the real and imaginary WS potentials were adjusted by an automatic search routine on the basis of a χ^2/F criterion to obtain best fit to the experimental cross sections.

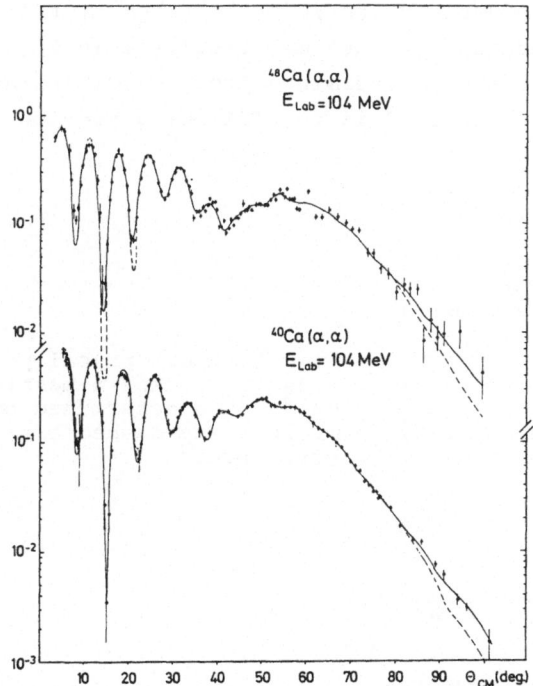

Fig. 1: Experimental and calculated differential cross sections.
Solid curves: Fourier-Bessel fits; dashed curves: folding model fits.

In Table 1 the results for the volume integral per nucleon pair J/4A, the root mean square (rms) radius and χ^2 per degree of freedom are compared to analyses using pure SW and SW squared potential, respectively. Comparing the χ^2 values it is obvious that the experimental data are

Table 1: Results of optical potential analyses of elastic α-particle scattering from ^{40,48}Ca using different parametrizations of the potential: SW = Saxon-Woods; $(SW)^2$ = Saxon-Woods squared; FB = Fourier-Bessel (eq. 2.1).

Target	Procedure	$J/4A$ (MeV·fm^3)	$<r^2>^{1/2}_{pot}$ (fm)	χ^2/F
^{40}Ca	SW	330 ± 1	$4.46 \pm .03$	5.4
	$(SW)^2$	327 ± 1	$4.32 \pm .03$	2.7
	FB	323 ± 2	$4.38 \pm .04$	1.5
^{48}Ca	SW	326 ± 1	$4.57 \pm .03$	4.3
	$(SW)^2$	318 ± 1	$4.24 \pm .03$	3.6
	FB	318 ± 2	$4.51 \pm .04$	3.0

distinctly better reproduced by the FB method than by the SW or SW^2 analyses. The FB potentials obtained are displayed in Fig. 2. The shaded areas are the error bands determined from the covariance matrix of the FB coefficients as described in the following section.

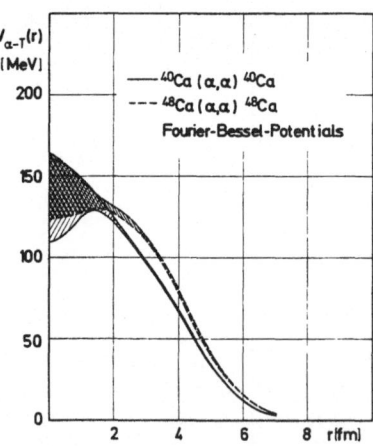

Fig. 2: Real optical potential for elastic 104 MeV α-particle scattering on ^{40,48}Ca determined by the FB method. The dashed areas are the error bands.

3. UNCERTAINTIES IN THE POTENTIAL

In the course of calculating the parameters of the potential by least-squares fit to experimental results the uncertainties in the parameters (for instance b_n) are usually determined from the diagonal elements of the covariance maxtrix (M^{-1})

$$(\Delta b_n)^2 = (M^{-1})_{nn} \, f_{min} \tag{3.1}$$

where (M^{-1}) is the inverse of the matrix M formed by the partial derivatives of $f(b_i) = \chi^2/F(b_i)$ at the minimum

$$M_{mn} = (\frac{\partial^2 f}{\partial b_n \, \partial b_m})_{min} \tag{3.2}$$

However, the uncertainties in the best-fit parameters in most cases are less interesting than the uncertainties in derived quantities such as the mean square uncertainty in the value of the potential $[\Delta V(r)]^2$ at a radius r. For the FB potentials this uncertainty is given by

$$[\Delta V(r)]^2 = \sum_{m,n=1}^{N} <\Delta b_m \, \Delta b_n> \; j_0(\frac{m\pi r}{R_c}) \; j_0(\frac{n\pi r}{R_c}) \tag{3.3}$$

where $<\Delta b_m \, \Delta b_n> = (M^{-1})_{mn} \, f_{min}$ are the correlations between the coefficients b_n. It should be noted that the correlations are very important in the present problem and this is an essential difference when compared with the electron scattering case. In addition, it should be emphasized that although the SW-term does not explicitly enter into the determination of the error band (eq. 2.1) the resulting errors are those of the <u>total</u> potential because only that can affect $f = \chi^2/F$. One can check this fact by using a <u>pure</u> FB series without SW-term. Choosing the same number of terms and the same cut-off radius, one obtains the same χ^2-value, volume integral, rms-radius and the same error band of the potential as compared to the case with SW-term included. Also the errors of the volume integral and rms-radius (c.f. Tab. 1) provided by corresponding formulae [2] remain unchanged when using the second approach.

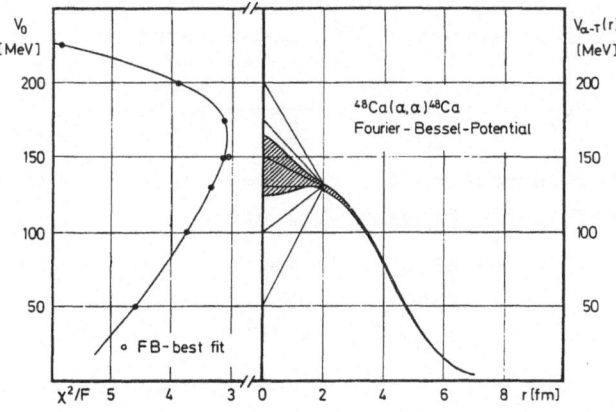

Fig. 3: Different linear approaches of the central potential region ($r \leq 2$fm) of the optical potential. The left hand part displays the the corresponding χ^2-value as function of $V(r=0)$.

In order to prove that also the interior of the potential (r<2fm) is well determined (as indicated by the error band), we approached the potential for $0 \leq r \leq 2$fm by several linear functions. For r > 2fm the best-fit FB potential was taken. The result of various calculations is demonstrated in Fig. 3. If the potential form chosen is distinctly outside of the error band, the χ^2-values increase considerably. Thus, the experimental scattering cross sections can only be well reproduced when the potential lies inside the error band.

4. FOLDING MODEL ANALYSES

As we are interested in more detailed information on the radial shape of the nuclear density distributions the experimental data have also been analyzed on the basis of folding model descriptions of the real potential. In such a model the real part of the optical potential is generated by folding an effective alpha-bound nucleon-interaction $v_{eff}^{\alpha N}(r,r_\alpha)$ into the nuclear matter distribution $\rho_m(r)$ of the target nucleus. The folding model in its simplest form is successful only over the diffraction region of the angular distribution. The results of the FB analyses over a wide angular region clearly indicate the presence of saturation effects [2]. We have, therefore, employed a modified folding model for the analysis of the present data over the full angular range. The method was to fix first the parameters of the interaction by requiring a fit to the ^{40}Ca data assuming a known density distribution, and then to obtain the parameters for the neutron density distribution of ^{48}Ca from a fit to ^{48}Ca data.

We used a simple Gaussian interaction modified by a density dependent factor determined from fits to the ^{40}Ca cross section

$$v_{eff}^{\alpha N} (|\vec{r}-\vec{r}_\alpha|) = V_o \exp \left[- \frac{|\vec{r}-\vec{r}_\alpha|^2}{\mu_o^2}\right] \left[1 - \gamma \rho_m^{2/3}\right] \qquad (4.1)$$

where the last factor introduces the required saturation effects. The parameter values used in the present work are:

$$V_o = 85.8 \pm 0.7 \text{ MeV}$$
$$\mu_o = 1.697 \pm 0.005 \text{ fm}$$
$$\gamma = 2.15 \pm 0.15 \text{ fm}^2$$

It is interesting to note that this phenomenological value for γ is very close to 2 fm^2 as found in more fundamental analyses by Jeukenne, Lejeune and Mahaux [5]. In order to remove the constraints on the densities ρ_m given by simple parametrizations as e.g. Fermi form, we des-

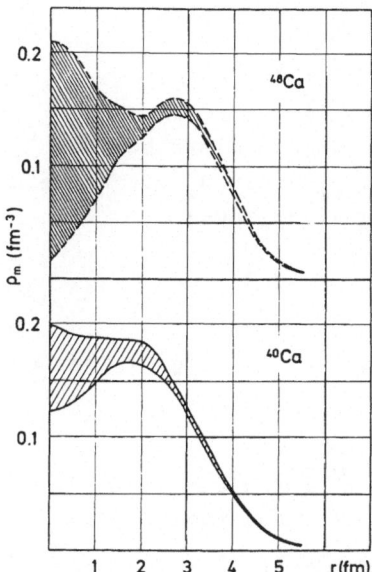

Fig. 4: Nuclear matter density distribution from a combined folding model and Fourier-Bessel analysis.

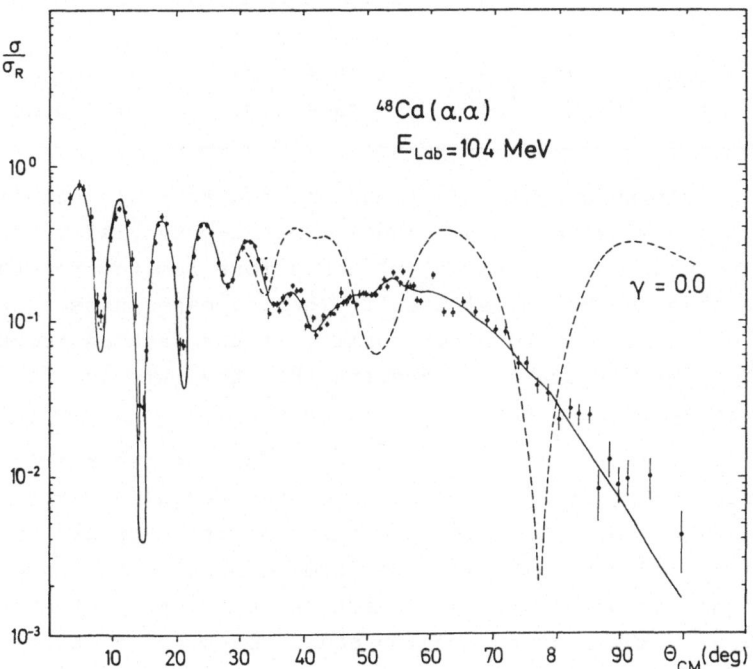

Fig. 5: Reproduction of the large angle scattering cross sections without the saturation term $(1-\gamma \cdot \rho_m^{2/3})$ (dashed line). Solid line: best fit FB-folding result.

cribed the nuclear matter distribution as

$$\rho_m(r) = \frac{A}{Z} \rho_p(r) + \sum_{n=1}^{N'} \beta_n j_0(\frac{n\pi r}{R_c'}) \qquad (4.2)$$

where $\rho_p(r)$ was derived from the well-known charge distributions [6,7]. In order to preserve the volume integral of ρ_m, only $\beta_2 \ldots \beta_N$ were varied in the χ^2 fits and β_1 was determined as follows:

$$\beta_1 = \sum_{n=2}^{N'} \frac{(-1)^n \beta_n}{n^2} \qquad (4.3)$$

Results for the folding model fits are also shown in Fig. 1 (dashed curves) and are summarized in Table 2.

Table 2: Results for folding model calculations

Target	χ^2/F	$J/4A$ (MeV fm^3)	potential r_{pot}	mass $r_m(48) - r_m(40)$
			rms radii (fm)	
^{40}Ca	3.6	315 $\pm$ 2	4.37 $\pm$ 0.04	
^{48}Ca	5.0	314 $\pm$ 2	4.49 $\pm$ 0.04	0.20 $\pm$ 0.08
^{48}Ca/V_{eff}	3.7	310 $\pm$ 4	4.43 $\pm$ 0.03	0.07 $\pm$ 0.06

Fig. 4 shows the resulting density distributions together with their uncertainties. These are larger than those in the FB-potential method because the folding procedure causes $\rho_m(r)$ to be more remote from the experimental data than the potential itself. The increased uncertainty in the case of ^{48}Ca is due to the minor quality of the data as compared to the case of ^{40}Ca. We venture the statement that the bump at r = 2.5 - 3.0 fm is reflecting the $f_{7/2}$-neutron orbit.

The importance of the saturation term $(1-\gamma \cdot \rho_m^{2/3})$ for the description of the large angle scattering cross sections is demonstrated in Fig. 5, where γ is set equal to zero. In this case only the very forward cross sections are well reproduced. Also when performing drastic and unreasonable changes in the parameters of the effective interaction no satisfactory description of the backward region could be obtained taking $\gamma=0$.

In view of the extreme simplicity of the folding model and in particular of the neglect of explicit exchange effects and non-locality, small differences between the strength of the interaction for ^{40,48}Ca could be expected. Thus, we have tried to relax this constraint yielding a reduction in the χ^2-value when changing the strength of $V_{\alpha-N}$ by about 9 %

(c.f. Table 2, line 3). Hereby the radius of the nuclear matter distribution in ^{48}Ca is remarkably decreased. The errors of both procedures, however, still overlap. Introducing the different results of the ^{48}Ca nuclear matter density into considerably refined folding model calculations [8], one may conclude that - due to the oversimplifications of the effective interaction - the radii obtained by the simple folding procedures may be extreme values to be deduced from the experimental data. Averaging the results of the various methods, we obtain

$$r_m(48) - r_m(40) = (0.12 \pm 0.06)\,fm.$$

5. CONCLUSION

From Tables 1 and 2 we conclude that all methods give results which are consistent with each other when considering the realistic errors given by the FB method.

Table 3: Comparisons between the present and previous results

Method	$r_m(48) - r_m(40)$ (fm)	$r_n(48) - r_p(48)$ (fm)	Reference
α-scattering			
104 MeV	0.12 $\pm$ 0.06	0.18 $\pm$ 0.10	present work
79 MeV	0.05 $\pm$ 0.04	0.03 $\pm$ 0.03	9
1.37 GeV	0.12 $\pm$ 0.03	0.17 $\pm$ 0.05	10
p-scattering			
1.04 GeV		0.16 $\pm$ 0.05	11
π-cross sections		0.08 $\pm$ 0.05	12

Table 3 compares the present results with previous ones, all of which were obtained using simple analytic functions for the nuclear matter distribution. If the uncertainties obtained in the present work are typical for the uncertainties that should have been quoted for previous results, then all apparently conflicting results are in agreement with each other.

292

6. REFERENCES

1) Z. Majka and T. Srokowski, Acta Phys. Pol. $\underline{B9}$ (1978) 53
    A. Budzanowski et al., Phys. Rev. $\underline{C17}$ (1978) 951

2) E. Friedman and C.J. Batty, Phys. Rev. $\underline{C17}$ (1978) 34

3) B. Dreher et al., Nucl. Phys. $\underline{A235}$ (1974) 219

4) H.J. Gils et al., to be published; E. Friedman et al., Phys. Rev.
    Lett. (in press)

5) J.-P. Jeukenne, A. Lejeune and C. Mahaux, Phys. Rev. $\underline{C16}$ (1977) 80

6) R.F. Frosch et al., Phys. Rev. $\underline{174}$ (1968) 1380

7) J.B. Bellicard et al., Phys. Rev. Lett. $\underline{19}$ (1967) 527

8) Z. Majka, H.J. Gils and H. Rebel, Report KfK 2622 (1978),
    Z. Physik (in press), and contributions to this conference

9) G.M. Lerner et al., Phys. Rev. $\underline{C12}$ (1975) 778

10) G.D. Alkhazov et al., Nucl. Phys. $\underline{A280}$ (1977) 365

11) G.D. Alkhazov et al., Phys. Lett. $\underline{57B}$ (1975) 47; Nucl. Phys. $\underline{A274}$
    (1976) 443; A. Chaumeaux et al. Phys. Lett. $\underline{72B}$ (1977) 33

12) M.J. Jakobson et al., Phys. Rev. Lett. $\underline{38}$ (1977) 1201

VALIDITY OF REFINED FOLDING MODEL APPROACHES
FOR LIGHT PROJECTILE SCATTERING

Z. Majka[+], H.J. Gils and H. Rebel
Kernforschungszentrum Karlsruhe GmbH
Institut für Angewandte Kernphysik
Postfach 3640, D-7500 Karlsruhe
Federal Republic of Germany

1. INTRODUCTION

The first term of a formal multiple scattering expansion of the real part of the optical potential has become popular in scattering analyses as folding potential. There are basically two folding procedures. A single folding takes an adequate (semi-empirical) nucleon-nucleus-1 potential and folds it into the density distribution of nucleus-2 [1]. A double folding starts with an effective nucleon-nucleon interaction folded into the density of both nuclei [2-4]. Actually the folding potentials in the simplest form do overestimate the real potential by about 30 % for elastic alpha particle scattering [2] and by as much as a factor two for heavier ions [5]. Recently the folding model procedure has been refined including the exchange term [4,6,7] and the density dependence of the effective nucleon-nucleon interaction [6,8].

In the present investigations we are concerned with the validity of different folding model approaches for description of the differential cross section for 104 MeV alpha-particle [9] and 156 MeV ^{6}Li ion scattering [10] from ^{40,48}Ca. Moreover a phenomenological parametrization of the folded potentials is derived.

[+] On leave from Institute of Physics, Jagellonian University, Cracow, Poland

2. REFINED FOLDING POTENTIAL MODELS

The simple double folding potential /S/ in form

$$U_{PT}^S(r) = \iint d\vec{z}_P \, d\vec{z}_T \, \rho_P(\vec{z}_P) \, \rho_T(\vec{z}_T) \, t^{NN}(\vec{r}_{NN}) \qquad (2.1)$$

establishes the starting point of our investigations. There are two single folding approaches:

i   Projectile-nucleon approach /P/, (target folding):

$$U_{PT}^P(r) = \int d\vec{z}_T \, \rho_T(\vec{z}_T) \, U_{P-N}(\vec{R}_P) \qquad (2.2)$$

ii Target-nucleon approach /T/, (projectile folding):

$$U_{PT}^T(r) = \int d\vec{z}_P \, \rho_P(\vec{z}_P) \, U_{T-N}(\vec{R}_T) \qquad (2.3)$$

Usually, a projectile-nucleon U_{P-N} or a target-nucleon U_{T-N} potentials are taken in phenomenological form from nucleon scattering analyses. In the present investigation, U_{X-N}/X = P(T)/ are calculated in a semi-microscopic way including one-nucleon exchange effects and density dependence of the nucleon-nucleon interaction:

$$U_{X-N}(\vec{R}_X) = \int d\vec{z}_X \, \rho_X(\vec{z}_X) \, t_\rho^{NN}(\vec{r}_{NN},\rho_X) + \int d\vec{z}_X \, \rho_X^{Mix}(\vec{z}_X,\vec{R}_X) t_\rho^{NN}(\vec{r}_{NN},\rho_X) \quad (2.4)$$

For a detailed discussion of eq. (2.4) see Ref.[13]. Considering the complex projectile-target potential one has to note that the exchange and the density dependence effects are included merely in target nucleon-projectile and projectile nucleon-target system, respectively so that the two procedures (eqs. 2.2 and 2.3) may differ in the results.

Finally, we calculate the projectile-target potential taking into consideration density dependence of the nucleon-nucleon interaction effect resulting from both colliding nuclei. Two models are considered:

iii Sudden approximation /SD/ where density in the overlap region is a sum of the density of colliding nuclei $\rho_{TOT} = \rho_P + \rho_T$

iv Intermediate approximation /IM/ where $\rho_{TOT} = m\rho_P + \rho_T$ (0<m<1).

In the sudden and the intermediate approximations we neglect antisymmetrization effects as will be motivated later, and the projectile-target potential takes form:

$$U_{PT}^{SD/IM}(r) = \iint d\vec{z}_P \, d\vec{z}_T \rho_P(\vec{z}_P) \, \rho_T(\vec{z}_T) \, t_\rho^{NN}(\vec{r}_{NN},\rho_{TOT}) \qquad (2.5)$$

The coordinates used in all former equations are presented in Fig. 1.

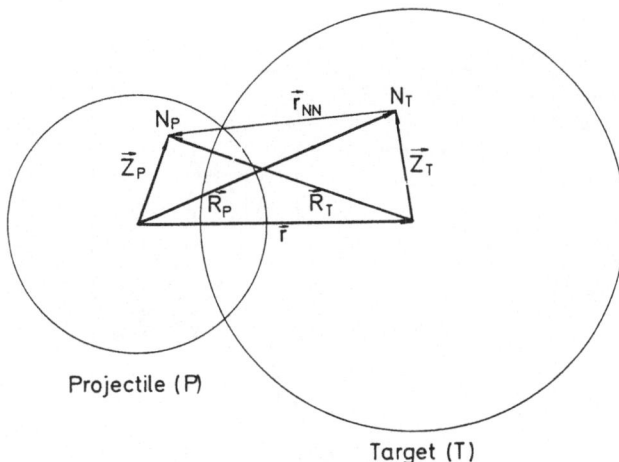

Fig. 1: Coordinates of the projectile-target-system

The effective nucleon-nucleon interaction is introduced by the form:

$$t_\rho^{np} = \frac{1}{8}\ (t_{SE} + 3t_{TE})$$

$$t_\rho^{pp} = \frac{1}{4}\ t_{SE} \qquad\qquad\qquad (2.6)$$

The singlet t_{SE} and triplet t_{TE} s-state effective interaction are more explicitly expressed by:

$$t_i(\vec{r}_{NN}) = g_i(\rho)\ \cdot\ v_{KK}^i(r_{NN})$$

$$= c_i[1+\alpha_i\rho_X^{2/3}]\ v_{KK}^i(r_{NN}) \qquad x = P,T,TOT \quad (2.7)$$

where v_{KK}^i are the singlet and triplet state potentials given by Kallio and Kolltveit [11] with Moszkowski-Scott separation distances 1.025 fm and 0.925 fm, respectively. The values of the parameters c_i and α_i are taken from Ref. [12] . The density independent effective nucleon-nucleon interaction in eq. (2.1) is given by eq. (2.7) with $g_i(\rho) = 1$ (simple Kallio and Kolltveit nucleon-nucleon interaction).

The quantities ρ_P and ρ_T are the matter point density distributions of the projectile and target nucleus, respectively, as extracted from measured charge density distribution by unfolding the charge density distribution of a single proton and ρ^{Mix} is a mixed density originating from the nonlocal exchange part of the potential resultung from anti-symmetrization requirements.

Fig. 2 displays the real part of the alpha-particle ^{40}Ca potential calculated using simple double folding procedure /S/ and approaches i and
ii.

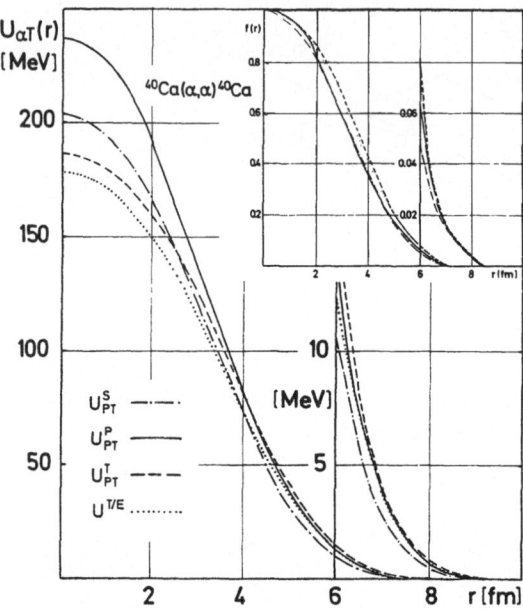

Fig. 2: Microscopic α-particle ^{40}Ca potentials calculated using
different approaches. Inset: Form factors of presented
potentials

Additionally, in order to notify and separate effects arising from density dependence and those from exchange the α-particle ^{40}Ca microscopic
potential without antisymmetrization term in eq. (2.4) using target-nucleon approach is calculated, too.

Applying the microscopic potentials for the description of the measured
differential cross sections of elastic scattering one has to add an imaginary part iW of the optical potential. We do this in a phenomenological way using volume and surface Woods-Saxon squared form. Additionally
a renormalization factor λ_R for the depth of the real potential has to
be introduced. Comparison of the experimental angular distributions for
α-particle ^{40}Ca scattering with those calculated using folded potentials
are presented in Fig. 3.

The microscopic potential is characterized by the central depth $U_{PT}(r=0)$,
halfway radius $R_{1/2}$, the 10-90 % distance t_{10-90}, mean square radius
$<r^2>^{1/2}$ and volume integral per nucleon J_N (see Table 1). Additionally,

Table 1 contains values of the normalization factor and χ^2/F for the best fits.

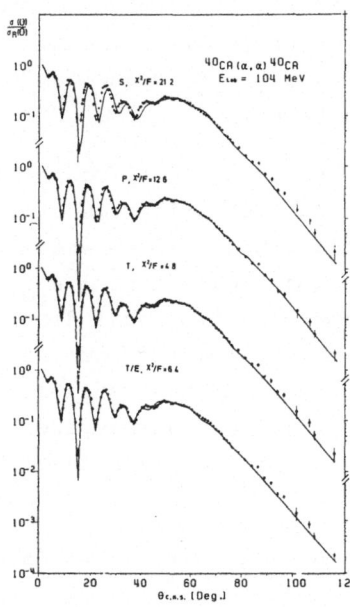

Fig. 3: Optical model fits to the $^{40}Ca(\alpha,\alpha)^{40}Ca$ cross sections using various approaches for the real folded potential

Table 1: Parameters of the real part of the α-particle ^{40}Ca potential

Approach	$U_{PT}(r=0)$ (MeV)	λ_R	$<r^2>^{1/2}$ (fm)	J_N (MeV fm^3)	$R_{1/2}$ (fm)	t_{10-90} (fm)	χ^2/F
S	203.3	0.791	4.084	346.9	3.352	3.980	21.2
P	237.6	0.685	4.015	400.4	3.327	4.089	12.6
T	186.2	0.791	4.246	393.9	3.726	4.093	4.8
TE	178.1	0.848	4.247	360.7	3.618	4.148	6.4

As can be seen from Fig. 2 and Table 1, the inclusion of the density dependence of the nucleon-nucleon interaction and antisymmetrization effects into folding procedure (target folding and projectile folding) significantly modify depth and shape of the simple double folding potential. Comparison with experimental data favors that procedure which includes the effects under consideration into the nucleon-target system (projectile folding) also indicating that antisymmetrization effects are of minor importance as compared to density dependence of the nucleon-nucleon interaction. Although the agreement with the experimental data is improved by use of folded potentials referred by apporaches i and ii, a renormalization factor λ_R smaller than 1 is required. Additionally, to show that renormalization factor of the depth of the folded potential

(ii) simulate effects resulting from complex projectile we have calculated the nucleon-^{40}Ca potential using eq. (2.4). The depth of the microscopic nucleon-^{40}Ca potential calculated for the energy considered
(E_N = 55 MeV corresponding to 104 MeV/4 + average kinetic energy of the
nucleons in the α-particle) agrees within 5 % to the phenomenological
nucleon-nucleus potential given by Fricke et al. [14].

The exchange term (second term of the r.h.s. in eq. 2.4) contributes
less than 5 % for energy under consideration. For this reason we neglect
the exchange term for the following calculation (eq. 2.5) where the
effective interaction depends actually on the local density ρ_{TOT} of the
overlapping system. Due to the Pauli distortion this local density has
to be assumed to be intermediate between the arithmetic sum of ρ_p and
ρ_T (sudden approximation: maximum compression) and the adiabatic case
$\rho_{TOT} \simeq \rho_T$ (minimum compression). For simplicity we parametrize by
$\rho_{TOT} = m\rho_p + \rho_T$ with m (0≤m≤1) accounting for the compression of nuclear matter in the overlap region.

Adjusting the quantity m (in addition to the imaginary potential) by
fitting the experimental α-^{40}Ca cross sections we found that for m ≈ 0.5
the best fit is obtained keeping renormalization factor λ_R = 1.

Similar calculations performed for α-particle ^{48}Ca scattering show that
significantly better fits to the experimental data have been obtained
introducing the neutron density distribution $\rho_n \neq N/Z \ \rho_p$.

For ^{6}Li ion scattering, however, even with the maximum value of the
compression of the nuclear matter in overlap region (sudden approximation the folded potential has to be renormalized by factor λ_R < 1.
The above result confirms earlier conclusions [15] that ^{6}Li ion scattering has a behavior quite different from the scattering of other complex particles.
Fig. 4 displays the experimental and theoretical differential cross
sections corresponding to the results presented in Table 2.

Table 2: Parameters of the real part of the microscopic α-^{40}Ca and
^{6}Li-^{40}Ca potentials

Potential	U (r=0) (MeV)	λ_R	m	$<r^2>^{1/2}$ (fm)	JN (MeV·fm^3)	$R_{1/2}$ (fm)	t_{10-90}	χ^2/F
^{40}Ca(α,α)/IM/	151.4	1.0	0.5	4.22	298.0	3.576	4.098	4.0
^{40}Ca(Li,Li)/SD/	201.3	0.76	1.0	4.62	310.0	3.611	4.60	5.9

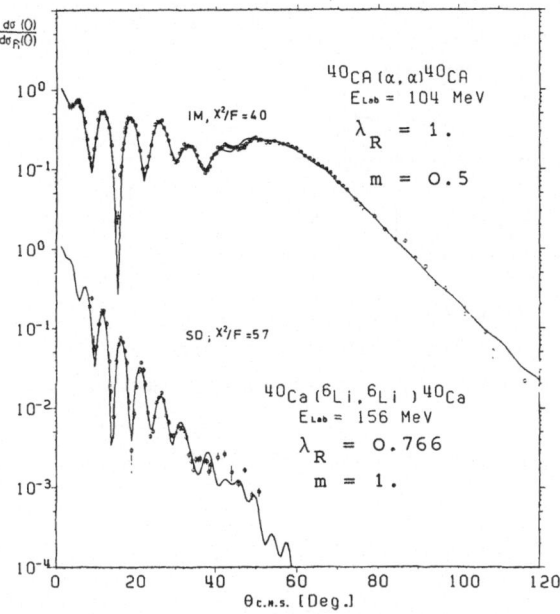

Fig. 4: Optical model analyses of ^{40}Ca(α,α) and ^{40}Ca$(^6$Li,^{6}Li)
scattering on the basis of intermediate and sudden approximation
of the real potential, respectively.

3. CONCLUSIONS

Concluding, we would like to emphasize that it is possible to extend the
folding model description of 100 MeV alpha-particle scattering to large
angle scattering without any renormalization of the real potential when
the density dependence of the nucleon-nucleon interaction is adequately
included. The experimental cross sections are described with merely the
same quality as by use of phenomenological potentials [16]. For ^{6}Li-
scattering good representation of the experimental data can only be obtained by requiring a normalization factor of the microscopic optical
potential smaller than 1.

Finally, we tested different phenomenological parametrization of the
microscopic potentials. As can be seen from Table 3, popular Woods-
Saxon and squared Woods-Saxon parametrizations do not represent microscopic potentials satisfactorily*.

---

*
  Angular distributions were calculated using phenomonological imaginary potential with parameters for the best fits presented in Fig. 4.

Table 3: Comparison of the χ^2/F values* for different parametrization of the microscopic $^{40}Ca(\alpha,\alpha)$ /IM/ and $^{40}Ca(^6Li,^6Li)$ /SD/ potentials.

Potential	χ^2/F		
	WS	WS2	MF
$^{40}Ca(\alpha,\alpha)$	70.5	14.5	4.8
$^{40}Ca(^6Li,^6Li)$	69.5	11.9	7.6

WS = Woods-Saxon parametrization

WS2 = squarea Woods-Saxon form

MF = modified Fermi parametrization (eg. 3.1)

In contrast, we found that the modified Fermi parametrization [13] in form

$$U_{P-T}(r) = U\left[1 + \omega\left(\frac{r}{C}\right)^2\right]\left[1 + \exp\left(\frac{r-C}{a}\right)\right]^{-m} \qquad (3.1)$$

with the parameters presented in Table 4 fits microscopic $^{40}Ca(\alpha,\alpha)$ /IM/ and $^{40}Ca(^6Li,^6Li)$ potentials adequatly.

Table 4: Values of the best fits parameters of the phenomenological representation of the microscopic $^{40}Ca(\alpha,\alpha)$ /IM/ and $^{40}Ca(^6Li,^6Li)$ /SD/ potentials

Potential	U (MeV)	$C \cdot A_T^{-1/3}$ (fm)	a (fm)	ω	m
$^{40}Ca(\alpha,\alpha)$	183.3	1.605	1.628	−0.166	3.032
$^{40}Ca(^6Li,^6Li)$	200.3 +	1.601	1.945	−0.153	2.726

One of the authors (Z.M.) would like to thank the Kernforschungszentrum Karlsruhe GmbH for a grant and Prof. G. Schatz for kind hospitality in the Institut für Angewandte Kernphysik at the Kernforschungszentrum Karlsruhe.

+ The depth is renormalized by the best fit factor $\lambda_R = 0.766$

References

1) D.F. Jackson and V.K. Kembhavi, Phys. Rev. 178 (1969) 1626
   P.P. Singh et al., Phys. Lett. 59B (1975) 113

2) A. Budzanowski et al., Phys. Lett. 32B (1970) 431
   A. Budzanowski et al., Particle and Nuclei 5 (1973) 97

3) J.P. Vary and C.B. Dover, Phys. Rev. Lett. 31 (1973) 1510
   G.R. Satchler, Proc. Symp. on Macroscopic features of heavy-ion
   collisions, Argnonne, Illinois, Argonne National Laboratory
   Report ANL-PIIY-76-2 (April 1976) unpublished

4) Z. Majka et al., Phys. Rev. C18 (1978) 114

5) G.R. Satchler, Phys. Lett. 59B (1975) 121

6) B. Sinha, Phys. Rev. C5 (1975) 1546

7) Y. Eisen and B. Day, Phys. Lett. 59B (1976) 253

8) F. Petrovich et al., Phys. Lett. 71B (1977) 259
   W.G. Love, Phys. Lett. 72B (1977) 4

9) H.J. Gils et al., to be published

10) J. Buschmann et al., Verhandlg. DPG 4(1978) 945

11) A. Kallio and K. Kolltveit, Nucl. Phys. 33 (1964) 8

12) A.M. Green, Phys. Lett. 24B (1967) 384

13) Z. Majka, Phys. Lett. 76B (1978) 161

14) M.P. Fricke et al., Phys. Rev. 156 (1967) 1207

15) G.R. Satchler and W.G. Love, Preprint 1978
    G.R. Satchler, private communications

16) H.J. Gils et al., Proc. XVI Int. Winter Meeting on Nucl. Phys.,
    Bormio (Italy) Jan. 16-20, 1978
    E. Friedman et al., Phys. Rev. Lett., in press
    Z. Majka et al., Report KFK 2622 (1978), Z. Phys., in press

Energy Dependence of the Phenomenological α-^{90}Zr Optical Potential

L.W. Put

Kernfysisch Versneller Instituut, Rijksuniversiteit Groningen

Groningen, The Netherlands

1. Introduction.

In this contribution a survey is given of analyses in terms of phenomenological and microscopic optical potentials of a set of data on elastic scattering of α-particles from ^{90}Zr over a wide energy range. This data set consists of angular distributions at E_α= 40, 59, 80, 100 and 118 MeV [1,2] taken at the K.V.I. variable energy cyclotron, supplemented with the 142 MeV data taken at Maryland [3]. The 118 and 142 MeV data subtend to sufficiently large scattering angles to satisfy the condition [4] to resolve the discrete ambiguities in the well depth of the real part of the optical potential.

The principal aims of this study were: 1) to determine which family of real potentials at low energies corresponds with the unique real potential at high energy; 2) to determine the energy dependence of the optical model parameters; 3) to investigate the applicability of the simple folding model over a large range of energies; 4) to stimulate theorists to improve on the microscopic calculations of the optical potential.

In this contribution the emphasis is put on the results obtained on the energy dependence of the form factors of the potential. Results obtained by other authors in analyzing this set of data are also discussed in this contribution.

2. Results of analyses.

2.1. Phenomenological potentials.

The Woods-Saxon parameters obtained from six-parameter fits to the data are given in the upper part of table 1 and the fits, except for E_α= 142 MeV, are shown in fig. 1 (solid curves). Only those potentials are presented which correspond to the unique potentials at 118 and 142 MeV (see for discussion ref. [2]). At energies $\gtrsim$ 80 MeV the shape parameters are almost constant with energy whereas at low energies, i.e. 40 and 59 MeV, the parameters, particularly those of the real potential, deviate considerably from the high-energy values. The use of the high-energy real and imaginary form factors for fitting the low-energy data, i.e. searching on V and W only, results in a bad description of the amplitudes of the oscillatory pattern at forward angles (fig. 1, dashed curves). When the parameters of the imaginary form factor are also searched upon some improvement is obtained but the deviations at forward angles remain, thus indicating that the data cannot be described with an energy-independent real WS form factor. Actually we have shown [2] that within this parametrization the main energy dependence of the optical potential form factors is in the real form factor. It is worth mentioning that similar energy dependences in the form factor have been observed

Table 1
Best fit optical model parameters for scattering of α-particles from ^{90}Zr as a function
of bombarding energy. Results are given for a Woods-Saxon and for a square of a Woods-
Saxon parametrization of the real potential. In both cases a Woods-Saxon imaginary
potential was used.

E_α (MeV)	V (MeV)	r_R (fm)	a_R (fm)	W (MeV)	r_I (fm)	a_I (fm)	χ^2/F
Woods-Saxon							
40.0	112.3	1.477	0.542	15.28	1.556	0.344	6.0[a]
59.1	120.6	1.351	0.676	20.60	1.516	0.572	14.9
79.5	141.2	1.225	0.821	18.49	1.575	0.565	9.2
99.5	133.3	1.235	0.805	19.63	1.571	0.562	5.1
118	124.4	1.255	0.792	20.55	1.570	0.566	5.2
141.7	118.3	1.263	0.787	20.84	1.564	0.573	7.3
square of Woods-Saxon							
40.0	136.6	1.533	0.963	15.03	1.647	0.332	12.4[a]
59.1	153.5	1.434	1.151	18.13	1.564	0.530	8.2
79.5	167.5	1.374	1.241	18.98	1.563	0.590	4.1
99.5	170.9	1.347	1.289	19.71	1.551	0.619	2.7
118	150.7	1.388	1.212	20.34	1.554	0.629	2.1
141.7	141.6	1.396	1.196	21.66	1.527	0.667	1.2

a) fit was made to data for $\theta_{cm} < 142°$.

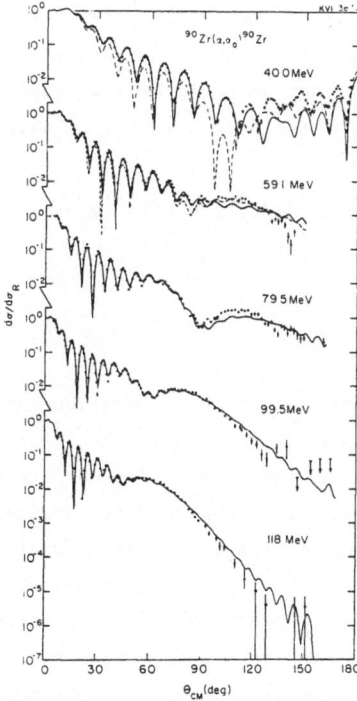

Fig. 1. Differential cross sections for elastic scattering from ^{90}Zr as a ratio to
Rutherford scattering. Solid curves are optical model fits to the data with Woods-
Saxon parameters as given in table 1. Dashed curves are fits for WS optical potentials
with geometrical parameters as determined from fitting the 118 MeV data. For the three
highest energies dashed curves coincide with solid curves.

for α-scattering from other nuclei (see ref. [2] for references).

In spite of the bad fits obtained at low energy by using the average high-energy form factor (r_R = 1.245 fm, a_R = 0.801 fm, r_I = 1.570 fm and a_I = 0.567 fm) a linear energy dependence of the depths of the real potential was observed over the whole energy range studied. It can be represented by

$$V(E_\alpha) = V_0(1 + \alpha_V E_\alpha)$$

with

$$V_0 = 155.2 \pm 5.8 \text{ MeV},$$

$$\alpha_V = -0.0016 \pm 0.0003 \text{ MeV}^{-1} \tag{1a}$$

This value of α_V is in excellent agreement with the prediction α_V = -0.0016 MeV^{-1} of Jackson and Johnson [5]. Below E_α = 80 MeV W increases rapidly with E_α, between 80 and 142 MeV the behaviour of W with E_α can be represented by a linear relation with

$$W_0 = 19.17 \pm 1.43 \text{ MeV},$$

$$\alpha_W = 0.0003 \pm 0.0006 \text{ MeV}^{-1} \tag{1b}$$

It has been argued by Goldberg [6] that the fits to data at $E_\alpha \approx$ 140 MeV give an indication that the Woods-Saxon representation cannot simultaneously provide the optimum shape for the tail and the surface region of the potential and he showed that other parametrizations of the real potential result in improved fits to data at 140 MeV. To find out whether a wrong parametrization could be the origin of the observed energy dependence of the real form factor the data were fit by using several other representations of the real potential

$$V_2(r) = V\{1 + \exp((r^2 - r_R^2 A^{2/3})/2a_R r_R A^{1/3})\}^{-1}, \tag{2}$$

$$V_3(r) = V\{f(r)\}^2, \tag{3}$$

$$V_4(r) = V\{f(r)\}^{2.65}, \tag{4}$$

$$V_5(r) = V\{f(r) + \tfrac{1}{4} f''(r)\}. \tag{5}$$

where f(r) represents the WS form factor. As we have concentrated on the study of the real potential the standard WS shape was used for the imaginary potential in all further analyses. Only the results for the square of a WS potential (eq. 3) will be presented here as these are very similar to the results obtained for eqs. (2), (4) and (5). From the lower half of table 1 the improvement in quality of the fit to the high-energy data is evident. It can also be seen that a similar behaviour is found as for the WS representation: the low-energy form factor is different from the almost constant high-energy form factor, this is confirmed by the failure of the high-energy (WS)^2 form factor in giving a detailed description of the low-energy data, particularly at the forward angles. It is noteworthy that the parametrization eq. (4), successful in the case of low-energy (20-50 MeV) α-scattering from ^{40}Ca [7], does not give an energy-independent form factor over the wide energy range spanned by the ^{90}Zr data.

We also applied a method somewhat similar to the model independent methods used in

Table 2.
Values of χ^2/F obtained in various fits to the ^{90}Zr data. Details are given in the text.

E_α (MeV)	full spline	100 MeV spline	energy dep. imag. f.f. [b]	folding [c]	spline + folding [d]	folding [e]	folding + add. [f]
40.0 [a]	3.2	24.9	16.1	34	8.4	31	10.6
59.1	2.5	25.5	69.4	23	9.4	13	10.6
79.5	2.8	4.8	13.4	22	3.9	23	4.7
99.5	1.5	1.5	9.8	33	1.6	39	2.1
118	1.1	2.7	13.2	25	2.5	12	2.5
141.7	0.6	5.4	–	96	3.5	26	1.7

a) fit was made to data for $\theta_{cm} < 142^\circ$.
b) results from ref. [8]; global fit with real $(WS)^2$ and imaginary volume plus derivative $(WS)^2$ form factors.
c) effective α-nucleon interaction folded over target matter distribution.
d) same as c) for $r \gtrsim 5.5$ fm; spline for $r < 5.5$ fm.
e) results from ref. [11]; nucleon-nucleus potential folded over the α-particle.
f) same as e) + additional phenomenological term (eq. (6)).

determining charge distributions from electron scattering data. Instead of representing the real potential by the parameters V, r_R and a_R the values of the real potential at ten radii were chosen as the parameters to search for. The potential values at other radii were obtained by constructing a cubic spline through the values of V(r) at these ten radii. This is not a model-independent method in the strict sense of the word, the choice of the number of radii and their position and the use of a WS shape for the imaginary potential will of course have some effect on the final result. But this method relieves us of the constraints imposed by the two-parameter parametrizations of the form factor.

With this method excellent fits to the data were obtained (fig. 2, solid curves; table 2). In fig. 3 the splines for 40 and 100 MeV are compared with the best-fit WS potential. At 40 MeV no unambiguous result was obtained, fig. 4 shows three different splines which give essentially the same value of χ^2/F. The origin of the undulations in the 40 MeV splines and the dip at r=0 fm is not yet understood, efforts to force the splines to have a more smooth character always resulted in a considerably worse fit.

The splines obtained at high energies were very similar in shape. In fact using the shape of the 100 MeV spline to fit the data at other energies (fig. 2, dashed curves; table 2) gave good results for $E_\alpha \gtrsim 80$ MeV, the results at low energies, however, indicate that we did not succeed in finding an energy-independent form factor of the real potential. But it was found that the 40 MeV data could be reasonably described with a potential which partly consisted of the 100 MeV spline, namely for $r \gtrsim 6$ fm, and an optimized spline for $r < 6$ fm. The same was true for a spline which was restricted to have the shape of the 100 MeV spline for $r < 6$ fm. These ambiguities in determining the 40 MeV potential will be further discussed in section 3.

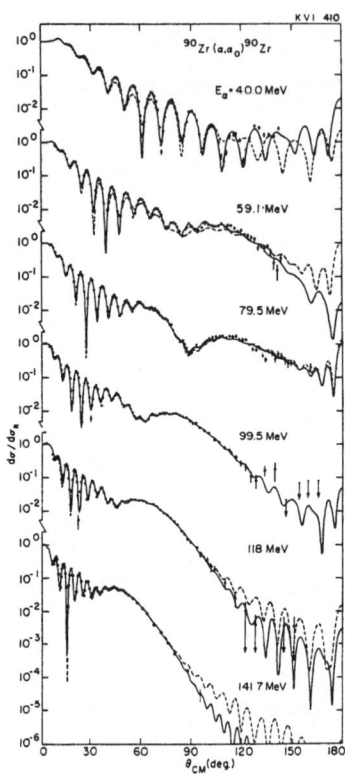

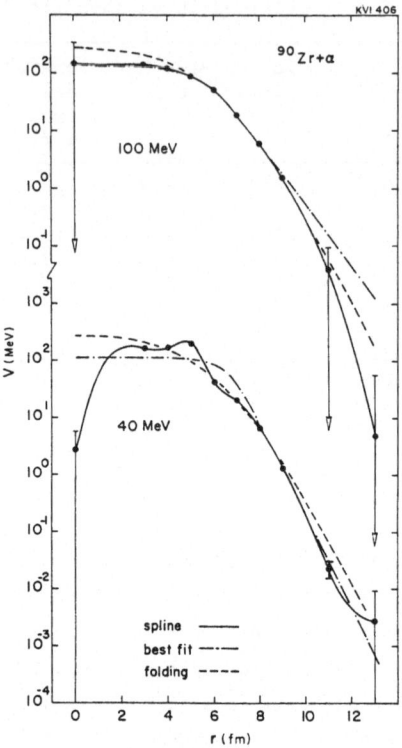

Fig. 2. Fits to the data obtained for potentials with the real part given by a
spline function. Solid curves are optimized for each energy. Dashed curves are
for splines which have the same shape as
the spline obtained from the fit to the
100 MeV data.

Fig. 3. Radial dependence of the real potential as represented by a spline function. For comparison best-fit Woods-Saxon
potentials and potentials with the energy-
independent form factor as obtained by
folding an effective α-nucleon interaction
over the target nucleus are also shown.

Majka *et al*. [8,9] have analysed the ^{90}Zr data in a different way. Whereas they
used an energy-independent $(WS)^2$ form factor for the real potential the imaginary potential was taken as a sum of a $(WS)^2$ term and the derivative of a $(WS)^2$. As the
strenghts of these two terms vary in a different way the form factor of the imaginary
potential changes with energy. Also with this approach a reasonable description of the
data could be obtained, values of χ^2/F obtained in their global fit are given in the
4th column of table 2. As the use of a WS imaginary potential did not lead to an energy-independent $(WS)^2$ real potential (see table 1) Majka's result might be a manifestation of the correlation between the real and the imaginary potential. Further discussion will be given in section 3.

2.2. Microscopic potentials.

The analyses with a potential obtained by folding an effective α-nucleon interaction over the density distribution of the target nucleus (details in ref. [2]) re-

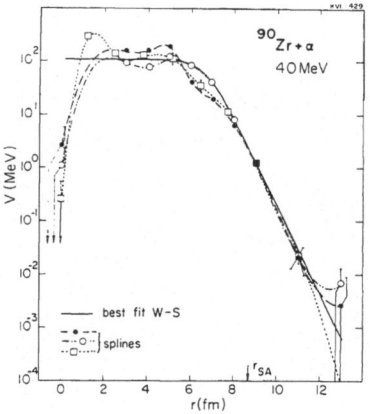

Fig. 4. Three different spline potentials which give identical fits to the 40 MeV data. The arrow indicates the position of the strong absorption radius.

sulted in bad fits to the data. Some of the fits are shown in fig. 5, χ^2/F values are given in the 5th column of table 2. This is not at all surprising as in such calculations one neglects that the α-particle is not a free α-particle inside the nucleus, i.e. antisymmetrization, density dependence of the interaction, coupling to other reaction channels, etcetera are neglected. In fact, it is only expected that this model gives a good prediction of the tail of the potential. Actually the tails of the high-energy spline potentials were found to be rather similar to the tail of this folding potential (see fig. 3). And further it was found that not only the high-energy data but also the low-energy data could be described by potentials which for $r > 5.5$ fm had the shape of the folding potential and consisted of a spline for $r < 5.5$ fm (6th column of table 2).

Singh et al. [10] and Cowley and Wall [11] fitted the data with a potential obtained by folding a phenomenologically determined nucleon-nucleus potential over the α-particle. Although this method, which also results in an energy-independent form factor of the α-nucleus potential, gives better results (7th column in table 2), than folding of the α-nucleon potential over the target nucleus it fails in reproducing the details of the forward angular maxima of the low-energy angular distributions [10]. Cowley and Wall [11] also added a phenomenological term to the potential as calculated with the just described method and then obtained very good fits to the data (8th column in table 2). The three parameters in this additional repulsive term, which is given by

$$V_m(r) = V_0\{1 + \exp - (\frac{r-R}{a_0})^2\}, \qquad r \leqslant R \qquad (6)$$

$$= 2V_0 \quad \exp - (\frac{r-R}{a_0})^2 , \qquad r > R$$

vary with energy in a similar way as the WS parameters: strongly decreasing or increasing below $E_\alpha = 80$ MeV and about constant above this energy. Consequently, the

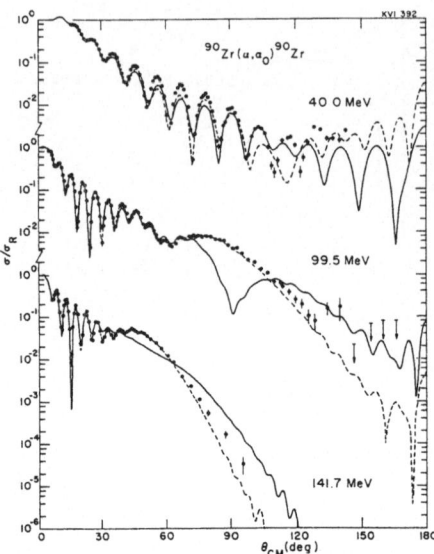

Fig. 5. Fits to some of the data with potentials obtained by folding an effective α-nucleon potential over the target nucleus. Solid curves are for potentials calculated for nuclear matter distribution as determined from electron scattering. Dashed curves were obtained for phenomenologically adjusted parameters of the matter distribution (see ref. [2]).

total potential has an energy-dependent shape, in fact the tails of the total potentials are very similar to the tails of the best-fit WS potentials. Whether this additional term in the potential is not just a convenient parametrization of the energy dependence of the potential or that the authors interpretation in terms of Pauli effects and of coupling of pick-up and stripping channels is correct has to be found out from detailed microscopic calculations.

The microscopic calculations of Perkin *et al.* [12] contain a number of corrections to the simple folding model calculations. Energy dependence of the nucleon-nucleus potential, three-body terms and Pauli effects were taken into account. To obtain good fits to the data, however, a substantial part of the calculated potentials, i.e. beyond about 6 fm, had to be replaced by phenomenological potentials. This failure also calls for a more detailed calculation!

3. Discussion of the results.

The observation of the difference in the form factor of the phenomenological potentials at low and high energy raises the question whether this difference can be attributed to a specific radial range. To answer this question it is necessary to know how well the potentials have been determined at each energy.

From a comparison of many of the potentials obtained in fitting the 100 MeV data (see fig. 10 in ref. [2]) it was found that all real potentials that give a good fit approximately coincide over the radial range from 4 fm to 8 fm. But at radii less than 4 fm and beyond 8 fm the potentials differ considerably. This suggests that for this energy the potential is well determined in the range 4-8 fm and that the scattering is very insensitive to the interior and the tail of the potential. This is confirmed by studies in which either the interior or the tail of the best-fit real WS potential at each energy was distorted by multiplication with a factor $\exp[-c(r_D-r)^2]$, with $c = 0.4$ fm^{-2}. The effect on the quality of the fit to the data is shown in fig. 6, for high energies as well as for low energies an effect is seen to be present only over the range 4-9 fm. A very similar result was obtained by Cowley and Wall [11] who applied a different type of distortion, i.e. a multiplication with a factor $[1-0.1 \exp\{-((r-RT)/AT)^2\}]$ with AT = 0.5 fm.

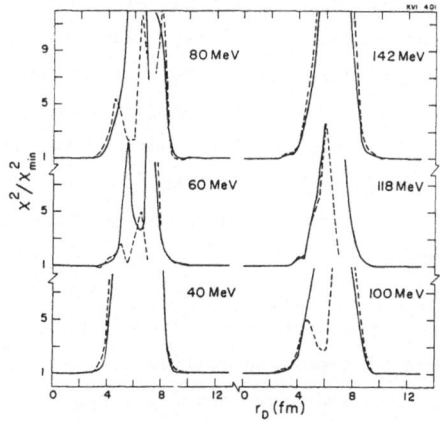

Fig. 6. Ratio χ^2/χ^2_{min} as a function of r_D, the radius below or above which the best-fit WS potential is distorted by multiplication with a factor $\exp(-0.4(r_D-r)^2)$. Solid curves are for data over the full angular range. Dashed curves are for truncated data sets, the truncation angles correspond to a momentum transfer $q = 3.5$ fm^{-1}.

All the analyses have shown that the data for energies between 80 and 142 MeV can be well described with form factors which are independent of E_α. We can thus conclude that in this energy range the real form factor does not depend on E_α and that it is well determined between 4 and 8 fm. This radial range seems to be dominant in determining the high-energy scattering.

At 40 MeV the situation is quite different. An extensive comparison with the WS potential is not very meaningful since the parametrizations of eqs. (2)-(5) gave worse fits to the data. While the various splines which give excellent fits to the 40 MeV data distinctly differ from each other for $r < 8$ fm the tails of these splines closely follow the tail of the best-fit WS potential (see fig. 4). This seems to confirm the strong absorption picture that at low energies the potential value at the strong absorption radius and the tail of the potential are very important and that the scattering is insensitive to the surface and interior part of the potential [13, 14]. This would also explain that the folding potential and the WS potential with the high-energy form factor, which both have tails which differ from the good-fit potentials of fig. 4 (see fig. 3), fail in giving a good description of the data. But this is contradicted by the fact that good fits can be obtained with potentials that have been constructed by suitably replacing the inner part (r < 6 fm) of potentials which do not give good fits to the data. This was found (see sect. 2) for the folding potential and for the 100 MeV spline potential, for both the inner part was replaced by an optimized spline potential. This must mean that at low energies the scattering is not only sensitive to the tail but also to the potential at radii less than 6 fm, and furthermore that changes in the tail can be compensated by changes in the interior or the surface part of the potential. The negative effect on the quality of fit of efforts to smooth the splines of fig. 4 also indicates a sensitivity to the surface and the interior of the potential. From fig. 6 one might even conclude that at low energies the scattering is insensitive to the tail and sensitive to the same radial part of the potential as at high energies. Whether this apparent contradiction with the strong absorption picture should be interpreted in terms of important contributions to the scattering from the interior of the nucleus awaits calculations of the type Brink and Takigawa made for α-scattering from ^{40}Ca [15].

It is clear from the presented results that although the determination of the unique potential at high energies has enabled us to resolve the discrete ambiguity at low energies, and even in spite of the indications for sensitivity to the surface part of the potential, we are faced with ambiguities in determining the detailed shape of the real potential. Because of these ambiguities it is not possible to draw conclusions about the difference between high- and low-energy potentials. Apart from the ambiguities in the real potential there is the additional ambiguity whether to ascribe the failure of the high-energy form factor at low energies to an energy-dependence of the real form factor [2,11,12,16] or to an energy-dependent imaginary form factor [8] or may be to a combination of both. Clearly, the guidance of theoretical calculations is required to remove these ambiguities!

At present no microscopic calculations of the real potential exist that can reproduce the ^{90}Zr data and the phenomenologically determined potentials. As mentioned already in section 2 and as shown in fig. 3 the potential obtained by folding the α-nucleon interaction over the target nucleus is only correct for r > 5.5 fm. Folding over the α-particle does not give a detailed description of the data [10,11] either. More sophisticated calculations in which effects are taken into account which are neglected in the simple folding model calculations predict a real form factor which changes with energy [9,12,16,17] albeit that the calculations of Sinha [17] and of Majka et al. [9] only predict a very minor effect. The potential calculated by Sinha[17] which includes density dependence effects does not account for the 142 MeV data. The calculation of Perkin et al. [12] (see sect. 2.2) could only reproduce the data when the calculated potential beyond about 6 fm, i.e. in the dominant part of the potential, was replaced by a phenomenological potential. Majka *et al.* [9] corrected for one-nucleon exchange, but the radius of the calculated potential came out to be too small. The calculation of Chang and Ridley [16], in which binding energy effects are treated explicitly, also does not result in a detailed description of experimental data (no calculations made for ^{90}Zr).

In view of the results of Majka *et al.* [8,9] in analyzing the data with an energy-dependent imaginary form factor calculations of the type made for ^{40}Ca by Vinh Mau [18], which resulted in only a slight energy dependence of the real form factor and an appreciable variation of the imaginary form factor with energy, would be of interest for ^{90}Zr.

4. Conclusions.

The real and imaginary part of the optical model potential describing elastic α-particle scattering from ^{90}Zr were found to have form factors which over the range $E_\alpha = 80-142$ MeV are independent of the bombarding energy. The shape of the real potential in this energy range seems to be determined within close limits over the range 4-8 fm. Low-energy data cannot be described by simultaneously using the same real and imaginary form factors as at high energy. There is strong, although not conclusive, evidence that at low energies the form factor of the real potential has a different shape than at high energies. Some indications for an energy-dependent form factor of the imaginary potential are also present. Ambiguities in the determination of the real potential and the correlation between real and imaginary potential prohibit to draw definite conclusions about the shape of the real and imaginary potential at these low energies.

The available microscopic calculations do not account for the phenomenologically determined potentials and for the details of the experimental data. More detailed microscopic calculations may help us to improve on our understanding of α-particle scattering.

References.

1) L.W. Put and A.M.J. Paans, Phys. Lett. 49B (1974) 266.
2) L.W. Put and A.M.J. Paans, Nucl. Phys. A291 (1977) 93.
3) D.A. Goldberg, S.M. Smith and G.F. Burdzik, Phys. Rev. C10 (1974) 1362.
4) D.A. Goldberg and S.M. Smith, Phys. Rev. Lett. 29 (1972) 500.
5) D.F. Jackson and R.C. Johnson, Phys. Lett. 49B (1974) 249.
6) D.A. Goldberg, Phys. Lett. 55B (1975) 59.
7) F. Michel and R. Vanderpoorten, Phys. Rev. C16 (1977) 142.
8) Z. Majka and T. Srokowski, Acta Phys. Pol. B9 (1978) 53.
9) Z. Majka, A. Budzanowski, K. Grotowski and A. Strzalkowsky, Phys.Rev.C18(1978)114.
10) P.P. Singh, P. Schwandt and G.D. Yang, Phys. Lett. 59B (1975) 113.
11) A.A. Cowley and N.S. Wall, Phys. Rev. C17 (1978) 1322.
12) D.G. Perkin, A.M. Kobos and J.R. Rook, Nucl. Phys. A245 (1975) 343.
13) D.F. Jackson and C.G. Morgan, Phys. Rev. 175 (1968) 1402.
14) P.P. Singh and P. Schwandt, Phys. Lett. 42B (1972) 181.
15) D.M. Brink and N. Takigawa, Nucl. Phys. A279 (1977) 159.
16) H.H. Chang and B.W. Ridley, University of Colorado, preprint.
17) B. Sinha, Phys. Rev. C11 (1975) 1546.
18) N. Vinh Mau, Phys. Lett. 71B (1977) 5.

Accurate Optical Potentials for Elastic α-Particle Scattering from Nuclei Around A=40

R. CEULENEER

Université de l'Etat, B-7000 Mons, Belgium

In recent years, much effort has been devoted to microscopic calculations of the nucleus-nucleus interaction. At the present time, the results obtained for physical observables by these approaches are too inaccurate to be compared directly with the data. Therefore, accurate phenomenological otpical potentials are essential links between such calculations and experiment. Moreover, it is interesting to investigate the feasibility of the optical model description of some nuclear processes in order to clarify their physical interpretation. In this respect, it is worth mentioning the controversy raised by the "anomalous large angle scattering" of α-particles from medium-weight nuclei[1]. Finally, precise phenomenological optical potentials might allow the extraction of reliable values for physical quantities such as nuclear matter radii[2].

Most of the optical model analyses are carried out using analytical form factors. Their lack of flexibility introduces spurious constraints and, as a matter of fact, it has been shown that a significant improvement of the fits to the data is obtained using either a spline function representation[3] or a Fourrier-Bessel expansion[4] for the real potential.

In this note, recent results obtained by Michel and Vanderpoorten[5] in their analysis of elastic α-particle scattering from ^{40}Ca between 22 and 142 MeV and from ^{36,40}Ar, ^{42,44,48}Ca at 29 MeV using a spline function representation for the real potential are presented. The parameters determined by least-squares fit to the data are:

1°) 10 values V_n for the real potential ranging from 0 to 9 fm in 1 fm steps.

2°) 3 parameters for the imaginary potential whose form factor was chosen to be a squared Woods-Saxon.

3°) A normalization factor multiplying the cross section, introduced in order to take into account the large experimental uncertainties in the absolute normalization of the data. This parameter was found to remain within acceptable limits, given that discrepancies between cross sections from different laboratories at neighbouring energies might amount to 50%.

The calculated cross-sections for α-particle scattering from ^{40}Ca at indicated incident energies E_α and from ^{36,40}Ar, ^{40,42,44,48}Ca at E_α = 29 MeV are displayed in fig. 1 and fig. 2. The real potentials corresponding to these cross sections are presented in fig. 3 and fig. 4. It is found that, with increasing incident energy, their volume integral decreases at a rate compatible with Jackson's prediction[6], while the volume integral of the imaginary part increases smoothly. Although these very satisfactory features can be explained by means of simple physical arguments, it would be very interesting to understand them more firmly on the basis of microscopic calculations.

Acknowledgement

I am indebted to my colleagues F. Michel and R. Vanderpoorten for their collaboration in preparing this note.

References

1) F. Michel and R. Vanderpoorten, Phys.Rev. C16(1977)142 and references therein
2) H.J. Gils, contribution to this workshop
3) L.W. Put and A.M.J. Paans, Nucl.Phys. A291(1977)93 and contribution to this workshop
4) E. Friedman and C.J. Batty, Phys.Rev. C17(1978)34
5) F. Michel and R. Vanderpoorten, Preprint Mons
6) D.F. Jackson and R.C. Johnson, Phys.Lett. 49B(1974)249

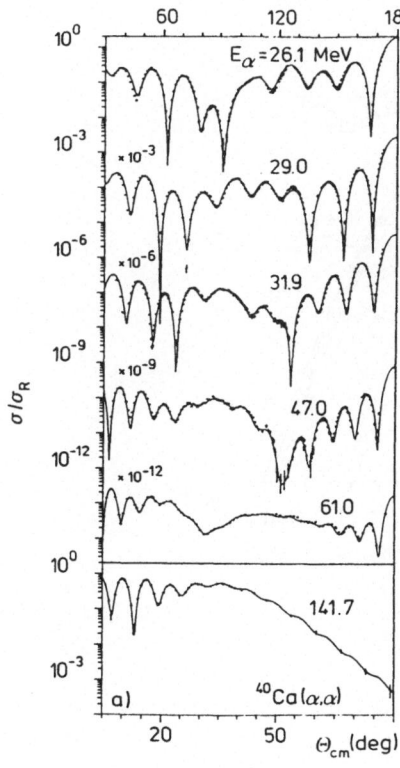

FIGURE 1

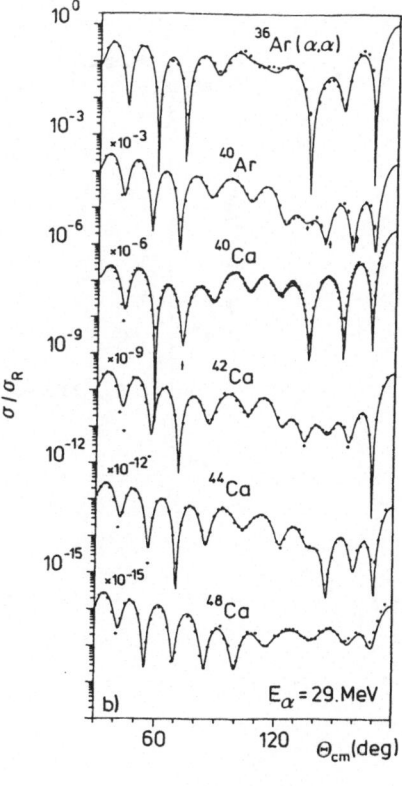

FIGURE 2

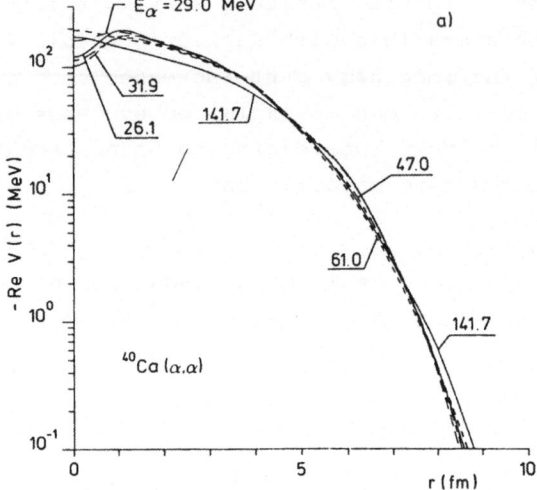

FIGURE 3

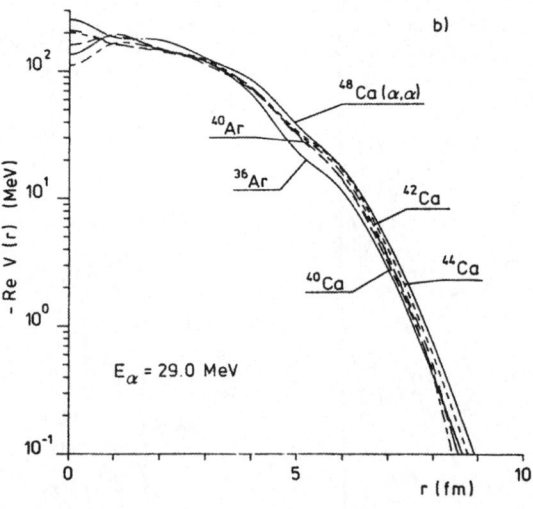

FIGURE 4

High Energy Alpha Scattering Used to Study the Uniqueness and Shape of the Optical Potential

S. Wiktor
Institute of Nuclear Physics, Cracow, Poland

C. Mayer-Böricke, A. Kiss, M. Rogge, P. Turek
Institut für Kernphysik, KFA, Jülich, BRD

The subject of this report is an analysis of the elastic alpha scattering at relatively high energies i.e. 120, 145 and 172.5 MeV on some light (^{12}C, ^{24}Mg, ^{27}Al) and heavier (^{58,60,62,64}Ni) nuclei. The experimental data have been obtained using the isochronous cyclotron in Jülich. The analysis was done in terms of the phenomenological optical potential.

The first attempt of the analysis was to examine the uniqueness of the potential parameters. It has been known for some time[1]) that the discrete ambiguity in potential parameters disappears if the energy of bombarding particles is sufficiently high and the measurements of angular distributions are extended to sufficiently large angles. The analysis of our data began with the potential of the form:

$$V(r) = V_c(r,r_c) - V \cdot f(r,r_v,a_v) - W \cdot f(r,r_w,a_w)$$

where V_c is the Coulomb potential, V and $\bar{W}$ are the depths of the real and imaginary parts of the nuclear potential and $f(r,r_i,a_i)$ is the standard Saxon-Woods form factor. Both volume and surface forms of the absorptive part of the potential have been used separately. The parameter V was varied in steps and the other five parameters were searched simultaneously to obtain optimum fits to the experimental data for every step of the parameter V. The results have been reported already[2]). As a reminder of the essential results, two pictures are shown. From Fig. 1 it is seen that the volume absorption gives definitively better agreement with experimental data than the surface absorption. Moreover, since in each case there appears only one absolute minimum, it is to be concluded that the potential corresponding to the minima in χ^2 vs. parameter V is unique. In the case of the lowest incident energy (120 MeV) the plots in fig. 1 do not exclude the occurance of additional minimum in the region of $V = 190$ MeV, however, this last one is not as deep as the main minimum in the region of $V = 110$ MeV.

The differential cross-sections calculated from the best fit potentials are displayed together with the experimental data on fig. 2. For

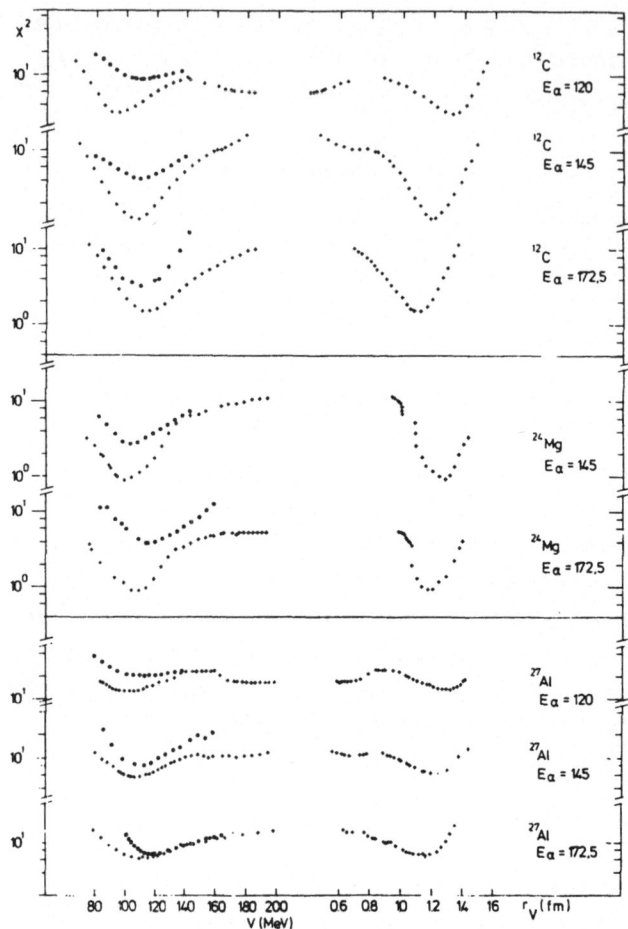

Figure 1 - The minimized values of χ^2 versus V (left side) and vs. r_v (right side) for volume absorption (crosses) and surface absorption (dots).

comparison, the best fits with surface absorption only are also indicated by the broken lines. For some range of V in the neighborhood of the minima in χ^2 the volume integrals are characterized by approximately constant values. This indicates the existence of the continuous ambiguity, or, in other words, correlations between the parameters. Of special significance is the correlation between the parameters V and r_v. The values of the volume integrals decrease systematically with increasing incident energy.

Having found the best fit parameters for different incident energies of the projectile, the analysis with fixed geometry was undertaken. The geometrical parameters, averaged over the energies, have been kept constant

Table 1

Target, E	6 - param. search								2 - param. search		
	V	r_v	a_v	W	r_w	a_w	$\dfrac{\chi^2}{F}$	V	W	$\dfrac{\chi^2}{F}$
(MeV)	(MeV)	(fm)	(fm)	(MeV)	(fm)	(fm)		(MeV)	(MeV)	
^{12}C 120	110	1.19	0.79	15.0	1.86	0.54	7.62	116	15	9.66
" 145	115	1.16	0.81	15.5	1.89	0.49	2.04	114	15	2.56
" 172	116	1.12	0.83	16.9	1.82	0.53	1.44	110	16	2.04
^{24}Mg 145	100	1.29	0.77	20.4	1.75	0.48	0.89	103	21	1.55
" 172	104	1.22	0.83	23.9	1.63	0.58	3.99	103	22	7.04
^{27}Al 120	102	1.31	0.75	20.3	1.67	0.56	12.5	111	20	14.0
" 145	105	1.24	0.78	19.8	1.63	0.60	6.44	105	19	6.61
" 172	111	1.16	0.81	20.9	1.58	0.67	5.86	104	21	7.81

where n and m are variable powers. At the beginning the powers n and m
were taken equal. It is found that the growth of the power implies immediately the growth of the potential depth and consequently also the
growth of remaining parameters. This is understandable since the increasing power makes the potential in the peripheral region shallower and the
demand to restore the fit to experimental data forces the depth of the
well to increase. This leads to a new ambiguity i.e. the correlation between V and n. Figure 3 presents the behavior of χ^2 vs. the power n for
different nuclei and different energies. In every point all six optical
model parameters were searched and the resulting χ^2 corresponds to optimal sets of these parameters. On the basis of these results one cannot
claim that the SW^2 form is generally more appropriate than the standard
SW form. A tendency can be noticed to change the shape of the potential
with the mass of the target nucleus and the energy of projectile.

In the next stage of the analysis different powers of the real and
imaginary parts n ≠ m were used. The dashed lines of fig. 3 correspond
to the case with constant value of m and varied the n-value only. From
this figure one can draw the conclusion that the shapes of real and imaginary parts of potentials do not have to be the same and they may vary
when changing the energy of projectile and mass target nuclei.

The question arises what influence the experimental error has on the
behavior of these curves. This question has been tested on the example
of ^{64}Ni by changing the normalization of experimental data. The results
are presented in fig. 4, where the χ^2 value is displayed vs. the power n
for a few values of normalization. As seen from fig. 4, the error in

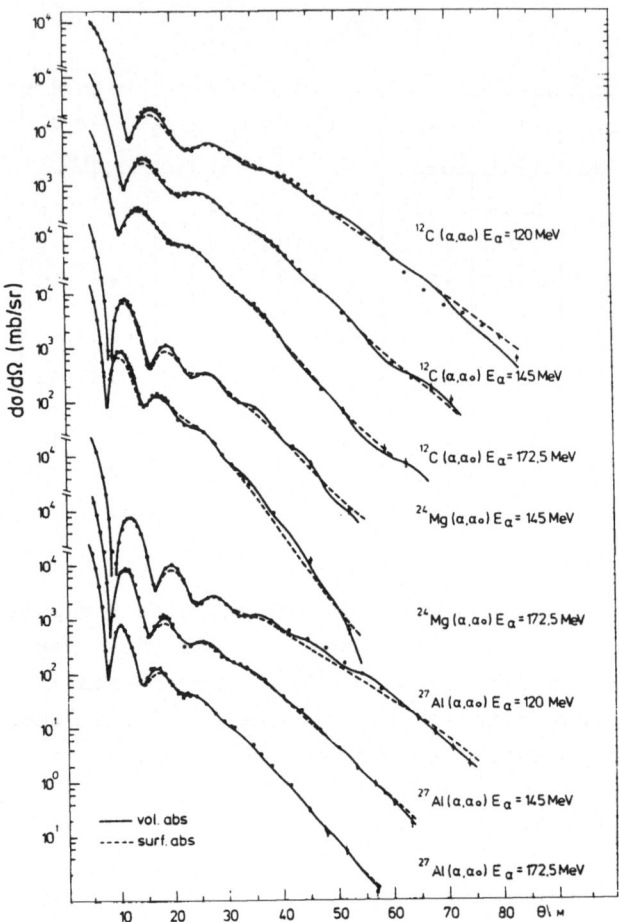

do/dΩ (mb/sr)

^{12}C (α,α_0) E_α = 120 MeV

^{12}C (α,α_0) E_α = 145 MeV

^{12}C (α,α_0) E_α = 172.5 MeV

^{24}Mg (α,α_0) E_α = 145 MeV

^{24}Mg (α,α_0) E_α = 172.5 MeV

^{27}Al (α,α_0) E_α = 120 MeV

^{27}Al (α,α_0) E_α = 145 MeV

—— vol. abs
----- surf. abs

^{27}Al (α,α_0) E_α = 172.5 MeV

Figure 2 - Elastic alpha scattering angular distributions for ^{12}C, ^{24}Mg and ^{27}Al for different incident energies between 120 and 172.5 MeV. The solid curves are the best optical model fits with volume absorption, the dashed curves those with surface absorption. Error bars indicate statistical uncertainities.

and only the parameters V and W were searched. In this approach, however, the best fit values of χ^2 became large by about 20 percent or more. Such change in χ^2 is reflected in markedly worse fits to the experimental data. In table 1 the best fit parameters and the χ^2 values for scattering on light nuclei are given.

From microscopic calculations, which were developed in recent years[3]), it appears that the potential form can deviate substantially from that of the Saxon-Woods type. Therefore an attempt has been made to reanalyze the data in terms of a modified potential form:

$$V(r) = V_c(r,r_c) - V \cdot f(r,r_v,a_v)^n - W \cdot f(r,r_w,a_w)^m$$

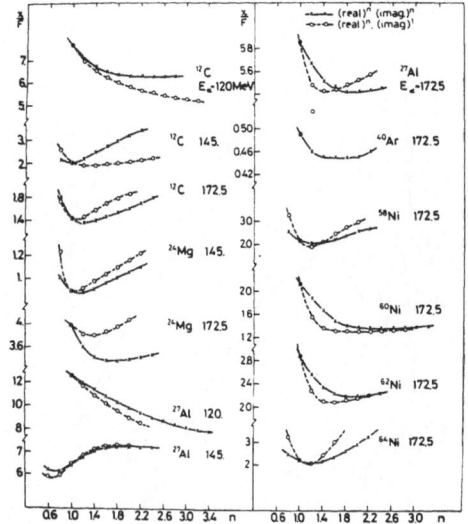

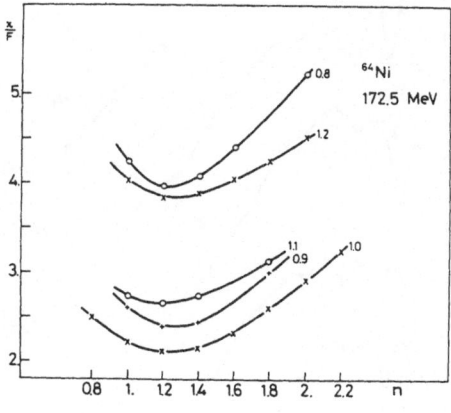

Figure 3 - The dependence of χ^2 on the power n. The solid line corresponds to n = m and the dashed line corresponds to m constant (m = 1).

Figure 4 - The dependence of χ^2 on the power n in the case of ^{64}Ni. Various curves correspond to various normalizations.

absolute value of the cross-section has no influence on the shape and position of the minimum in the curves. Hence, only the relative errors in cross-section may exert the influence on the shape of the curves shown in fig. 3. The main uncertainty of this kind is inherent in the error of the angular position of the detector, especially at those points where the cross-section changes rapidly with the angle.

Of course the curves presented in fig. 3 cannot be treated as crucial for making decisive conclusions concerning the discussed forms, since small changes in experimental data may seriously affect them. Our aim was only to bring attention to the examination of other than SW potential shapes.

In fig. 5 the influence of the above discussed various factors on calculated angular distribution is illustrated. Generally the fits, judged by eye, are not sensitive enough to examine such fine effects as the magnitude and shape of the assumed potential. The numerical values of χ^2, as are shown in figs. 1 and 3, provide more sensitive criteria.

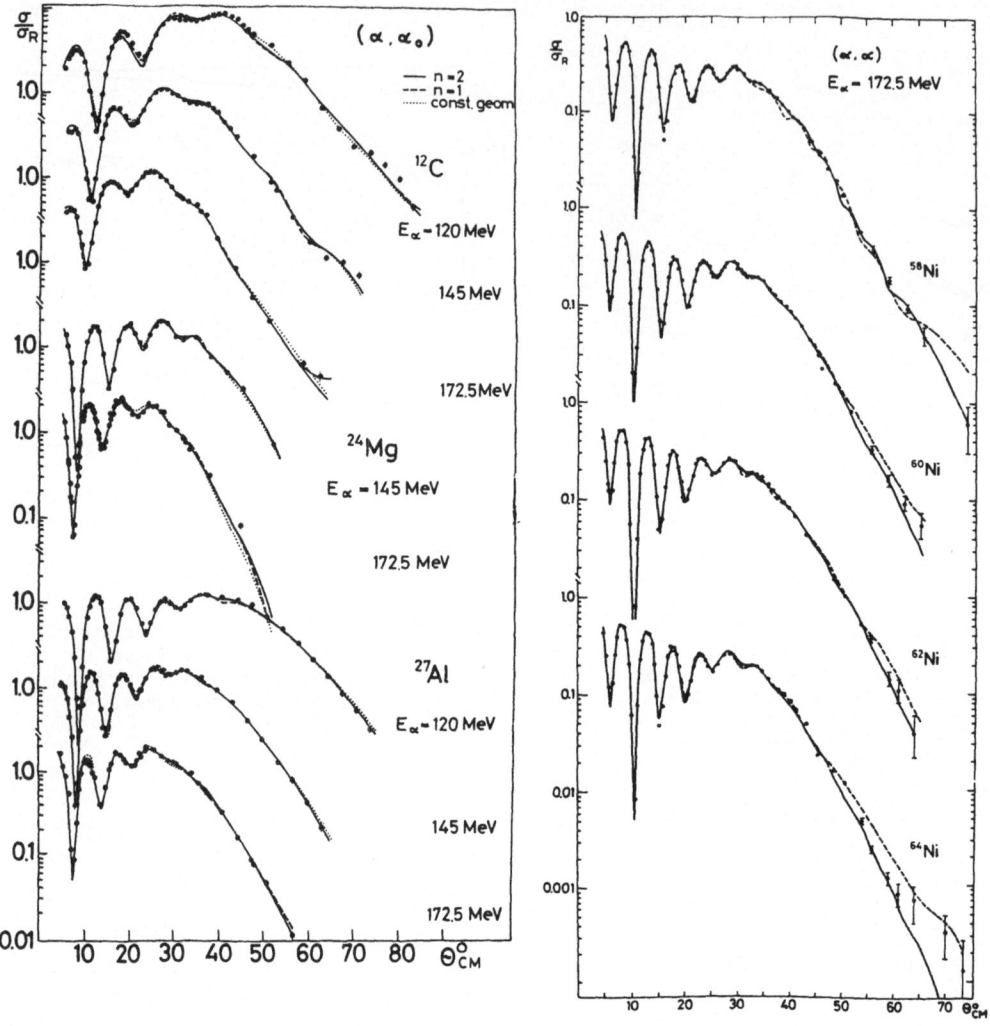

Figure 5 - Elastic α-scattering cross-section reduced to Rutherford
cross section. The solid lines represent the 6 - parameter
(SW)² fits, the dashed lines represent the 6 - parameter SW
fits and the dotted lines represent the 2 - parameter SW fits.

References:

1) B. Tatischeff, I. Brissaud, Nucl.Phys. $\underline{A155}$(1970)89
   S.M. Smith, G. Tibell, A.A. Cowley, D.A. Goldberg, H.G. Pugh,
   W. Reichart and N.S. Wall, Nucl.Phys. $\underline{A207}$(1973)273
   D.A. Goldberg, S.M. Smith, H.G. Pugh, P.G. Ross and N.S. Wall,
   Phys.Rev. $\underline{C7}$(1973)1938
   D.A. Goldberg, S.M. Smith and G.F. Burdzik, Phys.Rev. $\underline{C10}$(1974)1362
   A. Budzanowski, H. Dabrowski, L. Freindl, K. Grotowski, S. Micek,
   R. Planeta, A. Strzalkowski, M. Bosman, P. Leleux, P. Macq,
   J.P. Meulders and C. Pirart, Phys.Rev. $\underline{C17}$(1978)951
   Z. Majka, A. Budzanowski, K. Grotowski and A. Strzalkowski,
   Institute of Nuclear Physics, Cracow, Report No. 940/PL 1977

2) S. Wiktor, A. Kiss, C. Mayer-Böricke, M. Rogge and P. Turek,
   Annual Report 1976 of the Institute of Nuclear Physics, KFA Jülich
   A. Budzanowski, C. Alderliesten, J. Bojowald, C. Mayer-Böricke,
   W. Oelert, P. Turek and S. Wiktor, Annual Report 1977 of the Institute
   of Nuclear Physics, KFA Jülich

3) F.A. Brieva and J.R. Rook, Nucl.Phys. $\underline{291A}$(1977)299,317
   L.W. Put and A.M.J. Paans, Nucl.Phys. $\underline{A291}$(1977)93
   N. Vinh Mau, Phys.Lett., $\underline{B71}$(1977)5

ODD-EVEN DEPENDENCE OF THE OPTICAL POTENTIAL

Y.C. Tang

Institut für Theoretische Physik der Universität Tübingen, BRD

and

School of Physics, University of Minnesota, USA

1. INTRODUCTION

The conventional optical model, which employs a local, ℓ-inde-pendent potential for its real central part, has been quite success-ful in explaining the essential features of light-ion scattering by medium-and heavy-weight nuclei. As has been frequently demonstrated, even a simple folding prescription [1] for this model can lead to very satisfactory agreement with experimental data, except perhaps at extreme backward angles [2]. On the other hand, for the analysis of scattering problems such as ^{3}He + α [3], α + ^{6}Li [4], and ^{12}C + ^{13}C [5] where the incident and target nuclei have similar mass, the con-ventional model has been found to be rather inadequate. This indi-cates, therefore, that there must be certain basic features of the internuclear interaction which are not properly contained in the op-tical potential of such a model and these features become especially important when the nucleon-number difference of the colliding nuclei is small.

In the conventional optical model, the Pauli principle is not explicitly considered and internuclear antisymmetrization is only approximately taken into account by adjusting the depth parameters and form factors of the optical potential. The question thus natur-ally arises: are these adjustments flexible enough to properly repre-sent antisymmetrization effects? To answer this question, we have recently undertaken a project [6,7] to examine the structure of ker-nel functions in resonating-group calculations where totally anti-symmetric wave functions are used. As will be discussed below, the result of this examination does reveal that the basic shortcoming of the conventional model lies in the lack of an ℓ-dependence in its real central part and, for a reasonable description of the effects of internuclear antisymmetrization, the optical potential must gene-rally contain at least an odd-even ℓ-dependent or parity-dependent component.

In sect. 2, we give a brief description of the resonating-group formulation and show that antisymmetrization effects are represented by various nucleon-exchange terms in the exchange-normalization and exchange-Hamiltonian kernel functions. Section 3 is devoted to a discussion of the general features of the effective local potentials which are constructed to yield the same Born scattering amplitudes as these exchange terms. Explicit resonating-group calculations in ^{3}He + α and α + ^{16}O systems, where contributions from individual nucleon-exchange terms are investigated, are then discussed in sect. 4. From these calculations, one sees that the conclusions reached from the simple Born-approximation study of sect. 3 are in fact generally valid even at lower energies. Finally, in sect. 5, we summarize the results, and discuss the situation under which exchange effects are particularly significant and the introduction of a Majorana component becomes very important if a local-potential analysis of experimental data is to be successfully made.

2. RESONATING-GROUP FORMULATION AND EXCHANGE KERNEL FUNCTIONS

We discuss here the general case of A + B scattering, where the nuclei A and B contain N_A and N_B ($N_A > N_B$) nucleons, respectively (for simplicity, we assume the spins of the nuclei to be zero and the charge of the proton to be infinitesimally small). In the simplest, one-channel resonating-group formulation (for details, see refs. [7,8]), the trial wave function Ψ is written as

$$\psi = \mathcal{A}\left[\phi_A \phi_B F(\vec{R}) Z(\vec{R}_{cm})\right] , \qquad (1)$$

where $\mathcal{A}$ is an operator for the total antisymmetrization of the wave function and $Z(\vec{R}_{cm})$ is any normalizable function describing the center-of-mass motion of the entire system. The functions ϕ_A and ϕ_B represent the internal structures of the clusters; they are chosen to be translationally-invariant products of single-particle functions of the lowest configuration in harmonic-oscillator wells of width parameters α_A and α_B, respectively. The function $F(\vec{R})$ describes the relative motion between the clusters; it is obtained by solving the projection equation

$$\langle \delta\psi \mid H - E_T \mid \psi \rangle = 0 , \qquad (2)$$

where E_T is the total energy of the system composed of cluster internal energies E_A and E_B, and the relative energy E in the c.m. system.

The Hamiltonian H is a Galilean-invariant operator, given by

$$H = \sum_{i=1}^{N} T_i + \sum_{i<j=1}^{N} V_{ij} - T_{cm}$$ (3)

with $N = N_A + N_B$ being the total number of nucleons, T_{cm} being the kinetic-energy operator of the total center-of-mass, and V_{ij} being a nucleon-nucleon potential chosen to fit the two-nucleon scattering data especially in the low-energy region.

By using eqs. (1) - (3) one obtains, after some straight-forward manipulation, the following integrodifferential equation for the relative-motion function $F(\vec{R}')$:

$$\left[-\frac{\hbar^2}{2\mu} \nabla_{\vec{R}'}^2 + V_D(\vec{R}') - E \right] F(\vec{R}') + \int K(\vec{R}',\vec{R}'') F(\vec{R}'') d\vec{R}'' = 0 \quad,$$ (4)

where V_D is a direct potential and $K(\vec{R}', \vec{R}'')$ is an energy-dependent kernel function given by

$$K(\vec{R}',\vec{R}'') = H_E(\vec{R}',\vec{R}'') - E_T N_E(\vec{R}',\vec{R}'')$$ (5)

with H_E and N_E being the exchange-Hamiltonian and the exchange-normalization kernels, respectively; these kernels have the forms

$$H_E(\vec{R}',\vec{R}'') = \langle \phi_A \phi_B \delta(\vec{R}-\vec{R}')z \,|\, H \,|\, \mathcal{A}'' [\hat{\phi}_A \hat{\phi}_B \delta(\vec{R}-\vec{R}'')z] \rangle$$ (6)

and

$$N_E(\vec{R}',\vec{R}'') = \langle \phi_A \phi_B \delta(\vec{R}-\vec{R}')z \,|\, \mathcal{A}'' [\hat{\phi}_A \hat{\phi}_B \delta(\vec{R}-\vec{R}'')z] \rangle \quad.$$ (7)

In eqs. (6) and (7), $\hat{\phi}_A$ and $\hat{\phi}_B$ are antisymmetric functions obtained by antisymmetrizing the cluster internal functions ϕ_A and ϕ_B, and the operator $\mathcal{A}''$ is defined by the equation

$$\mathcal{A}'' = \mathcal{A}' - 1 \quad,$$ (8)

with $\mathcal{A}'$ being an antisymmetrization operator which interchanges nucleons in different clusters. From equ. (4) one sees that if the two clusters are considered as structureless, then the effective interaction between them must be both nonlocal and energy-dependent.

From the above discussion, one also sees that the effects of intercluster antisymmetrization are contained in the exchange kernel functions H_E and N_E. If such effects are omitted by setting the anti-symmetrization operator A' as unity, then these kernel functions will vanish. In this crude approximation, the effective intercluster potential will, therefore, just be the direct potential V_D (i.e., the usual double-folding potential) which is a simple ℓ-independent local potential if a purely central nucleon-nucleon force, such as the one used in ref. [9], which contains specifically no Majorana component, is employed.

Upon performing the integration over nucleon spatial coordinates, the expression for N_E in eq. (7) can be reduced to

$$N_E(\vec{R}',\vec{R}'') = \sum_X N_E^X(\vec{R}',\vec{R}'') \qquad , \qquad (9)$$

where

$$N_E^X(\vec{R}',\vec{R}'') = P_{XN}\exp\left(-A_{XN}\vec{R}'^2 - C_{XN}\vec{R}'\cdot\vec{R}'' - A_{XN}\vec{R}''^2\right) , \qquad (10)$$

with X ($X \geq 1$, with its largest value equal to N_B in most cases) being the number of nucleons interchanged between the clusters and P_{XN} being a polynomial in $\vec{R}'^2, \vec{R}'\cdot\vec{R}''$, and $\vec{R}''^2$. By using a complex generator-coordinate technique recently developed [7,10], one can derive general expressions for the coefficients A_{XN} and C_{XN} [11]. These expressions are

$$A_{XN} = \frac{\mu_0^2}{4X}\frac{X^2\left[(\alpha_A-\alpha_B)^2 + \frac{2}{\mu_0}(N_A+N_B)\alpha_A\alpha_B\right] + N_A N_B\left(1-\frac{X}{\mu_0}\right)(\alpha_A+\alpha_B)^2}{N_A N_B(\alpha_A+\alpha_B) - X(N_A\alpha_A+N_B\alpha_B)}$$
$$(11)$$

and

$$C_{XN} = -\frac{\mu_0^2}{2X}\frac{X^2(\alpha_A-\alpha_B)^2 + N_A N_B\left(1-\frac{X}{\mu_0}\right)(\alpha_A+\alpha_B)^2}{N_A N_B(\alpha_A+\alpha_B) - X(N_A\alpha_A+N_B\alpha_B)} \qquad (12)$$

with

$$\mu_0 = \frac{N_A N_B}{N_A+N_B} \qquad (13)$$

being the reduced nucleon number of the two clusters.

In a similar manner, one obtains for H_E the form

$$H_E(\vec{R}',\vec{R}'') = \sum_X \sum_q H_{Eq}^X(\vec{R}',\vec{R}'') \qquad , \qquad (q = a,b,c,d,e) \qquad (14)$$

where

$$H_{Eq}^{X}(\vec{R}', \vec{R}'') = P_{xq} \exp(-A_{xq}\vec{R}'^2 - C_{xq}\vec{R}'\cdot\vec{R}'' - B_{xq}\vec{R}''^2)$$

$$+ \text{hermitian conjugate}, \tag{15}$$

with P_{Xq} (q = a,b,c,d,e) being again a polynomial in $\vec{R}'^2, \vec{R}'\cdot\vec{R}''$, and $\vec{R}''^2$. Here one sees that, in contrast to the exchange-normalization case, there appear now five types of exponential factors for each X-value (this number is reduced to four in the special case where $\alpha_A = \alpha_B$; also, for X = N_B, there are only three types). In these factors, the analytic expressions for A_{Xq}, B_{Xq}, and C_{Xq} can also be derived by using the complex-generator-coordinate technique mentioned above. These expressions are, however, quite lengthy in the general case where $\alpha_A \neq \alpha_B$; hence, we shall not list them here, but refer the interested readers to ref. [12] for details.

In the kernel function H_E, the type-α term arises from the kinetic-energy operator and those internucleon potential-energy operators in which both nucleons belong to the same cluster and both are either involved or not involved in any intercluster nucleon-exchange process (for a diagrammatical representation of each type in H_E, see ref. [12]). For this particular type, the exponential factor for each value of X is exactly the same as that in the corresponding term of the exchange-normalization kernel. In addition, it should be noted that in the case where the nucleon-nucleon potential has a rather long range (i.e., the range parameter K in eq. (16) of ref. [9] takes on a value appreciably smaller than the width parameters α_A and α_B), the exponential factors involved in types b,c,d, and e become quite similar to the exponential factor which appears in eq. (10). Thus, there are reasonable indications that the structures of H_E and N_E are in fact not greatly different (see ref. [12] for an explicit verification of this assertion); in the following, we shall therefore mainly discuss the properties of the exchange-normalization kernel and mention the specific features of the exchange-Hamiltonian kernel only in a relatively brief manner.

3. EFFECTIVE LOCAL POTENTIALS

To assess the relative importance of the various nucleon-exchange terms in the exchange-normalization and exchange-Hamiltonian kernel functions, we adopt the following procedure. We construct effective local energy-dependent "exchange" potentials which yield, in the Born approximation, the same scattering amplitudes as these nucleon-exchange terms, and then examine the energy and spatial dependence of these exchange potentials. For the general case where α_A is not equal to α_B, this procedure can in fact be straightforwardly carried out, but the resultant expressions for the exchange potentials will be rather complicated. Hence, for clarity in discussion, we adopt the assumption

$$\alpha_A = \alpha_B = \alpha . \tag{16}$$

This assumption does enable us to considerably simplify our presentation, but will not affect the conclusion in any essential way.

3.1 Study of exchange-normalization kernel

By using a prescription given previously [13], we find that the effective local potential $\tilde{V}_{XN}(\vec{R})$, which yields the same Born scattering amplitude as the kernel term N_E^X has the form

$$\tilde{V}_{XN}(\vec{R}) = \tilde{P}_{XN} \exp\left[-(k/k_{XN})^2\right] \exp\left[-(R/R_{XN})^2\right] (1 \text{ or } P^R) , \tag{17}$$

where k denotes the wave number given by $(2\mu E)^{1/2}/\hbar$, $\tilde{P}_{XN}$ is a polynomial in k^2 and R^2, and P^R is a Majorana operator interchanging the position coordinates of the two nuclei, treated now as structureless point particles. This effective potential, which is characterized by a characteristic wave number k_{XN} and a characteristic range R_{XN}, can be shown to have the following general properties:

(i) $X < \mu_0$. In this case, C_{XN} has a negative value and the effective potential $\tilde{V}_{XN}$ is a Wigner-type potential which yields large Born scattering amplitudes only at forward angles. Also, it is easily shown that both k_{XN} and R_{XN} decrease monotonically with increasing X and have their largest values when X = 1.

(ii) $X > \mu_0$. In this case, C_{XN} has a positive value and the effective potential $\tilde{V}_{XN}$ is a Majorana-type potential which yields large Born scattering amplitudes only at backward angles. Also, one readily finds that both k_{XN} and R_{XN} increase monotonically with increasing X and have their largest values when $X = N_B$.

In situations where grazing collisions are dominant, such as heavy-ion scattering, α-scattering by medium-and heavy-weight nuclei, and so on, it is of course evident that longer-ranged effective potentials will have larger influence. Therefore, our analysis indicates that, among all exchange terms, the one-exchange term (X = 1) with characteristic values k_{1N} and R_{1N}, and the core-exchange term (X = N_B) with character-istic values k_{CN} and R_{CN} are the most important.

It is our belief that, even when barrier and absorption effects are not too important, the above statement concerning the dominance of the one-exchange and core-exchange terms is still generally valid. By examining the depths of the effective potentials for all values of X in ^{3}He + α and other relatively simple systems where resonating-group kernel functions are entirely known (not just their exponential factors), we have found that the range of the effective potential is in fact the dominant factor in determining its relative importance. We must emphasize, however, that our belief is based merely on results obtained from examining specific systems rather than on any general consideration. Therefore, it is important that further investigation must still be made in order to establish a more solid basis for the validity of our statement.

To continue our discussion, we list below the expressions for the various relevant characteristic wave numbers and characteristic ranges:

$$k_{IN} = \left[\mu_0 \left(2\mu_0 - 1 \right) \alpha \right]^{1/2} , \tag{18}$$

$$R_{IN} = \left[\frac{4(\mu_0 - 1)}{2\mu_0 - 1} \frac{1}{\alpha} \right]^{1/2} , \tag{19}$$

$$k_{CN} = \left(\frac{N_A N_B}{N_A - N_B} \alpha \right)^{1/2} , \tag{20}$$

$$R_{CN} = \left(\frac{4}{N_A - N_B} \frac{1}{\alpha} \right)^{1/2} . \tag{21}$$

Also, for the sake of comparison, we give here the direct potential V_D which, in the case of a central nucleon-nucleon potential having a Gaussian spatial dependence with a range parameter K (see eq.(16) of ref.[9]), has the form

$$V_D(\vec{R}) = \tilde{P}_D \exp\left[-(R/R_D)^2 \right] , \tag{22}$$

where

$$R_D = \left[\frac{\mu_0 \alpha + (2\mu_0 - 1)K}{\mu_0 \alpha K} \right]^{1/2} \tag{23}$$

and $\widetilde{P}_D$ is a polynomial in R^2. In addition, it is important to note that in the polynomial factors $\widetilde{P}_{1N}$ and $\widetilde{P}_D$ the highest powers of R^2 are the same. As for the polynomial factor $\widetilde{P}_{CN}$ occuring in the core-exchange potential $\widetilde{V}_{CN}$, its highest power of R^2 has recently been determined by Baye et al. [14]; for the interesting case where N_A and N_B are nearly equal (see the discussion below), this highest power is again approximately the same as that appearing in $\widetilde{P}_{1N}$ and $\widetilde{P}_D$. Therefore, since the polynomial factors in $\widetilde{V}_{1N}$, $\widetilde{V}_{CN}$, and V_D have similar values for their highest powers in R^2, it is appropriate to simply examine the exponential factors in order to decide the situations under which the effective potentials $\widetilde{V}_{1N}$ and $\widetilde{V}_{CN}$ make important contributions.

Let us now study the spatial dependence of $\widetilde{V}_{1N}$ and $\widetilde{V}_{CN}$. By comparing the values of R_{1N} and R_{CN} with the value of R_D, we can make the following general remarks:

(i) The ratio R_{1N}/R_D is smaller than but close to 1. For example, in the realistic case where K is close to α and μ_0 is appreciably larger than 1, the value of R_{1N}/R_D is approximately equal to 0.8. This indicates, therefore, that the one-exchange term may be generally important, which is consistent with the results obtained in a number of previous investigations [15]. In these investigations, the purpose was to see if the phase-shift values calculated with the resonating-group method (with central nucleon-nucleon potential only) can be reasonably reproduced by a potential model in which one solves, instead of the integro-differential equation (4), but a simpler equation

$$\left[-\frac{\hbar^2}{2\mu}\nabla^2 + \widetilde{V}(\vec{R}) - E\right] F(\vec{R}) = 0 \quad, \tag{24}$$

where $\widetilde{V}(\vec{R})$ is an effective internuclear potential assumed to have the form

$$\widetilde{V}(\vec{R}) = V_D(R) + V_W(R) + V_M(R)P^R \quad, \tag{25}$$

with the terms $V_W(R)$ and $V_M(R)P^R$ introduced to represent the main effects of antisymmetrization. Indeed, these investigations have invariably shown that the V_W term must have a non-negligible magnitude in comparing with the V_D term. In addition, the fact that R_{1N} is less than R_D is also in agreement with an empirical finding [16], obtained by potential-model analyses of p, ^{3}He, and α scattering by ^{16}O, that the range of V_W tends to be somewhat shorter than that of the direct potential V_D.

(ii) The characteristic range R_{CN} decreases with increasing value of the nucleon-number difference

$$\delta = N_A - N_B \qquad (26)$$

between the nuclei A and B. This means that one expects the core-exchange effect to become less important as δ increases. Indeed, we have reached a similar conclusion based on the results of many resonating-group calculations [15,17]. There it was found that the degree of odd-even ℓ-dependence, exhibited by the calculated phase shift, turns out to be quite strong in scattering systems involving two Δ-shell nuclei where δ is small, and weak in systems such as α + ^{16}O and n + ^{40}Ca where δ takes on much larger values. In addition, of course, the finding that core-exchange effects are important in α + ^{6}Li, ^{12}C + ^{13}C, and ^{12}C + ^{16}O scattering [4,5,18] supports the assertion reached by our present analysis.

The situation in α + ^{40}Ca scattering at relatively low energies of a few MeV/nucleon needs some clarification. The successful fitting of experimental data by Kondo et al. [19] employing a potential containing an odd-even ℓ-dependence would seem to indicate that core-exchange effects are important in this system. Recently, however, it has been found [20] that the use of parity-independent potentials can similarly lead to excellent agreement with experiment over a wide energy range. The fact that the same experimental result can be explained by different sets of potentials is just a manifestation of potential-model ambiguities, as has been pointed out especially by Wall [21]. Based on our study here (see also ref.[22]), we are firmly of the opinion that the essential features of α + ^{40}Ca scattering can be properly accounted for without the incorporation of an odd-even ℓ-dependent component into the effective potential.

It should be remarked, however, that even when the core-exchange potential has a small magnitude, one may still observe significant effects in situations where partial wave scattering amplitudes strongly cancel one another. Generally, these occur at backward angles when the scattering energies are relatively high. For instance, in a phenomenological potential-model study of nucleon scattering by ^{40}Ca at about 30 MeV [2], it was found that when a small Majorana potential (V_M = -0.01 V_D in eq.(25)) is introduced, the scattering behaviour at angles larger than about 150° is appreciably affected and the differential cross section at 180° is increased by a factor of around 3.

3.2 Study of exchange-Hamiltonian kernel

The study of the exchange-Hamiltonian kernel $H_E(\vec{R}',\vec{R}'')$ is conducted in exactly the same way, resulting in effective exchange potentials $\tilde{V}_{Xa}, \tilde{V}_{Xb}, \tilde{V}_{Xc}$, and $\tilde{V}_{Xd}$ (as mentioned above, the type-a term has the same exponential factor for each value of X as the corresponding exchange-normalization term and, hence, will not be further considered; also, in the case where $\alpha_A = \alpha_B = \alpha$, type-d and type-e terms contain the same exponential factor). The expressions for these potentials are quite lengthy and, therefore, will not be given here (see ref.[12] for details). The properties of these exchange potentials can, however, be briefly summarized as follows:

(i)   For each type, the one-exchange and the core-exchange terms are again the most important ones among all exchange terms.

(ii) The one-exchange and core-exchange potentials are generally (with very few exceptions) Wigner-type and Majorana-type potentials, respectively.

(iii) For the characteristic wave number, there appears a major difference between the results obtained from the exchange-Hamiltonian study and the exchange-normalization study. This occurs in the one-exchange case for type c, arising from internucleon potential-energy operators in which the two nucleons belong to different clusters and both are involved in an intercluster nucleon-exchange process (see ref.[12] for a detailed explanation). For instance, when K is chosen as equal to α, one finds that the characteristic wave number has now the expression

$$k_{1c} = g_\mu k_{1N} , \tag{27}$$

where the multiplicative factor g_μ , given by

$$g_\mu = \left[\frac{6\mu_0 - 1}{2\mu_0 - 1} \right]^{1/2} , \tag{28}$$

can assume a value almost equal to 2. For all other one-exchange and core-exchange characteristic wave numbers, the values are, however, not greatly different from those of k_{1N} and k_{CN} given by eqs.(18) and (20), respectively.

(iv) —— The values of the characteristic ranges R_{1q} and R_{Cq} (q = b,c,d) are either smaller or slightly larger (by 10 % or less for $K = \alpha$) than the corresponding values in the exchange-normalization case. In particular, the characteristic range R_{1c} has a magnitude comparable to R_{1N} given by eq.(19).

Because of item (iv) above, the discussion in subsect. 3.1 concerning the spatial dependence of effective potentials remains essentially valid. For the energy dependence, on the other hand, one must of course take proper consideration of the factor g_μ given by eq.(28). This will be discussed below in subsect. 3.3.

3.3 Energy dependence of effective potentials

In this subsection, we examine the energy dependence of exchange effects by studying the expressions for k_{1N}, k_{CN}, and k_{1C} given by eqs.(18),(20), and (27). As has been discussed above, the major characteristic which determines the importance of an effective exchange potential is the range. However, even when it has a range comparable to that of the direct potential, the effects of this potential will still be relatively minor if its depth has a small magnitude. Therefore, for our present purpose, we shall make a reasonable, though somewhat arbitrary, assumption that the one-exchange and core-exchange potentials will become rather ineffective when their energy-dependent exponential factors (see eq.(17)) acquire values less than $e^{-\beta_1}$ and $e^{-\beta_C}$, respectively, with $\beta_1 = 2$ and $\beta_C = 4$ (note that we choose β_C to be larger than β_1 because, at relatively high energies, the core-exchange potential, in distinct contrast to the one-exchange potential, contributes mainly in an angular region where the direct potential has a rather small influence). Adopting this criterion, one can then easily find that the one-exchange potential has generally a small depth when E/μ_0 (i.e., the incident energy per nucleon in the laboratory system, regardless of whether A or B is the incident nucleus) is larger than $\tilde{E}_1$, where

$$\tilde{E}_1 = \frac{\hbar^2}{2M} g_\mu^2 \frac{2(2\mu_0 - 1)}{\mu_0} \alpha \tag{29}$$

with M being the nucleon mass, and the core-exchange potential has generally a small depth when E/μ_0 is larger than $\tilde{E}_C$ where

$$\tilde{E}_C = \frac{\hbar^2}{2M} \frac{4(2-\xi)^2}{1-\xi} \frac{1}{\delta} \alpha \tag{30}$$

with $\xi = \delta/N_A$ $(0<\xi<1)$.

The value of $\tilde{E}_1$, as determined from eq.(29), is in the range of about 50 to 100 MeV/nucleon as α varies from 0.2 to 0.4 fm^{-2} (we assume $K = \alpha$). Together with our discussion in subsect.3.1 concerning the range of the one-exchange effective potential, we can therefore conclude that in all scattering systems the one-exchange term has generally an important influence over a wide range of energies.

Table 1

Values of $\tilde{E}_C$ in various systems

System	δ	$\alpha \, (\text{fm}^{-2})$	$\tilde{E}_C$ (MeV/nucleon)
$n + \alpha$	3	0.52	90
$^3H + \alpha$	1	0.45	152
$\alpha + {}^{16}O$	12	0.36	16
$^{16}O + {}^{17}O$	1	0.32	106
$^{16}O + {}^{20}Ne$	4	0.30	25
$^{16}O + {}^{40}Ca$	24	0.27	5

The situation with the core-exchange term is rather different. In table 1, we list the values of $\tilde{E}_C$ for some representative systems. Here it is seen that, because of the factor $1/\delta$ occurring in eq.(30), the core-exchange term has generally a slow energy dependence only when δ is relatively small.

4. EXPLICIT STUDY OF ^{3}He + α and α + ^{16}O SYSTEMS

The considerations given in sect.3 are made in the Born approximation; hence, one might expect that the results obtained should have only semi-quantitative significance at relatively high energies. In this section, however, we shall show by an explicit study of the ^{3}He+α and α+^{16}O systems that these results may in fact have general validity even at lower energies where present-day nuclear-physics experiments are mostly concerned with.

4.1. ^{3}He + α system

The procedure we use to study the importance of various nucleon-exchange terms ($X = 1,2,3$) appearing in the ^{3}He+α kernel function, given in ref.[23], is as follows. We choose a definite energy, and compare the differential cross sections and phase shifts obtained by

solving the integrodifferential equation (4) with the full kernel
(resonating-group or r-g calculation) and with different nucleon-exchange terms turned off. In fig. 1, we show such a cross-section comparison at a c.m. energy of 60 MeV (i.e., 35 MeV/nucleon). To obtain
the results given in this figure, we have omitted all charge effects
and used the nucleon-nucleon potential of ref.[9] with a Serber mixture. As is seen, the differential cross sections obtained with the
resonating-group calculation (solid circles) and with the two-exchange
term turned off (solid line) are almost identical at all angles, thus
indicating that the two-exchange term is not important, in agreement
with the Born-approximation prediction given in sect.3.

The importance of the one-exchange and the core exchange (i.e.,
three-exchange with X = 3) terms is also shown in fig. 1. Here one
sees that if the one-exchange term alone is included (dashed line),
the cross-section behaviour in the forward angular region can be reasonably reproduced but the strong rise in the backward direction does
not at all materialize. In a different way one can also see from this
figure that the one-exchange term must necessarily be considered in
the calculation. If one includes only the three-exchange term (dot-
dashed line), then one finds that the calculation yields a back-angle
rise in the cross section; however, the cross sections at forward
angles are severely underestimated and the over-all agreement with the
resonating-group result is quite poor.

Individual contributions of the one-exchange and three-exchange
terms at 60 MeV can be seen more clearly from fig. 2 (remember that
the two-exchange term has little influence). With the one-exchange

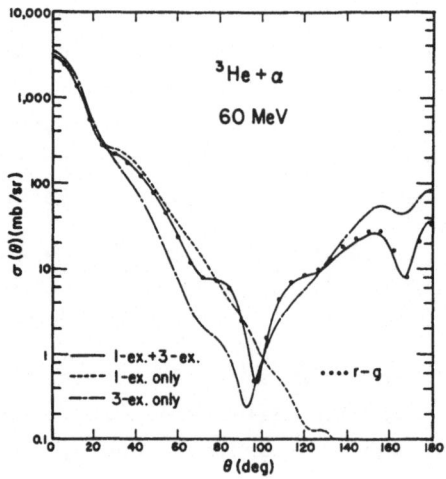

Fig. 1:
Comparison of ^{3}He+α differential cross sections at
60 MeV calculated with the
full resonating-group kernel
and with various nucleon-exchange terms omitted. The
common width parameter used
for the cluster internal
functions is α = 0.46 fm^{-2}.

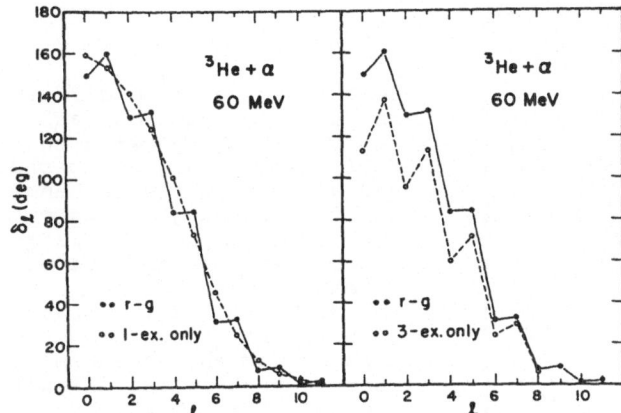

Fig. 2:
Comparison of ^{3}He+α phase shifts at 60 MeV calculated with the full resonating-group kernel and with various nucleon-exchange terms omitted.

term alone, it is shown in the left side of this figure that the calculated phase-shift points (open circles) lie on a rather smooth curve which cuts right across the zigzag line joining the resonating-group phase-shift points (solid circles). This indicates that the three-exchange effective potential is essentially a Majorana potential, exactly as claimed in sect.3. The situation is quite different when only the three-exchange term is considered. The right side of fig. 2 shows that the dashed line joining the calculated phase-shift points has now a zigzag behaviour, but in all partial waves the phase-shift values are smaller than those from the resonating-group calculation. This is an indication that the one-exchange effective potential is essentially an attractive Wigner potential, again in agreement with the discussion presented in sect.3.

4.2. $\alpha + {}^{16}O$ system

A similar procedure is used to study the importance of different nucleon-exchange terms (X = 1,2,3,4) in the $\alpha+{}^{16}O$ system. For this system, the value of μ_0 is 3.2, and the discussion in sect.3 shows that the one-, two-, and three-exchange terms should yield progressively weaker contribution. In addition, because the nucleon-number difference δ has a rather large value in this case, it is anticipated that the core-exchange term (i.e., four-exchange term with X = 4) should also have little influence. That these expectations are indeed borne out even at a relatively low energy of 18 MeV (about 5.6 MeV/nucleon) is shown in fig. 3. Here one sees that the resonating-group cross-section values at all angles can be well accounted for by a calculation in which only the one-exchange and two-exchange terms are included.

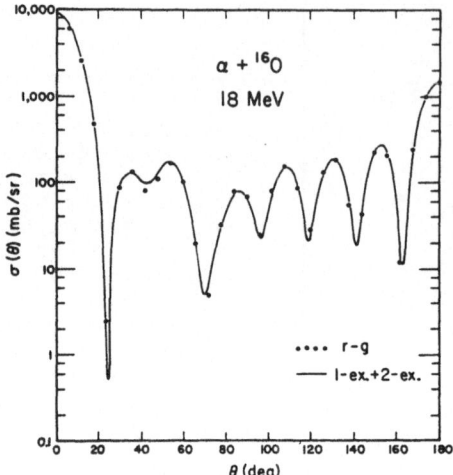

Fig. 3 :
Comparison of $\alpha + {}^{16}O$ differential cross sections at 18 MeV calculated with the full resonating-group kernel and with the three-exchange and four-exchange terms omitted. The common width parameter used for the cluster internal functions is $\alpha = 0.32$ fm^{-2}. All charge effects are omitted and the nucleon-nucleon potential used is that of ref.[9] with $w = 0.334$, $m = 0.481$, $b = 0.076$, and $h = 0.109$.

Table 2

$\alpha + {}^{16}O$ phase shifts δ_0 and δ_1, in degrees, at 18 MeV calculated with the full kernel and with various nucleon-exchange terms omitted

	col.1 r-g calc.	col.2 V_D+1-ex.	col.3 V_D+2-ex.	col.4 V_D+3-ex.	col.5 V_D+4-ex.	col.6 V_D only
δ_0	633.4	682.4	412.5	474.4	471.8	468.7
δ_1	623.9	667.0	405.7	465.3	458.3	461.7

To demonstrate the effects of the various nucleon-exchange terms in a more detailed manner, we list in table 2 phase shifts δ_0 and δ_1 calculated with the full resonating-group kernel and with some of these exchange terms omitted. From this table, we note the following salient features:

(i) From columns 1 and 2, one finds that, for an approximate calculation, it is not sufficient to take only the one-exchange term into account; the two-exchange term must also be included (see also ref.[22]).

(ii) By comparing entries in columns 2,3,4, and 6, one notes that the one-exchange term has a larger influence than the two-exchange term which, in turn, has a much larger influence than the three-exchange term (see also ref.[24]). Also, there appears the interesting feature that the one-exchange and three-exchange effective potentials are on-the-average attractive, while the two-exchange effective potential is on-the-average repulsive.

(iii) The one-, two, and three-exchange effective potentials are ess-
      entially Wigner-type potentials, while the four-exchange (core-
      exchange) effective potential is essentially a Majorana-type
      potential. In addition, it is noted from columns 5 and 6 that
      the four-exchange term has a small contribution in this case.
All these features are consistent with the discussion given in sect.3,
based on the Born approximation.

5. DISCUSSION AND CONCLUSION

By using the procedure of constructing effective local potentials,
we find that the essential features of internuclear antisymmetrization,
represented by various nucleon-exchange terms in the resonating-group
kernel function, can be briefly summarized as follows:
(i)   At least at relatively high energies, the one-exchange and the
      core-exchange terms are the most important ones among all ex-
      change terms. Especially this is so, when absorptions effects
      are important and, hence, grazing waves become dominant.
(ii) In all scattering systems, the one-exchange term is generally
      important over a wide energy range. The core-exchange term, on
      the other hand, has generally an important influence only when
      the nucleon-number difference of the interacting nuclei is rather
      small.
(iii) The effective local potentials corresponding to one-exchange and
      core-exchange terms are essentially Wigner-type and Majorana-
      type potentials, respectively.
      Based on these features, one may then come to the following use-
ful conclusions:
(i)   In a scattering calculation where the nuclei involved have a
      large difference in mass, such as $\alpha + ^{40}Ca$ scattering, it may be
      a reasonable approximation to consider only those exchange terms
      in which a small number of nucleons is interchanged between the
      clusters (e.g., one- and two-exchange terms in the $\alpha + ^{40}Ca$ case),
      thus substantially reducing the computational difficulty associ-
      ated with a many-nucleon resonating-group investigation.
(ii) If a local-potential model is constructed to analyze experimental
      scattering results, then the effective potential in this model
      (i.e., the real central part of the optical potential) must, in
      general, contain a Majorana exchange component (see eq.(25));
      in other words, the effective potential must have an odd-even
      ℓ-dependent or parity-dependent character. In this respect, it

is interesting to note that odd-even potential models have already been successfully used to fit experimental data of $p+^{3}He$, $p+\alpha$, $^{3}He+\alpha$, $\alpha+^{6}Li$, and $^{12}C+^{13}C$ scattering [3,5,25-28].

From our discussion it also becomes clear why the conventional optical model, in which the effective potential contains no Majorana component, works quite well for light-ion scattering by medium- and heavy-weight nuclei. The main reason for this is evidently that, in these cases, the nucleon-number difference of the interacting nuclei is sufficiently large such that, except for fitting scattering data at extreme backward angles, the core-exchange term has little influence and, therefore, can be reasonably omitted from the calculation.

To achieve a better understanding of the core-exchange effect, one should conduct both theoretical and experimental investigations for scattering systems involving nuclei of similar mass, such as ^{16}O scattering by ^{17}O, ^{18}O, and other heavier targets. For this purpose, the important point to note is that the experiment should cover a large angular region and be performed at energies of about 20 to 50 MeV/nucleon where resonance effects are small. At present, experimental data of this type exist in light-ion systems (see, e.g, ref.[4]) but are not yet available for heavy-ion scattering.

ACKNOWLEDGMENTS

I wish to thank Professor K. Wildermuth for his kind hospitality extended to me at the Institut für Theoretische Physik der Universität Tübingen. Also, it is my sincere pleasure to acknowledge the Alexander von Humboldt-Stiftung for financial assistance in the form of a Senior U.S. Scientist Award.

References:

1. G.W. Greenlees, G.J. Pyle, and Y.C. Tang, Phys. Rev. 131 (1968) 1115.
2. G.W. Greenlees, W. Makofske, Y.C. Tang, and D.R. Thompson, Phys. Rev. C6 (1972) 2057.
3. P.E. Frisbee, Ph.D.thesis, University of Maryland (1972); B. Buck, H. Friedrich, and C. Wheatley, Nucl. Phys. A275 (1977) 246.
4. D. Bachelier, M. Bernas, J.L. Boyard, H.L. Harney, J.C. Jourdain, P. Radvanyi, M. Roy-Stephan, and R. DeVries, Nucl. Phys. A195 (1972) 361.
5. W. von Oertzen, Nucl. Phys. A148 (1970) 529.
6. M. LeMere and Y.C. Tang, in Proceedings of the Third International Conference on Clustering Aspects of Nuclear Structure and Nuclear Reactions, Winnipeg, Canada (1978), and to be published in Phys. Rev. C.
7. Y.C. Tang, M. LeMere, and D.R. Thompson, will appear in Phys. Reports.
8. K. Wildermuth and Y.C. Tang, A Unified Theory of the Nucleus (Vieweg, Braunschweig, 1977).
9. D.R. Thompson and Y.C. Tang, Phys. Rev. C4 (1971) 306.
10. W. Sünkel and K. Wildermuth, Phys. Lett. 41B (1972) 439; D.R. Thompson and Y.C. Tang, Phys. Rev. C12 (1975) 1432, C13 (1976) 2597.
11. M. LeMere, Ph.D. thesis, University of Minnesota (1977).
12. M. LeMere, D.J. Stubeda, H. Horiuchi, and Y.C. Tang, to be published.
13. G.W. Greenlees and Y.C. Tang, Phys. Lett. 34B (1971) 359.
14. D. Baye, J. Deenan, and Y. Salmon, Nucl. Phys. A289 (1977) 511; D. Baye and P.H. Heenen, Fizika (Suppl.3) 9 (1977) 1.
15. D.R. Thompson, M. LeMere, and Y.C. Tang, Nucl. Phys. A270 (1976) 211.
16. M.LeMere, R.E. Brown, Y.C. Tang, and D.R. Thompson, Phys. Rev. C15 (1977) 1191.
17. Y.C. Tang, Fizika (Suppl.3) 9 (1977) 91.
18. D.Baye, Nucl. Phys. A272 (1976) 445.
19. Y. Kondo, S. Nagaka, S. Ohkubo, and O. Tanimura, Progr. Theoret. Phys. 53 (1975) 1006.
20. F. Michel and R. Vanderpoorten, Phys. Rev. C16 (1977) 142; H.P. Gubler, U. Kiebele, H.O. Meyer, G.R. Plattner, and I. Sick, Phys. Lett. 74B (1978) 202.
21. N.S. Wall, Phys. Rev. C14 (1976) 2326.
22. K. Langanke, J. Leutenantsmeyer, M. Stingl, and A. Weiguny, in Proceedings of the Third International Conference on Clustering Aspects of Nuclear Structure and Nuclear Reactions, Winnipeg, Canada (1978).
23. Y.C. Tang, E. Schmid, and K. Wildermuth, Phys. Rev. 131 (1963) 2631.
24. T. Matsuse, M. Kamimura, and Y. Fukushima, Progr. Theoret. Phys. 53 (1975) 706.
25. L.G. Votta, P.G. Roos, N.S. Chant, and R. Woody, III, Phys. Rev. C10 (1974) 520.
26. H. Bohlen, N. Marquardt, W. von Oertzen, and Ph. Gorodetsky, Nucl. Phys. A179 (1972) 504.
27. W. von Oertzen, in Proceedings of the INS-IPCR Symposium on Cluster Structure of Nuclei and Transfer Reactions induced by Heavy Ions, Tokyo, Japan, 1975 (IPCR Cyclotron Progress Report Supplement 4), p. 337.
28. D. Clement, E.J. Kanellopoulos, and K. Wildermuth, Acta Phys. Austr. 42 (1975) 29.

The Imaginary Part of the Heavy Ion Optical Potential

D. M. Brink

Dept. of Theor. Physics, Oxford.

Abstract

A number of different methods have been proposed for calculating the imaginary part of the Heavy-Ion Optical potential. In some of these a folding type procedure is used. Real and imaginary parts of the optical potential come from folding of the real and imaginary parts of some effective nucleon-nucleon t-matrix with nuclear densities. Another procedure calculates the loss of flux from the incident channel in second order perturbation theory. Green's function methods give a way of improving the perturbation approach.

In this lecture a different approach based on Feynman's path integral method is proposed. It is known that semi-classical methods are very useful for calculating heavy-ion scattering. The Feynman method provides a natural link between a complete quantal theory and semi-classical approximations. By writing a path integral expression for the scattering amplitude one obtains a formula relating the imaginary part of the optical potential to the coupling of the elastic channel to various inelastic and reaction channels. In a perturbation approximation this formula is quite analogous to Feshbach's formula which has been used recently to obtain a long-range optical potential describing the effects of Coulomb excitation on elastic scattering.

1.   Introduction.

In a complete treatment of the interaction of two heavy-ions the scattering equations include coupling between elastic and all reaction channels. This coupling allows for the possibility of polarization and excitation during the scattering process. The elastic scattering is affected by virtual transitions to excited states. In the optical model we choose to include these polarization effects in the optical potential rather than considering them explicitly. The imaginary part of the optical potential takes into account the loss of flux from the elastic channel into various reaction channels.

Most investigations of heavy-ion optical potentials have concentrated on calculating the real part of the potential. Some of the methods proposed for calculating the imaginary potential are reviewed in sections 3 and 4 of this talk. A new approach based on Feynman path integrals is introduced in section 5. The idea of this method comes from a recent application of the path integral method to deep-inelastic scattering of heavy ions.[1] In Ref. [1] the path integral method is used to obtain expressions for friction and diffusion coefficients describing transfer of energy between relative and internal degrees of freedom in a heavy-ion collision . Beck and Gross[2] have suggested that there should be a relation between the friction tensor and the imaginary part of the optical potential. The path integral approach enables this relation to be studied in more detail.

2. Experimental Data.

Before considering theories I will discuss briefly the information about optical potentials which can be extracted from experimental measurements of elastic scattering. It is necessary to distinguish between several different cases.

Strong Absorption: In this case any waves which penetrate into the interior of the optical potential are absorbed completely. The elastic angular distribution is diffractive and the ratio $\sigma(\theta)/\sigma_R(\theta)$ of the elastic cross-section $\sigma(\theta)$ to the Rutherford cross-section $\sigma_R(\theta)$ resembles a Fresnel diffraction pattern[3] when the Sommerfeld parameter n >> 1. When n < 1 there is a Fraunhofer-type angular distribution. The transition between these two kinds of angular distribution[4] occurs for n ∿ 1, and both Fraunhofer and Fresnel features coexist in the transition region.

When the interacting nuclei have zero spin the elastic scattering is determined by partial-wave scattering amplitudes S_ℓ or phase shifts δ_ℓ where $S_\ell = \exp(2i\delta_\ell)$. For strong absorption $|S_\ell|^2 \to 1$ for large relative angular momentum ℓ because then there is no interaction; $|S_\ell|^2 = \exp(-4\text{Im}\delta_\ell) \to 0$ for small ℓ because of absorption. A grazing angular momentum ℓ_g can be defined[5] as the value of ℓ for which the reflection coefficient $|S_\ell|^2 = \frac{1}{2}$, and a strong absorption radius d_g as the distance of closest approach for a classical Rutherford orbit with angular momentum ℓ_g.

According to Ball et al.[5] the elastic scattering data determine the strength of the real and imaginary potentials at the strong absorption radius d_g. The data also place some constraints on the surface diffuseness parameter; typically a ∿ 0.6 fm for a Woods-Saxon shape. These conclusions were drawn from a study of scattering of O and C by Pb. In these cases the strong absorption radius lies outside the Coulomb barrier of the heavy ion potential and d_g (or ℓ_g) is determined mainly by $\text{Im}V_N$.

The situation is different for lighter systems, for example α-scattering with lab energies $\lesssim$ 50 MeV or $^{16}O + ^{16}O$ scattering. In these cases the absorption occurs inside the Coulomb barrier and is relatively weak at the barrier. Then ℓ_g is the angular momentum for which the top of the Coulomb plus centrifugal barrier coincides with the incident energy and is determined mainly by $\text{Re}V_N$. The elastic scattering is not sensitive to $\text{Im}V_N$ provided it is strong enough to absorb any incident flux passing over the Coulomb barrier.

In both these cases elastic scattering gives no information about the heavy-ion optical potential for small separations between the interacting nuclei and limited information in the surface region.

Refractive Scattering: In certain cases it seems that $\text{Im}V_N$ is weak enough to allow the interior region of the optical potential to influence the elastic scattering. There is a large back-angle cross-section observed in $\alpha + {}^{40}\text{Ca}$ scattering for $E_{lab} \lesssim$ 40 MeV. In a semi-classical description this effect is produced by orbits deflected to large angles[6,7] by the real part of the $\alpha - {}^{40}\text{Ca}$ potential. At higher α-energies the nuclear rainbow phenomenon gives rise to characteristic structures in elastic cross-sections. This phenomenon has also been observed in ${}^6\text{Li} + {}^{28}\text{Si}$ elastic scattering[9] at for $E_{lab} = 135$ MeV. These effects are often referred to as refractive phenomena. When they are present it may be possible to get some information about the interior regions of the optical potential. Similar [8] effects occur in the scattering of ${}^{16}\text{O}$ by ${}^{28}\text{Si}$.

3. Folding Models.

Several theories of heavy-ion elastic scattering lead to a folding prescription of the optical potential. If two interacting nuclei have matter densities $\rho_1(\underset{\sim}{r}_1)$ and $\rho_2(\underset{\sim}{r}_2)$ respectively then the optical potential in a folding model is

$$V_{opt}(r) = \int d^3\underset{\sim}{r}_1 \int d^3\underset{\sim}{r}_2 \; \rho_1(\underset{\sim}{r}_1)\rho_2(\underset{\sim}{r}_2) \; g(\underset{\sim}{r}_{12})$$

(3.1)

where $\underset{\sim}{r}_{12} = \underset{\sim}{r} + \underset{\sim}{r}_2 - \underset{\sim}{r}_1$ and $g(\underset{\sim}{r}_{12})$ is some effective G-matrix describing the interaction of a pair of nucleons. Folding models have the nice feature that they provide a unified description of the real and imaginary parts of V_{opt}. In general g is complex and the real and imaginary parts of V_{opt} come from the real and imaginary parts of g in the folding integral (3.1).

For the limiting case of high energy nucleus-nucleus scattering g becomes the t-matrix for free space nucleon-nucleon scattering

$$g(r_{12}) = -\frac{2\pi \hbar^2 f(0)}{M} \; \delta(r_{12})$$

(3.2)

where M is the nucleon mass and f(0) as a complex forward nucleon-nucleon scattering amplitude evaluated at the incident lab. energy per nucleon. It is an appropriate average over spin and isospin variables which according to Dover and Vary[10] is

$$f(0) = \xi f_{np}(0) + (1-\xi) f_{pp}(0)$$

(3.3)

where $\xi = (N_1 Z_2 + N_2 Z_1)/A_1 A_2$ and $f_{np}(f_{pp})$ are the neutron-proton (proton-neutron) forward amplitudes.

Dar and Kirzon[11] obtain the high energy limit (2) by starting with Glaubers'[12]

amplitude for elastic scattering of two composite nuclei A and B. Using the impact
parameter representation one writes

$$f(Q) = \frac{iK}{2\pi} \int e^{i\, Q\cdot b} \left[1 - e^{i\, X_{AB}(b)} \right] d^2 b \qquad (3.4)$$

where b is the impact parameter, K is the incident momentum and $\underset{\sim}{Q}$ is the momentum
transfer. The Glauber phase is given by

$$X_{AB} = -i\, \ell n \left[\left\langle f_A f_B \right| \prod_{i=1}^{A} \prod_{j=1}^{B} \left\{ 1 - \Gamma(\underset{\sim}{b} - \underset{\sim}{s}_i + \underset{\sim}{v}_j) \right\} \left| f_A f_B \right\rangle \right] \qquad (3.5)$$

where f_A and f_B denote the ground state wave-functions of nuclei A and B composed of
A and B nucleons respectively, $\underset{\sim}{s}_i$ and $\underset{\sim}{v}_j$ are the projections of the coordinates of
the i-th and j-th nucleons on a plane perpendicular to K, and Γ is the nucleon-nucleon
profile function. This profile function can be expressed in terms of nucleon-nucleon
scattering amplitudes. Dar and Kirzon[11] are concerned with the strong absorption
case. Then only the tails of the density distributions ρ_A and ρ_B are important and
they are able to simplify eq. (3.4) to obtain an effective optical potential which
has the folding form (3.1). The Glauber model is a high energy approximation. To be
valid it requires that V/E << 1, where E is the incident energy per nucleon and V is
a typical nucleon-nucleon interaction strength.

Dover and Vary[10] have considered modifications of the high energy formula which
allow it to be used for energies down to about 10 MeV per nucleon. They use an effective g-matrix with a finite range, average over the Fermi motion and consider off shell
corrections. Their theory gives a reasonable description of α-nucleus scattering in
the energy range 10-40 MeV per nucleon. A similar approach has been used by Saloner
et al.[13]

4. Dynamic Potentials.

Folding models like the ones described in the last section take into account the
finite size of interacting nuclei through their densities but they do not take into
account specific effects due to their finiteness such as the possibility of collective
oscillations or of transfer of nucleons between one nucleus and the other. The approach discussed in this section is to assume that the basic effective interaction is
known but then to explicitly take into account the non-elastic excitations which are
available to the interacting nuclei. If one used the complete g-matrix for finite
nuclei then the folding model would automatically include specific effects of the
excitation spectrum of those nuclei. Using a g-matrix for nuclear matter in a local
density approximation includes excitation effects, but in some averaged way.

Following Feshbach[14] we denote the complete Hamiltonian for the colliding pair
of nuclei by

$$H = H_o + T_R + v \qquad (4.1)$$

where H_o denotes the internal Hamiltone for the isolated nuclei, T_R is the kinetic energy of relative motion and is the coupling interaction between the nuclei. The ground-state wave-functions of the isolated nuclei are denoted by ϕ_o and ψ_o. The generalized optical potential for elastic scattering at a relative energy E is then

$$V_{opt} = \langle \phi_o \psi_o | v | \phi_o \psi_o \rangle + \langle \phi_o \psi_o | v Q \frac{1}{E - H_{QQ} + i\varepsilon} Q v | \phi_o \psi_o \rangle$$

$$= V_F + \Delta V .$$

$$(4.2)$$

Here the first term in eq. (4.2) is the folded optical potential V_F and ΔV is the correction due to excitation effects. The operator Q projects off the ground states of the two nuclei.

The imaginary part of the optical potential takes account of the loss of flux from the elastic channel into various reaction channels. Eq. (4.2) provides a formula for calculating this effect. It has been used by various groups for obtaining the optical potential for nucleon-nucleus scattering[15,16]. Recently a long range imaginary potential for heavy-ion scattering has been calculated from the same starting point [17,18]. This potential takes account of the loss of flux from the incident channel due to the direct process of Coulomb excitation.

5. Path Integral Approach.

In this section we use Feynman's path integral method[19] in the version formulated by Pechukas[20] to obtain a formula for the optical potential. The method is to write two expressions for the scattering amplitude, one in terms of an optical potential and the other in terms of a detailed coupled channels description. Comparison of the two expressions gives a procedure for calculating the optical potential.

Feynman's expression for a transition amplitude K for scattering by an optical potential $V_{opt}(r)$ is

$$K = \int \mathcal{D} r(t) \, \exp \frac{i}{\hbar} S[r(t)] .$$

$$(5.1)$$

Here $r(t)$ is a path for the relative coordinate satisfying end point conditions appropriate for a scattering problem and the path integral extends over all possible paths satisfying these conditions. The action S is the time integral of the Lagrangian along the path $r(t)$

$$S[r(t)] = \int \left(\frac{1}{2} \mu \, \dot{r}^2 - V(r) \right) dt$$

$$(5.2)$$

where μ is the reduced mass of the interacting nuclei.

Next we write another expression for K from a microscopic point of view. The intrinsic motion of the fragments is described by a Hamiltonian H_o with eigenstates $|a>$, $|b> \dots$ and energies ε_a, $\varepsilon_b \dots$ The interaction between the relative motion $\underset{\sim}{r}(t)$ and the internal coordinates ξ is described by a potential $v(r,\xi)$ and the time evolution operator U of the system obeys the equation.

$$i\hbar \frac{\partial U}{\partial t} = \left[H_o + v(r(t),\xi) \right] U$$

(5.3)

For a given path $\underset{\sim}{r}(t)$ the solution with U = 1 at $t = t_o$ is denoted by $U[\underset{\sim}{r},t,t_o]$. According to Pechukas[20] one can write an expression for the amplitude K as a path integral

$$K = \int \mathcal{D}\underset{\sim}{r}(t) \, exp\{\tfrac{i}{\hbar} S_o[r(t)]\} \langle o| \, U[\underset{\sim}{r}, t_1, t_o]| o\rangle$$

(5.4)

where $S_o[\underset{\sim}{r}(t)] = \int_{t_o}^{t_1} \tfrac{1}{2}\mu \dot{\underset{\sim}{r}}^2 \, dt$.

Eqs. (5.1) and (5.4) give identical results for K provided

$$\int_{t_o}^{t_1} V_{opt}(r(t)) \, dt = i\hbar \ln\left[\langle o| U[r(t), t_1, t_2]| o\rangle \right]$$

(5.5)

for all paths $\underset{\sim}{r}(t)$.

In general it is impossible to satisfy the relation (5.5) for a local optical potential, because the structure of the right hand side of eq. (5.5) is much more complicated than the structure of the left hand side. This should not be unexpected because the optical potential (4.2) coming from Feshbach's formalism is non-local and energy dependent. Let us therefore ask for a less stringent requirement; namely that eq. (5.5) should hold for a certain class of "important paths".

For heavy-ion scattering the stationary phase method should be a reasonable approximation for evaluating the integrals (5.1) and (5.4). The stationary orbits are classical orbits in some appropriate potential $V_o(r)$. This suggests that the important paths referred to in the last paragraph should be classical orbits in $V_o(r)$ at the particular energy of the scattering experiment. An approximation would be to require that eq. (5.5) should hold for Rutherford orbits or, in the case where the incident energy is high above the Coulomb barrier, for straight line orbits tangential to Rutherford orbits at the point of closest approach. As there is a one parameter family of orbits corresponding to different impact parameters it should be possible to find V_{opt} to satisfy eq. (5.5).

How can the right hand side of eq. (5.5) be calculated? The simplest method is to take the leading term in time-dependent perturbation theory. This should be a good approximation for cases when there is strong absorption. In the tail of the potential the interaction $v(r,\xi)$ is small and perturbation theory is valid. For closer separations there is almost complete absorption and deviations from the perturbation result would not make much difference. The perturbation result gives

$$\int_{-\infty}^{\infty} V_{opt}[\underline{r}(t)] \, dt = \int_{-\infty}^{\infty} \langle 0| \, v(\underline{r}(t),\xi)|0\rangle \, dt$$
$$- \frac{i}{\hbar} \sum_{n\neq 0} \int_{-\infty}^{\infty} dt \int_{-\infty}^{t} dt' \, \langle 0|v(r(t),\xi)|n\rangle \langle n|v(r(t'),\xi)|0\rangle e^{\frac{i}{\hbar}(\varepsilon_0-\varepsilon_n)(t-t')}$$

$$(5.6)$$

The first term in eq. (5.6) is the average folded potential of Feshbach's theory (eq. (4.2)). The second term gives the polarization potential ΔV. The imaginary part of the optical potential is contained in the second term. Separating it leads to the relation

$$\int_{-\infty}^{\infty} Im \, V_{opt}[r(t)] \, dt = -\frac{1}{2\hbar} \sum_{n\neq 0} \left| \int_{-\infty}^{\infty} dt \, \langle n| \, v(r(t),\xi)|0\rangle e^{\frac{i}{\hbar}(\varepsilon_n-\varepsilon_0)t} \right|^2 .$$

$$(5.7)$$

Eq. (5.6) has a similar structure to the Feshbach's formula (4.2).

6. Long Range Polarization Potential.

Love et al.[17] and Baltz et al.[18] have calculated a long-range absorption in the heavy-ion optical potential due to Coulomb excitation of a low lying collective quadrupole state. These calculations are based on the Feshbach formula (4.2) and yield an imaginary part to the optical potential whose leading term has a radial dependence proportional to r^{-5}. The same problem can be studied by the path integral method introduced in section 5.

For this example eq. (5.7) reduces to

$$\int_{-\infty}^{\infty} Im \, V_{opt}[r(t)] \, dt = -\frac{\hbar}{2} P_2[r(t)]$$

$$(6.1)$$

where $P_2[r(t)]$ is the probability of exciting the collective 2^+ state calculated by perturbation theory. Using the results of Alder et al.[22] we can write

$$P_2 = K_0 \frac{9}{16} \left[|I_{20}(0,\xi)|^2 + 3|I_{22}(0,\xi)|^2 \right]$$

$$(6.2)$$

Here the $I_{2\mu}$ are the usual Coulomb excitation integrals[22], θ is the classical scattering angle for the Rutherford orbit involved in eq. (6.1), and ξ is the adiabaticity parameter of Coulomb excitation theory. Also

$$K_o = \frac{16\pi}{225} \frac{k^4}{n^2} \left[\frac{B_T(E2)\uparrow}{Z_T e^2} \right]$$

(6.3)

where k is the wave-number of relative motion of the nuclei, n is the Sommerfeld parameter. Eqs. (6.2) and (6.3) assume that only the target is excited. If the projectile is also excited another term must be added to eq. (6.2).

Inverting to find the long range behaviour of the optical potential gives V_{opt} $\propto r^{-5}$ for large r. This long range part has the same strength as the corresponding term in the potential of Baltz et al.[18] The long range part in Love et al.[17] is stronger by a factor of 4/3.

7. Relation with Dissipative Processes.

In this section we give another application of the ideas developed in section 5 by finding a relation between the imaginary part of the optical potential describing scattering of heavy-ions in excited states and a friction coefficient for deep inelastic scattering.

In the work of Agassi, Ko and Weidenmüller (AKW)[21] on a description of deeply inelastic collisions in terms of a transport equation the Hamiltonian of the system is assumed to have the form (4.1). Matrix elements of the interaction $v_{ab} = <a|v|b>$ between eigenstates of H_o are taken to be random variables with mean value zero and second moment

$$\overline{v_{ab}(r)\, v_{cd}(\tilde{r})} = \left(\delta_{ac}\delta_{bd} + \delta_{ad}\delta_{bc}\right)\left(D_a D_b\right)^{\frac{1}{2}} W_o\, f\{\tfrac{1}{2}(r+\tilde{r})\}$$

$$\times \exp\{-(\varepsilon_a - \varepsilon_b)^2/2\Delta^2\}\, \exp\{-(r-\tilde{r})^2/2\sigma^2\}$$

(7.1)

A justification of this form can be found in ref.[21] In eq. (7.1) D_a denotes the mean level spacing at energy ε_a, W_o is a strength parameter and f(r) is a form factor. Typical values of the correlation length σ and the correlation energy Δ are σ = 3.5fm and Δ = 7MeV.

Eq. (5.7) corresponds to a weak coupling formula in the terminology of AKW. This should be adequate for a discussion of the optical potential. When strong coupling effects become important for deep inelastic scattering the optical model absorption is so large that its numerical value is irrelevant. Substituting eq. (7.1) in eq.

(5.7) gives an expression for the optical potential when the interacting nuclei are in an excited state |a> as

$$\operatorname{Im} V_{opt}(r) = -\pi W_o f(r).$$ (7.2)

It is interesting that eq. (7.2) does not depend on the excitation energy of the target or projectile i.e. the optical potential should be the same in every excited state. This result depends on the validity of eq. (7.1) and this equation makes sense only if the excitation energy is large enough for the level spacing D_a to have a meaning. We should therefore not use eq. (7.2) for the optical potential when target and projectile are in their ground states. On the other hand as eq. (7.2) is independent of the excitation energy an extrapolation could give a reasonable estimate of the ground state optical potential.

A friction coefficient for deep inelastic scattering in weak coupling is calculated in ref. [1] as $c = \hbar\beta\pi W_o f/\sigma^2$ where β is the inverse of a nuclear temperature defined in terms of the level spacing $D(\varepsilon)$ as

$$\beta = -\frac{d}{d\varepsilon}\left[\ln D(\varepsilon)\right].$$

Equation (7.2) suggests a relation

$$c = \frac{\beta\hbar}{\sigma^2}\left|\operatorname{Im} V_{opt}(r)\right|.$$ (7.3)

Beck and Gross give a formula for c, but their formula is written in terms of the mean energy loss ΔE in a transition. Perturbation theory gives this as

$$\Delta E = \tfrac{1}{2}\beta\left(\hbar v/\sigma\right)^2$$ (7.4)

provided the relative velocity v is not too large (i.e. provided $\hbar v < \Delta\sigma$). Eliminating β from eq. (7.3) gives

$$c = \frac{2\Delta E}{\hbar v^2}\cdot\left|\operatorname{Im} V_{opt}\right|$$

which is exactly the result of Beck and Gross. The interpretation is somewhat different. In ref. [2] ΔE is taken as the spacing between single particle shell model states, i.e. $\Delta E \simeq 8 \text{MeV}$. In the present calculation ΔE is given by eq. (7.4) and depends on the energy of relative motion and on the nuclear temperature. The choice in ref. [2] is more appropriate to the early stages of a deep inelastic process when energy can be transferred only from relative motion to internal motion. In later stages energy can be transferred in both directions. On the average energy is lost from relative motion because of the increase in density of states $D^{-1}(\varepsilon_a)$ with energy. Then the expression (7.4) is more appropriate.

References

1. D. M. Brink, J. Neto and H. A. Weidenmüller (to be published).
2. R. Beck and D. H. E. Gross, Phys. Lett. 47B(1973) 143.
3. J. S. Blair, Phys. Rev. 95 (1954) 1218, Phys. Rev. 115 (1959) 928.
4. W. E. Frahn, Extended Seminar on Nuclear Physics Trieste 1973 (IAEA, Vienna, 1975) Vol 1.157.
5. J. B. Ball, C. B. Fulmer, E. E. Gross, M. L. Halbert, D. C. Hensley, C. A. Ludemann, M. J. Saltmarsh and G. R. Satchler, Nucl. Phys. A252 (1975) 208.
6. F. Michel and A. Vanderpoorten, Phys. Rev. C16 (1977) 142.
7. D. M. Brink and N. Takigawa, Nucl. Phys. A279 (1977) 159.
8. P. Braun-Munzinger, G. M. Berkowitz, T. M. Cormier, C. M. Jachcinskc, J. W. Harris, J. M. Barrette and M. J. Levine, Phys. Rev. Lett. 38 (1977) 944.
9. R. M. DeVries, D. A. Goldberg, J. M. Watson, M. S. Zisman and J. G. Cramer, Phys. Rev. Lett. 39 (1977) 450.
10. J. P. Vary and C. B. Dover, Proceedings of Second High Energy Heavy Ion Summer Study, Lawrence Berkeley Lab. (July 1974).
11. A. Dar and Z. Kirson, Phys. Lett. 37B (1971) 166, Nucl. Phys. A237 (1975) 319.
12. R. J. Glauber, High Energy Physics and Nuclear Structure, ed. G. Alexander (N.H. 1967).
13. D. A. Saloner and C. Toepffer, Nucl. Phys. A283 (1977) 108
    D. A. Saloner, C. Toepffer and B. Fink, Nucl Phys. A283 (1977) 131.
14. H. Feshbach, Ann. of Phys. 19 (1967) 287.
15. N. Vinh Mau and A. Bouyssy, Nucl. Phys. A257 (1976) 189.
16. P. W. Coulter and A. R. Satcher, Nucl. Phys. A293 (1977) 269.
17. W. G. Love, T. Terasawa and G. R. Satchler, Phys. Rev. Lett. 39 (1977) 6, Nucl. Phys. A291 (1977) 183.
18. A. J. Baltz, S. K. Kauffmann, N. K. Glendenning and K. Pruess.
19. R. P. Feynman and A. R. Hibbs, Quantum Mechanics and Path Integrals (McGraw Hill 1965).
20. P. Pechukas, Phys. Rev. 181 (1969) 174.
21. D. Agassi, C. M. Ko and H. A. Weidenmüller, Ann. Phys. (N.Y.) 107 (1977) 140.
22. K. Alder, A. Bohr , T. Huus, B. Mottelson and A. Winther, Rev. Mod. Phys. 28 (1956) 77.

HEAVY ION FOLDING POTENTIALS

W. G. Love

Department of Physics and Astronomy
University of Georgia
Athens, GA 30602, USA

Abstract: The calculation and application of folding-model potentials
for heavy-ion scattering are discussed for bombarding energies of the
order of 10 MeV/nucleon. Both single- and double-folding methods are
considered with an emphasis on using realistic internucleon interactions
based on G-matrices in a double-folding model. The desirability of a
consistent folding-model description of both light-ion and heavy-ion
scattering is stressed and progress in this direction and some residual
problems are discussed. An *empirical* assessment of the G-matrix folding
model is made by applying the model to a number of cases of heavy-ion
elastic and inelastic scattering.

1. INTRODUCTION

Although phenomenological heavy-ion (HI) optical potentials are
often used to describe elastic scattering, the use of *calculated* HI
potentials is appealing in that such potentials require as input, nuclear information from other processes such as nucleon-nucleon and
electron-nucleus scattering. When appropriate, the incorporation of
such nuclear information into HI potentials is probably carried out
most simply within the folding-model approximation.

To define folded potentials for heavy-ion collisions, we sketch
briefly the framework in which these potentials appear. The physical
picture and relevant coordinates are shown in Fig. 1. The complete
Hamiltonian for the colliding pair is given by:

$$H = H_1 + H_2 + T + V \qquad\qquad (1)$$

where H_1 and H_2 denote the internal Hamiltonians for the isolated ions,
T is the kinetic energy of *relative* motion in the center-of-mass system
(assumed to be ~10MeV/nucleon in accordance with the bulk of available
experimental data) and V is the coupling interaction between the ions.
We ignore antisymmetrization between nucleons in different ions, but

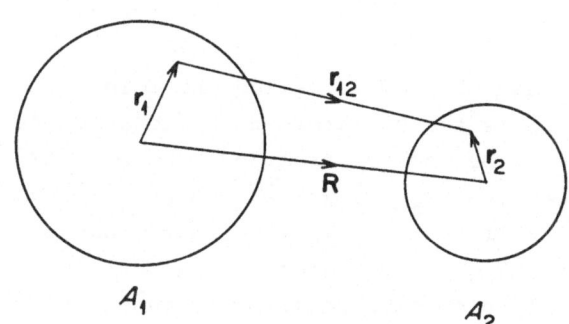

Figure 1. Coordinates used for the collision of two nuclei A_1 and A_2.

will return to this point later. We define a complete set of states for the *internal* Hamiltonians by

$$(H_1 - \varepsilon_{1i}) \psi_i = 0$$
$$(H_2 - \varepsilon_{2i}) \phi_i = 0 \qquad (2)$$

If the complete wave function is denoted by $\psi^{(+)}$ we note that measurements of elastic or inelastic scattering provide us with information about the projection of the complete wave function onto the channel subspace defined by the particular measurement. We denote these $\vec{R}$-dependent projections by

$$\chi_{ij}(\vec{R}) \equiv \langle \psi_i \phi_j | \psi^{(+)} \rangle \rightarrow \psi^{(+)} = \sum_{k,\ell} \chi_{k\ell}(\vec{R}) \psi_k \phi_\ell \qquad (3)$$

where i and j denote the final states selected for measurement. A set of coupled equations [1] governing the $\chi_{ij}(\vec{R})$ can be found by inserting the above expansion into the Schrödinger equation. This yields

$$[T + (\psi_i \phi_j | V | \psi_i \phi_j) - E_{ij}] \chi_{ij}(\vec{R}) = - \sum_{[k,\ell \neq i,j]} (\psi_i \phi_j | V | \psi_k \phi_\ell) \chi_{k\ell}(\vec{R}) .$$

$$(4)$$

$$E_{ij} = E - \varepsilon_{1i} - \varepsilon_{2j} \qquad (i,j = 0 \ldots \infty)$$

The rounded brackets indicate that the matrix elements remain a function of $\vec{R}$.

An alternative to eq. 4 is provided by Feshbach's formalism [2] in which all *explicit* couplings to channels other than the one of interest are eliminated. For that channel in which both nuclei remain in their ground states, for example, this procedure results in a one-body Schrödinger equation for the $\chi_{oo}(\vec{R})$,

$$[T + U_{op} - E] \chi_{oo} = 0 \qquad (5a)$$

with the optical-potential operator U_{op} given by

$$U_{op} = (\psi_o \phi_o |V| \psi_o \phi_o) + (\psi_o \phi_o |VQ \frac{1}{E-H_{QQ}+i\varepsilon} QV| \psi_o \phi_o)$$

$$= U_F + \Delta U. \tag{5b}$$

U_F is defined as the first term in eqn. 5b and is called the folded potential for elastic scattering. Q is the operator which projects off the ground states of the two nuclei. By its construction, the elastic scattering amplitude may be obtained from χ_{oo} as $R \to \infty$. Note that this same folding potential and its generalization to excited states and off-diagonal couplings appear in eq. 4. The folded potential is real (provided V is real), but ΔU is generally a complex operator representing excitations to all states of the system. We refer to ΔU as a dynamic polarization potential. In practice V is an effective interaction (G-matrix) which itself includes some real polarization corrections.

2. FOLDING MODELS

If we assume V to be a local two-body operator the double-folded potential U_F becomes

$$U_F(\vec{R}) = \int d\vec{r}_1 \int d\vec{r}_2 \rho_1(\vec{r}_1) \rho_2(\vec{r}_2) V(\vec{r}_{12} = \vec{R}+\vec{r}_2-\vec{r}_1). \tag{6}$$

Here ρ_i denotes the ground state density of A_i and an appropriate average over spin and isospin is implied. The associated off-diagonal folded potentials (couplings) are obtained from eq. 6 by replacing ρ_1 and ρ_2 by the appropriate *transition* densities. It should be noted that U_F may also be regarded as the classical interaction energy between two ions with "frozen configurations." When neither target nor projectile possesses spin, U_F becomes a function of the magnitude of R only. More generally U_F will contain spin-orbit, tensor etc. terms [3]. Usually, these extra terms are relatively unimportant [3] when only cross sections are measured due to the relatively small number of participating nucleons. Although the Coulomb interaction is contained in both V in eq. 1 and U_{op} in eq. 5b, it is usually treated on a slightly different but equivalent footing so that U_F will denote only the nuclear contribution to the potential.

Before looking in more detail at the real-folded potential U_F let us briefly examine ΔU. Although it is impractical to calculate ΔU in detail, some of its gross features may be noted. Its most conspicuous feature [4] (learned empirically) is that of providing strong absorption

of elastic flux for encounters in which the distance of closest approach
is less than a distance known as the strong absorption radius, $D_\frac{1}{2}$. $D_\frac{1}{2}$
is defined as the distance of closest approach for that Rutherford orbit
for which the transmission coefficient is $\frac{1}{2}$. Inside this radius the
scattering is known [4] to be relatively insensitive to the refractive
(real) part of the potential due to almost complete absorption. Slightly
outside this radius the rapidly decaying nuclear interaction is swamped
by the Coulomb potential with the result that HI scattering is only
sensitive to the real part of the nuclear interaction very near $D_\frac{1}{2}$.
Such considerations have led to the adoption of a local imaginary potential as a model for ImΔU. Basically two types of geometry for the
imaginary potential (W) have emerged. One is a Woods-Saxon (WS) shape
or some variant thereof. The other, suggested by Vary and Dover [5],
is obtained by taking ImΔU to have the same shape as U_F. With only a
few exceptions [6] (long-range absorption due to the Coulomb force) one
of the above types of geometries have proven adequate.

Assuming we can find a reasonable phenomenological imaginary potential to represent ImΔU, eq. 5a becomes:

$$[T + V_{Coul}(R) + U_F(R) + iW(R) + Re\Delta U - E]\chi_{oo}(\vec{R}) = 0 \qquad (7)$$

If we set ReΔU = 0 in eq. 7 and are prepared to adjust W we get what is
usually meant by the folding model for elastic scattering. If ReΔU is
either negligible or geometrically similar to U_F, its effects may be
incorporated in a phenomenological way by renormalizing U_F. If the renormalization factor is near unity and the fit to the data is acceptable
then the folding model works. If the renormalization is large or the
data is poorly represented, we must look for explicit corrections. A
rough estimate [6] suggests that (ReΔU/ImΔU)<<1 when both are calculated
to second order.

The derivation of eq. 7 neglects antisymmetrization between nucleons
in different ions. Hence a number of additional corrections [7] are
necessary when the nuclei have appreciable overlap. Unfortunately most
estimates [7] (but see ref. 8) of these effects are made in an adiabatic
limit and it is not clear how applicable these are to elastic scattering.
Even near $D_\frac{1}{2}$ the relative angular kinetic energy is usually quite large
and the calculations of Zint [8] indicate a very weak dependence of
ReU on the *direction* of relative velocity. Moreover, there appears to
be considerable cancellation [7] amongst the various corrections.

Since we are interested in the real potential in a region where the total density is ~5% of central density [4] we might expect single-nucleon exchange (SNE) to be the leading correction arising from anti-symmetrization (multi-particle exchange terms involve exchange without interaction and in the region of small overlap these become small). The importance of these SNE terms have been estimated [9,10] for a common force with the result that SNE can be included approximately by replacing $V(r_{12})$ in eq. 6 by

$$V(r_{12}) + \hat{J}(E)\delta(\vec{r}_{12}) \tag{8}$$

where the strength $\hat{J}$ of the psuedopotential depends sensitively on the spin-space-isospin character of the force but very weakly on the energy E as long as $5 \lesssim E(MeV)/A \lesssim 15$. The t-matrix technique of Dover and Vary [5] also includes SNE terms but also in a very approximate way.

2.1 Double Folding Model Interactions

To evaluate the double-folding integral in eq. 6 we have to specify the interaction. We believe it is desirable to use an effective interaction based on a realistic nucleon-nucleon force since one of our long-range goals is to obtain a unified description of nucleon-nucleon, nucleon-nucleus and nucleus-nucleus scattering. In order to appreciate some of the results to be discussed later, let us look at the nucleon-nucleon dynamics in some detail.

Since the real part of the HI potential is only important for ion separations corresponding to a very small overlap of the two densities, the range of the interaction is extremely important. This is illustrated in Fig. 2 where the volume integral required to produce the correct HI potential at the strong

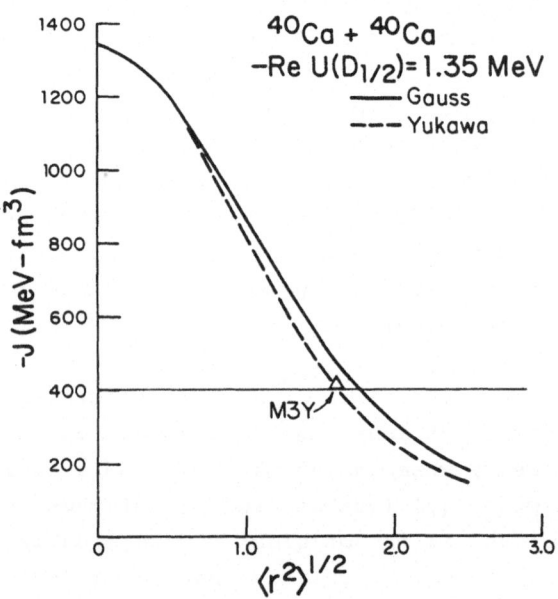

Figure 2. Volume integral vs. rms radius of nucleon-nucleon interaction to yield correct HI potential at the strong absorption radius.

absorption radius is shown for ^{40}Ca + ^{40}Ca at a few MeV/nucleon. The value of -400MeV-fm^3 is roughly that required for nucleon-nucleus scattering.

For HI collisions in which E/A~10MeV the *nucleon-nucleon* interaction is dominated by scattering in relative s-states due to the centrifugal barrier in the two-body subsystem. Consequently only those parts of realistic *odd*-state forces which possess a long range will be effective in HI collisions. Since the calculations are not actually done in relative coordinates, this manifests itself in almost complete cancellation between the direct and exchange (SNE) contributions arising from eq. 8 for the *short-range part* of odd-state forces. This extreme cancellation has been noted [10] in a number of cases and can be understood by noting that the forces constructed in relative coordinates for use in odd states are not restricted to odd states when *only* the direct (or exchange) terms are included. Thus when short-range odd-state forces are used both direct and exchange terms must be included.

Although odd-state forces are believed to possess a long-range term arising from the OPEP, most HI collisions (due to the scalar and/or isoscalar nature of at least one of the colliding partners) sample only the scalar and/or isoscalar part of V in which the OPEP contribution acting in odd-states exactly cancels that acting in even states. This cancellation does not occur in the SNE terms but long-range interactions are very ineffective in inducing nucleon exchange anyway.

Based on the above considerations we can now understand why early calculations [11] of U_F which employed purely even-state forces with rather long ranges overestimated the HI potential at $D_{\frac{1}{2}}$ (by roughly a factor of 2) and frequently had an unacceptably small slope. Many of them included the OPEP contribution either implicitly or explicitly without an odd-state counterpart to cancel it. Figure 3 illustrates this point where the even-state OPEP contribution is shown for the ^{16}O + ^{60}Ni system.

Two basic approaches have emerged for choosing the nucleon-nucleon effective interaction V to be used in double folding. The approach of Dover and Vary [5] is based on the idea that nucleon-nucleon collisions at very *low densities* should be amenable to an impulse approximation description. In our notation they take

$$V(r) = \bar{V} e^{-(r/r_o)^2} \tag{9}$$

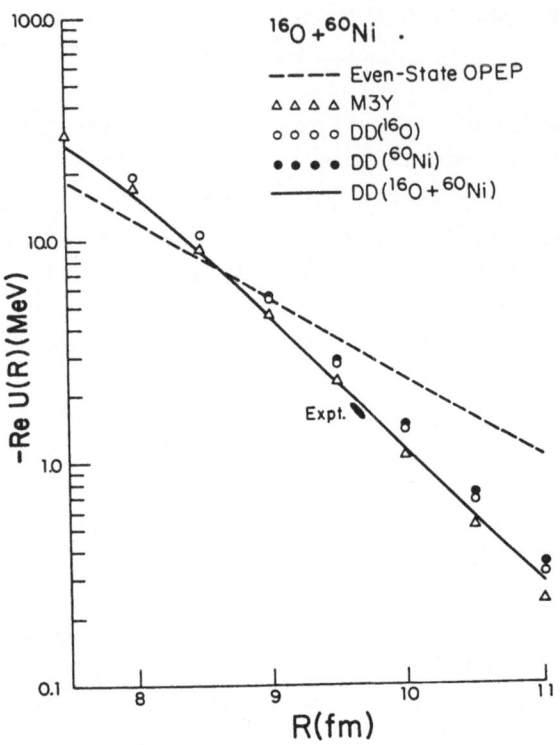

Figure 3. Folding potential for ^{16}O + ^{60}Ni scattering. DD denotes density-dependent G-matrix of Day; M3Y denotes ρ-independent G-matrix from eq. 10. The dashed curve is for that part of the even-state interaction ascribed to the OPEP.

where $\bar{V}$ (complex) is proportional to the forward nucleon-nucleon scattering amplitude averaged over the internal motion of the nucleons in the target and projectile, and corrected for off-shell propagation and the Pauli principle in intermediate states. The range r_O has been taken to be 1fm [12] or 1.4fm [12]. From Fig. 2 we see that such a large change in r_O will necessitate a considerable change in the volume integral.

The other approach is based on the *low energy* of the nucleon-nucleon collisions and takes V to be a local interaction which describes a selected set of G-matrix elements for two nucleons either bound or moving in nuclear matter. Since the bound G-matrix is real, the imaginary part of the optical potential has to be treated phenomenologically. Two distinct types of effective interactions based on G-matrices have been used. One [13] includes the density dependence of G in an average way and leads to a density-independent V. The other [14] represents V as a function of density through its dependence on k_F, the Fermi momentum. The density-independent V is based on representing a selected set of s- and p-wave G-matrix elements in an oscillator basis with the matrix elements of a sum of three Yukawa terms. One Yukawa was taken to be the OPEP; the second one was chosen to have a range of 0.4fm which roughly simulates multiple-pion exchange processes, and the third one of range 0.25fm was chosen for calculational convenience. The strengths for the latter two ranges were adjusted to fit the G-matrix elements. The complete interaction is given in ref. 13. For many HI calculations where only the scalar-isoscalar part of V is important we find, using the even-state Reid

G-matrix supplemented by the odd-state force of Eiilot *et al.* [15],

$$V(r) = \left[7999 \frac{e^{-4r}}{4r} - 2134 \frac{e^{-2.5r}}{2.5r} \right] \text{MeV}$$

and   (10)

$$\hat{J} = -262 \text{ MeV-fm}^3.$$

This interaction is, for HI calculations, essentially equivalent [10] to that in which only the OPEP part of the odd-state force in included. This alternate form of the interaction is:

$$V(r) = \left[6315 \frac{e^{-4r}}{4r} - 1961 \frac{e^{-2.5r}}{2.5r} \right] \text{MeV}$$

and   (11)

$$\hat{J} = -81 \text{ MeV-fm}^3.$$

The other low-energy technique used by Day and coworkers [14] uses the defining equation for the G-matrix ($G\phi = V_R\psi$) where ϕ (ψ) is the uncorrelated (correlated) relative wave function in nuclear matter, and V_R is a modified Reid soft-core potential. In coordinate space a trivially local equivalent G is given by

$$G(r,k_F) = \frac{V_R(r)\,\psi(r,k_F)}{\phi(r)}$$ (12)

averaged over the *relative* momenta. In the local-density approximation k_F depends on the *total* local density, and the integral in eq. 6 is substantially more complicated.

2.2 Single Folding Model

Numerous authors [17] have used the single-folding model to describe HI scattering. This model may be obtained from eq. 6 by replacing the integration over *either* of the two nuclear densities (say #2) by the phenomenological optical model for the scattering of a nucleon from that nucleus. In particular, the single folding potential becomes,

$$U_{SF}(\vec{R}) = \int d\vec{r}_1 \rho_1(\vec{r}_1)\, U_{n2}(|\vec{R} - \vec{r}_1|, E_n)$$ (13)

where U_{n2} describes the scattering of a nucleon from the nucleus A_2 at a bombarding energy $E_n \approx E/10$. Although in principle the imaginary part of the HI optical potential could be calculated this way, this has received less attention. Satchler [11] has pointed out that this method leads to a systematic overestimate of the HI potential $U_{SF}(D_\frac{1}{2})$ by roughly a factor of two independent of the choice U_{n2} vs. U_{n1}. A number of reasons for this discrepancy have been suggested [18,20]. First, the phenomenological U_{ni} contains the effects of coupling to other channels and this coupling is likely to be quite different when the incident nucleon is bound in another nucleus relative to when it is free. Secondly, nucleon-nucleus scattering is not nearly as sensitive to the tail of U_{ni} as is HI scattering so that a relatively small "error" in U_{ni} may translate into a large deviation in the HI optical potential at the strong absorption radius. Finally, if the underlying effective two-body interaction is density dependent, $(V \rightarrow V(r,\rho_1+\rho_2))$, eq. 13 will not include the effect ρ_1 has on V. The first of these effects is difficult to unravel. The latter two alternatives have received some attention recently [18-20].

Rickertsen and Satchler [18] have shown that the overestimate of the real part of the HI potential can be nearly eliminated by using the square of a Woods-Saxon shape factor for the nucleon-nucleus potential. This form is much smaller in the tail region and provides a simultaneous description of both nucleon-nucleus and nucleus-nucleus scattering. It remains to be seen whether this form is consistent with the use of a realistic two-body interaction to obtain the nucleon-nucleus potential.

Use of a density-dependent interaction has been made by Sinha [21] in the double-folding model. An examination of the role of density-dependent interactions with respect to the single-folding anomaly has recently been carried out [19-20] using both the density dependence predicted by the Day [14] G-matrix and that calculated in ref. 22. The density dependence found by both groups can be expressed to a good approximation in the form

$$V(r,\rho) = V_1(r) + V_2(r)e^{-\alpha\rho} , \qquad (14)$$

which is especially convenient for short-range forces where $\rho \approx \rho_1(r_1) + \rho_2(r_2)$ and the convolution form of eq. 6 is preserved for each term in eq. 14. By setting $\rho_1=0$ in the exponential in eq. 14, for example, we can get an estimate of the error made in evaluating U_{SF} in eq. 13 where the dependence of $V(r,\rho)$ on ρ_1 is not included. The results are shown in Fig. 3 for $^{16}O + ^{60}Ni$ using the G-matrix derived by Day [14]. It is

seen that omitting either density, corresponding roughly to using alternatively an $^{16}O+n$ or $^{60}Ni+n$ phenomenological potential, leads to an overestimate of the HI potential at $D_{\frac{1}{2}}$ by ~25%. Although this correction
does not resolve the discrepancy, it is a non-negligible correction and
should be borne in mind as an uncertainty in single-folding calculations.

2.3 Densities for Folding Models

The most direct measure we have of the densities to be used in the
folding model (eqs. 6 and 13) comes from electron scattering. This only
yields information about the nuclear *charge* density, which tells us primarily about the proton distribution. To get the point proton density
the finite size of the proton ($<r^2>$~0.76fm^2) must be unfolded. Generally
this is an essential correction. For example, for $^{40}Ca + ^{40}Ca$, failure
to include the finite size of the proton (in both nuclei) increases $U(D_{\frac{1}{2}})$
by 54% using the interaction in eq. 10. Since our information about
neutron distributions is indirect it is not clear how to best get these.
For N=Z nuclei it is likely adequate at this time to simply take $\rho_n=\rho_p$.
For N≠Z one can use either the independent-particle model [16] with the
well adjusted to yield the correct binding for the nucleons in the outer-
most shells or the results of those Hartree-Fock calculations which
correctly predict the proton distribution. For $^{40}Ca + ^{40}Ca$ a shell-model
density [23] which yields $<r^2>_n^{\frac{1}{2}}-<r^2>_p^{\frac{1}{2}} = -.09$ fm yields $U_F(D_{\frac{1}{2}}\approx10.6$fm)
smaller by 8% than by assuming $<r^2>_n^{\frac{1}{2}} = <r^2>_p^{\frac{1}{2}}$ and using the results [24]
of electron scattering. For $^6Li + ^{40}Ca$ the analogous difference in
$U_F(D_{\frac{1}{2}})$ is 1%. For the $^{40}Ca + ^{40}Ca$ system, correction for the charge
distribution of the *neutron* leads to an 8% increase in $U_F(D_{\frac{1}{2}})$. Recoil
corrections have also been estimated [23] for ^6Li in the $^6Li + ^{40}Ca$
system and are negligible *provided* the mean-square radius is unaltered
(which requires a readjustment of the well).

3. APPLICATIONS OF THE FOLDING MODEL

Although a formal justification of the folding model is unavailable,
the model is relatively simple and easy to apply. At this point we will
be content with an empirical assessment of this model.

To emphasize the important role of the value of the real part of
the HI optical potential at the strong absorption radius, we examine
$^{16}O + ^{63}Cu$ at $E(^{16}O) \simeq 42$MeV. Figure 4 shows the results of work recently
completed by the Florida State group [25]. For each of three different
models used to generate ReU, the best fits to the elastic scattering

each require ReU($D_\frac{1}{2}$)=−.68±.06MeV. The three models shown here are: 1)
the G-matrix folding model (GMFM) of eq. 10 with a slightly different
contribution from exchange, 2) the adiabatic model of ref. 26 with 'a'
denoting the range in this model and 3) a family of Woods-Saxon potentials having different diffusivities which span the shaded region.

Although the value of U($D_\frac{1}{2}$) is the best defined characteristic of
HI potentials, ReU($D_\frac{1}{2}$) is invariably only a few percent of the Coulomb
potential at that radius. Nevertheless, its role in describing the
elastic scattering is quite important [4]. A representative case which
illustrates this importance is shown in Fig. 5 where the folding model
results for ^{16}O + ^{63}Cu are compared with the experimental data of ref. 25.
The numbers in parentheses are (N,α) where U=N(1+iα)U_F.

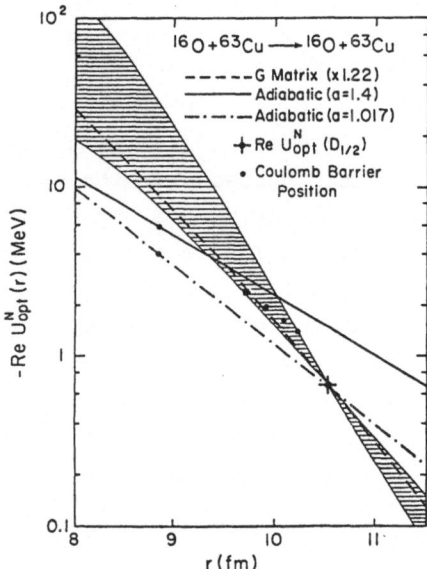
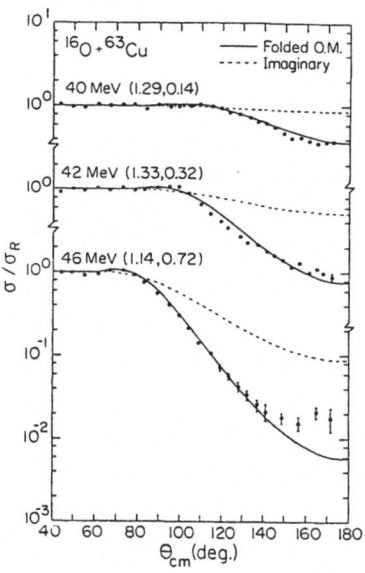

Figure 4. Tails of real nuclear
potentials from an adiabatic model,
a G-matrix folding model (GMFM) and
a phenomenological model.

Figure 5. ^{16}O + ^{63}Cu elastic scattering compared with the GMFM. The
solid curve denotes the results
where both ReU and ImU are included.
The dashed curve illustrates the
effects of setting ReU=0.

The G-matrix in eq. 10 (or one of its variants) has been applied
[10,23,25] to a large number of cases of HI elastic and inelastic scattering. Here we discuss a few recent applications of this model. When
the folding model is used, the optical potential is usually parameterized

as

$$U_{opt}(R) = NU_F(R) \tag{15}$$

or

$$U_{opt}(R) = NU_F(R) + iWS(R). \tag{16}$$

In the first form N is taken to be complex; in the second form N is real and WS denotes a phenomenological Woods-Saxon imaginary potential. ReN≈1.0 denotes empirical success of the folding model.

Figure 6 shows the results of the GMFM applied to the ^{40}Ca + ^{40}Ca data of Doubre *et al.* [27]. An earlier report [27] of these data found the folding-model predictions to deteriorate significantly with increasing bombarding energy when the form (15) was used and N was constrained to be 1.0+i0.7. The fits shown here use form (16) with N=1.18 for each center-of-mass energy. Although the fits to the lower energy data are still superior, the folding-model fit at each energy is now comparable to that obtained using a phenomenological WS potential [27]. The G-matrix of eq. 11 was used without exchange (M3YD).

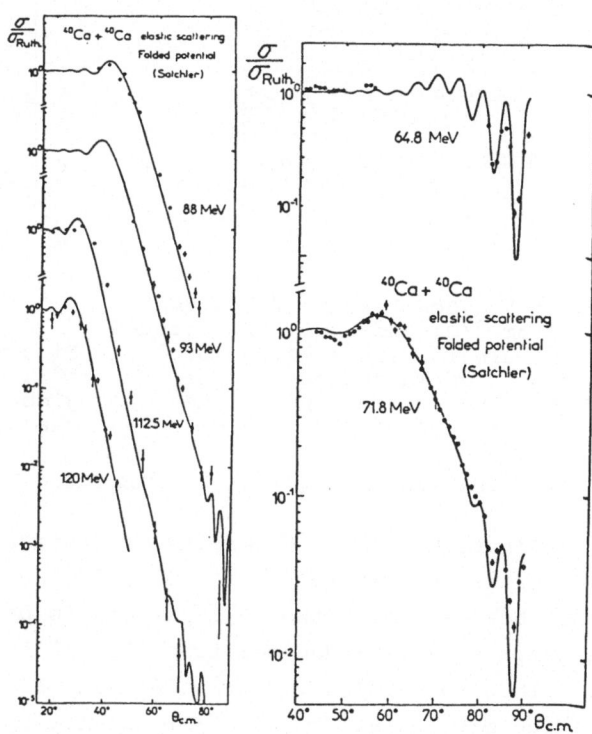

Figure 6. ^{40}Ca + ^{40}Ca elastic scattering compared with the calculations of the GMFM. The interaction M3YD was used with N=1.18.

The same interaction (M3YD) has also been applied [23] to the elastic and inelastic scattering of ^{16}O by ^{60}Ni at $E(^{16}O) = 141.7$MeV and the results are shown in Fig. 7. For either form of $W(r)$, ReN is found to be 1.00 ± 0.03. The Tassie model [28] was used for the inelastic (2^+) transition density which was normalized to the observed B(E2) and the inelastic scattering was treated in the distorted-wave approximation.

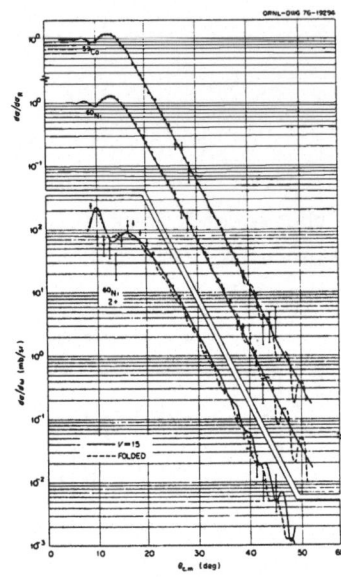

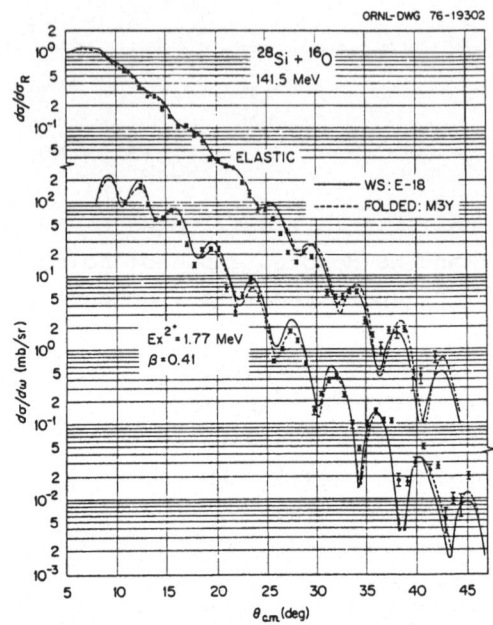

Figure 7. Scattering of 141.7MeV ^{16}O from ^{59}Co and ^{60}Ni. The dashed curves are for the folding model as discussed in the text. The solid curves correspond to the shallow WS potential.

Figure 8. Comparison between ^{16}O + ^{28}Si data and the predictions of global potentials as discussed in the text.

Similar results hold [23] for bombarding energies down to ~40MeV.

Figure 8 shows the results of similar calculations for the elastic and inelastic scattering of ^{16}O by ^{28}Si at E(^{16}O)=141.5MeV. In this case, however, both the WS(E-18) and folded model potentials are *global* potentials [23,29] deduced from ^{16}O + ^{28}Si elastic scattering from E(^{16}O)=33 to 215.2MeV. Form (16) with M3YD was used for the G-matrix folding model with N=0.89. For excitation of the 2^+ state in ^{28}Si the Tassie model [28] was used with a strength consistent with a B(E2↑) of 327e^2fm^4. Both types of calculation are in reasonable agreement with the data for elastic and inelastic scattering.

The double-folding model using eq. 10 (M3Y) has also been applied to the elastic scattering of several relatively light systems in the energy range 100-170MeV. The results are shown in Fig. 9 where the nucleus listed first is the projectile. The ^{16}O + ^{12}C data is that of Hiebert and Garvey [30]; the other data in Fig. 9 is that of Nair *et al.* [30]. A common imaginary WS geometry was taken for each of these systems. Reading down Fig. 9, ReN was found to be 1.03, 1.04, 1.03, 1.11 and 1.30

respectively. A folding-model description of the $^{14}N + ^{12}C$ and $^{14}N + ^{16}O$ data has also been reported by Moffa *et al.* [30]. Although only forward angle data are shown in Fig. 9, it has been shown [31] that the large-angle data for $^{12}C + ^{12}C$ scattering cannot be represented by a shallow WS potential as has been used to describe the scattering of $^{16}O + ^{16}O$. A reanalysis of the $^{16}O + ^{16}O$ data using the GMFM would be interesting.

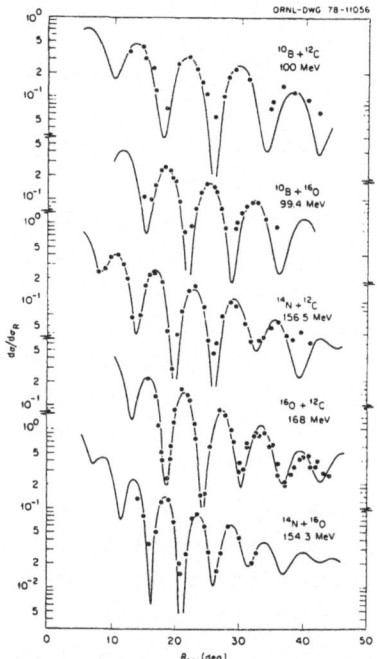

Figure 9. A comparison of relatively light-ion scattering with folded-model calculations.

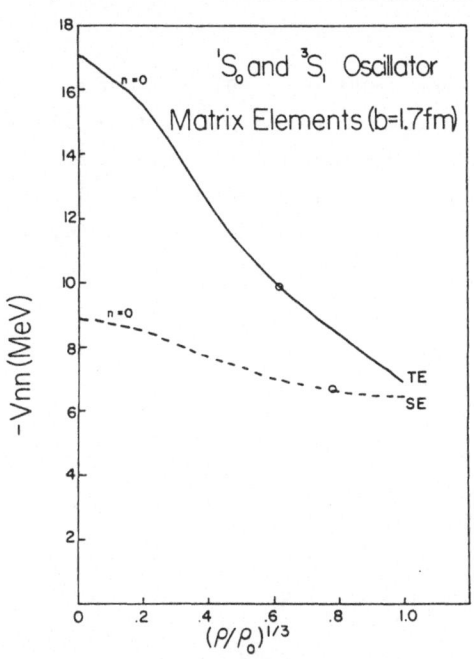

Figure 10. Oscillator matrix elements for the G-matrix of Day and the M3Y interaction of ref. 13 as a function of density.

The density-dependent G-matrix of Day [14] as parameterized in eq. 13 has also been applied with comparable success to a number of cases including $^{16}O + ^{60}Ni$ as shown in Fig. 3. A typical difference between the density-dependent and density-independent GMFM is $\lesssim 5\%$ in $U_F(D_{\frac{1}{2}})$. Figure 10 illustrates the density dependence of Day's G-matrix with the circle denoting the strength of the density-independent G-matrix associated with the interaction of eq. 10. Since the two G-matrices predict comparable values of $U_F(D_{\frac{1}{2}})$, the figure suggests that the dominant contributions to $U_F(D_{\frac{1}{2}})$ come from a region in which $\rho/\rho_o \approx \frac{1}{3}$.

4. LIGHTER IONS

Although Li and ^4He ions share many of the features characteristic of the scattering of heavier ions, it has been noted by DeVries *et al.*

[32] that there are significant differences in the scattering of those projectiles with $A \leq 7$ when compared with those having $A \geq 12$. The transition between these two mass regions has been described [32] in terms of the empirical optical potential which changes from moderately absorptive and refractive ($A \leq 7$) to strongly absorbing and diffracting. It is of interest to see whether the folding model is consistent with this observation.

The GMFM (eq. 10) has been applied [33] to ^{6}Li elastic scattering in the energy range $E(^6\text{Li}) \approx 20\text{-}135\text{MeV}$ for a number of targets. The values of the calculated U_F in the surface region are consistently too large with N in eq. 16 ranging from 0.38 to 0.87 with an average value of ~0.6. In some cases N is not well determined by a visual comparison of the scattering cross sections. Figure 11 shows a comparison between the calculated and observed [34] elastic scattering for ^{6}Li + ^{40}Ca at 30MeV using the GMFM. Although the correct fall-off with angle is predicted with N=1, the calculated cross section with N=1 is slightly out of phase with the data. An excellent fit is obtained with N=0.6. The failure of the GMFM to yield the correct strength of the real potential is not understood but is likely attributable to the weakly bound nature of ^{6}Li. It should be noted that this system is sensitive to ReU(R) for R<3fm and *only* a renormalization of U_F is required even though large angle data are available.

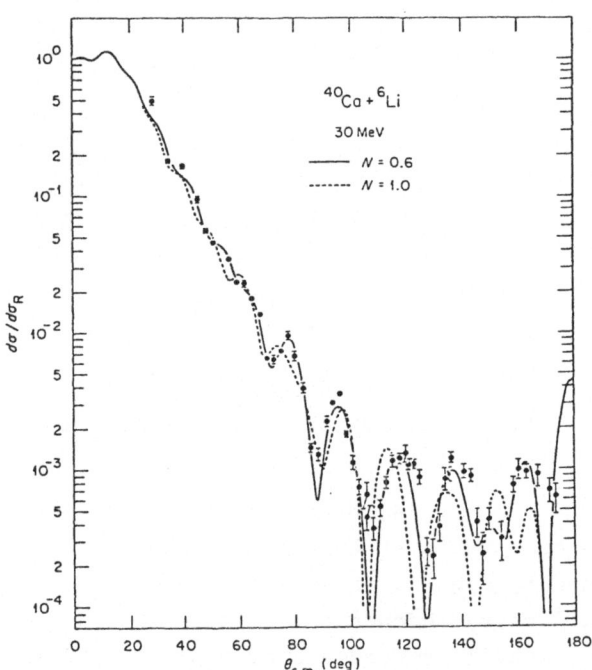

Figure 11. ^{6}Li + ^{40}Ca elastic scattering data compared with the GMFM potential calculations.

Numerous applications [35] of the folding model have been made for ^{4}He scattering based on different nucleon-nucleon or nucleon-nucleus interactions with reasonable success. Many of these calculations were based on a nucleon-nucleon force which proved to overestimate [11] $U(D_{\frac{1}{2}})$ by a factor of ~2 for projectiles with $A \geq 12$.

Eisen and Day [14] have satisfactorily described the elastic scattering of ^{4}He from ^{90}Zr and ^{208}Pb recently using a density-dependent GMFM (eq. 12) in the energy range between ~10 and ~25MeV. This ρ-dependent GMFM does not resolve the discrepancy found for ^{6}Li scattering.

Only a few applications of the ρ-independent (eq. 10) GMFM have been made to ^{4}He scattering. Calculations have been made, however, for the elastic scattering of ^{4}He from ^{40}Ca. These calculations [36] were motivated by the recent results of Michel and Vanderpoorten [37] who have shown that the longstanding anomaly associated with large angle ^{4}He scattering (ALAS) from ^{40}Ca can be explained by an optical potential using a common WS shape factor raised to the νth power for both the real and imaginary parts of the potential. After determining V(r) and W(r) by a search on U_{opt} at E_α=29MeV (ν=2.65), they then varied only the absorptive radius to fit the elastic scattering up to ~45MeV. The potential has now been calculated for this system using both the GMFM of eq. 10 and a slightly different representation [13,36] (M245) of the

G-matrix which was determined by fitting to the *same* G-matrix elements. The main difference between these two representations of the G-matrix is that M245 does not include an OPEP tail in *any* channel. It does, however, give a slightly better overall fit to the G-matrix elements than does M3Y in eq. 10.

In Fig. 12 the empirical real potential of Michel and Vanderpoorten is compared with the folded potentials, M3Y from eq. 10 and M245. Although the M3Y folded potential is ~25% smaller for R≈7fm, U_F predicted by M245 is in good agreement in the range R≈2-9fm. Calculations using both the full empirical potential and the folded potential U_F(M245) supplemented by the unadjusted

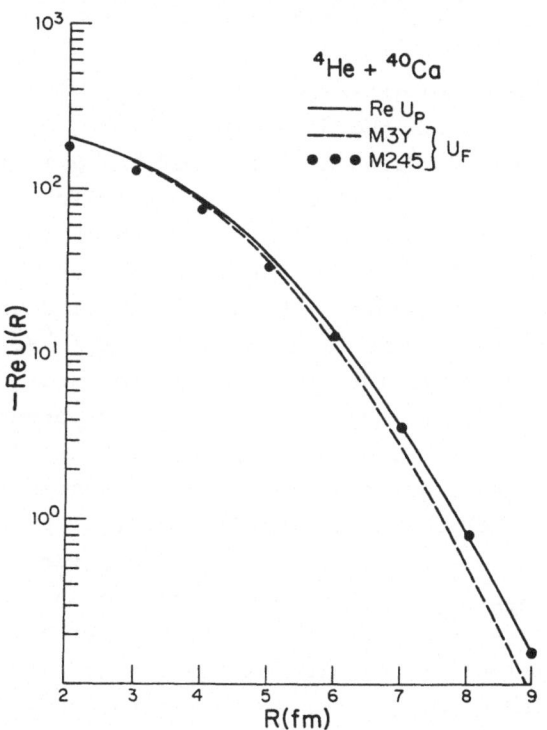

Figure 12. A comparison of empirical and GMFM potentials for ^{4}He + ^{40}Ca scattering. U_P denotes the empirical potential; M3Y and M245 denote folded potentials discussed in the text.

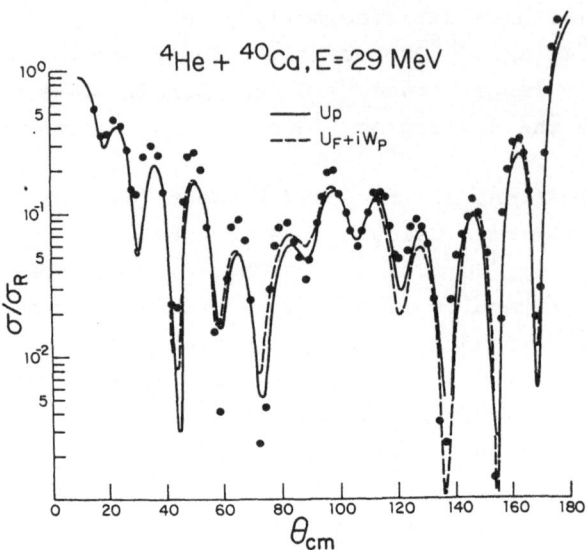

Figure 13. ^{4}He + ^{40}Ca scattering data at E_α=29MeV compared with the scattering predicted by the empirical potential and the M245 version of the GMFM as discussed in the text.

empirical imaginary potential were made. Figure 13 shows the results compared with the data with N=1.07. The GMFM is seen to give a fit to the data comparable to that of the phenomenological potential. The 7% adjustment is only necessary in order to fit the 90°-140° data and is well within the uncertainty of the folding model. Since both the folded and empirical real potentials are essentially energy independent, they should both explain the experimental data over the same energy range. Other folding-model calculations for ^{4}He + ^{40}Ca scattering have recently been reported [38] and these also describe the large angle scattering. The common feature of these folded potentials is a very deep real potential which has been shown by Brink and Takigawa [39] to be one of the two essential characteristics of a potential-model description of ALAS. The other characteristic is *relatively* weak absorption which allows the α-particle to penetrate into the nuclear interior. Although the GMFM (as applied here) does not provide an estimate of the absorption, it does yield a real potential consistent with the ALAS data. This same type of potential has also been shown [37] to significantly improve the description of the inelastic excitation of the 3$^-$ state(3.73MeV) in ^{40}Ca by α-particles.

It is interesting to note that the GMFM interaction of eq. 10 gives with no adjustment a ^{4}He + ^{4}He optical potential almost identical to that found by Buck *et al.* [40] which is known to describe ^{4}He + ^{4}He scattering up to $E_{cm} \simeq 40$MeV. Both the volume integrals and mean-square-radii of these potentials agree to within 2%. Curiously, the M245 interaction gives an α+α potential having a mean-square-radius ~16% larger than that found by Buck *et al.* [40].

To make connection with the scattering of nucleons, we show in Fig. 14 the scattering of n+^{60}Ni compared with the results [41] obtained

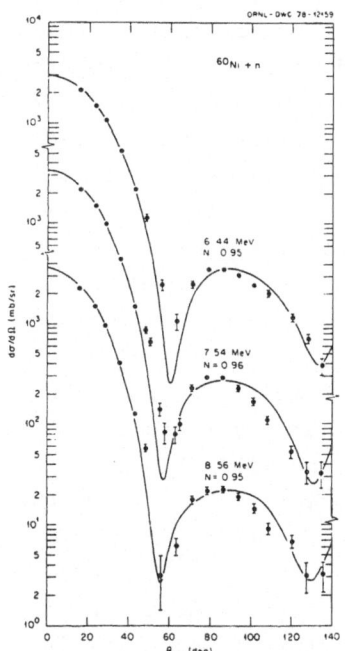

when the real part of the optical potential is calculated using the GMFM (eq. 10). Although the fits are not excellent by optical model standards, they are quite reasonable with only a 5% reduction in the strength of the interaction. A surface imaginary potential was used in these calculations.

5. HEAVY-ION SPIN-ORBIT POTENTIALS

There have been a number [3,42-49] of attempts to determine the spin-orbit (LS) potential acting between heavy ions. In particular, analyzing powers have been measured [42] for the elastic scattering of ^{6}Li ions from a number of light nuclei. Both single- [42-47] and double-folding [48,49] model estimates of the LS potentials have been made. Single folding based on an α-d cluster model [42], and double folding [49] based on the M3Y [13] G-matrix yield LS potentials which provide a reasonable description of the analyzing power for ^{6}Li scattering. The GMFM results are shown in Fig. 15.

Figure 14. A comparison of n+^{60}Ni scattering with the results of the GMFM.

These results are encouraging but there are uncertainties. For example, in the single-folding calculations of [42] an alternate choice [47] of the α-d cluster-model wave function resulted in analyzing powers much smaller than those observed. The difficulty with the GMFM for the *central* potential for ^{6}Li scattering has already been discussed. This difficulty was circumvented by Petrovich *et al.* [49] by using a phenomenological central potential.

There are other sources of the ion-ion LS potential which need to

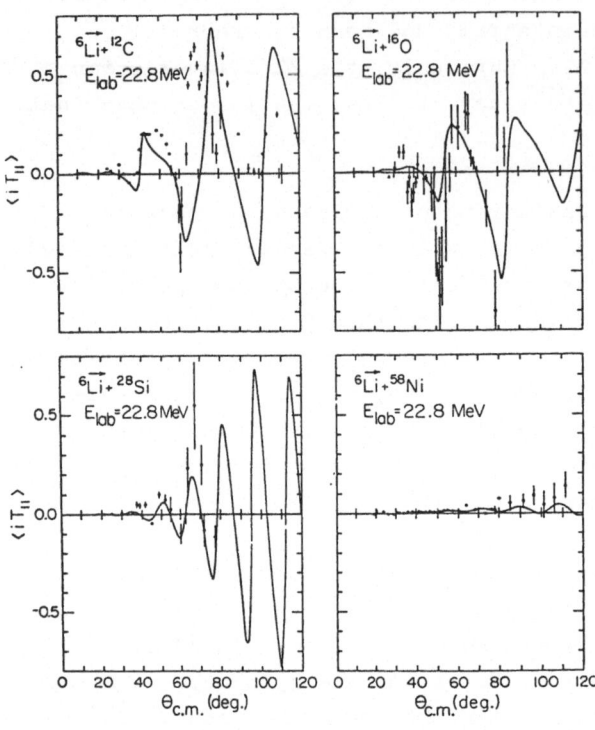

Figure 15. Analyzing powers [42] for
^{6}Li elastic scattering compared with
the results from ref. 49 using the
spin-orbit part of the G-matrix interaction.

be investigated. For example velocity-dependent forces, such as arise from central force exchange terms [3], contribute to the LS optical potential. A simple estimate of these (likely an upper limit) for ^{4}He + ^{9}Be and ^{4}He + ^{59}Co suggests that such terms may be larger than those obtained by simply folding the two-body LS interaction. For ^{6}Li projectiles, this correction should be negligible since the ^{6}Li wave function is believed to be dominated by the $^3S_1$ component [49] and these velocity-dependent corrections are proportional [3] to the matrix elements of $\vec{\ell}$. Higher-order dynamic polarization corrections [49] also need to be investigated since they can lead to *effective* LS potentials.

It would be quite interesting to see how well the *phase* of the LS part of the HI potential is pinned down. This might provide a more definitive clue as to its origin. In particular, asymmetry data as a function of bombarding energy might tell us if the LS potential arises predominatly from coupled channels effects. Also, if the LS potential arises from folding [3] those projectiles with the unpaired nucleon having $j=\ell+\tfrac{1}{2}$ should experience an LS potential of *opposite sign* from that of a projectile having an unpaired nucleon with $j=\ell-\tfrac{1}{2}$. If, however, the origin of the LS potential is a velocity-dependent force [3] no change in sign of the LS potential should be found. Clearly more "polarization" data using different projectiles at different energies must be obtained if we are to gain confidence in our understanding of the HI spin-orbit potential.

6. SUMMARY

In the past few years numerous calculations of the real part of the

optical potential have been made using the G-matrix double-folding model
based on a realistic nucleon-nucleon interaction. For projectiles with
E/A~10MeV and A≳10 both density-independent and density-dependent versions systematically yield HI potentials at the strong absorption radius
correct to within ~25% and more typically within 10-15%. Moreover, several inelastic transitions have been described with roughly the same
level of success when realistic transition densities are used. This
success stems from the much shorter range of the participating part of
the interaction than was used previously. This shorter range results
in both a reduced value of ReU at the strong absorption radius and a
steeper slope, both of which seem to be preferred by sufficiently detailed data. The difficulty [11] with the single-folding model remains,
but attempts to resolve this anomaly are being pursued.

For lighter ions (A≲10) the situation is less clear. A number of
double-folding model applications to ^{6}Li scattering indicate the need
for substantial renormalization while a few applications of the double-
folding model to ^{4}He scattering have been rather successful. A number
of single-folding calculations [35] have also been successful for ^{4}He
elastic scattering. The spin-orbit potential for ^{6}Li elastic scattering is predicted reasonably well by both the single-folding model
using the deuteron spin-orbit potential and the G-matrix double-folding
model.

Currently efforts are underway [22,50] to construct a complex (and
density dependent) G-matrix suitable for calculating both real and imaginary parts of the optical potential in a double-folding context. For
considerably higher bombarding energies (E/A≳100MeV) we may anticipate
the use of the free nucleon-nucleon t-matrix [51] as a complex effective
interaction for double folding. Some exploratory calculations [41] of
this kind have already been made.

The overall relative success of the double-folding model poses an
interesting theoretical problem: How do the numerous corrections to
this model cancel at the 20% level for such a variety of HI systems?
Although a completely satisfactory answer [7-8] to this problem is unavailable, the double-folding model remains a useful technique for calculating the real part of many HI optical and coupling potentials with
a minimal amount of input. In addition (within a folding-model context)
HI scattering has provided us with a relatively sensitive probe of the
tail region of the nucleon-nucleon interaction!

ACKNOWLEDGEMENTS

I am especially indebted to G. R. Satchler for numerous helpful discussions and for permission to use some of the material presented here. I am also indebted to F. Petrovich and the Florida State group for permission to use some of their results and to L. Rickertsen for a copy of his double-folding code. This work was supported in part by the National Science Foundation.

REFERENCES

1. T. Tamura, Rev. Mod. Phys. 37, 679 (1965).
2. H. Feshbach, Ann. of Phys. 19, 287 (1962).
3. G. H. Rawitscher, Phys. Rev. C6, 1212 (1972); W. G. Love, Nucl. Phys. A226, 319 (1974).
4. G. R. Satchler, Procs. Conf. of Reactions Between Complex Nuclei, Nashville, (North Holland Publishing Co., 1974), vol. 2, 171 (1974); N. K. Glendenning, Rev. Mod. Phys. 47, 659 (1975).
5. J. P. Vary and C. B. Dover, Phys. Rev. Lett. 31, 1510 (1973); C. B. Dover and J. P. Vary, Symposium of Classical and Quantum Mechanical Aspects of Heavy Ion Collisions, Heidelberg (Springer-Verlag, New York, 1975) vol. 33, 1 (1974).
6. W. G. Love, T. Teresawa and G. R. Satchler, Nucl. Phys. A291, 183 (1977); A. J. Baltz et al., Phys. Rev. Lett. 40, 20 (1978).
7. J. Fleckner and U. Mosel, Nucl. Phys. A277, 170 (1977); P. G. Zint and U. Mosel, Phys. Rev. C14, 1488 (1976).
8. G. H. Goritz and U. Mosel, Z. Physik A277, 243 (1976); P. G. Zint, Z. Physik A281, 373 (1977).
9. M. Golin, F. Petrovich and D. Robson, Phys. Lett. 64B, 253 (1976).
10. G. R. Satchler and W. G. Love, Phys. Lett. 65B, 415 (1976).
11. G. R. Satchler, Phys. Lett. 58B, 121 (1975).
12. C. B. Dover, P. J. Moffa and J. P. Vary, Phys. Lett 56B, 4 (1975); P. J. Moffa et al., Phys. Rev. Lett. 35, 992 (1975).
13. G. Bertsch, J. Borysowicz, H. McManus, and W. G. Love, Nucl. Phys. A284, 399 (1977); G. Bertsch, J. Borysowicz, and H. McManus, unpublished.
14. Y. Eisen and B. Day, Phys. Lett. 63B, 253 (1976).
15. J. P. Elliot et al., Nucl. Phys. A121, 241 (1968).
16. G. R. Satchler, in Procs. Symposium on Macroscopic Features of Heavy-Ion Collisions, 1976, Argonne National Laboratory Report ANL-PHY-76-2 (unpublished).
17. R. A. Broglia and A. Winther, Phys. Rep. 4C, 153 (1972); D. M. Brink and N. Rowly, Nucl. Phys. A219, 79 (1974); D. H. E. Gross and H. Kalinowski, Phys. Rev. 48B, 302 (1974).
18. L. D. Rickertsen and G. R. Satchler, Phys. Lett. 66B, 9 (1977).
19. W. G. Love, Phys. Lett. 72B, 4 (1977).
20. F. Petrovich, D. Stanley and J. J. Bevelacqua, Phys. Lett 71B, 259 (1977).
21. Bikash Sinha, Phys. Rev. Lett. 33, 600 (1974).
22. J. P. Jeukenne, A. LeJeunne and C. Mahaux, Phys. Rev. C16, 80 (1977).
23. G. R. Satchler, private communication; G. R. Satchler et al., Nucl. Phys. A298, 313 (1978).
24. C. W. DeJager, H. DeVries and C. DeVries, Atomic Data and Nuclear Data Tables, 14, 279 (1974).
25. H. Wojciechowski et al., Phys. Rev. C17, 2126 (1978); F. Petrovich, private communication.

26. H. J. Krappe and J. R. Nix, p. 159 in Procs. of the Third IAEA Symposium on Physics and Chemistry of Fission, Rochester, 1973 (IAEA, Vienna, 1974) Vol. 1; J. R. Nix and A. J. Sierk, Physica Scripta 10A, 94 (1974).

27. H. Doubre et al., Phys. Rev. C15, 693 (1977).

28. H. Uberall, Electron Scattering from Complex Nuclei (Academic Press, New York, 1971).

29. J. G. Cramer et al., Phys. Rev. C14, 2158 (1976).

30. J. C. Hiebert and G. T. Garvey, Phys. Rev. 135, B346 (1969); K. G. Nair et al., Phys. Rev. C12, 1575 (1975); P. J. Moffa et al., Phys. Rev. Lett. 35, 992 (1975).

31. R. M. Wieland, R. G. Stokstad, G. R. Satchler and L. D. Rickertsen, Phys. Rev. Lett. 37, 1458 (1976), and to be published; A. Gobbi et al., Phys. Rev. C7, 30 (1973); J. V. Maher et al., Phys. Rev. 188, 1665 (1969).

32. R. M. DeVries et al., Phys. Rev. Lett. 39, 450 (1977).

33. G. R. Satchler and W. G. Love, Phys. Lett. 76B, 23 (1978).

34. H. Bohn, K. A. Eberhard, R. Vandenbosch, K. G. Bernhardt, R. Bangert and Y-d. Chan, Phys. Rev. C16, 665 (1977).

35. See for example, A. M. Bernstein in Advances in Nuclear Physics, edited by M. Baranger and E. Vogt (Plenum Press, New York, 1969), Vol. 3; C. M. Lerner, J. C. Hiebert, L. L. Rutledge, Jr. and A. M. Bernstein, Phys. Rev. C6, 1254 (1972); P. Mailandt, J. S. Lilley and G. W. Greenless, Phys. Rev. C8, 2189 (1973); F. Michel, Phys. Rev. C13, 1446 (1976).

36. W. G. Love, in Procs. Symposium on Heavy-Ion Elastic Scattering, University of Rochester, 25-26 Oct. 1977, ed. R. M. DeVries; W. G. Love, Phys. Rev. C17, 1876 (1978).

37. F. Michel and R. Vanderpoorten, Phys. Rev. C16, 142 (1977).

38. N. Vinh Mau, Phys. Lett. 71B, 5 (1977); H. P. Gubler et al., Phys. Lett. 74B, 202 (1978).

39. D. M. Brink and N. Takigawa, Nucl. Phys. A279, 159 (1977).

40. B. Buck et al., Nucl. Phys. A275, 246 (1977).

41. G. R. Satchler and W. G. Love, to be published.

42. W. Weiss et al., Phys. Lett. 61B, 237 (1976).

43. H. Amakawa and K. I. Kubo, Nucl. Phys. A266, 521 (1976).

44. C. Chasman, P. D. Bond and K. W. Jones, Bull. Am. Phys. Soc. 20, 55 (1975).

45. S. Kubono et al., Phys. Rev. Lett. 38, 817 (1977).

46. L. A. Parks et al., Phys. Lett. 70B, 27 (1977).

47. W. J. Thompson, p. 14 in Procs. Conf. of Reactions Between Complex Nuclei. Nashville, 1974 (North Holland Publishing Co., 1974) Vol. 1.

48. P. J. Moffa, Phys. Rev. C16, 1431 (1977).

49. F. Petrovich, D. Stanley, L. A. Parks and P. Nagel, Phys. Rev. C17, 1642 (1978).

50. F. Brieva, H. V. Geramb and J. R. Rook, to be published.

51. W. G. Love et al., Phys. Lett. 73B, 277 (1978).

A Microscopic Nucleus-Nucleus Optical Potential

Bikash Sinha
Bhabha Atomic Research Centre
Nuclear Physics Division, Bombay 40085, India

Over the last few years a considerable amount of work has been done to calculate the nucleus-nucleus interaction potential. Several authors, recently, have pointed out the importance of Pauli exchange effects in estimating the interaction potential[1,2]. The single and double folding models of constructing the potential have met with varied success. Satchler[3] has shown how a suitable two-body interaction, when folded in with the nuclear density distributions of the two nuclei reproduce the elastic scattering data reasonably well. Such a folding procedure does not seem to work too well for heavier systems.

In the first part of my talk I will concentrate on the first-order part of the optical potential using the simple Skyrme type of two-body interaction. The local density in the Skyrme interaction (the three-body contact term in Skyrme interaction reduces to a density-dependent interaction for even-even nuclei) is computed by taking into account the contribution from both the target and the projectile density distribution[4]. Although nuclear saturation is guaranteed in this method, the folding of the two density distributions with the Skyrme interaction would ignore Pauli exchange corrections viz. effects of antisymmetrisation of the interaction matrix element and the Pauli distortion of kinetic energy. Pauli distortion refers to the fact that due to blocking of intrinsic states of a nucleus, the internal kinetic energy of the system tends to increase with the increasing overlap of the densities of the two nuclei. Fleckner and Mosel[5] recently demonstrated the role of the above mentioned Pauli exchange effects in detail. In this talk I would like to demonstrate the effects of Pauli exchange by first calculating the direct term where the two density distributions are folded in with the Skyrme interaction whereas the exchange term is computed by folding in the density matrices of the colliding nuclei. There are at least two other methods of computing the Pauli distortion effect - the energy-density method using Brueckner's sudden approximation[6] and the so called Proximity Potential using the geometrical properties of leptodermous systems[7]. Recently Brink and Stancu[8] have shown under what limiting condition the Proximity potential reduces to the energy density method. In the first part of the talk I would like to compare our results with these methods.

It is generally recognised now, that the Pauli distortion effect is primarily responsible for a shallow potential, turning repulsive for small values of the distance between the centres of the two ions - a fact not of great practical importance. Even for large separation however the Pauli exchange turns out to be important. On the other hand the folding procedure necessarily gives rise to a deep potential and so the controversy of "deep" and "shallow" potential persists.

In the second part of my talk I would like to draw your attention to the second-order part of the interaction potential - the principal value of which (polarisation term) leads to an additional term in the interaction potential. For EcM not too high (see later) the polarisation term, is attractive and it turns out that around the touching radius the contribution is significant. The pole term which is the imaginary part of the interaction potential is also computed.

2. FIRST-ORDER POTENTIAL

The direct and the exchange terms are given by $U_D(R) = \sum_{\substack{i \in 1 \\ j \in 2}} \langle ij | \mathcal{V}_d | ij \rangle$

and $U_{EX}(R) = \sum_{\substack{i \in 1 \\ j \in 2}} \langle ij | \mathcal{V}_{ex} | ji \rangle$ where $|i\rangle$ and $|j\rangle$ refer to the

single-particle wave-function of the nucleus $\underline{1}$ and $\underline{2}$ respectively. For our work we shall represent $|i\rangle = \Psi_i = \varphi_i \, e^{i \underline{k}_1 \cdot \underline{r}_1}$ and similarly

$|j\rangle \equiv \Psi_j = \varphi_j \, e^{i \underline{k}_2 \cdot \underline{r}_2}$ where φ_i and φ_j are the intrinsic single-particle wave functions of the respective nucleus, $e^{i\underline{k}_1 \cdot \underline{r}_1}$ and $e^{i\underline{k}_2 \cdot \underline{r}_2}$ are plane-wave representation of the relative motion of each nucleon with respect to the other nucleus, $\underline{k}_1$ and $\underline{k}_2$ being the respective wave number. The co-ordinates $\underline{r}_1$ and $\underline{r}_2$ refer to any two interacting nucleons in the nucleus $\underline{1}$ and $\underline{2}$, $\underline{R}$ is the distance between the centres of the two nuclei. The interaction potential now looks like

$$U_D(R) = \sum_{i,j} \int \varphi_i^*(\xi_1) \varphi_j^*(\xi_2) \, \mathcal{V}_d^{SK}(|\underline{s}|) \, \varphi_i(\xi_1) \varphi_j(\xi_2) \, d^3\xi_1 \, d^3\xi_2 \qquad (1)$$

where ξ_1 and ξ_2 are the internal co-ordinates of the two nuclei and

$$|\underline{s}| = |\underline{r}_1 - \underline{r}_2| = |\underline{\xi}_1 - \underline{\xi}_2 + \underline{R}| \qquad ; \text{ similarly,}$$

$$U_{EX}(R) = \sum_{i,j} \int \varphi_i^*(\xi_1) \varphi_j^*(\xi_2) \, \mathcal{V}_{ex}^{SK}(|\underline{s}|) \, \varphi_i(|\underline{s}+\xi_1|) \varphi_j(|\underline{s}+\xi_2|) e^{i(\underline{k}_1+\underline{k}_2)\cdot|\underline{s}|} d^3\xi_1 \, d^3\xi_2$$
$$(2)$$

Using Skyrme interaction one gets

$$U_T(R) = U_D + U_{EX} = \frac{3}{4} t_0 \int \rho_1 \rho_2 \, d^3\xi \; + \; \frac{1}{32}(5t_2 - 9t_1) \int (\rho_1 \nabla^2 \rho_2 + \rho_2 \nabla^2 \rho_1) \, d^3\xi$$

$$+ \frac{1}{16}(3t_1 + 5t_2) \int (\rho_1 \tau_2 + \rho_2 \tau_1) \, d^3\xi + \frac{1}{64}(5t_2 + 9t_1) \int (\rho_1 \nabla^2 \rho_2 + \rho_2 \nabla^2 \rho_1) d^3\xi$$

$$+ \frac{3}{16} t_3 \int \rho_L \rho_1 \rho_2 \, d^3\xi + \frac{1}{16}(3t_1 + 5t_2) \int (k_1^2 + k_2^2) \rho_1 \rho_2 \, d^3\xi \qquad (2a)$$

The derivation of eqn. (2a) is carried out exactly in the same manner as employed by Vautherin and Brink (see ref. 15) in their calculation of the Hartree-Fock single-particle energy of a nucleus using Skyrme interaction. The zero-range density-dependent term gives rise to an additional term

where ρ_L the local density, is evaluated at the mid-point of the interacting nucleons. Since the contact term is zero-range the local density is given by

$$\rho_L = \rho_1(|\underline{R} - \underline{\xi}|) + \rho_2(\underline{\xi}) \qquad (3)$$

In the above equations t_0, t_1, t_2 and t_3 are Skyrme parameters. The density ρ_1 and its derivatives are evaluated at $|\underline{R} - \underline{\xi}|$ whereas ρ_2 and its derivatives are evaluated at $\underline{\xi}$, the internal co-ordinate of any of the nucleus.

The kinetic energy density τ_1 and τ_2 are given by $\tau_{1,2} = \sum_{i,j} |\nabla \Psi_j|^2$ for our representation of $\Psi_{i,j}$, we get,

$$\tau_{1,2} = \tau_{1,2}^{s} + k_{1,2}^{2} \, \rho_{1,2} \qquad\qquad (4)$$

where $\tau_{1,2}^{s} = \sum_{i,\delta} |\nabla \varphi_{i,\delta}|^2$ is the static component of the kinetic energy density and $k_{1,2}^{2} \rho_{1,2}$ is an additional dynamical component arising due to relative motion. Combining eqns (2) (3) and (4) we get

$$U_{T}(R) = \frac{3}{4} t_0 \int \rho_1 \rho_2 \, d^3 \xi \;+\; \frac{1}{16}\left(3t_1 + 5t_2\right) \int \left(\rho_1 \tau_2^{s} + \rho_2 \tau_1^{s}\right) d^3 \xi$$
$$+ \frac{3}{16} t_3 \int \rho_L \, \rho_1 \rho_2 \, d^3 \xi \;+\; \frac{1}{64}\left(5t_2 - 9t_1\right) \int \left(\rho_1 \nabla^2 \rho_2 + \rho_2 \nabla^2 \rho_1\right) d^3 \xi$$
$$+ \frac{1}{8}\left(3t_1 + 5t_2\right) \int \left(k_1^2 + k_2^2\right) \rho_1 \rho_2 \, d^3 \xi \qquad\qquad (5)$$

except the last term the potential $U_{T}(R)$ looks exactly the same as obtained by Fleckner and Mosel[5] without the Pauli distortion.

As the nuclei approach each other, each nucleon in a nucleus experiences the single-particle field of the other nucleus, consequently, the energy of relative motion would be modified. Generally, one can write in a W. K. B. approximation

$$k_{1,2}^{2} = \frac{2m}{\hbar^2}\left(E_{1,2} - U_{12,21}\right) \qquad\qquad (6)$$

where U_{12} or U_{21} is the optical potential experienced by a single nucleon of kinetic energy E_1 or E_2 with respect to the other nucleus. The above relation for the local wave number has been used quite extensively for incident protons[9]. Ignoring the term $U_{12,21}$ in eq. (6) implies that there is no change in the relative motion energy due to interaction and therefore such an assumption implicitly ignores Pauli distortion effect. As the potential $U_{12,21}$ builds up its resistence for the ions to interact i. e. turns weaker, even repulsive, the internal kinetic energy increases at the expense of the relative kinetic energy. Indeed we shall demonstrate in section 4 that the dynamical component in eq. (5) agrees quite well in magnitude with the Pauli distortion effect, calculated by using energy-density method.

The major difficulty in evaluating eq. (6) is of course the uncertainty in $U_{12,21}$. Since the single nucleon in our case is not free but bound and embedded in the nuclear medium of its own nucleus, all kinds of Pauli correlation would come into play and it would be manifestly wrong to use a standard one-body optical potential. To overcome this difficulty we shall define U_{12} or U_{21} such that, the identity $U_{T}(R) = \int \rho_1 \, U_{12} \, d^3 \xi = \int \rho_2 \, U_{21} \, d^3 \xi$ is exactly satisfied. It is observed that eq. (5) can be written in the above manner without any loss of generality. Such a defination, further more guarantees "symmetry" in evaluating $U_{T}(R)$ and also, it would be possible to find out about the properties of U_{12} and U_{21} and how they deviate from a free nucleon-nucleus optical potential. It is simple to show that

$$U_{12} = \left\{ V_2 + \alpha \rho_2 (E_2 + E_1) - \alpha \rho_2 U_{21} \right\} \left(1 + \alpha \rho_2\right)^{-1} \qquad\qquad (7)$$

where

$$\alpha = \frac{2m}{\hbar^2} \frac{1}{8}\left(3t_1 + 5t_2\right)$$

$$V_2 = \frac{3}{4} t_0 \, \rho_2 + \frac{1}{16} (3t_1 + 5t_2) \left\{ \rho_2 \left(\tau_1^s / \rho_1 \right) + \tau_2^s \right\}$$

$$+ \frac{1}{64} (5t_2 - 9t_1) \left\{ \rho_2 \left(\nabla^2 \rho_1 / \rho_1 \right) + \nabla^2 \rho_2 \right\} + \frac{3}{16} t_3 \, \rho_L \, \rho_2 \qquad (8)$$

Similarly, one can define U_{21} and V_1. Substituting the relevant expression for U_{21} and eq. (7), it is straight forward to show that

$$U_{12} = \left\{ V_2 + \alpha \rho_2 \left(E_2 + E_1 \right) \right\} \left\{ 1 + \alpha \left(\rho_1 + \rho_2 \right) \right\}^{-1} \qquad (9)$$

The term in the denominator of eq. (9) can be visualised as a non-locality effect which tends to dampen the potential for small R. The familiar energy dependence of the optical potential is also evident. It should be noted however that the energy at which eq. (9) is evaluated has two components: each nucleon, in either of the nucleus, has a translational energy of relative motion and also the internal Fermi-motion energy; the energy at which the one-body potential $U_{12, 21}$ should be evaluated is thus given by $E_{1,2} \Rightarrow \bar{E}_{1,2} + \frac{\hbar^2}{2m} \, 0.6 \, k_F^2$. In this expression we have used the Thomas-Fermi approximation to evaluate the average kinetic energy of a nucleon. The local Fermi-momentum k_F is given by $k_F^3 = 1.5 \, \pi^2 \rho_L, \rho_L$ being the local density as defined before and $\bar{E}_{1, 2}$ is the translational kinetic energy per nucleon. With the change in the local wave-number the local energy of a nucleon changes and Pauli distortion sets in.

Now, in eq. (8) the kinetic energy density $\tau_{1,2}^s$ can in principle be computed by operating the kinetic energy operator on the antisymmetrised single-particle wave functions $\varphi_{i,j}$. Brink and Stancu[6] and more recently Fleckner and Mosel[5], have peformed such a calculation using two-centered Harmonic oscillator model for $^{16}O + ^{16}O$ system. One of the conclusions of the work of Brink and Stancu[6] later substantiated by Fleckner and Mosel[5] is that τ^s can be very well approximated by

$$\tau^s = 0.6 \, k_F^2 \, \rho + \frac{1}{2} \nabla^2 \rho \qquad (10) a$$

and the error is never more than 20%. Using eq. (9) and collecting all the terms, we get, finally,

$$U_T(R) = \int \rho_1 \, \bar{U}_{12} \, d^3 \xi = \int \rho_2 \, \bar{U}_{21} \, d^3 \xi$$

$$\bar{U}_{12} = \left\{ \frac{3}{4} t_0 \rho_2 + \frac{3}{8} (3t_1 + 5t_2) \, 0.6 \, k_F^2 \, \rho_2 + \frac{3}{64} (5t_2 - t_1)(\nabla^2 \rho_2 + \rho_2 \nabla^2 \rho_1 / \rho_1) \right.$$

$$\left. + \alpha \left(\bar{E}_1 + \bar{E}_2 \right) \rho_2 + \frac{3}{16} t_3 \, \rho_L \, \rho_2 \right\} \left(1 + \alpha \left(\rho_1 + \rho_2 \right) \right)^{-1} \qquad (10) b$$

a similar expression for U_{21} can be obtained by replacing $\rho_2, \nabla^2 \rho_2$ by

ρ_1 and $\nabla^2 \rho_1$ respectively. The functions $\bar{U}_{12}$ and $\bar{U}_{21}$ are essentially the functions U_{12}, U_{21}; the assumptions indicated by equs. (8) and (9) are however implicitely incorporated in eq. (10).

3. THE SECOND-ORDER POTENTIAL

The calculation of the second-order potential is motivated by the work of Vinh Mau[10]. One can write the second-order potential in general as

$$U_{II} = \sum_{m \neq 0} \lim_{\eta \to 0} \int d\epsilon_n \left\langle \psi_0 \left| \sum_i U_{12}(i) \, \psi_m \varphi_n \right\rangle \frac{\left\langle \varphi_n \psi_m \left| \sum_j U_{12}(j) \right| \psi_0 \right\rangle}{\epsilon_0 - \epsilon_n - \epsilon_m + i\eta} \right. \quad (11)$$

In eqn. (11) $U_{12}(i,j)$ is the one-body interaction potential between one nucleon of (say) the target and the projectile; ψ_0 is the ground-state wave function of the target, ψ_m and ϵ_m the wave-functions and excitation energies of its excited states, φ_n and ϵ_n are the wave functions and energies of the projectile in the intermediate states leave blank and is the (ϵ_0) centre of mass energy of the incident projectile. The principal value of the above equation is the polarisation term, contributing to the real part of the interaction potential whereas the pole term gives rise to the imaginary potential. In the calculation we shall assume that (1) the wave-function of the projectile in the field of the target nucleus can be approximated by a plane wave with a wave number $k_n^2 = (2\mu/\hbar^2)(\epsilon_n - U_R)$; $\mu = A_1 A_2/(A_1 + A_2)$ where U_R is the interaction potential. In principle, U_R should be the sum of both the first and the second-order term.

(2) The energy ϵ_0 is high enough so that all the energies ϵ_m would be such that $\epsilon_m < \epsilon_0$; this is justified for nucleus-nucleus collision because the average energy of excitation is around $\epsilon_m \sim 16.0$ MeV which is quite small compared to the usual laboratory energy at which experiments are done, above the Coulomb barrier, $(\sim 12 \, MeV; \, {}^{16}O + {}^{16}O)$ One can now write eq. (11) as

$$U_{II} = \sum_{m \neq 0} F(K, S_0) \left\langle \psi_0 \left| \sum_i U_{12}(i) \right| \psi_m \right\rangle \left\langle \psi_m \left| \sum_j U_{12}(j) \right| \psi_0 \right\rangle \quad (12)$$

where

$$F(K, S_0) = \frac{-\mu}{2\pi \hbar^2} \left[\cos(K S_0) + i \sin(K S_0) \right] \frac{1}{S_0} \frac{k}{K} \quad (13)$$

$$K^2 = \frac{2\mu}{\hbar^2}(\epsilon_0 - U_R - \epsilon_m) \quad : \quad k^2 = \frac{2\mu}{\hbar^2}(\epsilon_0 - \epsilon_m) : \quad S = |R - R'| \quad (14)$$

Since one is primarily interested in the surface region of interaction where $U_R \approx -5$ MeV it is expected that k and K are not very different. Further, we assume that ϵ_m is replaced by an average value $\langle \epsilon_m \rangle \sim$ 16 MeV; there are many good reasons to believe this is so[11, 12] and the energy of excitation in the entrance channel remains more or less constant.

We now apply 'closure' in summing over intermediate states and obtain,

$$U_{II} = F(K, S_0) \left[\left\langle \psi_0 \left| \sum_i U_{12}(i) \sum_j U_{12}(j) \right| \psi_0 \right\rangle - \left\langle \psi_0 \left| \sum_i U_{12}(i) \right| \psi_0 \right\rangle \left\langle \psi_0 \left| \sum_j U_{12}(j) \right| \psi_0 \right\rangle \right] \quad (15)$$

The $U_{12}'s$ we shall use to compute eq. (15) is the self-consistent one-body potential which leads on to the first-order interaction potential, as derived in section 2. Such a consistency, we feel, is necessary.

Using the U_{12} 's we now get,

$$U_{II}(R_0, S_0) = F(K, S_0) \left[\int U_{12}\left(|\underline{R}_0 - \underline{r} + \frac{1}{2}\underline{S}|\right) U_{12}\left(|R_0 - \underline{r} - \frac{1}{2}\underline{S}|\right) P_T(r) d^3r \right.$$
$$\left. - \frac{1}{4} \int U_{12}\left(|\underline{R}_0 - \underline{R} + \frac{1}{2}S_0 - \frac{1}{2}\underline{S}|\right) U_{12}\left(|\underline{R}_0 - \underline{R} - \frac{1}{2}S_0 + \frac{1}{2}\underline{S}|\right) P_T^2(R, s) d^3R d^3s \right]$$

(16)

which is certainly a "tough" integral.

The density-matrix is approximated in the Negele-Vautherin Campi approximation[13] which reads as follows:

$$P_T\left(\underline{R} + \frac{1}{2}\underline{s}, \underline{R} - \frac{1}{2}\underline{s}\right) = P_T(R) \hat{j}_1(ks)$$

$$k(R) = \left[\frac{5 \cdot 0}{3 P_T(R)}\left(\tau(R) - \frac{1}{4}\nabla^2 P(R)\right)\right]^{1/2}$$

(17)

$$\hat{j}_1(x) = 3!! \, j_1(x)/x \quad \text{and} \quad \tau(R) \quad \text{is the kinetic-energy}$$

density. For computation purpose eq. (16) is re-written

$$U_I(R_0, S_0) = F(K, S_0) \left[\int P_T\left(|\underline{R}_0 - \underline{x}|\right) d^3x \int U_{12}\left(|\underline{x} + \frac{1}{2}\underline{S}|\right) U_{12}\left(|\underline{x} - \frac{1}{2}\underline{S}|\right) d^3x \right.$$

$$\left. - \frac{1}{4}\int P_T^2\left(|\underline{R}_0 - \underline{x}|\right) d^3x \int U_{12}\left(|\underline{x} + \frac{1}{2}\underline{y}|\right) U_{12}\left(|\underline{x} - \frac{1}{2}\underline{y}|\right) \hat{j}_1\left\{k\left(|\underline{R} - \underline{x}|\right) |\underline{S}_0 - \underline{y}|\right\} d^3y \right. \quad (18)$$

and to obtain the local equivalent potential

$$\widetilde{U}_{II}(R_0) = \int U_I(R_0, S_0) \, j_0(KS_0) \, d^3S_0$$

(19)

It turns out that the exchange integral in eq. (18) is unmanageble unless one further assumes that $k\left(|\underline{R}_0 - \underline{x}|\right) \sim k_F^c$ where k_F^c is a constant, the local Fermi momentum at a certain radius. Since we are interested only in the peripheral region we chose a set of values of k_F^c, starting from the radial point at which the density falls off by half the central density, by one third and by one fourth. Fortunately the results do not change appreciably. The two integrations in the exchange integral "evens" out the sensitivity of k_F. It would be worthwhile to evaluate eq. (18) properly however, second-order excitation due to the coulomb field is calculated recently by Love et. al. [17] and Baltz et. al. [18].

4. RESULTS AND DISCUSSION

In this part we shall employ our results for ^{16}O incident on ^{16}O at various incident energies. In Fig. 1, the results obtained using eq. (10) are presented, nomenclatured as MFLD. The interaction potential as obtained by using the energy-density method using Skyrme interaction is also presented. It is evident that the exchange terms give rise to substantial repulsion in the potential as compared to just the direct term in eq. (10) shown in Fig. 1. In Table 1, the results are compared in more detail; the phenomenological results are obtained from Ngo[14]. The Pauli distortion effect as obtained from energy density method and the results obtained from our work correspond to the following expressions:

$$\Delta E_k^1 = \frac{\hbar^2}{2m}\int (\tau_t - \tau_1 - \tau_2) \, d^3\xi \qquad \text{(Energy-Density)}$$

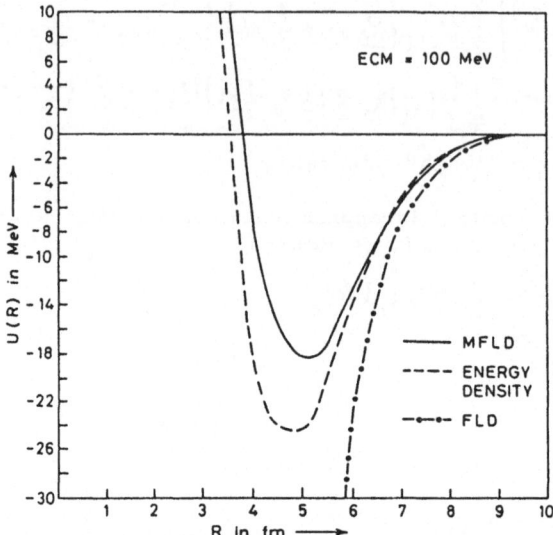

Fig. 1: The first-order interaction potential (MFLD)
as a function of intranuclear distance

$$\text{and} \qquad \Delta E_k^2 = \frac{1}{8} (3t_1 + 5t_2) \int (k_1^2 + k_2^2) \rho_1 \rho_2 \, d^3 \xi \quad (\text{Present Model})$$

in the first equation $\mathcal{T}_t$ corresponds to the kinetic-energy density of the combined system. As indicated earlier the two results seem to agree quite well. A detailed discussion on the results of the first-order potential is given ref. 15, we conclude our discussion with the following two observations.

(1) The kinetic energy term proportional to $0.6\ k_F^2$ is the single most important term for Pauli exchange correction, as suggested by other authors previously[6]. The energy dependence of our model is entirely due to knock-on exchange, reflecting the weaking of the two-body force with increase in energy. The <u>increase</u> in the interaction potential with the <u>increase</u> in energy (low energy region) due to the relaxation of Pauli blocking is not included in our model.

(2) The one-body potential prescribed in the text is manifestly different from a free nucleon-nucleus potential, as expected. A detailed discussion is presented in ref. 15, we note here that the one-body potential U_{12} between a nucleon (say) in the projectile and the target is halved when the projectile density becomes around 0.05 fm^{-3}. Nuclear saturation and Pauli exchange effects tend to weaken the interaction potential, the diminished strength of U_{12} is precisely because the projectile nucleon is embedded in its own nuclear medium and not free. It is important therefore that while computing (e.g.) second-order potential, as shown below, the driving potential for excitation should be U_{12} and not a free nucleon-nucleus potential.

In Fig. 2 the results obtained for the second-order virtual excitation are presented for ECM = 100.0 MeV. It is evident that beyond the touching

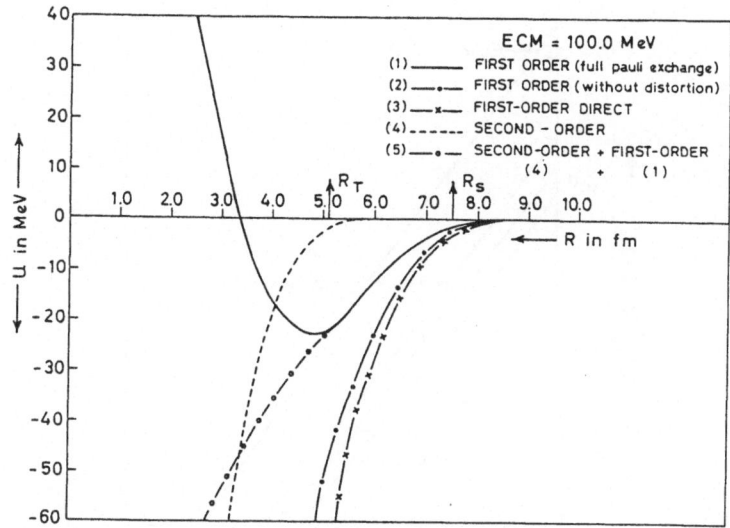

Fig. 2: The second-order interaction potential compared with
the first-order potential, ECM = 100.0 MeV

radius the polarisation term has virtually no contribution, but at lower
energy, Fig. 3, ECM = 50.0 MeV the polarisation term has significant contribution upto 5.5 fm. It turns out that the second-order term tends to
decrease with increasing energy and for ECM $\sim$ 500.0 MeV it turns
repulsive. In Fig. 4, the polarization term as a function of energy is
presented. Evidently, there is a nagging suspicion about the choice of   ,
the Fermi-momentum, but fortunately, as remarked earlier, the results are

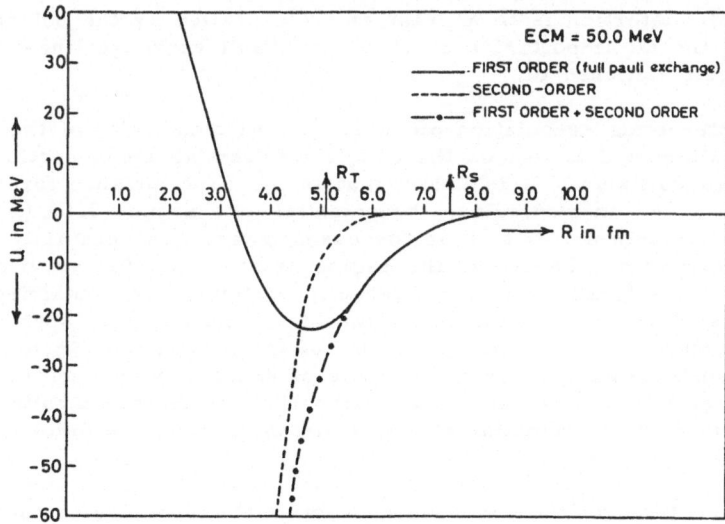

Fig. 3: The second-order and the first-order potential for
ECM = 50.0 MeV

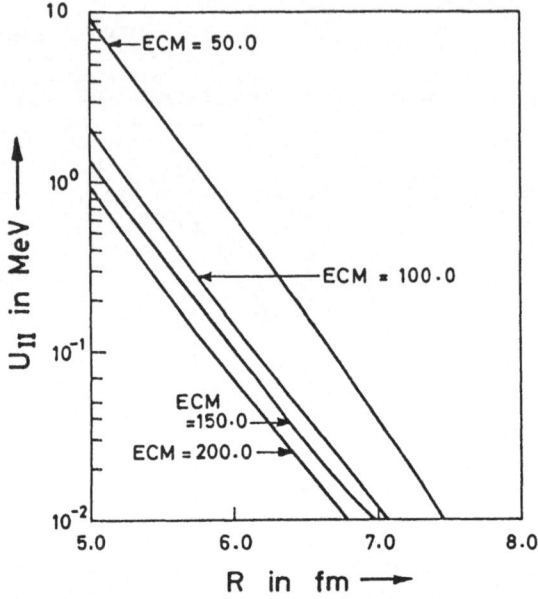

Fig. 4: Energy Dependence of the second-order potential

not sensitive to values of k_F at least around the interesting region. In Fig. 2 the direct term and the direct term plus the Pauli exchange terms arising solely due to the antisymmetrisation of the matrix elements and not from Pauli distortion effect are also shown. It seems that although at small distances Pauli distortion is to a great extent cancelled by the polarization term, around the all important surface region Pauli exchange lingers on but polarization gets switched off.

The interesting speculation ofcourse is the sensitivity of these effects on ion-ion scattering data and on the double differential cross-sections for heavily damped collision. It is known that in the first instance for light ion systems, $^{12}C, + ^{12}C$ in particular, the potential upto around 0.66 $(A^{1/3} + A_2^{1/3})$ [3] is determined. If that be the case, polarization certainly is going to be important. Secondly, the raging debate about the sensitivity of the restoring force (gradient of the interaction potential) for obtaining the solution of the classical equation of motion of two ions throws open an interesting speculation about the sensitivity of the polarization term specially at small distances for these kinds of experimental results. It is conjectured that for heavily damped collision phenomena the magnitude and shape of the interaction potential at small distances could be quite important.

The pole term of the second-order potential gives rise to the imaginary part of the optical potential. In Fig. 5 the results obtained from the present work are compared with phenomenological results[16]. In terms of form-factor and the general characteristics, the deep energy independent phenomenological form-factor looks quite similar to the results obtained theoretically. It is significant that for ECM = 50.0 MeV the theoretical

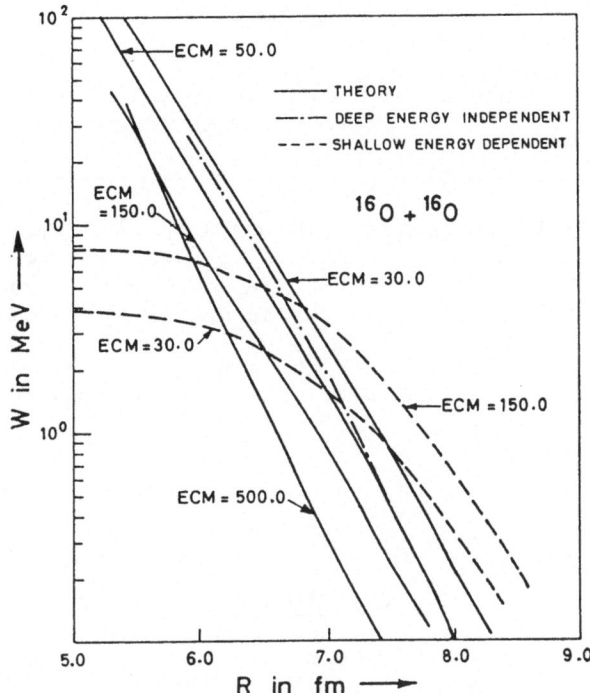

Fig. 5: Imaginary part of the interaction potential compared
with phenomenological results for various energies

form-factor agrees remarkably well with the phenomenological form-factor,
especially around 7.5 fm, the strong obsorption radius for $^{16}O + ^{16}O$
system. The significance of these results can be better understood by
fitting data. The imaginary potential tends to decrease slowly with increase
in energy, in Table II, the energy dependence is presented with other
results. A detailed application of the results for analysis of scattering data
and possibly heavily damped collision process are postponed for the future.

It is concluded that the polarisation term could be significant for
small distances between the centres of the nuclei but not of any practical
importance at large distances the imaginary potential calculated self-
consistenly seem to agree rather well with phenomenological results.

The author would like to thank his collegues at the Bhabha Atomic
Research Centre for many useful discussion and B.K. Jain for many
critical suggestions.

Table I

The Interaction Potential in MeV as a
function of R using different Models
(16_O + 16_O)

R in fm	U_T (Phenom)	U_T (MFLD)	U_T (Energy Density)
5.0	- 27.0	- 18.5	- 24.0'
6.0	- 24.8	- 14.0	- 14.2
7.0	- 4.7	- 5.1	- 5.0
8.0	- 1.0	- 1.0	- 1.1

Table II

R in fm	ECM in MeV	Ph. W in MeV (a)	W Theo. in MeV	Ph. W in MeV (b)
6.0	50.0	6.6	14.48	21.50
7.0	"	3.17	1.6	1.94
7.5	"	1.53	0.45	0.42
8.0	"	0.63	0.087	0.09
6.0	30.0	-	12.6	2.3
7.0	"	-	2.75	1.5
7.5	"	-	0.80	0.80
8.0	"	-	0.2	0.32

(a)   Energy independent, "deep"

(b)   Energy dependent, "Shallow"

REFERENCES

1. D. M. Brink and Fl. Stancu, Nucl. Phys. A243 (1975) 175
2. B. Sinha, Phys. Rev. C11 (1975) 1546
3. G. R. Satchler, Proc. Conf. macroscopic aspects of heavy ion collisions, Argonne 1976
4. B. Sinha Phys. Rev. Letts. 33 (1974) 600
5. J. Fleckner and U. Mosel, Nucl. Phys. A 277 (1977) 170
6. Fl. Stancu and D. M. Brink, Nucl. Phys. A270 (1976) 236
7. J. Blocki et al, Ann. Phys. 105 (1977) 427
8. D. M. Brink and Fl. Stancu Nucl. Phys.
9. B. Sinha, Phys. Rep. 20C (1975) 1
10. N. Vinh Mau, Phys. Lett. 71B (1977) 5
11. R. A. Broglia et al, Phys. Lett. 61B (1976) 113
12. B. Sinha, Phys. Lett. 71B (1977) 243, S. C. Phatak and B. Sinha (unpublished)
13. X. Campi and A. Bouyssy, Phys. Lett. 73B (1978) 263
14. C. Ngo, Private communication
15. B. Sinha and S. Moszkowski, Phys. Lett. to be published
16. R. H. Siemssen in Nuclear Spectroscopy and Reactions Part B, ed. J. Cerny Academic Press (1974) 233
17. W. G. Love et. al. Nucl. Phys. A 291 (1977) 183
18. A. J. Baltz et. al. Lawrence Berkeley Report

LONG RANGE ABSORPTION AND OTHER DIRECT REACTION COMPONENTS IN THE OPTICAL POTENTIAL[*]

A.J. Baltz
Brookhaven National Laboratory
Upton, New York 11973

and

N.K. Glendenning, S.K. Kauffmann and K. Pruess
Lawrence Berkeley Laboratory
Berkeley, California 94720

The effect of a strongly coupled inelastic excitation upon elastic scattering is represented as an optical potential component. In particular, a long range imaginary optical potential approximating the effects of quadrupole Coulomb excitation has been derived in closed form. An analytical closed form for sub-Coulomb elastic scattering is obtained by inserting this potential into a weak-absorption model, and connection is made with the semi-classical theory of Coulomb excitation. Above the Coulomb barrier, the long range absorptive potential may be incorporated into an optical model code. Alternatively a more elaborate analytical formulation has been made of the cross section itself in the weak absorption model.
The potential component arising from nuclear excitation of an inelastic state may be evaluated numerically on a computer. Two examples computed (50 MeV α scattering on ^{154}Sm and 60 MeV ^{16}O scattering on ^{40}Ca) exhibit strong ℓ-dependence in the potential component.

The effects of a strongly coupled direct inelastic transiton upon the elastic scattering cross section has been shown experimentally to be at times quite important both due to transitions that are primarily nuclear, such as ^{154}Sm(α,α') at 50 MeV,[1] and also due to transitions exhibiting strong Coulomb excitation effects such as ^{184}W$(^{18}$O,^{18}O$')$ at 90 MeV.[2] In such cases, analyses in terms of coupled channels calculations have provided a satisfactory description of the data.

An alternative theoretical description is the construction of an optical model component arising from an excited state's strong coupling to the ground state.[1,3] The possible advantages of such an approach are computational tractability and the fact that the physical nature of a complex optical potential is perhaps more transparent than the coupling between channels in a computer code.

For the present we confine ourselves to a simple set of two coupled equations whose effect is to be represented by an optical model component

$$(E_1-H)\chi_1 = V_{12}\chi_2 \tag{1}$$

$$(E_2-H)\chi_2 = V_{21}\chi_1. \tag{2}$$

---
[*]Research supported by the U.S. Department of Energy.

V_{ij} is the inelastic transition form factor, H is an optical model Hamiltonian, and χ_1, χ_2 are the ground and excited state scattering wave functions. For tractability we ignore reorientation couplings. Eq. (2) may be written as an integral equation

$$\chi_2 = G_2^{(+)} V_{21} \chi_1 \tag{3}$$

where $G_2^{(+)}$ is the outgoing boundary condition distorted-wave Green's function operator $(E_2-H)^{-1}$, and this result can be substituted into Eq. (1) to obtain

$$(E_1-H)\chi_1 = V_{12} G_2^{(+)} V_{21} \chi_1. \tag{4}$$

The elastic channel is thus formally uncoupled,[1] with the non-local potential operator $V_{12} G_2^{(+)} V_{21}$ bringing in the effects of coupling to all orders upon the elastic channel.

The non-local potential component to be evaluated may be written in coordinate space

$$V(r,r') = V_{12}(r)\ G_2^{(+)}(r,r')\ V_{21}(r') \tag{5}$$

where V_{12} and V_{21} are the multipole operators connecting ground and excited state, i.e.,

$$V_{12} = V(r)\sum_M Y_{LM}^*(\hat{r}) \tag{6}$$

$$V_{21} = V(r')\sum_{M'} Y_{LM'}(\hat{r}'). \tag{7}$$

A partial wave expansion of $G_2^{(+)}$ may be made in coordinate space

$$G_2 = \frac{-2\mu}{rr'k\hbar^2}\sum_{\ell'm'} f_{\ell'}(r_<)h_{\ell'}^{(+)}(r_>) Y_{\ell'm'}(\hat{r}) Y_{\ell'm'}^*(\hat{r}') \tag{8}$$

where $f_{\ell'}(r_<)$ and $h_{\ell'}(r_>)$ are optical model wave functions with regular and outgoing boundary conditions respectively. We may project out the ℓ-dependent non-local radial potential component

$$U_\ell(r,r') = -\frac{2\mu}{k\hbar^2}\ V(r)V(r')\sum_{\ell'} f_{\ell'}(r_<)h_{\ell'}^{(+)}(r_>)$$

$$\times \frac{2L+1}{4\pi}\langle \ell 0 L 0|\ell'0\rangle^2. \tag{9}$$

This is the ℓ-dependent, non-local optical potential component corresponding to the effects of the inelastic excitation upon the elastic channel. By incorporating this potential component into our solution of the partial wave version of Eq. (4) we obtain a result still completely equivalent to the coupled channels solution with no reorientation.

A local equivalent potential[4] may now be defined for U (r,r')

$$U_\ell(r) = \frac{1}{X_\ell(r)} \int dr' U_\ell(r,r') X_\ell(r').$$ (10)

Of course to evaluate this potential exactly one must know the solution of the Schrödinger equation which includes its effect. We have chosen to solve this problem numerically by iteration, a procedure which is efficient when it converges, as it does in the specific cases we consider.

But before considering numerical evaluation of Eq. (10) we will treat a case tractable by analytic means, namely quadrupole Coulomb excitation. This case has particular interest because of recent data such as the 90 MeV ^{18}O on ^{184}W scattering.[2] As is seen in Figure 1 the usual Fresnel pattern (e.g., such as with a ^{208}Pb target) is damped below the Rutherford cross section in the region of the grazing angle. The angular distribution is well reproduced by a coupled channels calculation which includes Coulomb excitation of the 111 keV 2^+ rotational state in ^{184}W. An optical model description of the contribution of the 2^+ state to the elastic scattering was carried out by Love, Terasawa, and Satchler.[4] Their approximation was to use plane waves for the intermediate state and ground state in an equation analogous to Eq. (10). A classical correction was then made for the Coulomb braking. The potential obtained was dominantly negative-imaginary, and apart from finite size corrections, has a radial dependence of $r^{-5}[1-(z_1 z_2 e^2/rE_{cm})]^{-\frac{1}{2}}$. This potential successfully reproduced the angular distribution in the ^{184}W case with a curve very similar to the coupled channel calculation without a long rang absorptive potential.

We have derived a more exact expression for this long range potential by making use of a Coulomb-distorted Green's function in Eq. (9) and a Coulomb-distorted wave functions for the X_ℓ's in Eq. (10). That is in these expressions we let $X_\ell \rightarrow F_\ell$, $h_{\ell'}^{(+)} \rightarrow H_{\ell'}$ and $f_{\ell'} \rightarrow F_{\ell'}$ where $F_{\ell'}$ and $H_{\ell'}$ will be taken to the regular and outgoing boundary Coulomb wave functions, respectively. Recalling that the outgoing wave function can be expressed in terms of the regular and irregular Coulomb wave functions

$$H_{\ell'}(r_>) = G_{\ell'}(r_>) + i F_{\ell'}(r_>)$$ (11)

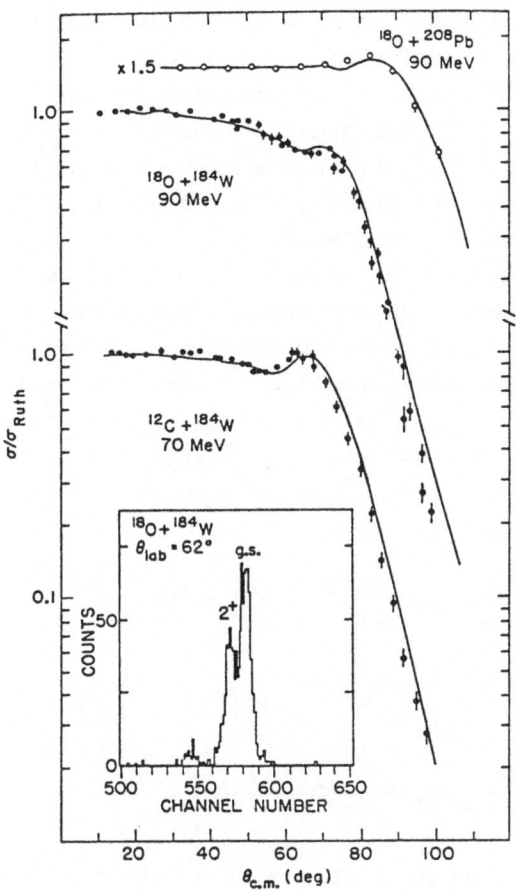

Fig. 1. Elastic scattering of ^{18}O and ^{12}C on ^{184}W. The usual Fresnel
pattern of ^{18}O + ^{208}Pb elastic scattering is shown for comparison.

and putting in the Coulomb quadrupole excitation operator for V(r) the
local potential from Eq. (10) takes the form

$$U_\ell(r) = -\frac{2\mu}{kh^2}\frac{4\pi}{25}z_p^2 e^2 B(E2)\uparrow\sum_{\ell'}\langle\ell 020|\ell'0\rangle^2\frac{1}{r^3}$$

$$\times\left[i\frac{F_{\ell'}(r)}{F_\ell(r)}\int_0^\infty dr'\,F_{\ell'}(r')\frac{1}{r'^3}F_\ell(r')\right.$$

$$+\frac{F_{\ell'}(r)}{F_\ell(r)}\int_r^\infty dr'\,G_{\ell'}(r')\frac{1}{r'^3}F_\ell(r')$$

$$\left.+\frac{G_{\ell'}(r)}{F_\ell(r)}\int_0^r dr'F_{\ell'}(r')\frac{1}{r'^3}F_\ell(r')\right]. \qquad (12)$$

We ignore the real components of this optical potential because they oscillate in sign as a function of r and merely serve to put "hair" on top of the real Coulomb potential. This corresponds to keeping only the on-energy-shell part of the Green's function for the intermediate state, a not unreasonable ansatz in the case of a smooth quasi-classical reaction such as Coulomb excitation. On the other hand, we can evaluate the imaginary component in closed form.

For the sake of simplicity we assume no energy loss in the quadrupole transition. However an approximate semiclassical energy loss factor $g_2(\xi)$ may be applied to our results at the end.[4,6] Making use of the closed forms for the $\frac{1}{R^3}$ Coulomb integrals[7,6] and the Coulomb wave recursion relations[8] we simplify the imaginary part of Eq. (12) with the additional assumption that either η or $\hat{\ell} = \ell + \frac{1}{2}$ is large, the usual semiclassical conditions. One obtains the long range imaginary potential for a given partial wave ℓ:

$$U_\ell(r) = - i \frac{2\mu}{k\hbar^2} \frac{\pi}{50} z_p^2 e^2 B(E2)\uparrow g_2(\xi)$$

$$\times \left[\left(\frac{\eta^2 k^2 (3\hat{\ell}^2 + \eta^2)}{\hat{\ell}^2 (\hat{\ell}^2 + \eta^2)^2} - \frac{\eta k^2}{\hat{\ell}^3} \arctan \frac{\hat{\ell}}{\eta} \right) \frac{1}{r^3} \right.$$

$$\left. + \frac{4\eta k \hat{\ell}^2}{(\hat{\ell}^2 + \eta^2)^2} \frac{1}{r^4} + \frac{2\hat{\ell}^4}{(\hat{\ell}^2 + \eta^2)^2} \frac{1}{r^5} \right] . \qquad (13)$$

This ℓ-dependent potential is compared with the ℓ-independent potential of Love, Terasawa, and Satchler in Figure 2. The LTS potential crosses our ℓ-dependent potential several fermis outside of the classical turning point for the small and intermediate ℓ values of interest. For the case in Figure 2 our formula has been compared with the results of a computer evaluation of the imaginary part of Eq. (12) and for all partial waves agreement is quite good (to within several percent except for computationally unstable points where $1/F_\ell(r)$ becomes large). This ℓ-dependent long range absorptive potential has been incorporated into an optical model code and the resulting cross section curve is practically indistinguishable from the corresponding calculations using the LTS potential (or from the original coupled channels calcuations) for $^{18}O + ^{184}W$ at 90 MeV in the angular region of experimental interest.

As an extension of this work we have found it possible to consider in a general way the effects of long range absorption upon the elastic scattering by deriving a cross section formula in closed form. Below the Coulomb barrier this cross section formula provides the most

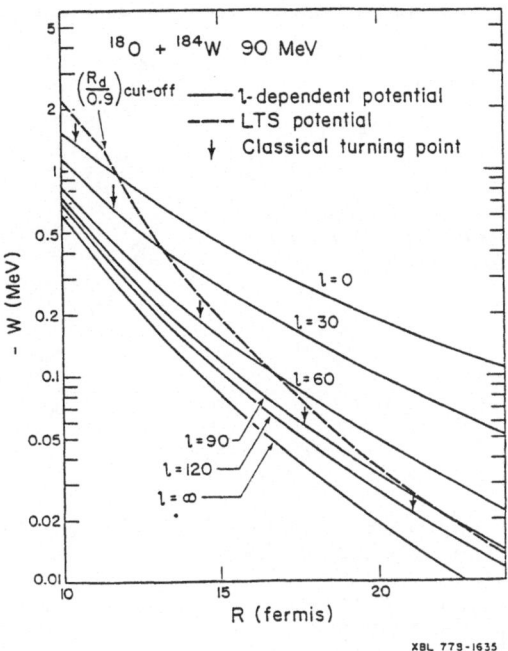

Fig. 2. ℓ-dependent imaginary optical potential obtained from Eq. (13)
compared with the LTS potential for $^{18}O + ^{184}W$ at 90 MeV.

concise way to compare our potential with the LTS potential as well as
showing the connection with the semiclassical theory of Coulomb excitation. For the general case, valid also above the Coulomb barrier we
have obtained a modified form of Frahn's strong absorption formula[9] by
using a perturbative JWKB integral evaluated along the Coulomb trajec-
tory[10] for the long range absorption contribution to the phase shift. The
detailed formula for the above barrier case is discussed elsewhere.[11,15]
A similar closed form has recently been independently proposed by Frahn
and Hill.[12] Here we will only show an example of the use of our general
cross section formula in fitting 90 MeV $^{18}O + ^{184}W$ elastic data (Fig. 3).
The fit of the formula is comparable to the optical model calculation
with long range absorption or to the coupled channels calculation.

Below the Coulomb barrier our result becomes independent of nuclear
surface parameters other than B(E2)↑ and we obtain a simple form for
the elastic scattering ratio to Rutherford cross section

$$\sigma(\theta)/\sigma_R(\theta) = \exp[-K\ f(\theta)], \qquad (14)$$

where all the specific parameters of the reaction are contained in the
constant

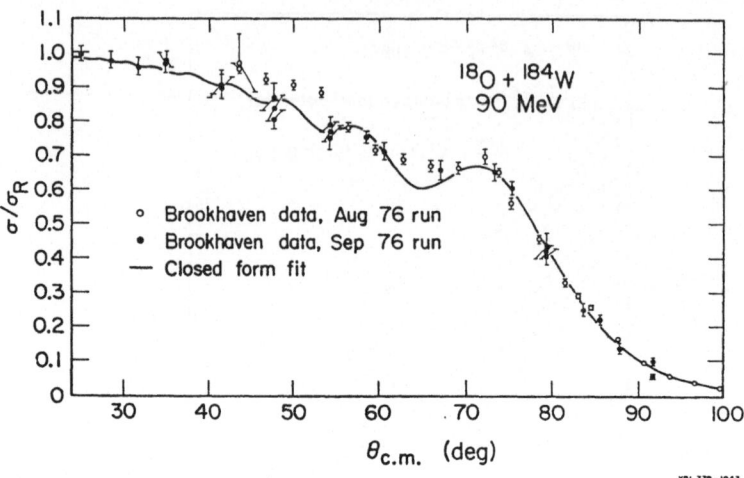

XBL 779-1943

Fig. 3. Comparison of our weak absorption model formula for the cross
section with the Brookhaven data.

$$K = \frac{16\pi}{225} \frac{k^4}{\eta^2} \left[\frac{B(E2)\uparrow}{z_T^2} \frac{g_2(\xi)}{e^2} \right] \tag{15}$$

and $f(\theta)$ is a universal function of angle

$$f(\theta) = \frac{9}{4} \left((\cos\frac{\theta}{2})^4 \, (\frac{4}{3}D^4 + \frac{104}{105}D^5) + (\sin\theta)^2 \right.$$

$$\times \left[\frac{\pi}{4}D^3 + (\frac{64 - 15\pi}{30})D^4 \right]$$

$$\left. + \left\{ \left[3 + (\tan\frac{\theta}{2})^2 \right] (\sin\frac{\theta}{2})^4 - (\tan\frac{\theta}{2})^3 (\frac{\pi-\theta}{2}) \right\} (D^2 + \frac{2}{3}D^3) \right) \tag{16}$$

with

$$D = (1 + \csc\frac{\theta}{2})^{-1}. \tag{17}$$

$f(\theta)$ has the smooth behavior exhibited in Fig. 4(a).

A similar expression may be obtained for the cross section produced by the LTS potential differing only in the form of the universal
function of angle $\bar{f}(\theta)$. The ratio of $\bar{f}(\theta)/f(\theta)$ has been plotted in
Fig. 4(b). Clearly at intermediate angles of about 40° to 110° the
ratio deviates little from unity, implying excellent agreement for the
prediction of the two potentials. However, beyond 110° (corresponding
to LTS cutoff of the Coulomb correction factor at $R_d/0.9$) there is no

theory from the LTS potential but only a possible prescription. For the sake of tractability we have merely ignored the cutoff in the ratio calculation. In Fig. 4(c) we show the elastic cross section in a sub-Coulomb case with small energy loss for which data exists at two angles, $^{16}O + {}^{162}Dy$ at 48 MeV.

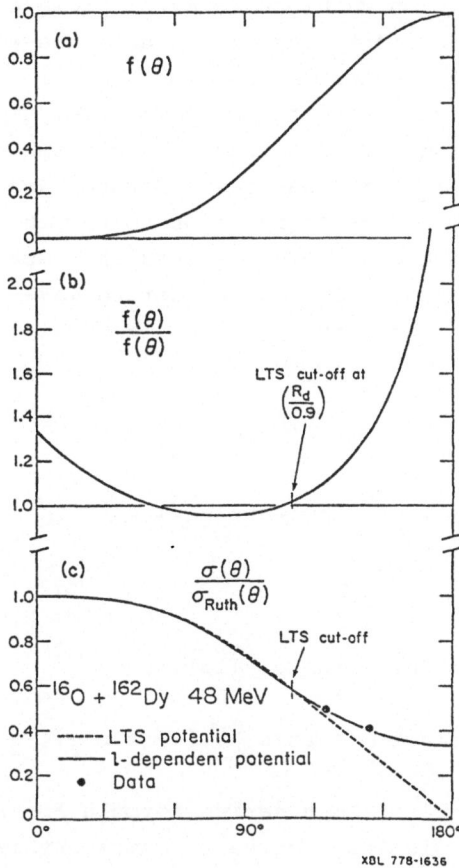

XBL 778-1636

Fig. 4. (a) Universal function of angle, $f(\theta)$. (b) Ratio of $\bar{f}(\theta)$ for the LTS potential to $f(\theta)$ for our potential. (c) Elastic-scattering cross section for $^{16}O + {}^{162}Dy$ at 48 MeV calculated from Eq. (14) incorporating $f(\theta)$ for our potential and $\bar{f}(\theta)$ for the LTS potential. Data are from Lee and Saladin (Ref. 16).

Recently more complete angular distributions have been obtained for sub-Coulomb 70 MeV ^{20}Ne scattering on Sm isotopes.[13] In Figure 5 we see comparison of data with our formula which includes here also a term for excitation of the 2^+ state in ^{20}Ne (dashed curve). While the qualitative agreement is good, at backward angles discrepancies occur especially for ^{148}Sm and ^{150}Sm. These discrepancies may be at least partially attributed to the larger energy loss factors ξ, which are only described approximately by the angle independent factor $g_2(\xi)$. At angles farther forward and especially for cases with a very low lying 2^+ state (small ξ) we expect both our potential and cross section formula to have greater validity. Furthermore at more forward angles (corresponding to a greater distance of closest approach) there is less multiple Coulomb excitation to higher states. However multiple Coulomb excitation will have an effect less direct upon the elastic scattering than on the inelastic 2^+ scattering in general.

Connection can be made with the semiclassical theory of Coulomb excitation[6] by exploiting the fact that our on-shell approximation for the Green's function makes it separable. Cotanch and Vincent have recently used a separable Green's function to sum the distorted wave series.[14] In our Coulomb case we use a separable Green's function for the ground state and then the Coulomb distorted wave series can be summed.[15] As we did in obtaining Eq. (14), a quasi-classical substitution is made

$$\hat{\ell} = \eta \cot \frac{\theta}{2} \tag{18}$$

in the scattering amplitudes. We obtain finally

$$\frac{\sigma(\theta)}{\sigma_R(\theta)} = \left(\frac{1 - \frac{1}{4} K g(\theta)}{1 + \frac{1}{4} K g(\theta)} \right)^2 \tag{19}$$

where K is identical to the K in Eq. (15), and

$$g(\theta) = \frac{9}{4} \left[\frac{1}{3}(\sin \frac{\theta}{2})^4 + (\tan \frac{\theta}{2})^4 (1-(\tan \frac{\theta}{2})(\frac{\pi-\theta}{2}))^2 \right] . \tag{20}$$

This is the on-shell Coulomb Born series formula for sub-Coulomb elastic scattering. It is instructive to compare this formula with the JWKB formulation based on the long range absorptive potential,

$$\frac{\sigma(\theta)}{\sigma_R(\theta)} = \exp[-K\, f(\theta)] . \tag{14}$$

At 180° the formulas agree exactly to second order in K (which is equivalent to fourth order in the interaction). At other angles the same correspondence is broken only by a small deviation (less than 4 percent anywhere) between $g(\theta)$ and $f(\theta)$. Thus to a very good approximation the JWKB optical model approach is equivalent to summing the scattering series on-energy-shell for the case of sub-Coulomb elastic scattering.

In a parallel manner the Coulomb Born series may be summed for the amplitude of inelastic Coulomb excitation to the 2^+ state. The result is

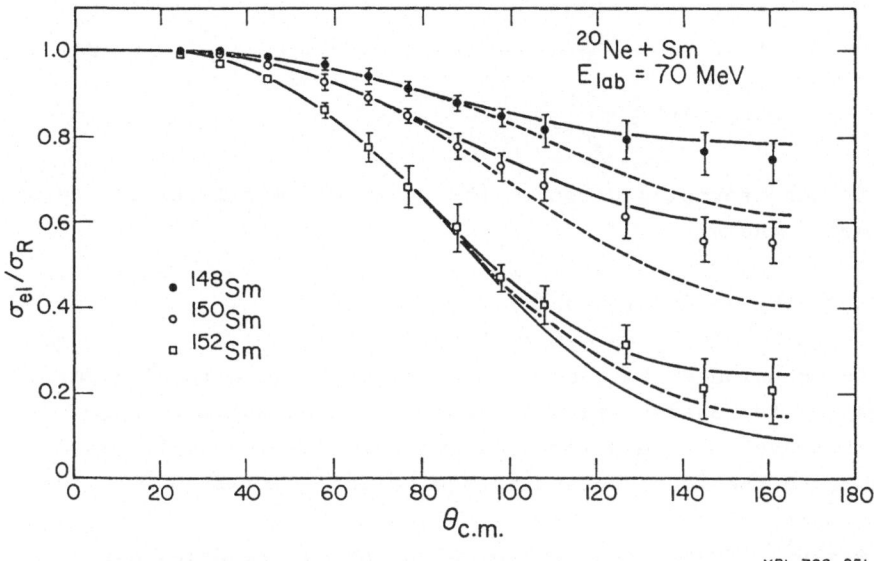

XBL 782-251

Fig. 5. Angular distributions from elastic scattering of ^{20}Ne on
samarium nuclei. Dashed curves show calculations using Eq.
(14) with a term for the ^{20}Ne 2^+ excitation added in. Solid
curves show coupled channel calculations with both 2^+ states
and reorientation included. The lower solid curve for ^{152}Sm
shows the calculation without reorientation, a significant
effect for this isotope.

$$\frac{\sigma^{2^+}(\theta)}{\sigma_R(0)} = \frac{K\,g(\theta)}{[1 + \frac{1}{4}K\,g(\theta)]^2} \; . \tag{21}$$

Now we note that the first order semiclassical result for Coulomb
excitation is just

$$\frac{\sigma^{2^+}_{(1)}(\theta)}{\sigma_R(\theta)} = K\,g(\theta)\,. \tag{22}$$

Thus we may rewrite the equations for elastic and inelastic scattering
in terms of the first order semiclassical Coulomb excitation cross
section

$$\frac{\sigma^{0^+}(\theta)}{\sigma_R(\theta)} = \left(\frac{1 - \frac{1}{4} \dfrac{\sigma^{2^+}_{(1)}(\theta)}{\sigma_R(\theta)}}{1 + \frac{1}{4} \dfrac{\sigma^{2^+}_{(1)}(\theta)}{\sigma_R(\theta)}} \right)^2 \tag{23}$$

$$\sigma^{2^+}(\theta) \;=\; \frac{\sigma^{2^+}_{(1)}(\theta)}{\left(1+\dfrac{1}{4}\dfrac{\sigma^{2^+}_{(1)}(\theta)}{\sigma_R(\theta)}\right)^2} \;. \qquad (24)$$

Note that these formulas preserve the quasi-classical unitary relationship at every angle

$$\sigma^{0^+}(\theta) \;+\; \sigma^{2^+}(\theta) \;=\; \sigma_R(\theta). \qquad (25)$$

Having concluded the general discussion of long range absorption,
we now return to consideration of the optical potential component
arising from nuclear inelastic coupling. In the cases discussed we
have evaluated the local potential equivalent $U_\ell(r)$ from Eq. (10) by
numerical means.

It was previously suggested by Glendenning, Hendrie, and Jarvis[1]
that the effect of inelastic scattering could be represented by an
optical potential component and these authors considered the case of
50 MeV α scattering on ^{148,150,152,154}Sm. The coupling is dominantly
nuclear and it reflects the change from a spherical vibrational nucleus
^{148}Sm to a rotational nucleus ^{154}Sm. It was found that a single optical
potential could describe both spherical and deformed Sm isotope elastic
scattering data when the strongly coupled excited states were treated
explicitly. However, in the absence of explicit coupling to excited
states, the optical potential that reproduces the elastic scattering is
quite different in the two cases. Moreover, while the optical model
parameters for ^{148}Sm differ little from the coupled channels parameters,
the optical model parameters for ^{152}Sm are quite different from the
coupled channels parameters.

We have calculated the component of the optical potential from the
direct rotational coupling of the 2^+ state to the ground state. The
real part of the optical potential component is exhibited in Figure 6
as a function of orbital angular momentum ℓ. Clearly it is highly
ℓ-dependent, repulsive in the low partial waves, increasing in magnitude
to the surface, changing sign and becoming attractive, and then decreasing in magnitude for high partial waves. The empirical optical model
component (the difference between optical model and coupled channels
parameters from Ref. 1) is ℓ-independent and repulsive as is also seen
in Figure 6 (dashed line).

The imaginary part of the optical model component is shown in
Figure 7. It is ℓ-dependent but absorptive for all partial waves. The

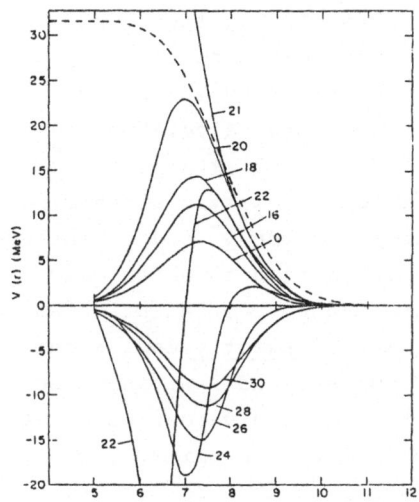

Fig. 6. Real potential component for 50 MeV α + ^{154}Sm scattering.

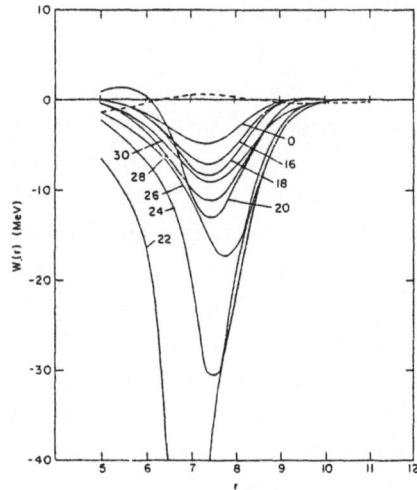

Fig. 7. Imaginary potential component for 50 MeV α + ^{154}Sm scattering.

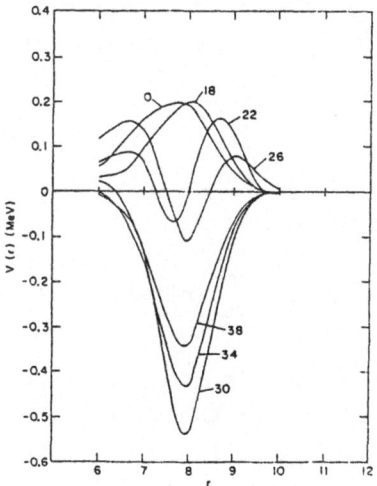

Fig. 8. Real potential component for 60 MeV ^{16}O + ^{40}Ca scattering.

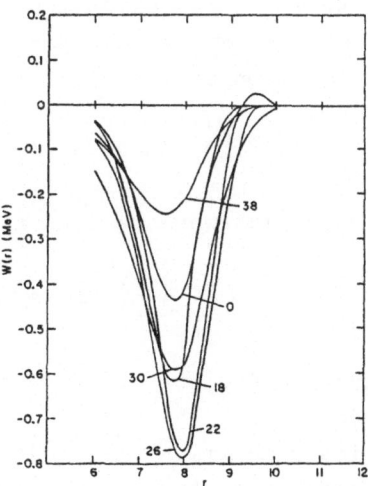

Fig. 9. Imaginary potential component for 60 MeV ^{16}O + ^{40}Ca scattering.

empirical imaginary potential component is of small magnitude, but relatively diffuse in its small absorption outside the surface.

We have investigated a second case of recent interest, 60 MeV ^{16}O scattering on ^{40}Ca.[17] In this case a coupled channels calculation was able simultaneously to reproduce the 0^+, 3^-, 5^- and 2^+ states in ^{40}Ca, while DWBA calculations using parameters fitted to elastic scattering failed to reproduce the angular distributions for the 3^- and 5^- inelastic scattering. For this coupled channel calculation in which only the 3^- state was coupled to the ground state and no reorientation was assumed, our optical model formulation Eq. (4) is <u>exactly</u> equivalent to the coupled channels formulation. In Figures 8 and 9 we show the real and imaginary parts of the ℓ-dependent local equivalent potential component which exactly represent the effect of the coupling of the 3^- state upon the elastic scattering. The general pattern is similar to the α-Sm case of Figures 6 and 7: the real potential component is repulsive for low partial waves and attractive for high partial waves; the imaginary potential component is dominantly absorptive with an ℓ-dependence of strength peaking in the surface partial waves. In both cases the ℓ-dependence of the imaginary potential seems to reflect the ℓ-window of a direct reaction in the presence of a strongly absorptive background potential; flux is lost from the elastic channel into the inelastic channel primarily in the surface partial waves.

If we wish to look at the amplitude for inelastic scattering to the 3^- excitation in this particular case we can obtain it directly from an equation of the form of Eq. 3,

$$\chi_{3^-} = G_{3^-}^{(+)} V_{3^-0^+} \chi_{0^+} . \tag{26}$$

Taking a partial wave we find the asymptotic form

$$\chi_\ell^{3^-}(r \to \infty) = h_\ell^{3^-}(r \to \infty) \sum_{\ell'} C_{\ell'} \int_0^\infty f_\ell^{3^-}(r') V(r') \chi_{\ell'}^{0^+}(r') dr' . \tag{27}$$

Since the coefficient of the outgoing wave function is the scattering amplitude we have the coupled channels equivalent for the inelastic scattering transition amplitude in the form of DWBA. All 0^+-3^- coupling effects enter through the ground state wave function $\chi_{0^+}(r')$, and the excited state wave function $f_{3^-}(r')$ is just an optical model wave function without the effect of the strong coupling to the 0^+ ground state. We have in fact incorporated the ground state wave functions $\chi_{0^+}(r')$ into a DWBA code, and very good numerical agreement is obtained with calculations using the coupled channels code CHUCK[18] for the 3^- cross section in the ^{16}O + ^{40}Ca case. This optical potential method for calculating coupled elastic and inelastic cross sections may be straightforwardly generalized to a sum of excited states coupled only to the ground state.

Finally we recall the strong ℓ-dependence of our calculated potentials arising from direct inelastic channels. In contrast, the usual optical model prescription for fitting elastic scattering makes use of an ℓ-independent potential. However one must question the usefulness of wave functions generated by this procedure when large direct reaction strength is present. The heavy ion DWBA angular distribution anomalies may not be unrelated to the use of an ℓ-independent optical potential even when a large percentage of flux is going into direct channels.

References

1. N.K. Glendenning, D.L. Hendrie, and O.N. Jarvis, Phys. Lett. 26B, 131 (1968).
2. C.E. Thorn, M.J. LeVine, J.J. Kolata, C. Flaum, P.D. Bond, and J.C. Sens, Phys. Rev. Lett. 38, 384 (1977).
3. H. Feshbach, Ann. of Phys. 19, 286 (1962).
4. W.G. Love, T. Terasawa, and G.R. Satchler, Phys. Rev. Lett. 39, 6 (1977); Nucl. Phys. A291, 183 (1977).
5. A.J. Baltz, S.K. Kauffmann, N.K. Glendenning, and K. Pruess, Phys. Rev. Lett. 40, 20 (1978).
6. K. Alder, A. Bohr, T. Huus, B. Mottelson, and A. Winther, Rev. Mod. Phys. 28, 432 (1956).
7. L.C. Biedenharn and C.M. Class, Phys. Rev. 98, 691 (1955).
8. Milton Abramowitz and Irene A. Stegun, Handbook of Mathematical Functions (Dover, New York, 1970).
9. W.E. Frahn, Generalized Fresnel Model for Very Heavy Ion Scattering. III Dynamic Polarization Effects (University of Cape Town preprint, 1977).
10. S.K. Kauffmann, Relation of Phase Shifts to Potential Parameters in the Elastic Scattering of Very Heavy Ions (University of Cape Town, Department of Physics preprint, Cape Town, 1976); C.E. Porter, Phys. Rev. 99, 1400 (1955).
11. A.J. Baltz, S.K. Kauffmann, N.K. Glendenning, and K. Pruess, Lawrence Berkeley Laboratory preprint LBL 6588.
12. W.E. Frahn and T.F. Hill, Zeitschrift für Physik A 285, 315 (1978).
13. P. Doll, M. Bini, D.L. Hendrie, S.K. Kauffmann, J. Mahoney, A. Menchaca-Rocha, D.K. Scott, T.J.M. Symons, K. Van Bibber, M.P. Viyogi, H. Wieman, and A.J. Baltz, Phys. Lett. 76B, 566 (1978).
14. S. Cotanch and C.M. Vincent, Phys. Rev. C14, 1739 (1976).
15. A.J. Baltz, N.K. Glendenning, S.K. Kauffmann, and K. Pruess, to be published.
16. I.Y. Lee and J.X. Saladin, Phys. Rev. C9, 2406 (1974).
17. K.E. Rehm, W. Henning, J.R. Erskine, and D.G.Kovar, Phys. Rev. Lett. 40, 1479 (1978).
18. P.D. Kunz, CHUCK, coupled channels code (unpublished).

ADIABATIC AND DYNAMIC POLARIZATION EFFECTS
IN SUBCOULOMB ELASTIC SCATTERING[*]

G. Baur, Institut für Kernphysik der KFA Jülich,
D-5170 Jülich, Germany

F. Rösel and D. Trautmann, Institut für theoretische Physik
der Universität Basel, CH-4056 Basel, Switzerland

I. Introduction

Even well below the Coulomb barrier, where the short range nuclear
forces between two ions are negligible, there are deviations in
elastic scattering from the pure Rutherford cross section. These
deviations may be due to various effects of long range interactions,
like electron screening (or the formation of electronic quasimole-
cules), vacuum polarization, relativistic effects, electromagnetic
excitation of nuclear levels and radiative corrections (e.g. brems-
strahlung). Many of these effects can be reliably described by appropriate local potentials[1].

Because nuclei are extended objects, they can be polarized in the
electric field of the other, which can also be interpreted as virtual
transitions to excited intermediate states. These transitions can be
characterized by the adiabaticity parameter $\xi_{if} = \eta_f - \eta_i$ where η_i and
η_f are the usual Coulomb parameters. In the case, where the virtual
excitation takes place via very high-lying states ($\xi_{if} > 1$), the transition is adiabatic and the effect can be represented by a local real
potential (see e.g. ref. 2)). This will be discussed in the first part
of this talk. However, if $\xi_{if} < 1$, real excitations can become large and
the cross section is well given in terms of the excitation probability
P_o, which can be calculated reliably in the semiclassical approximation[1,2,3]. Such calculations are compared to recently developed
optical model approaches[4,5] and discussed in the second part of this
talk.

II. Polarization Effects in the Subcoulomb Elastic Scattering of Heavy Ions

a) Adiabatic Case ($\xi_{if} \geq 1$)

The most important intermediate states which have to be taken into account are the strongly collective E1 transitions to the giant dipole states. Because of their high excitation energy the condition $\xi_{if} > 1$ is usually well fulfilled and it can be shown (see e.g. [1,2]) that the virtual excitations of these states can be represented to a good approximation by the real polarization potential

$$V_{pol}(R) = -\frac{1}{2}\frac{e^2}{R^4}(\alpha_1 z_2^2 + \alpha_2 z_1^2) \qquad (1)$$

where z_i and α_i are the charge and the polarizability of nucleus i, i=1,2. This dipole polarizability can be independently estimated by the photo-sum σ_{-2}:

$$\alpha \overset{\leq}{=} \frac{\hbar c}{2\pi^2}\sigma_{-2} = \frac{\hbar c}{2\pi^2}\int_0^\infty \frac{\sigma(E)}{E^2}\,dE \qquad (2)$$

where $\sigma(E)$ is the total photo-absorption cross section. The long range polarization potential (eq. (1)) will change the Rutherford orbit, which can be calculated quantum-mechanically, and more simply (to avoid the extended integration of the radial Schrödinger equation and the sum over many partial waves) in the classical approximation (all orders in V_{pol} or only first order). In the latter case, the deviation $\Delta(\theta) = (\sigma(\theta)-\sigma_{Ruth}(\theta))/\sigma_{Ruth}(\theta)$ is given by the simple scaling law[2]

$$\Delta_{classical}^{(first\ order)}(\theta) = -\frac{E^3}{e^6(z_1 z_2)^4}(\alpha_1 z_2^2 + \alpha_2 z_1^2)g(\theta) \qquad (3)$$

where $g(\theta)$ is a universal function of θ; E denotes the bombarding energy. In Fig. 1 the effect of the electric dipole polarizability is shown together with the competing effects of vacuum polarization and nuclear effects. (We neglect the influence of the quadrupole moment of the deuteron on the elastic scattering cross section, for such effects see ref. 6).) It is seen that the effect of vacuum polarization is of the same order of magnitude as the polarization effects, but the angular dependence is rather different. At backward angles the nuclear effects have some influence even far below the Coulomb barrier.

An especially suited estimation of the influence of a complex optical potential of the Woods-Saxon type on the elastic scattering cross section is given in ref. 7), eqs. (19-23). From these formulae it can

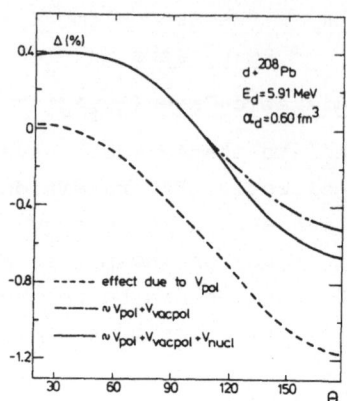

<u>Fig. 1:</u> Deviation $\Delta = \dfrac{\sigma - \sigma_R}{\sigma_R}$ from Rutherford scattering due to deuteron polarization (the polarization of ^{208}Pb is negligible here), nuclear interactions (described by a standard optical model) and vacuum polarization.

be seen that the deviation Δ behaves essentially like $\exp(-R_c/a)$ ($R_c = \dfrac{Z_1 Z_2 e^2}{E}$ ... classical turning point, a ... diffuseness of the optical potential). We see that nuclear effects tend extremely strongly to zero with decreasing bombarding energy; on the other hand, the deviation due to polarization, eq. (3), depends only on the third power of E. Therefore we can always find an energy region where polarization effects dominate completely over nuclear effects. (It should be kept in mind that phenomenological optical potentials, extrapolated from higher bombarding energies, can only serve as a rough guide for the actual nuclear effects below the Coulomb barrier.) Vacuum polarization and screening effects depend even less strongly than the polarization effects on the bombarding energy.

A similar behaviour is found for the reactions $\alpha + ^{208}$Pb and ^{16}O$ + ^{208}$Pb, as can be seen in Fig. 2. We feel that a direct measurement of the polarizability of a nucleus, independent of the photo absorption method, is possible, since the vacuum polarization contribution can be reliably subtracted. A favourable case could be the system $d + ^{208}$Pb (see Fig. 1) with the rather easily polarizable deuteron and the rather "stiff" ^{208}Pb.

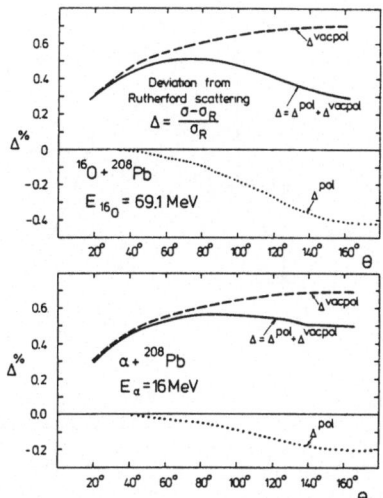

<u>Fig. 2:</u> Deviation Δ from the Rutherford cross section due to vacuum polarization Δ^{vacpol} (dashed line taken from ref. 8)), polarization Δ^{pol} (dotted line; we use $\alpha({}^4He) = 0.07$ fm^3 and for ${}^{16}O$ and ${}^{208}Pb$ the polarizabilities were calculated with the formula[1,9]: $\alpha = 3.5 \cdot 10^{-3}$ A$^{5/3}$ fm^3). The sum is given by the continuous line.

b) "Dynamic" Case, $\xi_{if} \leq 1$

In this case, flux can be taken out of the elastic channel because of the appreciable excitation probability of P_i of certain (low lying, strongly collective) levels i. Neglecting now to a good approximation the change of the Rutherford orbit, wen can express the elastic cross section in terms of the excitation probability P_o of the ground state[1,2,3]

$$\frac{d\sigma}{d\Omega}(\theta) = \frac{d\sigma_{Ruth}(\theta)}{d\Omega} \cdot P_o(\theta),\tag{4}$$

where P_o can be reliably calculated with the semiclassical coupled channels method[1]. Let us now contrast such calculations to the recently developed optical model approach[4,5]. We consider the example[4] of elastic scattering of ${}^{40}Ar$ on ${}^{238}U$ at $E_L = 340$ MeV, disregarding nuclear interaction between the heavy ions. We use different coupling schemes, as is illustrated in Figs. 3(a)-3(c). Fig. 3(a) shows the two-state model used in refs. 4,5). In Fig. 3(b) the reorientation effect in the 2$^+$ state is also considered; in Fig. 3(c) the rotational band is included up to the 4$^+$ state. We only discuss target excitations in ${}^{238}U$, which is a very strongly collective

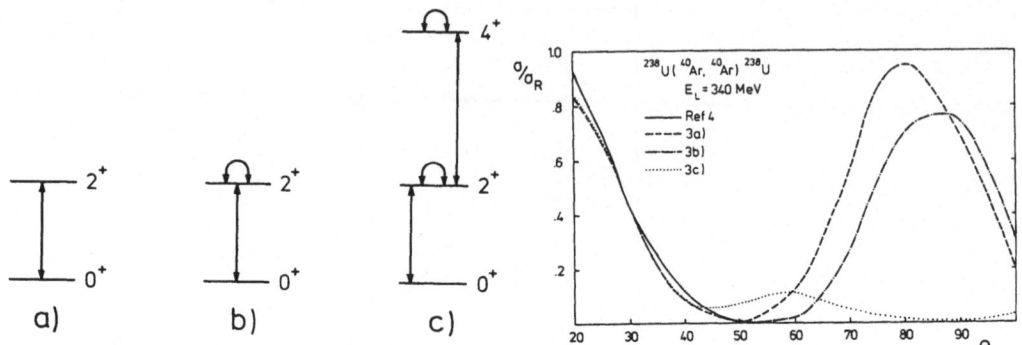

Fig. 3: Coupling schemes used in the semiclassical Coulomb excitation calculations. The coupling scheme of Fig. 3(a) corresponds to the one used in refs. 4,5).

Fig. 4: Comparison of semiclassical Coulomb excitation calculations using the coupling scheme of Fig. 3(a), 3(b) and 3(c), with the optical model calculations of ref. 4). Note that no nuclear effects are included in the Coulomb excitation calculations. This becomes unrealistic beyond the grazing angle $\theta \gtrsim 50^{\circ}$.

nucleus ($B(E2,0^{+} \to 2^{+}) = 13e^{2}b^{2}$, $E_{2+} = 44$ keV, $E_{4+} = 148$ keV). The results are shown in Fig. 4 together with the optical model calculations of ref. 4). It can be seen that the influence of the reorientation in the second state, Fig. 3(b), and the coupling to the 4^{+} state, Fig. 3(c), have little influence on the elastic scattering in the region below the grazing angle $\theta \lesssim 50^{\circ}$. For the (admittedly unrealistic) situation for $\theta \gtrsim 50^{\circ}$, however, an important difference arises between the three coupling schemes. There it would be no longer sufficient to consider only the first 2^{+} state in the construction of the optical model, as is done in refs. 4,5). The rise of P_{o} as a function of θ is characteristic for the two-state model, where the flux goes back from the 2^{+} to the 0^{+} state. In a realistic situation, more flux is distributed to the higher ($4^{+},6^{+}$, etc.) members of the rotational band. Therefore, the elastic cross section will become even smaller. As a criterion one can take the coupling strength parameter $\chi_{0 \to 2}^{eff}(\theta)$ of the ground state rotational band[1]. In our case we have $\chi_{0 \to 2} = 9.88$, which gives a $\chi_{0 \to 2}^{eff}(50^{\circ}) = 2.08$. For values larger than this, the two-state model can no longer be applied. For heavy ion scattering of strongly collective nuclei this value may well be exceeded. So we can confirm the assumption of refs. 4,5) that only second order contributions are important for the potential in the case studied here. However, in general, for very large couplings $\chi(\chi^{eff} \gtrsim 3)$ higher order terms are important. Then the construction of a local optical potential will become even more difficult. But, of course, the semiclassical theory of Coulomb excitation can still be applied.

III. Conclusion

The study of subcoulomb elastic scattering is an interesting field, where there is an interplay of different mechanisms of fundamental physical importance. The quantum electrodynamical effect of vacuum polarization can be of the same order of magnitude as the polarization effects studied here. The calculation of classical trajectories gives very accurate results. The importance of the different effects will depend on the bombarding conditions (energy, scattering angle) and on the specific structure of the colliding nuclei. The high-lying giant dipole state can usually be taken into account in the adiabatic approximation, whereas, if strongly collective low-lying states are present (like the ground state rotational band in strongly deformed nuclei) the elastic scattering is most reliably calculated in the semiclassical coupled channels approach. There is a close analogy to the study of the energy shifts of exotic atoms[9] where, in a classical interpretation, the Kepler ellipses are disturbed by the same kind of physical effects as the hyperbolic trajectories in the scattering case. Therefore, we believe that this very old kind of elastic scattering experiments, which has led Rutherford to the study of the subject of nuclear physics itself, can tell us now also new fundamental things.

References

1. K. Alder and A. Winther, Electromagnetic Excitation (North Holland, Amsterdam, 1975)
2. G. Baur, F. Rösel and D. Trautmann, Nucl.Phys. A288 (1977) 113
3. G. Baur, F. Rösel and D. Trautmann, Phys.Rev. C17 (1978) 2256
4. W.G. Love, T. Terasawa and G.R. Satchler, Phys.Rev.Lett. 39 (1977) 6; Nucl.Phys. A291 (1977) 183
5. A.J. Baltz, S.K. Kauffmann, N.K. Glendenning and K. Preuss, Phys. Rev.Lett. 40 (1978) 20
6. F. Rösel, K. Alder and U. Smilansky, Ann.Phys. 78 (1973) 518
7. A. Traber, D. Trautmann and F. Rösel, Nucl.Phys. A291 (1977) 221
8. J. Rafelski, Phys.Rev. C13 (1976) 2086
9. T.E.O. Ericson and J. Hüfner, Nucl.Phys. B47 (1972) 205

OPTICAL MODELS FROM EXPERIMENTS WITH ORIENTED HEAVY IONS

G. Tungate

Max-Planck-Institut für Kernphysik Heidelberg

and

D. Fick

FB Physik, Philipps-Universität, Marburg

The determination of the full optical potential is limited by the degree of completeness of the available experimental data. Angular distributions of differential cross sections can be used to determine the central parts of the potential but give only a rough indication of the non-central parts. Thus to determine the total optical potential for particles with non-zero spin, one must include angular distributions of the polarization observables in optical model fitting procedures. The development of the polarized heavy ion source at Heidelberg has provided a useful means for measuring polarization analysing powers for nuclear interactions using ^{6}Li and ^{7}Li projectiles.

Vector analysing power data for the elastic scattering of 22.8-MeV ^{6}Li from ^{12}C, ^{16}O, ^{28}Si and ^{58}Ni have provided data which has been used to test the heavy ion spin-orbit potential. The general features of the data have been reproduced by spin-orbit terms derived from folding models. Similar results have been given by a model based on an α-d structure [1] and a double folding model [2]. The data and fits from [1] are shown in Fig. I. Hill and Frahn have analysed the data using the closed formalism approach [3]. From a parameterized S-matrix which includes spin-orbit interactions they provide good fits to the data. This analysis suggests that a complex spin-orbit potential is needed with an imaginary part one quarter to one half as large as the real part.

So far only one angular distribution of vector data which also contains some third-rank components has been measured for ^{7}Li. This was taken for the elastic scattering of ^{7}Li from ^{58}Ni at 14 MeV and is small and rather featureless. More extensive vector analysing power measurements with the ^{7}Li beam are planned for the near future. It is hoped

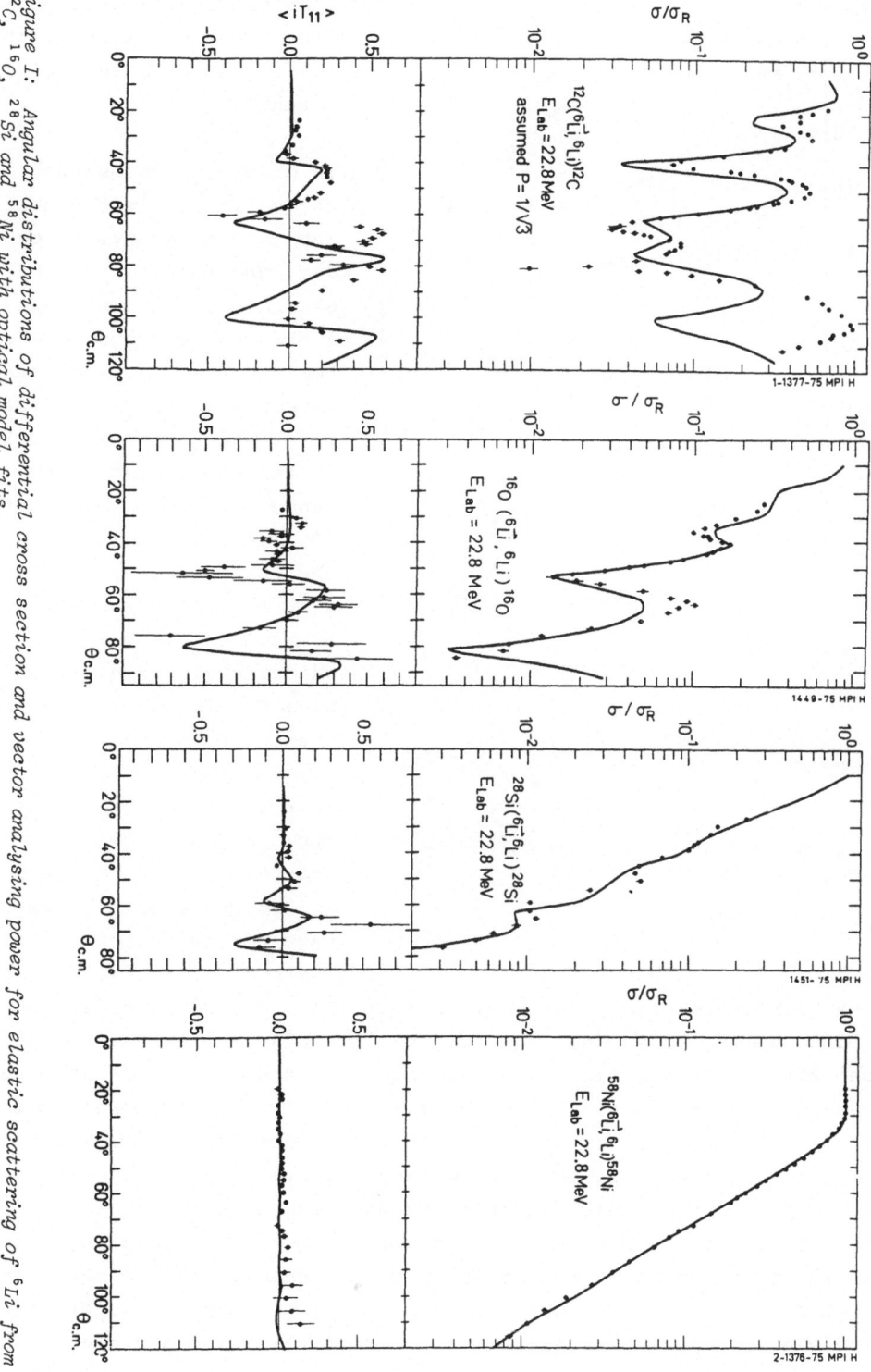

Figure 1: Angular distributions of differential cross section and vector analysing power for elastic scattering of ^6Li from ^{12}C, ^{16}O, ^{28}Si and ^{58}Ni with optical model fits.

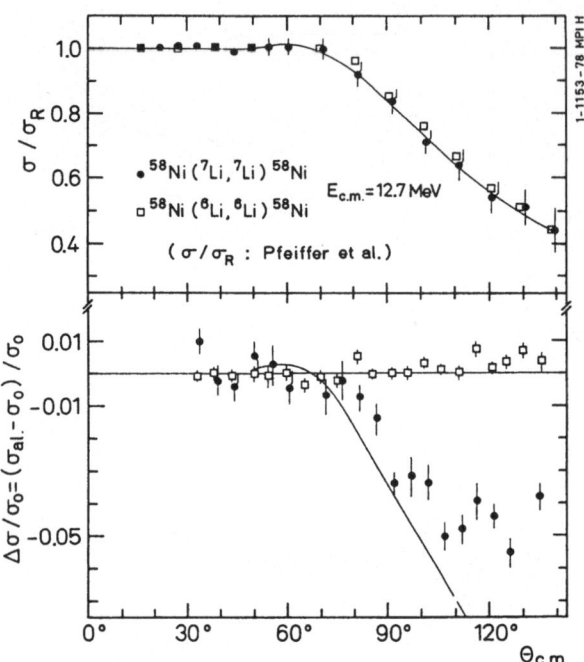

that these data will pro-
vide a further test of the
folding-model spin-orbit
potentials.

In order to demonstrate
shape effects in the second-
rank tensor analysing power
data and maintain a simple
experimental arrangement,
a transverse coordinate system was used for some measurements. In this system
the Z axis is perpendicular
to the scattering plane.
Hence for measurements of
$^{T}T_{20}$ (the small T indicates
the transverse system while
an H indicates the helicity
or Madison system), the spin
quantization axis of the
beam is also normal to the
scattering plane.

Fig.II Different cross section and tensor
polarization data for $^{58}Ni(^{6}Li,^{6}Li)^{58}Ni$ and
$^{58}Ni(^{7}Li,^{7}Li)^{58}Ni$ at $E_{c.m.} = 12.7$ MeV.
$\Delta\sigma/\sigma_{o} = {}^{T}T_{20}$. The experimental values of $\Delta\sigma/\sigma_{o}$
for ^{7}Li have a common uncertainty of 10%.
The curves are from an optical model calcu-
lation for the ^{7}Li scattering. |4|

Measurements of the $^{T}T_{20}$
for ^{6}Li and ^{7}Li elastically
scattered from ^{58}Ni at E_{cm}
= 12.7 MeV show a large effect for the deformed ^{7}Li
nucleus and a small effect

for the almost spherical ^{6}Li nucleus (Fig. II). From these data one
is led to believe that the $^{T}T_{20}$ for the ^{7}Li data is mainly due to a
shape effect. This belief is strengthened by the small odd-rank (vector and third rank) analysing power measured for ^{7}Li scattered from
^{58}Ni. Due to the lack of suitable computer codes for spin 3/2 systems
the T_{2q} data for the ^{7}Li scattering have received only a crude analysis.
This, however, gives qualitative representation of the ^{58}Ni tensor
data. The interaction radius for an aligned beam is assumed to change
by an amount ΔR which is proportional to the quadrupole moment of the
projectile. Then the situation for the $^{T}T_{20}$ for simple Fresnel scattering can be represented by Fig. III. From this simple picture an

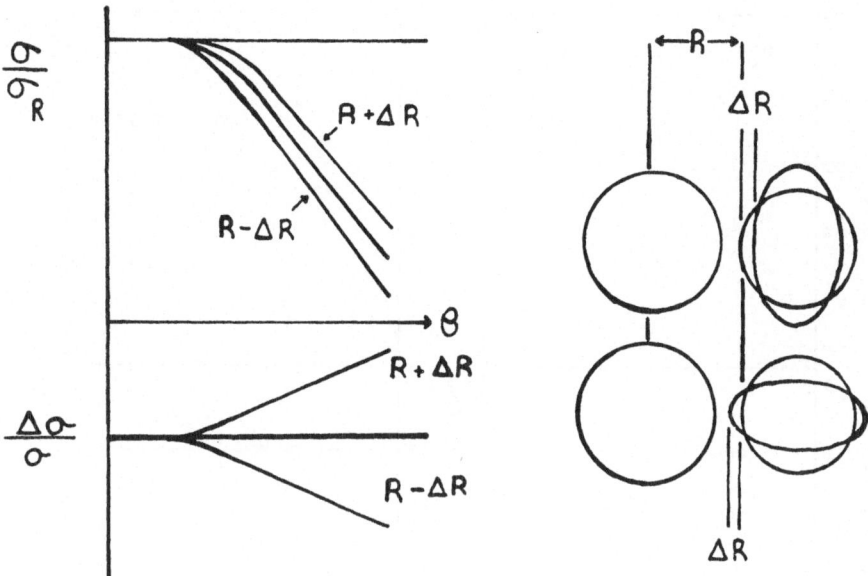

Figure III: *Pictorial representation of shape effects for Fresnel scattering.*
$\Delta\sigma = \sigma_{polarized} - \sigma_{unpolarized}.$

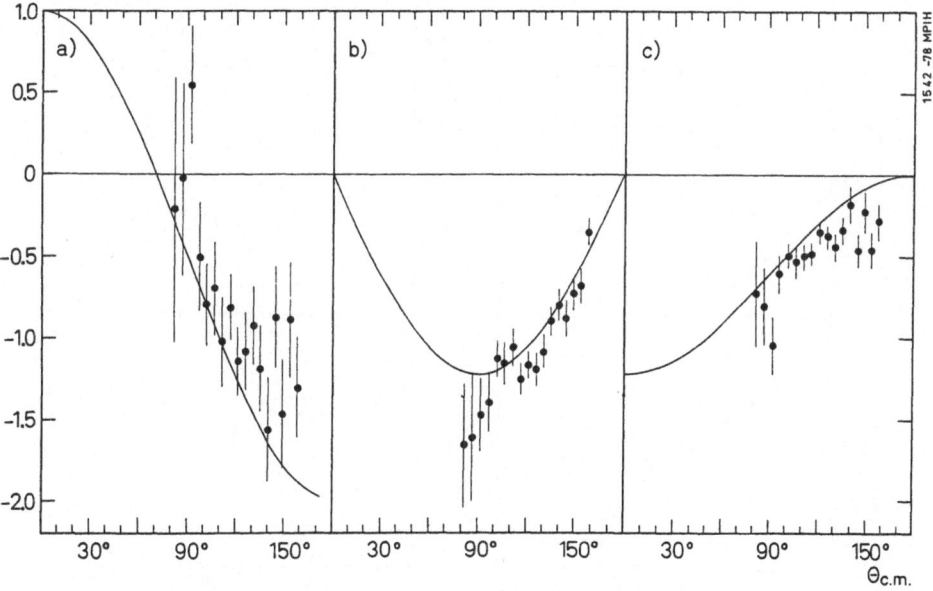

Fig.IV Ratio of tensor analysing powers in Madison frame to the $^{T}T_{20}$, *a)* $^{H}T_{20}/^{T}T_{20}$
b) $^{H}T_{21}/^{T}T_{20}$ *and c)* $^{H}T_{22}/^{T}T_{20}$. *The curves are from calculations based on a simple shape effect model and contain no adjustable parameters.*

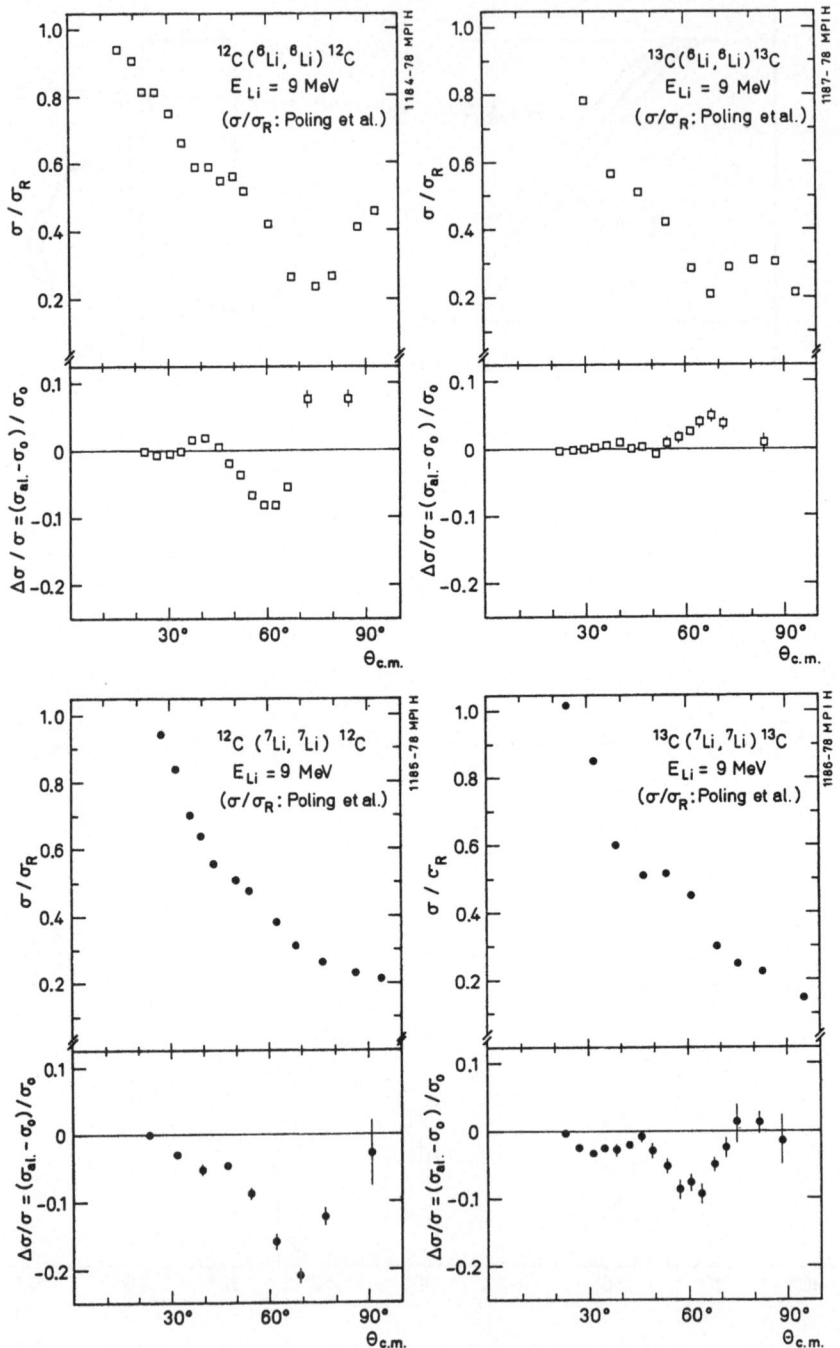

Figure V: Differential cross section and $^T T_{20}$ data for ^6Li and ^7Li scattered from ^{12}C and ^{13}C for E_{Li} = 9 MeV |4|.

estimate of the effect for the $^TT_{20}$ for ^7Li scattered from ^{58}Ni has been calculated and is shown by the solid curve in Fig. II. Assuming the absorption relative to the unpolarized case to be proportional to ΔR at closest approach, the model can be extended to predict the effect of the other second-rank analysing powers with respect to the $^TT_{20}$. The fits with this model are shown in Fig. IV. Here the fits contain no free parameters and considering the simplicity of the model agree remarkably will with the data. In taking ratios of analysing powers, the errors for the forward-angle data have become very large, hence the limited data at small scattering angles. This simple model also explains why in the deuteron optical model the forward-angle T_{22} data can almost always be fitted with a T_R potential, whereas the T_{20} data are difficult to fit.

Finally, $^TT_{20}$ data has also been taken for the elastic scattering of ^6Li and ^7Li from ^{12}C and ^{13}C at 9 MeV. Here the data (see Fig. V) show more structure and the magnitude of the effect for ^6Li suggests that the shape effect may no longer be dominant. Only at forward angles where the particles just feel the nuclear force can one see evidence for shape effects. This again reflects the situation found in deuteron scattering.

To end this discussion we would like to emphasize that the only way to understand these data fully is to perform a complete analysis using either an optical model or a parameterized S-matrix. Especially for the former method theoretically predicted folding potentials would be of great help. But for both methods computer codes applicable for spin 3/2 particles are essential.

REFERENCES

1. Weiss, W., et al.: Phys. Lett. 61B, 237 (1976)
2. Petrovich, F., et al.: Phys. Rev. C17, 1642 (1978)
3. Hill, T.F., Frahn, W.E.: contributed paper to Int. Conf. Nuclear Structure, Tokyo, September 1977
4. Draves et al.: Phys. Lett. B, to be published

Molecular States in Heavy Ion Potentials[*]

G. Terlecki, D. Hahn and W. Scheid

Institut für Theoretische Physik der Universität Gießen

and

R. Koennecke and W. Greiner

Institut für Theoretische Physik der Universität Frankfurt

1. Introduction

In the scattering of light heavy ions, e.g. $^{12}C+^{12}C$, $^{12}C+^{16}O$, resonance structures have been observed in various cross sections. These structures have been interpreted, first by Bromley et al.[1], as caused by nuclear molecules, and they are - in our mind - the most pronounced and richest cluster phenomena known. In this talk we want to focus the discussion on selected aspects for a theoretical description of quasimolecular resonance phenomena in heavy ion collisions. Review articles on the status of the field are collected in Ref. 2.

In the next sections we interpret certain structures in the cross sections of lighter heavy ions as molecular resonances. We discuss the conditions for the existence and excitation of nuclear molecules and the calculation of real and imaginary heavy-ion potentials. The following sections treat the nuclear molecule in a rotating coordinate system and serve to clarify the excitation mechanism of molecular collective and single-particle states. The last section deals with the application of parastatistics to nucleus-nucleus scattering, which yields some new points in connection with the antisymmetrization effects between nuclei.

---

[*] This work has been supported by the Bundesministerium für Forschung und Technologie, the Deutsche Forschungsgemeinschaft and the Gesellschaft für Schwerionenforschung (GSI).

2. Molecular Configurations

Structures in the cross sections, which are interpreted as molecular resonances, have been found especially in the $^{12}C-^{12}C$, $^{12}C-^{16}O$ and $^{16}O-^{16}O$ scattering. As example we show the elastic $^{16}O-^{16}O$ cross section in Fig. 1. Molecular configurations can be distinguished from the usual compound states that they are built up by clusters which keep their individuality to a large extent. There are several types of molecular configurations which give rise to different structures in the cross sections.

The first type of molecular configurations are the virtual states in the elastic nucleus-nucleus potential (dashed lines in Fig. 2). They cause the energy-dependent gross structures with widths of the order of 2 MeV in the $90°$-elastic cross sections of the $^{12}C-^{12}C$, $^{12}C-^{16}O$ and $^{16}O-^{16}O$ reactions above the Coulomb barrier (Figs. 1 and 10). The virtual resonances are the doorway states for all further molecular resonance phenomena in the various exit-channels.

A second type of molecular configurations are the quasibound states in the nucleus-nucleus potential (full lines in Fig. 2). They are excited directly for bombarding energies below the Coulomb barrier or via the virtual resonances in a double resonance mechanism[3]. One has observed the quasibound resonances most prominently in the elastic $^{12}C-^{12}C$ scattering below the Coulomb barrier with widths of the order of 100 keV. At incident energies above the Coulomb barrier they contribute to the intermediate structures superimposed over the gross structure in the elastic and inelastic cross sections.

A third type of molecular configurations happens in the α-particle transfer reactions. In these reactions, e.g. $^{16}O(^{16}O, ^{12}C)^{20}Ne$, cross sections have been measured which are comparable with the elastic and inelastic cross sections. Since systems consisting of $^{12}C-$, $^{16}O-$nuclei have α-cluster structures, molecular configurations can play an active role,in which α-particles constitute a homopolar binding in the sense of molecular chemistry. Such ideas were suggested by Michaud and Vogt to explain the transition between the resonating states in the entrance channel and the α-transfer exit channels. The α-cluster molecular configurations may generate intermediate structures in the cross sections which are not yet explored theoretically.

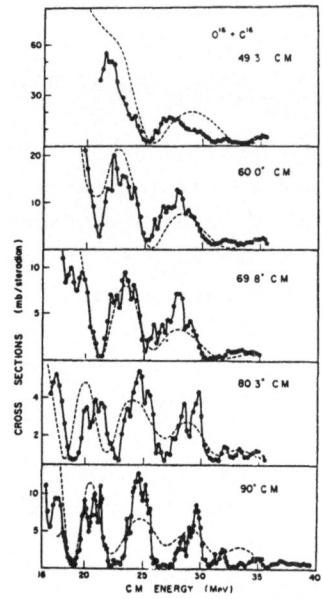

Fig. 1 Elastic $^{16}O+^{16}O$-scattering

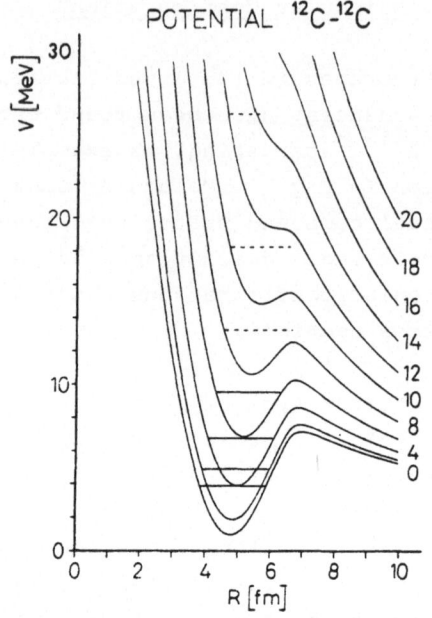

Fig. 2 Sudden $^{12}C-^{12}C$ potential

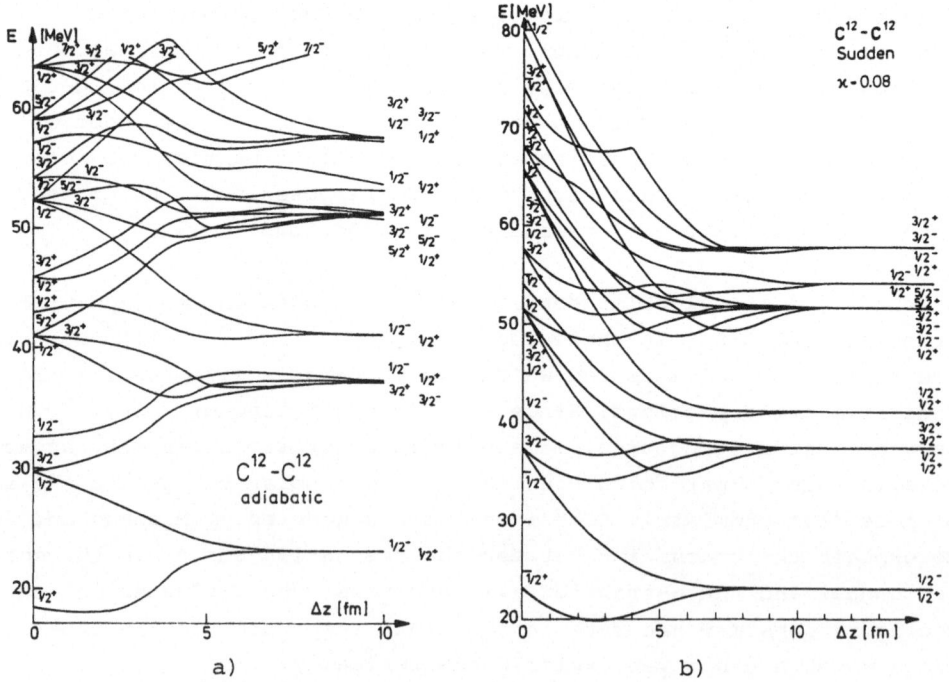

Fig. 3 Adiabatic and sudden TCSM level diagrams
(from Ref. 8)

Molecular resonances measured in the γ-, p- and α-exit channels with
widths of the order of 100 keV proceed probably through special compound states of the fused system. These compound states are produced
over the resonating virtual state in the entrance channel and, therefore, carry the same angular momentum as the gross structure in the
elastic cross section. This explanation is strongly supported by the
measured fusion cross sections which reveal the same gross structures
as the elastic cross sections.

All the above mentioned molecular configurations have in common that
they are excited through the virtual resonances in the elastic channel. For the appearence of resonance structures in the various exit
channels it is necessary that the virtual resonances are not absorbed
into complex compound configurations. Experimental signs for undamped
virtual resonances are the gross structures in the elastic and fusion
cross sections.

Since more and more experimental data give reliable evidence for molecular resonances, we like to remember on the situation before a few
years ago, when the intermediate structures in the cross sections
have been explained as statistical fluctuations by several experimental groups. Therefore, we are aware of the possibility that compound
elastic processes may also produce similar intermediate structures
which are not yet separated from those of molecular origin.

3. Theory of Molecular Resonances

In this section we discuss general problems connected with the excitation of molecular resonances.

3.1 Sudden and Adiabatic Real Potentials: The energies of the molecular resonances are mainly determined by the real nucleus-nucleus
potential. The potential depends on the coordinates describing the
dynamics of the system. These are the coordinate $\vec{r}$ for the relative
motion of the nuclear centers and coordinates for the intrinsic degrees of freedom of the nuclei. For example, we may choose quadrupole
coordinates $\alpha_{2\mu}$ for even-even nuclei with collective low energy spectra, which can be explained by vibration or rotation-vibration models.
The general expression for the real nucleus-nucleus potential may be
written in a rotational-invariant form as follows:

$$W(\vec{r},1,2) = V(r) + \sum_{L,M} Q_{LM}(1,2,r) \, Y_{LM}^{*}(\vartheta,\varphi) \qquad (1)$$

The numbers 1 and 2 abbreviate the whole set of intrinsic coordinates. The potentials are fixed by the condition that they vanish asymptotically.

The potentials in (1) are approximately obtained via the expectation value of the total energy for vanishing relative velocities. A useful method for the calculation of potentials is based on the two-center shell model (TCSM)[4] and the Strutinsky method. Figs. 3 and 13 give examples of level diagrams computed with the symmetric and asymmetric TCSM. Further methods to calculate real potentials can be found in literature in connection with folding potentials, Hartree-Fock calculations and generator-coordinate treatments using realistic nucleon-nucleon potentials.

The microscopic description of the scattering process within the TCSM is founded on the assumption that the nucleons move on molecular orbits. The molecular picture is justified, if the relative velocity of the centers is adiabatically slow compared with the orbiting velocities of the nucleons. In the considered systems, at bombarding energies slightly above the Coulomb barrier, the reaction and orbiting times are of the same order of magnitude, namely about $3 \cdot 10^{-22}$ sec. Therefore, the nucleons follow the time-evolution of the TCSM potential only approximately adiabatically.

The non-adiabaticity of the scattering process leads to velocity-dependent potentials. We distinguish two main types of nucleus-nucleus potentials, namely the adiabatic and sudden potentials[5]. Fig. 4 shows the two kinds of potentials for the case of $^{16}O+^{16}O$. They are defined as follows:

a) The adiabatic potential represents the minimum of energy for a fixed relative distance. Microscopically, the nucleons occupy the energetically lowest single particle levels up to the Fermi level in the level diagram of the adiabatic TCSM (see Fig. 3a). The parameters of the adiabatic TCSM potential are constrained by the condition, that the volume inside the equipotential surface at the nuclear surface is conserved for all relative distances.

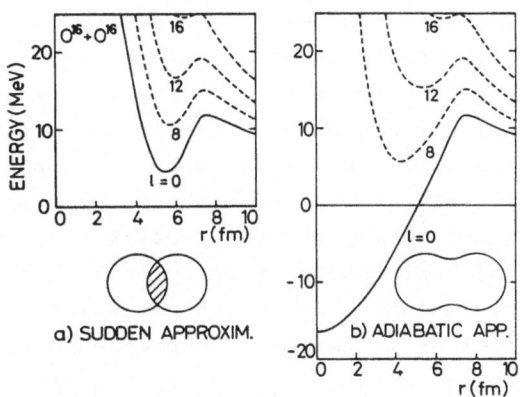

Fig. 4   Sudden and adiabatic potentials

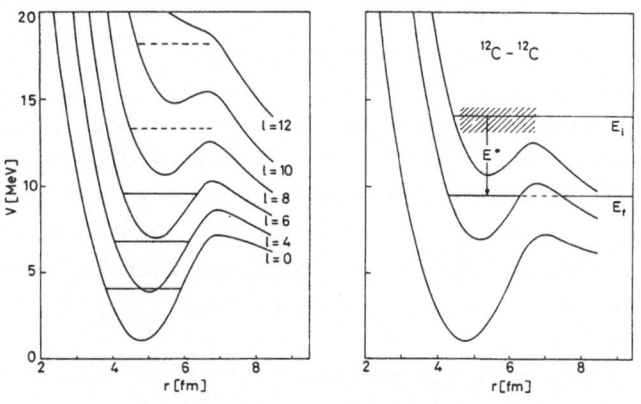

Fig. 5   Double resonance mechanism

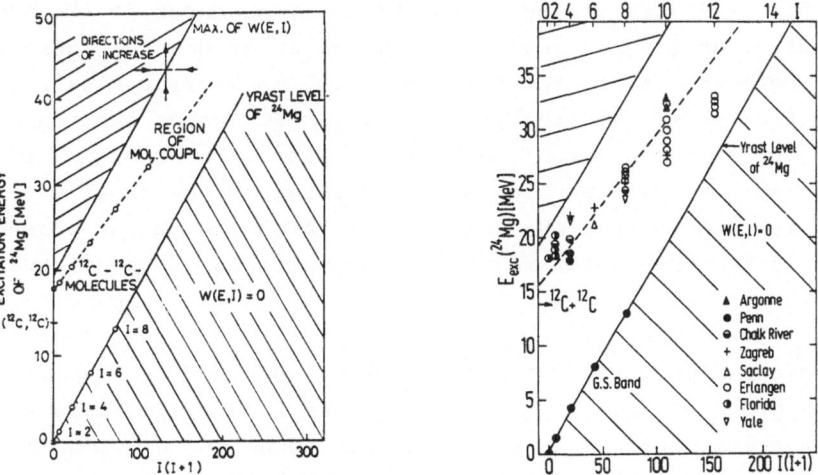

Fig. 6   Quasimolecular resonances in ^{24}Mg(righthand figure from Ref.17)

b) Non-adiabatic potentials, respectively sudden potentials, have a core-behaviour inside the overlap region, since the nucleons become more and more excited with increasing relative velocities. For lower relative velocities one may assume a non-adiabatic occupation of the levels in the adiabatic TCSM. The non-adiabatic occupation probabilities can be approximatively calculated with the velocity-dependent Landau-Zener formula,applied to avoided crossings of the TCSM levels where the excitation of nucleons is strongly enhanced[6]. Such a procedure leads to velocity-dependent heavy ion potentials. At very high relative velocities nuclear matter gets compressed during the reaction. Compression effects can be simulated by keeping the frequency of the harmonic two center oscillator fixed and equal to its asymptotic value ω_∞ for all relative distances[7]. The resulting single particle levels of the TCSM, shown in Fig. 3b, and the corresponding nucleus-nucleus potential rise steeply inside the overlap region.

Both types of potentials are equal near the Coulomb barrier and show the same attractive slope behind it. Therefore, in any case the potentials have quasibound and virtual resonances. The energies of the resonances become fairly independent of the used approximations for higher angular momenta as can be recognized from the dependence of the potentials on angular momentum shown in Fig. 4.

3.2 <u>Double-Resonance Mechanism</u>: The gross structure in elastic cross sections of the $^{12}C-^{12}C$, $^{12}C-^{16}O$ and $^{16}O-^{16}O$ scattering is generated by the virtual resonance states in the quasimolecular potential. A phase shift analysis yields widths of the resonances in the order of 2 MeV and about 2-3 MeV for their separation in energy. Therefore, at bombarding energies not too far above Coulomb barrier,always one surface partial wave resonates in the elastic 90°-cross section. According to our experience for the $^{16}O-^{16}O$ system, the energies of the virtual resonance states lie near the minima of the gross structures (interference effects), whereas the maxima of the fusion cross sections should be observable right at the positions of the virtual resonances.

Intermediate structures with widths of 0.1-0.3 MeV are superimposed over the gross structures (see Fig. 1). We interpret these structures with the indirect excitation of quasibound states. Imanishi[9] was the first who suggested the indirect excitation of the quasibound states in the $^{12}C-^{12}C$ scattering via the inelastic excitation of the

first 2^+-state of ^{12}C at 4.43 MeV, in order to explain the resonance
states near the Coulomb barrier. In Ref. 3 we have introduced the
double resonance mechanism for an enhanced excitation of the quasibound states and found that these states cause the intermediate structures in the experimental cross sections above the Coulomb barrier.

The double resonance model is depicted in Fig. 5 for the ^{12}C+^{12}C
scattering[10]. It can be applied to all direct reactions going from the
elastic channel to inelastic, transfer and fusion channels. For simplification let us assume the same potentials for the relative motion
in the elastic and excited channels. This approximation is given up
in the "coexistence" model which we discuss in one of the next sections. The transition strength between the elastic and a certain inelastic channel is largely enhanced, if the radial partial waves in
both channels resonate simultaneously with their corresponding virtual
and quasibound resonance states (see Fig. 5). The double resonance
effect requires necessarily that the difference in energy and angular
momentum between the virtual and quasibound states can be taken over
by the intrinsic configuration of the nucleus-nucleus system. As
example we mention the collective excitation of the first 2^+-state in
the ^{12}C+^{12}C-case (see Figs. 10 and 11).

The intermediate structures lie inside the width of their feeding virtual resonances, i.e. inside the corresponding gross structures of
the elastic cross section. Their widths are determined by the coupling
strengths between the elastic and excited channels. All intermediate
structures excited via the double resonance effect carry the same
total angular momentum as the gross structure of the elastic channel,
if the nuclei have ground state spin zero.

3.3 Conditions for the Appearence of Molecular Structures: Gross
structures in the elastic 90°-cross section as shown in Fig. 1
occur under the condition that the surface partial wave resonating
with a virtual state is only slightly absorbed. The molecular resonance states belong to very deformed cluster-configurations of the compound system and, therefore, have small widths for their decay into
configurations where the identity of the two clusters, e.g. of the
^{12}C- or ^{16}O-clusters, is completely resolved.

The small absorption of molecular configurations can be understood
when we consider the dependence of the imaginary potential on the

excitation energy and angular momentum[10]. As discussed in Sect. 4 the imaginary potential depends on the level density of those compound states which can be directly excited from the elastic channel. Fig. 6 gives a schematic plot of the resulting imaginary potential over the energy-angular momentum plane. As example we have chosen the $^{12}C-^{12}C$ system. The yrast line is obtained by extrapolating the ground state band of ^{24}Mg. Below this line the imaginary potential is exactly zero, because of the absence of compound states. The upper line marks the maximum of absorption. Between these two lines there exists a region with a small, nearly vanishing imaginary potential.

The range bordering the yrast line is called the molecular channel region with the molecular configurations therein. At a fixed bombarding energy the molecular region represents a window in angular momentum and allows the surface partial waves to enter the nuclear reaction zone. In that range all the molecular resonances lie which have been found by several groups in various reaction channels as pointed out in Fig. 6. The molecular configurations are damped mainly by the coupling between the molecular states having the same angular momentum. We note that the total angular momentum of the nucleus-nucleus system enters the absorption which usually differs from the angular momentum of the relative motion in the excited channels. This has the consequence that all molecular states, excited through the resonating virtual state in the elastic channels, feel the same small absorption as the doorway state.

In conclusion, the appearence of gross structures are experimental signatures for a weak absorption of the virtual resonance states. Molecular intermediate structures can only be observed if the gross structures remain umdamped.

3.4 The Coexistence Model: The dependence of the real potentials on the relative velocity of the nuclei leads to the coexistence model[11] depicted in Fig. 7.The model assumes a coexistence of sudden and adiabatic potentials for the elastic and excited channels.

According to this model the elastic scattering of nuclei is described in the sudden potential since the elastic scattering is a fast process and fulfills the requirements for the sudden approximation. Therefore, the position of the gross structures,i.e. the energies of the virtual resonances,have to be computed with the sudden potential. When the

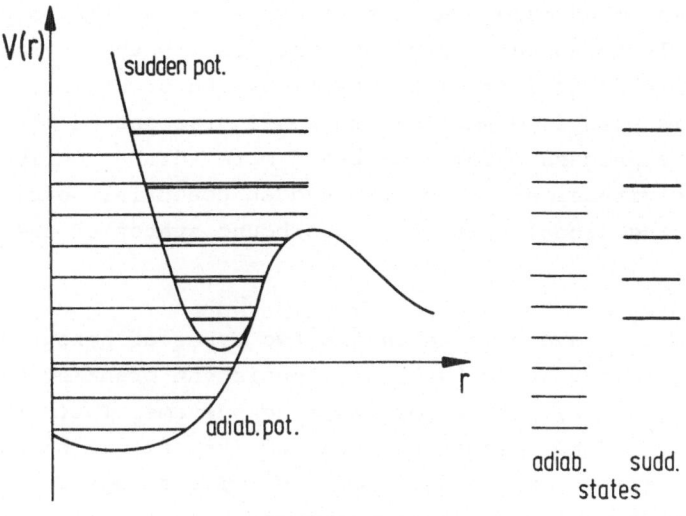

Fig. 7   The coexistence model

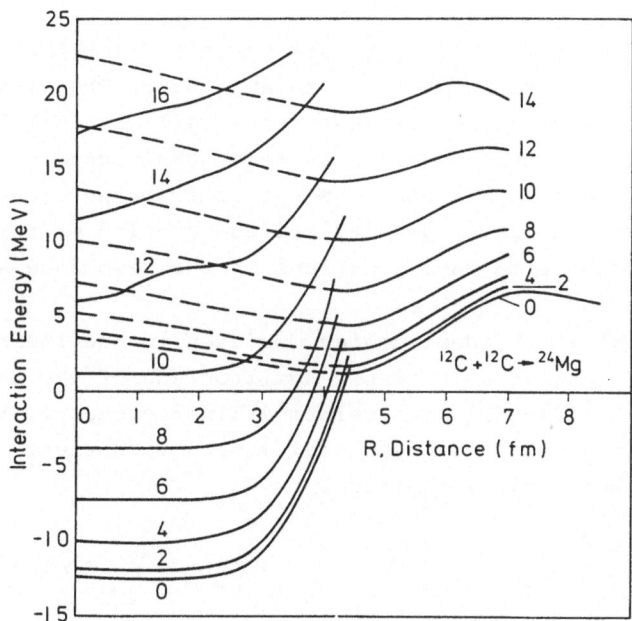

Fig. 8   Adiabatic and sudden potential for $^{12}+^{12}C$
based on a two center shell model calcu-
lation carried out by Chandra and Mosel [12].

long-living quasibound molecular states are excited, the potential
changes gradually to an adiabatic one. Whereas in the entrance channel the cluster-structure is well preserved, the structures in the
excited channels pass into shell-model configurations of the compound
nucleus when the process becomes more and more adiabatically slow. In
conclusion, the virtual states of the sudden potential, excited in
elastic scattering, couple with the quasibound states of the adiabatic potentials.

Chandra and Mosel[12] have calculated the two types of potentials for
the $^{12}C+^{12}C$ collision, using their selfconsistent cranking approximation and the Strutinsky method for rotating systems. Their results are
depicted in Fig. 8. The adiabatic potential (r<4 fm) is based on a
deformed ground state configuration of ^{24}Mg. The sudden potential
(r>4 fm) belongs to the molecular configuration of two ^{12}C-clusters,
which are distorted in the overlap region.

4.   The Elastic Imaginary Potential in the Theory of Molecular Reso-
nances

In this section we derive the imaginary potential for the elastic
scattering of two identical nuclei with spin zero. The idea is used
that the inelastic excitation is enhanced in all states, in which the
radial motion of the excited nuclei is in a quasibound molecular
state. As result we find that the absorption from the elastic channel
is proportional to the level density and the overlap of the elastic
radial wave function with wave functions of the resonance states.

The starting point is the Hamiltonian, written in the laboratory
system. It consists of the intrinsic Hamiltonians of the separated
nuclei, $H_o(1,2)=H_1(1)+H_2(2)$, the relative kinetic energy, the real
nucleus-nucleus potential V(r) and real interaction energy between
the nuclei and their relative motion:

$$H = H_o(1,2) + \frac{\vec{P}^2}{2\mu} + V(r) + \sum_{L,M} Q_{LM}(1,2,r) Y_{LM}^*(\vartheta,\varphi) \tag{2}$$

The numbers 1 and 2 represent the sets of intrinsic coordinates for
which we may choose the collective surface coordinates $\alpha_{2\mu}^{(i)}$.

The stationary scattering problem, $H\psi = E\psi$ is solved with two different types of basis states for the intrinsic degrees of freedom. At
first we take the states of the separated nuclei and, then, states
of molecular type. The later states depend on the relative distance
and are referred to a rotating coordinate system with its z-axis in
the direction of $\vec{r}$.

4.1 <u>Expansion in States of the Separated Nuclei</u>: The wave function
can be decomposed in the radial and orbital wave functions and the
eigenfunctions of the separated nuclei:

$$\Psi_{IM} = \sum_{\ell, J, n} R^I_{\ell J n}(r) \left[i^\ell Y_\ell(\vartheta, \varphi) \otimes \varphi_{Jn}(1,2) \right]^{[I]}_M$$

$$H_o(1,2) \, \varphi^{(1,2)}_{JnM} = E_{Jn} \, \varphi^{(1,2)}_{JnM}$$

(3)

In this ansatz we disregard the antisymmetrization of the relative
coordinate with the intrinsic coordinates. Since we consider identical
nuclei with spin zero, the wave function has to be symmetrized for the
exchange of the nuclei. The radial wave functions fulfill the following system of coupled channel equations:

$$\left(-\frac{\hbar^2}{2\mu r^2} \frac{d}{dr} r^2 \frac{d}{dr} + \frac{\ell(\ell+1)\hbar^2}{2\mu r^2} + V(r) + E_{Jn} - E \right) R^I_{\ell J n}(r)$$

$$= - \sum_{\ell', J', n'} M^I_{\ell J n, \ell' J' n'}(r) \, R^I_{\ell' J' n'}(r)$$

(4)

$$M^I_{\ell J n, \ell' J' n'}(r) = \sum_L i^{\ell'-\ell} (-)^{\ell'+I+J} \begin{Bmatrix} I & J & \ell \\ L & \ell' & J' \end{Bmatrix} (\ell \| Y_L \| \ell')(Jn \| Q_L \| J'n')$$

This system of equations can be used to derive an effective Hamiltonian for the elastic radial wave function R^I_{elast}. For nuclei with
ground state spin zero the elastic channel has spin J=0 and orbital
angular momentum ℓ=I. The inelastic radial functions are approximately
determined by the inhomogeneous equations:

$$\left(-\frac{\hbar^2}{2\mu r^2}\frac{d}{dr}r^2\frac{d}{dr}+\frac{\ell(\ell+1)\hbar^2}{2\mu r^2}+V(r)+\upsilon(r,E_{Jn})\right.$$

$$\left.+E_{Jn}-E\right)R^I_{\ell Jn}=-M^I_{\ell Jn,I00}(r)R^I_{elast}(r) \tag{5}$$

In Eq. (5) we have approximated the effect of all neglected coupling terms by an effective energy-dependent complex potential $v(r,E_{Jn})$ defined as:

$$\upsilon(r,E_{Jn})R^I_{\ell Jn}(r)\approx\sum_{\ell',J',n'\neq elast}M^I_{\ell Jn,\ell'J'n'}(r)R^I_{\ell'J'n'}(r) \tag{6}$$

For solving Eq. (5) we expand the inelastic radial functions in terms of the solutions of the homogeneous equation with outgoing waves (Kapur-Peierls-solutions):

$$\left(-\frac{\hbar^2}{2\mu r^2}\frac{d}{dr}r^2\frac{d}{dr}+\frac{\ell(\ell+1)\hbar^2}{2\mu r^2}+V(r)+\upsilon(r,E^*)\right.$$

$$\left.-E_{\ell\nu}(E^*)\right)\Psi_{\ell\nu}(r,E^*)=0 \;. \tag{7}$$

The real part of the eigenvalues $E_{\ell\nu}$ determines the energies of the resonances in the potential $V(r)+\mathrm{Re}\,v(r,E^*)$ and the imaginary part their widths. Inserting the solution of (5) into Eq. (4) we obtain an integro-differential equation for the elastic radial function:

$$\left(-\frac{\hbar^2}{2\mu r^2}\frac{d}{dr}r^2\frac{d}{dr}+\frac{\ell(\ell+1)\hbar^2}{2\mu r^2}+V(r)-E\right)R^I_{elast}(r)$$

$$=\sum_{\ell',J,n,\nu}M^I_{I00,\ell'Jn}(r)\frac{1}{E_{Jn}+E_{\ell'\nu}(E_{Jn})-E} \tag{8}$$

$$\times\Psi_{\ell'\nu}(r,E_{Jn})\cdot\langle\Psi_{\ell'\nu}(r',E_{Jn})|M^I_{\ell'Jn,I00}(r')|R^I_{elast}(r')\rangle.$$

For the further evaluation of the right hand side of Eq. (8) we introduce the density $\rho_J(E)$ of levels with channel spin J (without M-dege-
neracy) and replace the summation over n by an integration over the
excitation energy. We finally obtain an absorptive expression on the
right hand side of Eq. (8):

$$i\pi \sum_{\ell',J,\nu} \frac{1}{(2I+1)(2J+1)} (I \| Y_J \| \ell')^2 \overline{|(00 \| Q_J \| Jn)|^2}$$

$$\times \rho_J(\epsilon_{\ell'\nu}^*) \; \Psi_{\ell'\nu}(r, \epsilon_{\ell'\nu}^*) \; \langle \Psi_{\ell'\nu}(r', \epsilon_{\ell'\nu}^*) | R_{elast}^I(r') \rangle . \tag{9}$$

In this expression we have assumed that the transition matrix elements
vary so slowly in r that they can be taken out of the integration. The
energies $\epsilon_{\ell\nu}^*$ are defined as the zeros of the equations:

$$\epsilon_{\ell\nu}^* + Re \left(E_{\ell\nu}(\epsilon_{\ell\nu}^*) \right) - E = 0 . \tag{10}$$

Expression (9) describes the absorption from the elastic channel by
the inelastic excitation of compound states with channel spin J and
excitation energy $\epsilon_{\ell\nu}^*$. The excitation energies are discrete, since
the radial motion of the inelastic channels is assumed to be in the
resonance states of the potential $V(r)+Re \; v(r, E^* = \epsilon_{\ell\nu}^*)$. The resonance states, described by $\psi_{\ell\nu}$, play the role of doorway states for the
formation of the compound system. Their width consists of two parts,
namely of the spreading width for the transition into the compound
system, determined by the imaginary part of $v(r, E^*)$, and of the escape
width for the transition through the potential barrier. The absorption
from the elastic channel is proportional to the overlap of the elastic
radial function with the wave functions $\psi_{\ell\nu}$. The summation over ν is
restricted to a very small number of states, because the excitation
energies $\epsilon_{\ell\nu}^*$ have to be larger than some minimum excitation energy
above which it is first justified to apply formulas for statistical
level densities. This minimum excitation energy lies in the order
of 10 MeV in the considered light nuclei. For excitation energies
below the minimum value all channels have to be coupled explicitly
to the elastic channel.

The level density ρ_J is composed of the level densities of the individual nuclei. In a microscopic description of the scattering process the quantities Q_{LM} in Eq. (2) are the multipole terms of the two-body force acting between two nucleons in different nuclei. Such an interaction gives rise to 1p-1h excitations in both nuclei. Therefore, the level density ρ_J is approximately the product of the 1p-1h densities in the individual nuclei. In an actual calculation of ρ_J one has to take care of the symmetry of the states when the nuclei are identical.

A local imaginary potential is obtained when the kinetic energy of the radial motion is neglected in Eq. (4). Such an approximation means an infinite mass for the radial motion in the inelastic channels. It results for the right hand side of Eq. (8):

$$i \, W(r, I, E) \, R_{elast}^{I}(r) \tag{11}$$

$$W(r, I, E) = \pi \sum_{\ell', J'} \frac{(I \| Y_{J'} \| \ell')^2}{(2I+1)(2J'+1)} \overline{|(00 \| Q_{J'} \| J'n')|^2 \rho_{J'}(\epsilon_{\ell'}^*)}$$

$$\epsilon_\ell^*(r) = E - \frac{\ell(\ell+1)\hbar^2}{2\mu r^2} - V(r) - \mathrm{Re}\, v(r, \epsilon_\ell^*(r)) \tag{12}$$

The same formula for W is obtained, when we calculate the transition probability to the excited channels using Fermi's golden rule (see Refs. 10 and 13).

4.2 <u>Expansion in Molecular States</u>: The Hamiltonian H is transformed to the rotating coordinate system[14]:

$$H = H_o(1', 2') - \frac{\hbar^2}{2\mu r} \frac{\partial^2}{\partial r^2} r + \frac{(\vec{I} - \vec{J}')^2}{2\mu r^2} + V(r)$$

$$+ \sum_L Q_{LO}(1', 2', r) \sqrt{\frac{2L+1}{4\pi}} \tag{13}$$

The primes indicate the coordinates referred to the axes of the rotating system. We denote the total angular momentum by $\vec{I}$ and the intrinsic one by $\vec{J}$. In solving the scattering problem we first set up the molecular intrinsic wave functions as eigensolutions of:

$$\left(H_0(1',2') + \sum_L Q_{L0}(1',2',r)\sqrt{\frac{2L+1}{4\pi}} + \frac{\vec{J}_{x'}^2 + \vec{J}_{y'}^2}{2\mu r^2} \right) \varphi_{\lambda K}(1',2',r)$$

$$= \varepsilon_{\lambda K}(r)\,\varphi_{\lambda K}(1',2',r) \tag{14}$$

Here, the quantum number K denotes the projection of angular momentum on the intrinsic z-axis. When the nuclei are separated, the states $\varphi_{\lambda K}$ approach the states φ_{JnK}, defined in Eq. (3). Using the molecular wave functions $\varphi_{\lambda K}$ we expand the scattering solutions as follows (K=integer):

$$\Psi_{IM} = \sum_{K \geq 0, \lambda} R_{K\lambda}^{I}(r)\,\Phi_{IMK\lambda} \tag{15}$$

$$\Phi_{IMK\lambda} = \sqrt{\frac{2I+1}{16\pi^2(1+\delta_{K0})}} \left(D_{MK}^{I\,*}\,\varphi_{\lambda K}(1',2',r) + (-)^{I+J} D_{M-K}^{I\,*}\,\varphi_{\lambda-K}(1',2',r) \right)$$

The wave functions $\Phi_{IMK\lambda}$ are symmetric for the exchange of the nuclei and, therefore, have the same structure as those used in the strong-coupling model (Nilsson-model). The radial wave functions are obtained from the following system of coupled channel equations:

$$\left(-\frac{\hbar^2}{2\mu r}\frac{d^2}{dr^2}r + \varepsilon_{\lambda K}(r) + V(r) + \frac{(I(I+1) - K^2)\hbar^2}{2\mu r^2} \right.$$

$$\left. + \langle \varphi_{\lambda K} | -\frac{\hbar^2}{2\mu}\frac{\partial^2}{\partial r^2} | \varphi_{\lambda K}\rangle - E \right) R_{K\lambda}^{I}(r) =$$

$$-\sum_{K'\lambda' \neq (K,\lambda)} M_{K\lambda, K'\lambda'}^{I}(r)\, R_{K'\lambda'}^{I}(r) \quad . \tag{16}$$

$$M^I_{K\lambda,K'\lambda'}(r) = \langle \Phi_{IMK\lambda} | -\frac{\hbar^2}{2\mu}\frac{\partial^2}{\partial r^2} + \frac{1}{2\mu r^2}(I^+J^- + I^-J^+)|\Phi_{IMK'\lambda'}\rangle$$

$$-\frac{\hbar^2}{\mu}\langle\Phi_{IMK\lambda}|\frac{\partial}{\partial r}|\Phi_{IMK'\lambda'}\rangle\frac{1}{r}\frac{d}{dr}r \tag{17}$$

In contrast to the transition potentials in Eq. (4) the quantities M^I are differential operators. They are easily reduced to matrix elements with the molecular wave functions $\varphi_{\lambda K}$. Before we apply the previous formalism in order to derive the imaginary potential, we average the coupled equations over the molecular intrinsic states. For that we introduce a normalized distribution $\zeta(\varepsilon_{\lambda K}-\varepsilon)$ which is peaked about $\varepsilon=\varepsilon_{\lambda K}$, and define the level densities and the following quantities (see Mshelia et al.[15]):

$$\varrho_K(\varepsilon,r) = \sum_\lambda \zeta(\varepsilon_{\lambda K}(r)-\varepsilon) \tag{18}$$

$$\varrho_K(\varepsilon,r)\begin{cases} R^I_{K\varepsilon}(r) \\ t_{K\varepsilon}(r) \\ M^I_{K\varepsilon,00}(r) \end{cases} = \sum_\lambda \zeta(\varepsilon_{\lambda K}(r)-\varepsilon)\begin{cases} R^I_{K\lambda}(r) \\ \langle\varphi_{\lambda K}|-\frac{\hbar^2}{2\mu}\frac{\partial^2}{\partial r^2}|\varphi_{\lambda K}\rangle \\ M^I_{K\lambda,00}(r) \end{cases}$$

$$\varrho_K(\varepsilon,r)\varrho_{K'}(\varepsilon,r)M^I_{K\varepsilon,K'\varepsilon'}(r) = \sum_{\lambda,\lambda'}\zeta(\varepsilon_{\lambda K}(r)-\varepsilon)\zeta(\varepsilon_{\lambda'K'}(r)-\varepsilon)M^I_{K\lambda,K'\lambda'}(r)$$

In all the above sums the elastic channel ($K=\lambda=0$) has to be excluded. With these definitions we obtain a coupled equation for the radial wave function in the elastic channel, which is only approximately correct[15].

$$\left(-\frac{\hbar^2}{2\mu r}\frac{d^2}{dr^2}r + \varepsilon_{00}(r) + V(r) + \frac{I(I+1)\hbar^2}{2\mu r^2} + \langle\varphi_{00}|-\frac{\hbar^2}{2\mu}\frac{\partial^2}{\partial r^2}|\varphi_{00}\rangle\right.$$

$$\left.-E\right)R^I_{elast}(r) = -\sum_K \int d\varepsilon' \varrho_K(\varepsilon',r)M^I_{00,K\varepsilon'}(r)R^I_{K\varepsilon'}(r) \tag{19}$$

Analogously to Eq. (6) we introduce the approximation:

$$v_{K\epsilon}^I(r) R_{K\epsilon}^I(r) = \sum_{K'} \int d\epsilon' \, \rho_{K'}(\epsilon', r) \, M_{K\epsilon, K'\epsilon'}^I(r) \, R_{K'\epsilon'}^I(r) \qquad (20)$$

By means of exactly the same methods as applied in Sect. 4.1, we finally derive a nonlocal and then a local imaginary expression for the right hand side of Eq. (19). In the local approximation it results:

$$i\pi \sum_{K \geq 0} \rho_K(\epsilon(r), r) \, M_{00, K\epsilon(r)}^I(r) \, M_{K\epsilon(r), 00}^I(r) \, R_{elast}^I(r) \qquad (21)$$

with

$$\epsilon(r) = E - \left(V(r) + \frac{(I(I+1) - K^2)\hbar^2}{2\mu r^2} + t_{K\epsilon}(r) + Re\left(v_{K\epsilon}^I(r)\right) \right)$$

Further research work and numerical calculations with the imaginary potentials have now to be carried out in order to proof these theoretical ideas.

5. Excitation of Quasimolecular Collective States

In this section we discuss the inelastic excitation of identical nuclei with spin zero. Their spectra should be explainable by collective models like vibration or rotation-vibration models. In that case the intrinsic dynamics of the system can be described by surface multipole coordinates $\alpha_{\lambda\mu}^{(i)}$ which define the nuclear surface of the separated nuclei:

$$R^{(i)} = R_0 \left(1 + \sum_{\lambda\mu} \alpha_{\lambda\mu}^{(i)} \, Y_{\lambda\mu}^*(\vartheta_i, \varphi_i) - \frac{1}{4\pi} \sum_{\lambda\mu} |\alpha_{\lambda\mu}^{(i)}|^2 \right) \qquad (22)$$

with i = 1,2

The numbers 1 and 2 refer to nucleus 1 and 2. The asymptotic definition (22) of the nuclear shapes has to be extrapolated into the overlap region of the nuclei as indicated in Fig. 9.

5.1 The Real Potential Energy for Identical Nuclei: In the case of identical nuclei, e.g. $^{12}C + ^{12}C$ and $^{16}O + ^{16}O$, the general expression for the potential is given in the multipole deformation coordinates up to second order as follows (see Eq. (1)):

$$W(\vec{r},1,2) = V(r) + \sum_{L,M} Q_{LM}(1,2,r) Y^*_{LM}(\vartheta,\varphi) \qquad (23)$$

with
$$Q_{LM}(1,2,r) = I_L(r)\left((-)^L \alpha^{(1)}_{LM} + \alpha^{(2)}_{LM}\right)$$

$$+ \sum_{L_1 \leq L_2} \left\{ J_{L_1 L_2 L}(r)\left((-)^L\left[\alpha^{(1)}_{L_1} \otimes \alpha^{(1)}_{L_2}\right]^{[L]}_M + \left[\alpha^{(2)}_{L_1} \otimes \alpha^{(2)}_{L_2}\right]^{[L]}_M\right)\right.$$

$$\left. + K_{L_1 L_2 L}(r)\left((-)^L\left[\alpha^{(1)}_{L_1} \otimes \alpha^{(2)}_{L_2}\right]^{[L]}_M + \left[\alpha^{(2)}_{L_1} \otimes \alpha^{(1)}_{L_2}\right]^{[L]}_M\right)\right\}$$

The matrix elements of Q_L between asymptotic nuclear states can easily be reduced to electric multipole transition probabilities which may be taken from experiment or calculated in the framework of a collective nuclear model. The transition potentials I, J and K can be computed with the folding procedure. This procedure is based on the sudden approximation and assumes an effective nuclear two-body potential of Yukawa-type acting between the nuclear densities. A further commonly used method is a Taylor-expansion of the potential in powers of the multipole coordinates $\alpha_{\lambda\mu}$, which yields the following expression up to second order:

$$W(\vec{r},1,2) = V(r) - R_o \frac{dV}{dr} \sum_{\lambda\mu} \left((-)^\lambda \alpha^{(1)}_{\lambda\mu} Y^*_{\lambda\mu} + \alpha^{(2)}_{\lambda\mu} Y^*_{\lambda\mu}\right)$$

$$+ \frac{R_o^2}{2} \frac{d^2V}{dr^2} \left(\sum_{\lambda\mu}\left((-)^\lambda \alpha^{(1)}_{\lambda\mu} Y^*_{\lambda\mu} + \alpha^{(2)}_{\lambda\mu} Y^*_{\lambda\mu}\right)\right)^2$$

$$+ \frac{R_o}{4\pi} \frac{dV}{dr} \sum_{\lambda\mu}\left(|\alpha^{(1)}_{\lambda\mu}|^2 + |\alpha^{(2)}_{\lambda\mu}|^2\right). \qquad (24)$$

The last term in (24) is usually negligible. We note that the first order transition potential is independent of the multipole order in the above approximation:

$$I_L(r) = -R_o \frac{dV}{dr}. \qquad (25)$$

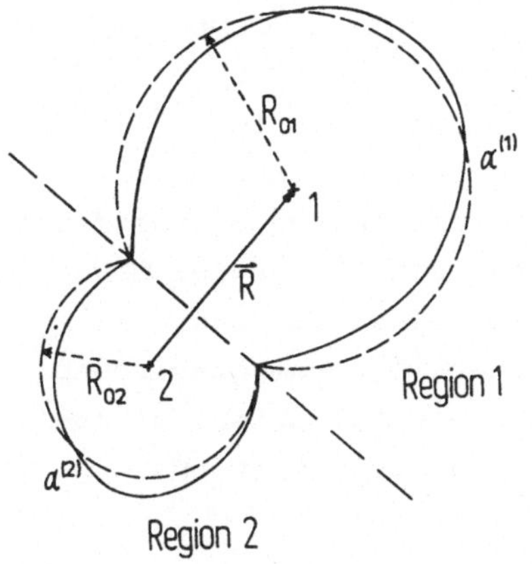

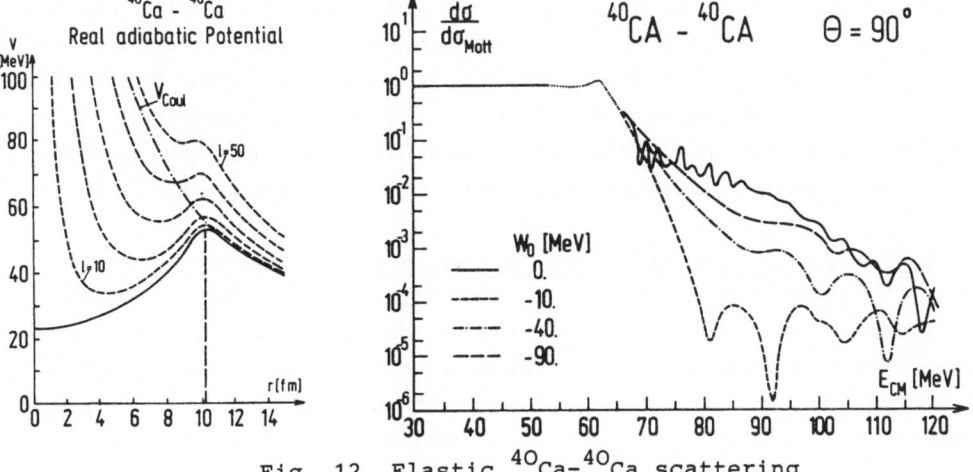

Fig. 9 Definition of the coordinates

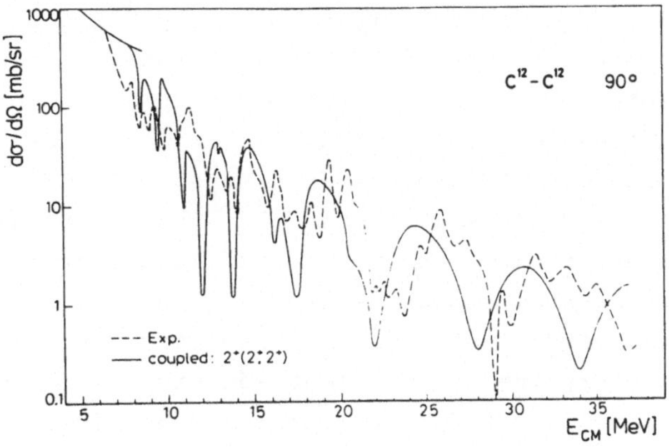

Fig. 10 Elastic ^{12}C-^{12}C scattering

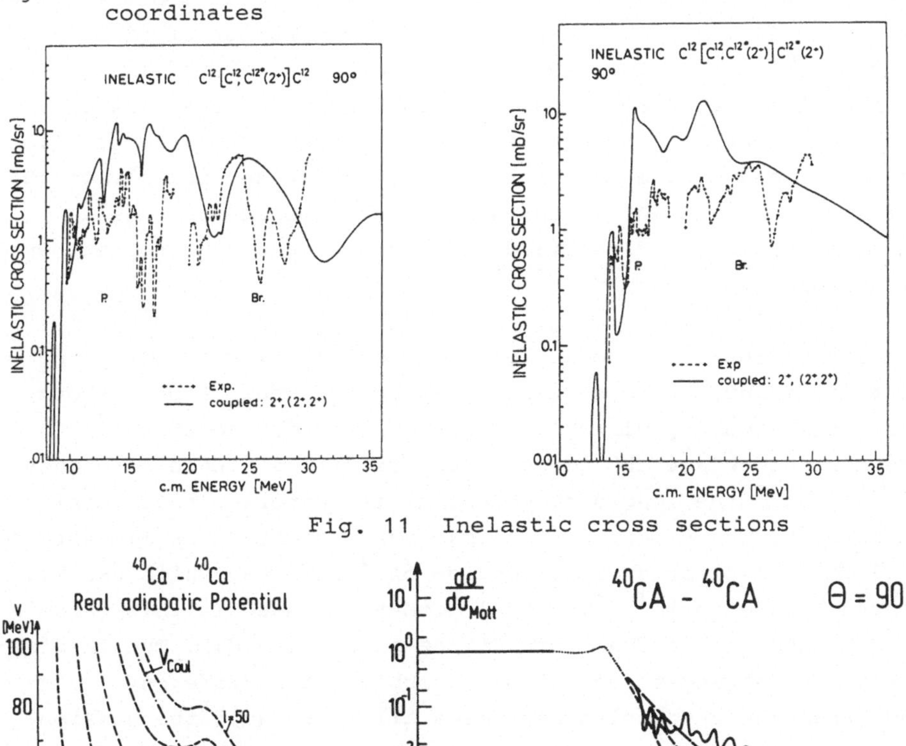

Fig. 11 Inelastic cross sections

Fig. 12 Elastic ^{40}Ca-^{40}Ca scattering

5.2 <u>Inelastic Excitation in the ^{12}C-^{12}C Scattering</u>: As example for
the excitation of collective states we choose the ^{12}C-^{12}C system. In
Ref. 10 we have carried out coupled channel calculations for ^{12}C+^{12}C
which are shown in Figs. 10 and 11. These calculations include the
ground state channel and the single and mutual excitation of the first
2^+-state at 4.43 MeV in ^{12}C. The potentials between the nuclei have
been obtained in an extended liquid drop model[5]. The real nucleus-
nucleus potential is depicted in Fig. 2 together with the quasibound
and virtual states. The absorptive part of the optical potential has
been related to the level density of the compound states in the
framework of the statistical model (see Sect. 4).

The gross and intermediate structures in the calculated cross sections
of Figs. 10 and 11 can be explained with the excitation of virtual and
quasibound resonances in the real ^{12}C-^{12}C potential shown in Fig. 2.
The intermediate structures appear at such bombarding energies at
which the conditions for the double resonance mechanism are fulfilled.
Since the positions of the structures depend quite sensitively on the
real potential, a phase shift analysis of the experimental inelastic
cross sections could yield detailed information about the correct energies of the molecular resonances.

The gross structures in the elastic cross section of Fig. 10 can be
assigned to individual virtual resonances up to about 30 MeV. Above
this energy neighbouring virtual resonances interfer so strongly that
we can no more correlate the gross structure with a single resonance.
We note, that also the imaginary potential may generate additional
gross structures. The effect of the imaginary potential is illustrated
in Fig. 12 for the elastic cross section of ^{40}Ca-^{40}Ca, which has been
investigated by Koennecke[16]. The real potential, used in this calculation, is obtained in the framework of the adiabatic TCSM. The imaginary potential is assumed to be volume-absorbing with different
strengths. When the absorption increases all effects of the quasimolecular real potential become flattened out as shown in Fig. 12.

5.3 <u>Discussion of Inelastic Excitation in the Rotating Coordinate
System</u>: To study the excitation of collective states of identical
nuclei closer, we transform the Hamiltonian to a rotating coordinate
system with its z-axis in the direction of the relative coordinate
$\vec{r}$ (see Fig. 9). The following Hamiltonian results if we restrict the
deformation coordinates to quadrupole surface coordinates only:

$$H = H_o(1,2) - \frac{\hbar^2}{2\mu r^2} \frac{\partial}{\partial r} r^2 \frac{\partial}{\partial r} + \frac{(\vec{I} - \vec{J}_{coll})^2}{2\mu r^2} + V(r) + i W(r)$$

$$+ \sqrt{\frac{5}{2\pi}} I_2(r) \alpha_o^+ + \sum_\mu \left(f_\mu^+(r) |\alpha_\mu^+|^2 + f_\mu^-(r)|\alpha_\mu^-|^2 \right) \qquad (26)$$

with

$$\alpha_\mu^\pm = \frac{1}{\sqrt{2}} \left(\alpha_{2\mu}^{'(1)} \pm \alpha_{2\mu}^{'(2)} \right) \qquad (27)$$

$$f_\mu^\pm(r) = \sum_{L=0,2,4} (-)^\mu (2\mu 2 - \mu | L 0) \sqrt{\frac{2L+1}{4\pi}} \left(J_{22L}(r) \pm K_{22L}(r) \right) \qquad (28)$$

Instead of the quadrupole coordinates $\alpha_{2\mu}^{'(1,2)}$ defined with respect to the rotating coordinate system, we have introduced the coordinates $\alpha_\mu^\pm$. In the overlap region the symmetric (antisymmetric) coordinates $\alpha_\mu^+(\alpha_\mu^-)$ describe quadrupole-type (octupole-type) excitations of the compound system. Therefore, the potential energy functions fulfill the relation $|f_\mu^+| < |f_\mu^-|$ for $r < R_{critical}$. For $r \to \infty$ they vanish: $f_\mu^+(r \to \infty) = f_\mu^-(r \to \infty) = 0$.

The intrinsic excitation of the nuclei can be described by two different sets of states, namely with the states of the separated nuclei and with molecular collective states, which depend on the relative distance. In the following we discuss both methods separately.

a) **Basis of states of separated nuclei**: The coupled channel calculations, presented in the previous section, have been carried out in the basis of the ^{12}C-states. In this basis the nuclei become mainly excited via the potential term proportional to α_o^+ in (26). The term excites ß-vibrations with K=0 along the intrinsic z-axis symmetrically in both nuclei. Therefore, in the rotating coordinate system the potential, linear in the quadrupole coordinates, causes the excitation of a single collective mode (ß-vibration) only. This simplified picture for the excitation of collective states is perturbed by the Coriolis-interaction in Eq. (26), proportional to $\vec{I} \cdot \vec{J}_{coll}$, which mixes the K-quantum numbers of the collective states in the rotating coordinate system.

b) <u>Basis of molecular collective states</u>: Out of the Hamiltonian (26) we isolate an intrinsic part H_{intr}, which describes the dependence of the collective modes on the relative distance between the nuclei. The intrinsic Hamiltonian consists of H_0 and the last two potentials in Eq. (26). For simplicity let us assume that the separated nuclei have an harmonic vibrator spectrum, i.e.

$$H_0(1,2) = \frac{B_2}{2} \sum_\mu \left(|\pi_\mu^+|^2 + |\pi_\mu^-|^2 \right) + \frac{C_2}{2} \sum_\mu \left(|\alpha_\mu^+|^2 + |\alpha_\mu^-|^2 \right) \quad (29)$$

Then the eigenstates of the intrinsic Hamiltonian can be analytically obtained as states of harmonic vibrators with r-dependent frequencies. For that purpose we rewrite H_{intr}:

$$H_{intr} = \frac{B_2}{2} \left\{ \sum_\mu \left(|\pi_\mu^+|^2 + \omega_\mu^{+2}(r) \, |\alpha_\mu^+ - \beta_0(r)\, \delta_{\mu 0}|^2 \right) \right.$$

$$\left. + \sum_\mu \left(|\pi_\mu^-|^2 + \omega_\mu^{-2}(r) \, |\alpha_\mu^-|^2 \right) \right\} - \frac{C_0^+}{2} \beta_0^2 \quad (30)$$

with the abbreviations

$$\omega_\mu^{\pm 2} = C_\mu^\pm / B_2 \quad ; \qquad C_\mu^\pm = C_2 + 2\, f_\mu^\pm(r) \, ; \quad (31)$$

$$\beta_0(r) = -\sqrt{\frac{5}{2\pi}} \, I_2(r) / C_0^+(r) \quad (32)$$

The interpretation of Eq. (3) is quite obvious: The excitation energies $\hbar\omega_\mu^\pm(r)$ of the molecular collective modes depend on r and $|\mu|$ like the molecular single-particle levels in the TCSM. For large separations they pass over into the asymptotic vibrator energy of the nuclei. The potential linear in α_0^+ in Eq. (26) gives rise to a static r-dependent deformation β_0. The static deformation belongs to the shape of the nuclear system with minimum energy; i.e. it describes the change of the nuclear shape during an adiabatic collision. Whereas sudden collisions are best described with states of the separated nuclei, the molecular states of collective or single-particle nature are most appropiate for all adiabatic situations.

The Taylor-expansion (24) yields the following expressions for the constants $c_\mu^\pm$ and β_0, where we have disregarded the monopole correction:

$$C_\mu^+ = C_2 + \delta_{\mu 0} \frac{5}{2\pi} R_0^2 \, d^2V/dr^2 \; ; \; C_\mu^- = C_2 \; ;$$

$$\beta_0 = \sqrt{\frac{5}{2\pi}} \, R_0 \, \frac{dV}{dr} \Big/ C_0^+ (r)$$

(33)

According to Eq. (33) the excitation energy $\hbar\omega_0^+$ is lowered because of $d^2V/dr^2 < 0$ at the barrier. Therefore, the symmetric β-vibrations are energetically favoured in the molecular basis.

Cindro[17] has suggested that the quasimolecular resonances shown in Fig. 6 can be interpreted as states of a very deformed ^{24}Mg-system and, therefore, may be classified according to the quantum numbers of the rotation-vibration model. With the Hamiltonian (26) we are able to describe a continuous transition from the rotation-vibration states of the compound nucleus to the quadrupole degrees of freedom of the nuclei. The five degrees of freedom of the rotation-vibration model are described by $\vec{r}$ and α_μ^+. We assume that the z-axis of the rotating coordinate system coincides with one of the principal axes of the compound system. Therefore, all excitations, described by α_1^+ and α_{-1}^+, have to be energetically raised by an additional potential. Both the α_0^+-vibration and the radial relative motion pass over into the β-vibration of the compound system.

6. Molecular Single-Particle Motion

As already stated, one assumes in the TCSM that the nucleons occupy molecular single-particle states during the collision. Molecular single-particle states are not yet well established in heavy ion collisions. Recently molecular wavefunctions were used to describe polarization effects in proton transfer reactions[18].

Specific signatures of molecular single-particle states should show up in inelastic and transfer cross sections. Especially the study of crossings of molecular levels can become an important tool for detecting molecular single-particle effects in heavy-ion collisions. At points of level crossings the excitation of nucleons becomes enhanced as shown by Fano and Lichten for the analogous excitation of electrons

in atomic collisions. Fig. 13 presents the single-particle levels calculated by Park[19] for the neutron-transfer in the reaction $^{13}C+^{16}O \rightarrow$
$^{12}C+^{17}O$. The 1p1/2-level occupied by the neutron in ^{13}C has an avoided
crossing (dashed lines in Fig. 13) with a second $\Omega=1/2$-level in the
overlap region. This latter level ends asymptotically in the 1d5/2-
and 2s1/2-levels of ^{17}O after a further crossing with a third $\Omega=1/2$-
level. The level-scheme demonstrates the importance of level crossings
for the selection of certain final states.

The scattering of nuclei which can be decomposed into cores and extra-
particles should reveal specific molecular single-particle effects.
The extra-nucleons move on molecular orbits in the mean field of all
the other nucleons. The mean field is best described with the potential
of the TCSM. Like the Nilsson-model, the two-center shell model is
referred to a body-fixed coordinate system in which the centers lie
on the z-axis. Therefore, the single particle motion has to be transformed to the rotating body-fixed coordinate system with its z-axis
defined by the direction of the relative coordinate $\vec{r}$ [14]. The relative
coordinate is set equal to the distance between the centers of TCSM
potential. The transformation of the Hamiltonian to the rotating coordinate system leads to the following expression [20]:

$$H = -\frac{\hbar^2}{2\mu r^2}\left(\frac{\partial}{\partial r}+D\right)r^2\left(\frac{\partial}{\partial r}+D\right) + \frac{(\vec{I}-\vec{J})^2}{2\mu r^2}$$

$$+ U_{C_1 C_2}(r) + \sum_{i=1}^{N} h_{TCSM}\left(\vec{P}'_{icm}, \vec{r}'_{icm}, \vec{S}'_i, r\right) \tag{34}$$

with the abbreviation:

$$D = \frac{1}{A}\left(A_2 \sum_{i=1}^{N_1} \frac{\partial}{\partial z'_{icm}} - A_1 \sum_{i=N_1+1}^{N} \frac{\partial}{\partial z'_{icm}}\right) \tag{35}$$

Here, we have assumed that nucleus 1(2) has C_1 (C_2) core-nucleons
and N_1 (N_2) extra-nucleons with $A_1=C_1+N_1$ ($A_2=C_2+N_2$) (see Fig. 14). The
total angular momentum is denoted by $\vec{I}$, the intrinsic angular momentum
of the extra-particles by $\vec{J}$ and the complex core-core potential by
$U_{C_1 C_2}(r)$.

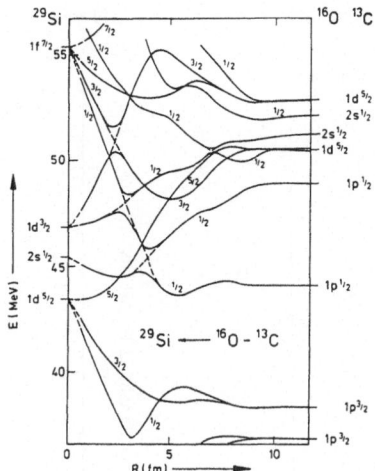

Fig. 13   Adiabatic level diagram of $^{13}C+^{16}O$

$$A_1 = N_1 + C_1 \qquad\qquad A_2 = N_2 + C_2$$

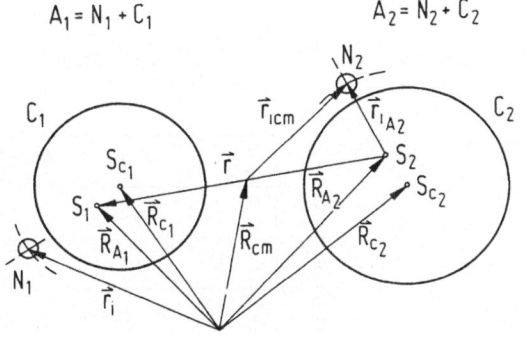

Fig. 14   Definition of coordinates in the dynamic TCSM

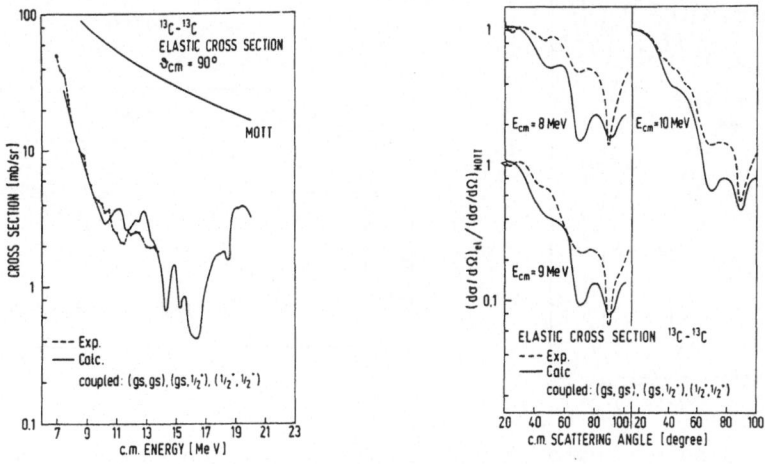

Fig. 15   Elastic $^{13}C-^{13}C$ scattering

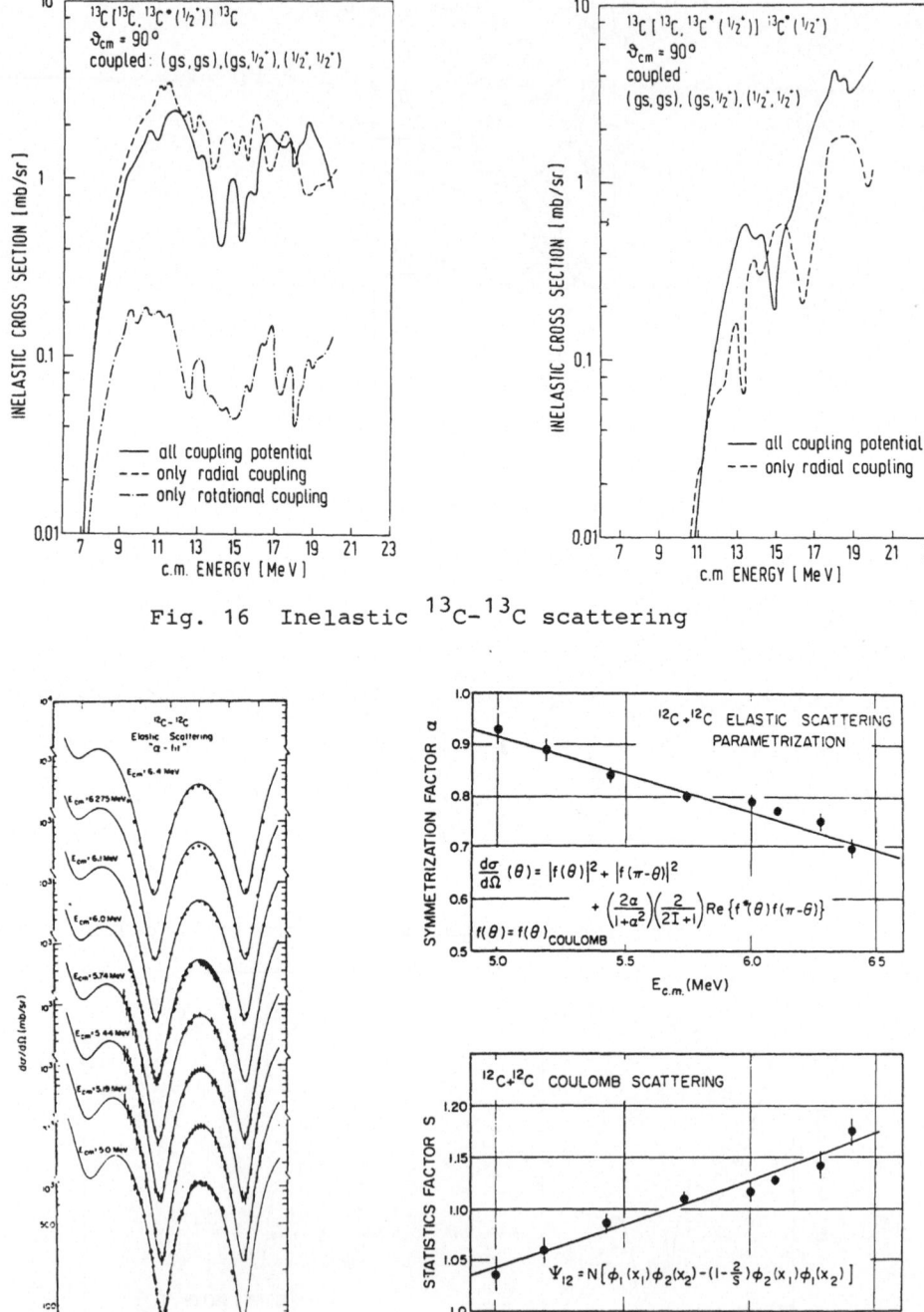

Fig. 16 Inelastic ^{13}C-^{13}C scattering

Fig. 17 Single parameter analysis of precision elastic scattering
in the Coulomb barrier region (from Ref. 23)

Terlecki [20] has carried out coupled channel calculations for the
^{13}C-^{13}C system using the Hamiltonian (34). In the ^{13}C-^{13}C scattering
two loosely-bound neutrons move in the potential generated by two
^{13}C-nuclei. In the coupled channel calculations the single and mutual
excitation of the first 1/2$^+$-state in ^{13}C at 3.09 MeV have been taken
into account. Figs. 15 and 16 show the obtained elastic and inelastic
cross sections. The elastic cross section is compared with the experimental data of Helb and Voit[21]. In Fig. 16 we investigate the influence of the various coupling terms of the kinetic energy in Eq. (34).
The Coriolis-coupling plays only a minor role. The radial coupling
($\sim\partial/\partial r$) operates on the r-dependent part of the molecular single-
particle wave functions and grows proportionally with the relative
volocity of the nuclei. As a future extension of this work we intend
to couple the transfer channels of the extra-particles leading to
^{12}C+^{14}C and the inelastic channels for the ^{12}C-core excitation to the
above considered channels.

7. Application of Parastatistics to Heavy Ion Scattering

The scattering of nuclei with integer spin below the Coulomb barrier
is described by a symmetrized wave function, reflecting the fact that
those nuclei behave like bosons. However, at higher incident energies
when the nuclear densities overlap in the reaction zone, we are deal-
ing with a system of fermions which obey the Pauli principle. This can
be effectively attributed to a loss of bosonity. Fink, Müller and Greiner[22] characterized the loss of bosonity by parastatistics of the
order s, where s is a continuous parameter. In this section we sketch
the application of the parastatistics for the description of the Coulomb scattering of identical nuclei.

According to the theory of parastatistics the relative wave function
of two nuclei with spin zero and obeying parabose-statistics can be
written as:

$$\psi(\vec{r}) = \varphi(\vec{r}) + \left(\frac{2}{s} - 1\right) \varphi(-\vec{r}) \tag{36}$$

The parameter s allows a continuous transtion from the Bose-statistics
(s=1) to the Fermi-statistics (s→∞) by varying s in the interval s $\geq$1.
The wave function (36) consists of a mixture of a symmetrized and antisymmetrized part:

$$\psi(\vec{r}) = \frac{1}{s}\left(\varphi(\vec{r}) + \varphi(-\vec{r})\right) + \left(1 - \frac{1}{s}\right)\left(\varphi(\vec{r}) - \varphi(-\vec{r})\right) \tag{37}$$

With Coulomb wave functions in Eq. (36) the following cross section results ($\alpha = \frac{2}{s} - 1$):

$$\frac{d\sigma}{d\Omega} = \left(\frac{e^2 Z^2}{4E}\right)^2 \left(\frac{1}{\sin^4 \vartheta/2} + \frac{1}{\cos^4 \vartheta/2} \right.$$

$$\left. + \frac{4\alpha}{1+\alpha^2} \cdot \frac{1}{\sin^2 \vartheta/2 \cdot \cos^2 \vartheta/2} \cdot \cos\left(\eta \ln \tan^2 \frac{\vartheta}{2}\right) \right). \tag{38}$$

This cross section deviates most in the minima and maxima from the usual Mott-cross section characterized by $\alpha = \pm 1$. Recently the Yale group[23] carried out very precise measurements of the elastic cross sections of the $^{12}C-^{12}C$ system in the vicinity of the Coulomb barrier (see Fig. 17). They found deviations from the Mott scattering which can be described by the parameter α depending linearly on the incident energy as shown in Fig. 17. This result points out that, in first order, the parastatistics parameter s can be used to simulate effects arising from nuclear contributions to the potential, and from the internal structure and Coulomb excitation of the nuclei. In a next step, experiments with fermion systems like the $^{13}C-^{13}C$ system should be carried out to test the applicability of the parafermi-statistics.

References:

1 D.A. Bromley, J.A. Kuehner and E. Almquist, Phys. Rev. Lett. $\underline{4}$ (1960) 365; Phys. Rev. $\underline{123}$ (1961) 878

2 Proceedings of the International Conference on Resonances in Heavy-Ion Reactions, Hvar (Yugoslavia, 1977), ed. by N. Cindro, North-Holland Publ. Co., Amsterdam 1978;
   H. Feshbach, in Proceedings of the European Conference with Heavy Ions, Caen 1976, J. Phys. $\underline{37}$ (1976) C5-177
   A. Richter and C. Toepffer, in "Heavy Ion Collisions", ed. by R. Bock, North Holland Publ. Comp.
   N. Cindro, in Nuclear Spectroscopy and Nuclear Reactions with Heavy Ions, Proceedings of the Int. School of Physics "Enrico Fermi", Course LXII, edited by H. Faraggi and R.A. Ricci (North Holland, Amsterdam 1976) p 271
   P. Taras, in Proceedings of the 3rd. Int. Conference on Clustering Aspects of Nuclear Reactions, Winnipeg (Canada), 1978; W. Scheid and W. Greiner, ibid.
   W. Greiner, in Proceedings of the Int. Conference on "Dynamical Properties of Heavy-Ion Reactions", Johannesburg 1978; R. Stokstad, ibid.

3 W. Scheid, W. Greiner and R. Lemmer, Phys. Rev. Lett. $\underline{25}$ (1970) 176

4 P. Holzer, U. Mosel and W. Greiner, Nucl. Phys. $\underline{A138}$ (1969) 241; D. Scharnweber, U. Mosel and W. Greiner, Nucl. Phys. $\underline{A164}$ (1971) 257; J. Maruhn and W. Greiner, Z. Physik $\underline{251}$ (1972) 431

5 W. Scheid, R. Ligensa and W. Greiner, Phys. Rev. Lett. $\underline{21}$ (1968) 1479; W. Scheid and W. Greiner, Z.Phys. $\underline{226}$ (1969) 364

6 D. Glas and U. Mosel, Phys. Lett. $\underline{49B}$ (1974) 301

7 K. Pruess and W. Greiner, Phys. Lett. $\underline{33B}$ (1970) 197

8 T. Morović and W. Greiner, Z. Naturforsch. $\underline{31a}$ (1976) 327

9 B. Imanishi, Phys. Lett. $\underline{27B}$ (1968) 267; Nucl. Phys. $\underline{A125}$ (1969) 33; see also Y. Abe, in Proceedings of the Conference on Resonances (Ref. 2), p. 211

10 H.J. Fink, W. Scheid and W. Greiner, Nucl. Phys. $\underline{A188}$ (1972) 259

11 W. Greiner et al., in Proceedings of the Conference on Resonances (Ref. 2), p 109

12 H. Chandra and U. Mosel, Nucl. Phys. $\underline{A298}$ (1978) 151

13 G. Helling, W. Scheid and W. Greiner, Phys. Lett. $\underline{36B}$ (1971) 64

14 J.Y. Park, W. Scheid and W. Greiner, Phys. Rev. $\underline{C6}$ (1972) 1565

15 E. Mshelia, W. Scheid and W. Greiner, Nuovo Cimento $\underline{30A}$ (1975) 589

16 R. Koennecke, Diploma Thesis, University of Frankfurt, 1977 and to be published

17 N. Cindro et al., Phys. Rev. Lett. $\underline{39}$ (1977) 1135; N. Cindro and B. Fernandez, in Proceedings of the Conference on Resonances (see Ref. 2), p 417

18 K. Pruess, Nucl. Phys. $\underline{A278}$ (1977) 124; K. Pruess and P. Lichtner, Nucl. Phys. $\underline{A291}$ (1977) 475

19 J.Y. Park, to be published

20 G. Terlecki, W. Scheid, H.J. Fink and W. Greiner, Phys. Rev. $\underline{C18}$ (1978) 265

21 H.D. Helb, P. Dück, G. Hartmann, G. Ischenko, F. Siller and H. Voit, Nucl. Phys. $\underline{A206}$ (1973) 385

22 H.J. Fink, B. Müller and W. Greiner, J. Phys. G (Nucl. Phys.) $\underline{3}$ (1977) 1119

23 D.A. Bromley, in Proceedings of the Int. Conf. on Resonances (Ref. 2), p. 3

QUANTUM CORRECTIONS TO OPTICAL POTENTIALS

P.-G. Reinhard*, Inst. f. Kernphysik, Univ. Mainz, W. Germany

and

K. Goeke, Inst. f. Kernphysik, Kernforschungsanlage Jülich
and Physik-Department, Univ. Bonn, W. Germany

1. Introduction

Often heavy ion (HI) scattering is
described in terms of one single coordinate q, the relative distance between the centers; i.e., one replaces
the complexity of the mutual interactions by one interaction between
the bulks and hopes to end up with a
fairly, simple Hamiltonian for the
motion described by this one degree-
of-freedom. Thus, we are faced with
a similar problem as accounted in
the microscopic description of col-

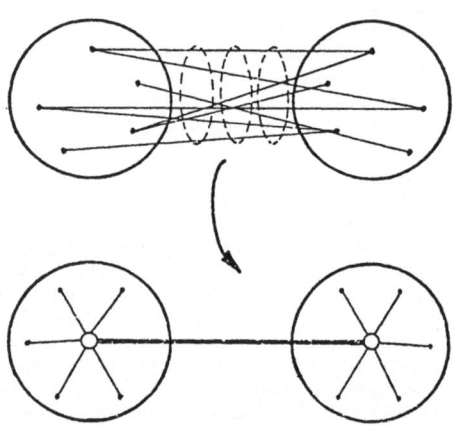

lective motion. Other examples are fission or surface vibrations where motion is
described by some shape parameters, like deformation, necking etc. (In fact, the
scattering description has to merge into the latter example if the nuclei approach
very closely, because the center separation becomes an inappropriate coordinate
there). It is the aim of the following talk to discuss some implications of this
collective approach, as used e.g., by selecting the center separation as a representative coordinate for the whole motion. Thereby, we concentrate on the problem
of systematically recovering the collective Hamiltonian, in particular, its static
part, which is, e.g., in HI-scattering the optical potential.

2. The deformation-path method

The ideal of a microscopic collective theory would be to have an explicit point
transformation from the many-particle coordinates x_i to one (or few) collective
operator $Q = Q(x_i)$ (in h.i. scattering, e.g., it becomes asymptotically the center
separation) and remaining intrinsic coordinates ξ_j and to transform the Hamiltonian
accordingly in order to obtain the explicit $H_{coll}(Q, \partial_Q)$, (which contains e.g., the
scattering potential as the ∂_Q^0 term), i.e.

$$\{ x_i \} \quad \rightarrow \quad Q, \{ \xi_j \}$$
$$\hat{H}(x_i, \partial_{x_i}) \quad \rightarrow \quad \hat{H}_{coll}(Q, \partial_Q) + \hat{H}_{intr.}(\xi, \partial_\xi) + \hat{H}_{coupl.} \tag{1}$$

This direct attack, however, is prohibitively complicated. One, therefore, prefers to extract $\hat{H}_{coll}$ in an indirect manner, having the idea of an explicit point trans-formation as a background.

As the tool for recovering $\hat{H}_{coll}$, one introduces the collective path $|q\rangle$. It is a set of wave functions in many body space labelled by the parameter q, which is to become the representative coordinate. In HI-scattering, e.g., one often uses the cluster basis

$$\Phi_q(x_1,\ldots x_A) = \alpha \{ \prod_{i_1}^{A_1} \psi_{i_1} (x_{i_1} - q/2) \cdot \prod_{i_2}^{A_2} \psi_{i_2} (x_{i_2} + q/2) \} \tag{2}$$

where q then describes the relative distance of the clusters. Once having the path, the collective (optical) potential is evaluated as the expectation value over the microscopic Hamiltonian, i.e., $\mathcal{V}(q) = \langle q| \hat{H}(x_i, \partial_{x_i})|q\rangle$.

In order to obtain the collective kinetic energy, we have to extend the static path $|q\rangle$ by adding a dependence on the collective momentum p. This leads to a dy-namical collective path $|q, p\rangle$. For the example of a scattering path, see eq. (2), a simple extension consists in adding plane wave currents $\exp(i x_{i_1} p/2)$, and $\exp(-i x_{i_2} p/2)$, respectively. A more refined choice is to evaluate[1] the dynamic features[2] of the path as the response to a "Cranking"-force $-q(i \partial_q)$. With the dynamic path given, we obtain the collective Hamiltonian as

$$\mathcal{H}(q, p) = \langle q, p | \hat{H}(x_i, \partial_{x_i}) | q, p \rangle . \tag{3}$$

This is a classical Hamiltonian function. It is still to be requantized in order to obtain the desired operator $\hat{H}_{coll}$.

The above sketched procedure, of course, is only one particular way of treating collective dynamics microscopically, as e.g., in h.i.-scattering. But it is the most widely used method, appearing in various forms and approximations, which em-brace the evaluation of potential-energy surfaces in a deformed shell model, fold-ing potential techniques, different stages of cranking, the vibrating-potential model and finally the adiabatic time-dependent Hartree-Fock theory (ATDHF)[1,2]. Recent theoretical developments have achieved an unification of all these models under the most general framework of ATDHF[2]. (Even the equivalence with the rather different Generator-coordinate method has been established[3]). A consistent theory of this type has to treat following problems:

1. Optimal choice of a collective path $|q\rangle$.
2. Reconstruction of $\hat{H}_{coll}$, using $|q\rangle$.
3. Interpretation of the collective wave function obtained with $\hat{H}_{coll}$.
4. Existence and limits of validity of a collective mode.
5. Relation to the Generator-coordinate method (GCM).

There has been much progress in the last years in studying the problems of this list by means of theories like ATDHF or a generalized Generator-coordinate method. We cannot outline all the details here. The main emphasis in this talk is on problem 2. We just assume the validity of a collective description (problem 4.) and we assume a properly chosen path $|q>$, or $|q, p>$, to be given (problem 1.) and ask then how to recover the quantum mechanical $\hat{H}_{coll}$. The point will be that $\mathcal{H}(q, p) = <q, p|H|q, p>$ is the classical limit of H_{coll} and thus, "covered" with some $\hbar^2$, $\hbar^4$,... terms. We have to trace these terms and to eliminate them. This leads to a systematic reconstruction of $\hat{H}_{coll}$.

3. The classical limit and its reversion

As we have seen above, the method starts with a (dynamical) collective path $|q, p>$ and leads to a classical Hamiltonian function $\mathcal{H}(q, p)$, see eq. (3), which determines the collective motion as an explicitly time-dependent process, i.e. $|\psi(t)> \sim |q(t), p(t)>$, where $q(t)$ and $p(t)$ are given by the classical equation-of-motion $\dot{q} = \partial_p \mathcal{H}$ and $\dot{p} = - \partial_q \mathcal{H}$. The important step is now, to realize that the states $|q, p>$ of the path represent collective wave packets; they have an average position $<q, p|\hat{Q}|q, p> = q$ and momentum $<q, p|\hat{P}|q, p> = p$ and finite spreading width in both, $<|(\hat{Q} - q)^2|> \neq 0$ and $<|(\hat{P} - p)^2|> \neq 0$. For the example of translations, using the center-of-mass as collective coordinate and a harmonic oscillator basis, the wave packet character can explicitly be displayed

$$
\Phi_q(x_1, \ldots x_A) = \mathcal{A} \{ \prod_i^A \psi_i(x_i - q) \}
$$

$$
= \exp(- \frac{\lambda}{2} (\frac{1}{A} \sum x_i - q)^2) \Phi_{intr.} \quad .
$$

(4)

In the general case, we can think of the wave packet being written in collective space,

$$
< Q | q, p > \propto \exp(ipQ/\hbar) \exp(- \frac{\lambda}{2} (Q - q)^2) ,
$$

(5)

where $\lambda = 2<P^2>/\hbar^2 \approx 1/2<Q^2>$ where $P = i\hbar \partial_q$. The ansatz (5) is appropriate up to order $\hbar^2$. But this is no principal restriction. The expansion can be extended systematically to higher orders by multiplying a polynomial in $(\hat{Q} - q)^n$ and fitting the coefficients with the higher moments $<(\hat{Q} - q)^n>$ and $<(\hat{P} - p)^n>$[2,5].

Altogether, we understand now that the $|q(t), p(t)>$ are moving wave packets, which are used to explore the collective dynamics, and the $\mathcal{H}(q, p)$ is the classical wave packet limit (not the WKB limit as usually discussed in quantum mechanics textbooks). This guarantees that $\mathcal{H}(q, p)$ reproduces H_{coll} $(Q \to q, P \to p)$ in order $\hbar^0$, but nothing more.

We are tracing the $\hbar^2$-terms by performing an analog calculation in Q-space, writing explicitly

$$H_{coll} = : \frac{P^2}{2M} : + V(Q) \tag{6}$$

$$: \frac{P^2}{2M} : = \frac{1}{4} \left(P^2 \frac{1}{2M} + 2P \frac{1}{2M} P + \frac{1}{2M} P^2 \right) ,$$

and integrating over the wave packet (5). This yields for the collective potential (the kinetic energy remains unchanged in order $\hbar^2$)

$$\mathcal{V}(q) = < q \mid \hat{H}_{coll} \mid q >$$

$$= \int dQ < q \mid Q > \left(: \frac{\hat{P}^2}{2M} : + V(Q) \right) < Q \mid q > \tag{7}$$

$$= V(q) + \frac{<Q^2>}{2} \partial_q^2 V + \frac{\hbar^2 <\partial_q^2>}{2M} + O(\hbar^4).$$

In the $\hbar^2$-terms we recognize the potential and kinetic zero-point energies (ZPE) of the collective wave packet[5]. In the spirit of a classical limit we have $\mathcal{V} = V$ since $\hbar \to 0$. But if one aims to obtain a quantized $\hat{H}_{coll}$, one has to look for the size and effects of the $\hbar^n$-terms and, if necessary, to remove it from $\mathcal{V}(q)$.

At this stage, we want to point out that the usual quantization-question, viz. the $\hat{P}$-ordering, has dissolved into the problem of ZPE subtraction. The form of the ZPE in eq. (7) is a consequence of the form of $: \hat{P}^2/2M :$ used in eq. (6). This particular form, however, is by no means necessary. One can choose any other $\hat{P}$-ordering, but then one has to consider a changed form for the ZPE. One only has to take care to treat $\hat{P}$-ordering and ZPE consistently. We furthermore see that finding out $\hat{H}_{coll}$ from $\mathcal{K}$ is more than simple "requantization". The term "reconstruction" describes things better. To summarize the procedure, we draw following schematic diagram:

	A-SPACE	Q-SPACE
(Assume given)	$\hat{H}(x_i, \partial_{x_i})$	$\hat{H}_{coll} = : \frac{\hat{P}^2}{2M} : + V(Q)$
introduce wave packet	$\phi_{q,p}(x_i)$	$\phi_{q,p}(Q)$
classical Hamiltonian	$<q,p\mid\hat{H}(x_i, \partial_{x_i})\mid q,p>_A = \mathcal{K}(q,p) \;=\; <q,p\mid\hat{H}_{coll}\mid q,p>_Q$	
trace folding terms		$= \frac{p^2}{2M} + V + \frac{\Delta Q^2}{2} V'' + \frac{1}{8M\Delta Q^2} + \cdots$
identify the moments	$<q\mid(\hat{Q}-q)^2\mid q>_A \quad = \quad \Delta Q^2 \;=\; <q\mid(\hat{Q}-q)^2\mid q>_Q$	
unfold		$V = \mathcal{V} - \frac{\Delta Q^2}{2}\mathcal{V}'' - \frac{1}{8\Delta Q^2 M}$

Here, we see explicitly the formal technique: Every step in A-space is simulated by the analog step in one-dimensional (collective) Q-space. This allows to analize systematically the effects of the collective wave packets |q, p> and to find corrections for it. The connection points are the expectation values $\mathcal{H}(q, p)$ and $<\hat{Q}^n>$, or $<\hat{P}^n>$, respectively. They allow to transfer the actual parameters, from A-space averages, to the formal parameters, used in Q-space. It is interesting to note that for reconstructing $\hat{H}_{coll}$ we need wave packet information, e.g., $<\hat{P}^2>$, in addition to the dynamical information $\mathcal{H}(q, p)$.

4. Influence of h^n-corrections

In the following figure, we try to illustrate the effects of the ZPE on the collective potential. We see that the ZPE modifies V(q) in essentially two ways. First, it raises the absolute energy; thus, for calculating ground state energies one has to subtract the ZPE, which allows the correlated ground state to come below the Hartree-Fock energy, i.e., $\mathcal{V}(q_{Min})$. Second, it changes the relative

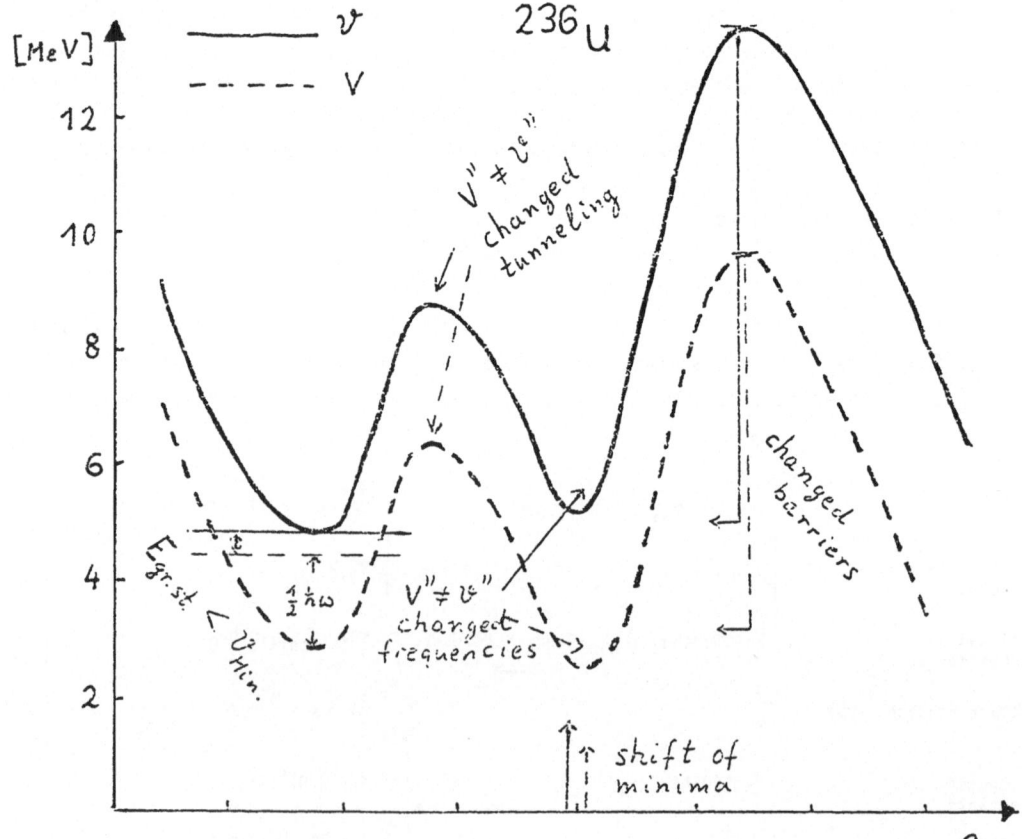

energies along the path if $\mathcal{V}$-V varies with q. This has effects on the collective dynamics; it may change positions of minima and maxima, curvatures there (thus, affecting frequencies) and barrier heights (thus, affecting phase shifts).

In fact, the above illustration represents also a realistic case, the symmetric fission, or scattering, path for ^{236}U. Three ZPE's have been included; the ZPE for motion along the path and the two ZPE's for rotation of the deformed (or two-centered) system as a whole. The effect on the total energy is 2 MeV at the first minimum. More interesting for collective dynamics is the effect on relative energies. It influences mostly the barriers; a lowering of 2 MeV, as observed above, is not negligible. The changes of frequencies or shifts of minima and maxima are small in this example. For lighter systems, the contribution of ZPE become even larger. For ^{32}S $\leftrightarrow$ ^{16}O + ^{16}O, e.g., we obtain a lowering of the barrier by about 6 MeV, which is quite a change. We ought to remind that just the barrier heights are important properties of scattering potentials and thus, the $\hbar^2$-corrections should be considered there.

By carrying the expansion one step further, we have also checked the $\hbar^4$-terms. They turn out to be very small in heavy systems; in ^{236}U, e.g., the contribution remains smaller than ±0.1 MeV. In light systems, however, it can happen in regions of strongly varying shell structure, e.g., sudden shell closures that the $\hbar^4$ contribution exceed the $\hbar^2$ terms. In such cases, the expansion of $\hat{H}_{coll}$ in orders $\hat{P}^n$, i.e., $\mathcal{V}$, $P^2/2M$, etc., which allows the expansion in orders $\hbar^n$ for the quantum corrections, breaks down. One has to use a complicated integral operator from the onset[6].

This leads into the domain of the generator-coordinate method (GCM). We cannot outline its details here. But at least, we want to communicate an important result concerning the interrelations: If $\hat{H}_{coll}$ can be expanded in orders p^n, or $\hbar^n$, respectively, the above method to reconstruct it from $\mathcal{K}(q, p)$ and the result for $\hat{H}_{coll}$ obtained from the GCM are completely identical[4]. It is only in case of an irreducible integral operator that the GCM may be superior[6]. In such a case, however, the concept of collective motion becomes doubtful anyway.

5. Summary

The purpose of this paper was to embed the concept of an optical potential for heavy ion scattering into the frame work of recently developed microscopic collective theories. This allowed to identify the usual h.i.-potential as a classical object depending on the center separation. A systematic expansion in powers of $\hbar$ could be performed, yielding quantum corrections in form of zero-point-energy subtractions. These corrections come out to lower the barrier heights by about 2 MeV for heavy systems and up to 6 MeV for light ones.

References:

1. F. Villars, Nucl. Phys. A285 (1977) 269
   M. Baranger, M. Veneroni, preprint 1978
2. K. Goeke, P.-G. Reinhard, Ann. of Phys. 112 (1978) 328
3. K. Goeke, P.-G. Reinhard, preprint 1978
4. P.-G. Reinhard, K. Goeke, Nucl. Phys., in press
5. P.-G. Reinhard, Nucl. Phys. A252 (1975) 120
6. P.-G. Reinhard, Nucl. Phys. A281 (1977) 221

* Supported by the Gesellschaft für Schwerionenforschung, Contract Number 06 MZ 709

THE OPTICAL MODEL IN ATOMIC PHYSICS

I.E. McCarthy

Institute for Atomic Studies
The Flinders University of South Australia
Bedford Park, S.A. 5042, Australia

The optical model for electron scattering on atoms has quite a short history in comparison with nuclear physics. The main reason for this is that until about five years ago there were insufficient data, since it was experimentally difficult to detect electrons with sufficient energy resolution to resolve states of the residual system. Angular distributions for elastic and some inelastic scattering have now been measured for the atoms which exist in gaseous form at reasonable temperatures. There are not many of these, inert gases, hydrogen, alkalies and mercury being the main ones. In addition, nearly every atomic ion can be produced in a fusion reactor, so it is very important to have a model that will accurately describe scattering and, even more important, total ionization cross sections for spherical atomic systems, although it is impossible to make targets for scattering experiments from the more highly-stripped ions.

It is interesting to fix the energy scale for atomic scattering. The separation energy of valence electrons is of the order of 20 eV. Hence the depth of the potential in the valence region is of order 20 eV - 50 eV. One would expect 1 eV in atomic scattering to correspond roughly to 1 MeV in nuclear scattering. Most atomic experiments are done in the intermediate-energy region of 100 eV - 1000 eV.

For theoretical considerations of the optical model we use the multichannel scattering formalism. The target states i, j, ... include continuum states which are made discrete for formal purposes by imposing periodic boundary conditions. The optical model potential coupling two channels i and j is written for energy E as

$$\tilde{V}_{ij} = V_{ij} + V_{iI}[E-K-V]^{-1}_{IJ} V_{Jj} \, , \tag{1}$$

where

$$V_{ij} \equiv <i|V|j> \quad .$$

The potential V is the real potential for the interaction between the incident electron and the rest of the atom. Exchange terms are implicitly included. The channels I,J (for which an implicit summation notation is introduced) do not include the elastic channel 0 or any inelastic channel that is to be explicitly treated in a coupled-channels treatment of a finite set of channels.

For illustration we will consider only the single-channel optical model potential $\tilde{V}_{00}$.

$$\tilde{V}_{00} = \Sigma_\mu <0|T_\mu|0> \tag{2}$$

where T_μ are the operators for scattering of the incident electron by the individual particles μ of the atom (antisymmetry is implied). We will not include the Coulomb potential of the nucleus in the discussion, since this is trivially added to the optical model potential of the electrons.

Making an independent-particle (Hartree-Fock) approximation for the atomic ground state, we write the elastic scattering (with obvious notation) using the momentum-space representation.

$$f(\underline{K},\underline{K}') = <\underline{K}|\tilde{V}_{00}|\chi^{(+)}(\underline{K}')> \quad , \quad (K=K')$$

$$= \int d^3q_1 \int d^3q_1' \, \delta(\underline{K}-\underline{q}_1) \tilde{V}_{00}(\underline{q}_1,\underline{q}_1') \chi^{(+)}(\underline{K}',\underline{q}_1')$$

$$= \Sigma_\mu \int d^3q_1 \int d^3q_2 \int d^3q_1' \int d^3q_2' \, \delta(\underline{K}-\underline{q}_1)$$

$$\times \phi_\mu{}^*(\underline{q}_2) T_\mu(\underline{q}_1,\underline{q}_2;\underline{q}_1'\underline{q}_2') \phi_\mu(\underline{q}_2') \chi^{(+)}(\underline{K}',\underline{q}_1') \tag{3}$$

The key to an optical model is the approximation made for T_μ. In practice it is impossible to proceed if we do not assume that T_μ is locally translationally invariant.

$$T_\mu(\underline{q}_1,\underline{q}_2;\underline{q}_1',\underline{q}_2') = t_\mu(\underline{k},\underline{k}') \, \delta(\underline{Q}-\underline{Q}') \tag{4}$$

where we use the following coordinate transformations

$$\underline{k} = \frac{1}{2}(\underline{q}_1-\underline{q}_2), \quad \underline{Q}=\underline{q}_1+\underline{q}_2, \quad \underline{q} = \underline{q}_1'+\underline{q}_2'-\underline{q}_1, \tag{5}$$

$$\underline{P} = \underline{k}-\underline{k}' = \underline{q}_1-\underline{q}_1' \; .$$

The optical model potential for the shell μ is

$$\tilde{V}_\mu(\underline{q}_1,\underline{q}_1') = \int d^3q \, \phi_\mu{}^*(q) \, t_\mu(\underline{k},\underline{k}') \phi_\mu(\underline{q}+\underline{P}) \tag{6}$$

The approximation of local translational invariance is familiar and known to work very well for calculating atomic orbitals in the form of the Thomas-Fermi or local atomic matter approximation. With this approximation we can proceed in the direction of nuclear physics[2] and calculate t_μ as a local Bethe-Goldstone G-matrix for positive energies.

The G-matrix approach has the advantage that t_μ reduces to the Coulomb potential

$$t_\mu(\underline{k},\underline{k}') = 1/2\pi^2 P^2 \equiv v(P) \tag{7}$$

if the shell separation energy ε_μ is well below the threshold for both electrons to have their total energy above the Fermi level. However the Fermi level depends on r and does not correspond directly to a physical threshold such as the ionization threshold for the shell μ. In atomic physics the physical thresholds are well-separated energetically and cross sections near the thresholds are important, particularly for the fusion impurity-ion application. We choose a different approximation in which the Pauli condition is replaced by a condition for the physical ionization thresholds.

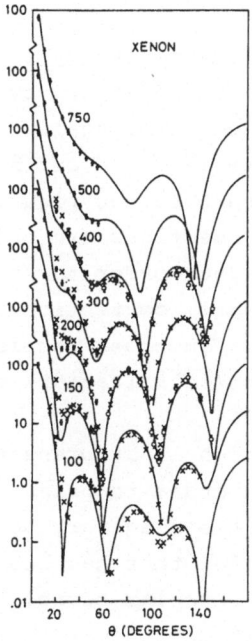

Fig. 1.   Elastic scattering of electrons from xenon at different energies (in eV). The differential cross section is in atomic units (a_0^2). The experimental data are represented by filled circles[5], open circles[6], and crosses[7]. The curve is the model of McCarthy et al.[4].

We first, however, show that the optical model makes sense in atomic physics by considering its recent history. So far no attempt has been made to introduce a physical threshold condition. Shells have been divided into valence and inert shells. For valence electrons the approximation has been made[3] that t_μ is the free Coulomb t-matrix calculated only up to second order, and with an arbitrary closure energy introduced to average the excitation energy of virtual channels I,J.

This parameter is fixed by requiring the long-range behaviour of the real part of the second-order term to correspond to the known induced-dipole polarization term $-\alpha/r^4$. There is thus one arbitrary parameter in the theory. The inert shells are treated with the approximation (7).

Angular distributions are quite well predicted by this model, but total reaction cross sections can be in error by about 30%.

The model of McCarthy, Noble, Phillips and Turnbull[4] is basically similar but contains another arbitrary parameter, the imaginary potential strength, which is adjusted to fit the total reaction cross section. Fig. 1 shows angular distributions for xenon at various energies with this model.

In atomic physics at intermediate energy the angular distribution shape for $\theta > 20°$ is quite well fitted simply by the first-order term (7) (Coulomb potential folding with exchange). A self-consistent equivalent-local exchange potential[8] based on the idea of Perey and Buck[9] produces phase shifts correct to two figures[10] even at energies below 1 eV. The absolute cross sections are only overestimated by about 10% by omitting complex higher-order terms, so what is needed in atomic physics is a relatively-small correction to the potential-folding term. Since the total reaction cross section depends entirely on the presence of the higher-order (polarization) term it is probably the most sensitive test of the model.

We now turn to the model which is at present being computed by A.T. Stelbovics and myself. In order to preserve the physical threshold condition for the appearance of the imaginary potential representing ionization we make an approximation that has been suggested by the results of (e,2e) experiments. Angular correlations can be fitted quite well by an averaged eikonal approximation in which the optical model wave function $\chi^{(+)}$ for each electron is assumed to be a plane wave in a constant real potential $\bar{V}$. Furthermore the value of $\bar{V}$ is very close to the separation energy ε_μ for the shell μ, since the experimental condition for zero recoil of the residual ion is very close to the 90° angular correlation for free electron collisions. We therefore approximate the single-particle potential by the separation energy ε_μ.

The Pauli principle is replaced by a physical threshold condition. Virtual momenta in the Green's function for the Bethe-Goldstone equation are excluded if both particles have energies below the continuum. We are thus excluding excited channels below the ionization continuum. For inert gases this is not a serious error, since the ionization cross section is about 80% of the total reaction cross section. The error

can be eliminated by treating discrete inelastic scattering explicitly
in a coupled-channels calculation.

The Green's function is thus similar to the Bethe-Goldstone
Green's function with the Fermi energy replaced by the shell separation
energy ε_μ and the single-particle potentials equal to ε_μ. Using a
spherically-averaged threshold (Pauli) operator $I_\mu(p)$ we have

$$G_\mu(p) = \frac{I_\mu(p)}{E^{(+)} - \varepsilon_\mu - E'} \quad ,$$

$$E' = \frac{1}{4} Q^2 + p^2 - 2\varepsilon_\mu \quad ,$$

$$I_\mu(p) = 0, \ a < 0 \quad ,$$

$$= a, \ 0 \leqslant a \leqslant 1 \quad ,$$

$$= 1, \ a > 1 \quad ,$$

$$a = (\frac{1}{4} Q^2 + p^2 - 2\varepsilon_\mu)/pQ \quad . \tag{8}$$

There is a pole in $G_\mu(p)$ at

$$p^2 = E + \varepsilon_\mu - \frac{1}{4} Q^2 \equiv Q_0^2 \quad . \tag{9}$$

The pole exists only for $E > \varepsilon_\mu$, since the threshold (Pauli) operator
$I_\mu(p)$ becomes zero at

$$\frac{1}{4} Q^2 + p^2 = 2\varepsilon_\mu \tag{10}$$

We now come to some problems that are specific to atomic physics
where the potential $v(P)$ has the Rutherford singularity (i.e. it is of
long range in coordinate space). The methods of nuclear physics do not
apply. For example the partial wave expansion is divergent on shell.
The Coulomb potential, however, has some simplifications of its own.
The Born expansion appears to be convergent, at least for a relevant
class of matrix elements, but more important is the fact that the
Coulomb t-matrix is known in closed form[11].

The approximation to be used for the shell scattering operator t_μ
is a solution of the integral equation

$$t_\mu(\underline{k},\underline{k}') = \langle \underline{k} | v + v \, G_\mu t_\mu | \underline{k}' \rangle \tag{11}$$

where G_μ is defined by (8).

This equation for t_μ may be considerably simplified by making the
approximation of Brieve and Rook[12].

$$G_\mu = G_R + G_F \quad ,$$

$$G_R(p) = C_R / (-\gamma^2 - p^2) \quad ,$$

$$G_F(p) = C_F (Q_o^2 - p^2) \quad . \tag{12}$$

The threshold function $I_\mu(p)$ is fitted with $Q_o^2 > 0$ according to

$$I_\mu(p) = C_F + C_R (Q_o^2 - p^2) / (-\gamma^2 - p^2) \quad , \tag{13}$$

$$C_F = I_\mu(Q_o) = (E - \varepsilon_\mu) / Q_o \Omega \quad . \tag{14}$$

For $Q_o^2 < 0$ there is no pole, so the Green's function is real. We obtain extra freedom for fitting by retaining the two-term approximation (12), but allowing two more free parameters Q_o' and C_F. In this case

$$I_\mu(p) = [C_F / (Q_o'^2 - p^2) + C_R / (-\gamma^2 - p^2)] (Q_o^2 - p^2) \tag{15}$$

where $Q_o'^2 < 0$. The quality of the approximation is illustrated by Table 1.

E(eV)	Q_o^2(a.u.)	C_F	γ^2(a.u.)	C_R	RMS
50	-1.7	-1.2	3.6	2.4	.17
100	-2.1	-1.4	4.0	2.6	.17
150	6.4	.36	23.0	.84	.16
200	7.4	.41	13.0	.74	.11
250	8.3	.48	10.0	.63	.086
300	9.2	.54	8.6	.55	.071
350	10.0	.60	7.6	.47	.054

Table 1. The parameters of the Green's function fitted for $\varepsilon_\mu = 100$ eV and $\Omega = K$. The column headed RMS gives the RMS deviation for a 30-point fit.

Because of the difficulties with the partial-wave expansion and the off-shell boundary conditions it is not possible to use the coupled-differential-equation method of Brieva and Rook for solving equation (11). We reduce (11) to two successive integral equations by defining an auxiliary negative-energy (real) t-matrix u.

$$u = v + v G_R u, \quad u \equiv v + w_R \quad . \tag{16}$$

We can now reduce (11) to the Lippmann-Schwinger equation for the potential u.

$$t_\mu = u + u G_F T_\mu \tag{17}$$

Equation (16) gives $C_R u$ as the Coulomb t-matrix for potential $C_R v$ and energy $-\gamma^2$. To solve (17) we need the fully-off-shell solution of the problem of scattering by a short-range potential w_R with a Coulomb tail.

Because of the smallness of the polarization potential w_R in comparison with v we make a multiple-scattering expansion for (17).

$$t_\mu = t + w_R + w_R G_0 t + t G_0 w_R + \ldots \; , \quad t \equiv v + w_F \; , \tag{18}$$

where $C_F t$ is the Coulomb t-matrix (complex) for potential $C_F v$ and energy Q_0^2.

Since the potential folding term for v may be simply evaluated, even though it contains the Rutherford singularity, we keep only the polarization part $t_\mu - v$ for the remainder of the discussion. This is a smooth function of k and k'. It is evaluated to first order in the multiple-scattering series.

$$t_\mu - v = w_R + w_F \; . \tag{19}$$

It is convenient to make an equivalent local and central approximation to $\tilde{V}_\mu(\underline{K},\underline{K}')$ as defined by equations (5) and (6). This potential will of course be energy dependent. In momentum space it depends only on K and P, whereas the potential (6) depends also on $\underline{K} \cdot \underline{P}$. We must therefore average over $\hat{\underline{K}} \cdot \hat{\underline{P}}$.

The averaged local central optical model potential is for the shell μ

$$\tilde{V}_\mu(K,P) + \frac{1}{2}\int d(\hat{\underline{K}} \cdot \hat{\underline{P}}) \int d^3 q \; \phi_i^*(\underline{q}) t_\mu(\underline{k},\underline{k}') \phi_i(\underline{P}+\underline{q}) \; , \tag{20}$$

where K is the incident momentum as shown by the delta function in equation (3). The coordinate transformations are

$$\underline{P} = \underline{K}-\underline{K}' = \underline{k}-\underline{k}' \; ,$$
$$\underline{Q} = \underline{K}+\underline{q} \; ,$$
$$\underline{k} = \frac{1}{2}(\underline{K}-\underline{q}) \; ,$$
$$\underline{k}' = \frac{1}{2}(\underline{K}'-\underline{P}-\underline{q}) \; . \tag{21}$$

The integral (21) would be extremely difficult to evaluate using the angular momentum expansions usual in nuclear physics, because of the coordinate transformations. However the first-order term is trivially evaluated in coordinate space, so we need only to evaluate the integral in $t_\mu - v$ given, for example, by (19). All the functions of the integrand are smooth, so we use a multiple-integration technique such as Monte Carlo to evaluate the integral using cartesian coordinates for the integrand. Multiple-integration techniques are not strongly dependent on the dimension of the integral.   It is certainly possible to go further in the multiple-scattering series (18) if necessary.   We have shown that 1 percent accuracy can be obtained in a reasonable time.

REFERENCES

1. I.E. McCarthy and E. Weigold, Phys. Reports $\underline{27C}$, 275 (176).
2. J. Hüfner and C. Mahaux, Ann. Phys. $\underline{73}$, 525 (1972).
3. F.W. Byron, Jr. and C.J.Joachain, Phys. Rev. $\underline{A15}$, 128 (1977).
4. I.E. McCarthy, C.J. Noble, B.A. Phillips and A.D. Turnbull, Phys. Rev. $\underline{A15}$, 2173 (1977).
5. R.H.J. Jansen and F.J. de Heer, J. Phys. B $\underline{9}$, 213 (1976).
6. J.F. Williams and A. Crowe, J. Phys. B $\underline{8}$, 2233 (1975).
7. S.J. Buckman, P.J.O. Teubner and H. Arriola (unpublished).
8. J.B. Furness and I.E. McCarthy, J. Phys. B $\underline{6}$, 2286 (1973).
9. F.G. Perey and B. Buck, Nucl. Phys. $\underline{32}$, 353 (1962).
10. M.E. Riley and D.G. Truhlar, J. Chem. Phys. $\underline{63}$, 2182 (1975).
11. J.C.Y. Chen and A.C. Chen, Adv. in Atom. and Molec. Phys. $\underline{8}$, 71 (1972).
12. F.A. Brieva and J.R. Rook, Nucl. Phys. A $\underline{291}$, 299 (1977).

THE KAON-NUCLEUS OPTICAL POTENTIAL FOR KAONIC ATOMS

R.C. Barrett

University of Surrey, Guildford, U.K.

I shall be concerned only with the K^--nucleus potential used in the calculation of kaonic atom energy levels. The K^+-nucleus potential can be calculated in
impulse approximation and the first order potential is accurate enough to investigate
nuclear matter densities (assuming we believe the K^+-nucleus interaction) (Dover and
Moffa, 1977).

The K^--nucleus interaction is much stronger and life is complicated by the
$\Lambda(1405)$ (or Y_0^*) resonance which is 27 MeV below the K^-p threshold and 35 MeV wide.
It is desirable to calculate the K^--nucleus interaction, however, to analyse the
measurements on kaonic atoms.

A kaonic atom is formed when a negative kaon is stopped in matter and cascades
firstly with Auger emission and then X-rays until the K^- is absorbed on the nucleus.
The energy and width of the last observed line are compared with the predictions with
purely electromagnetic forces to obtain the shift δE and width Γ due to the strong
interaction:

$$\delta E - i\Gamma/2 \equiv E_{n'\ell' \to n\ell}(\text{expt.}) - E_{n'\ell' \to n\ell}(\text{coul.})$$

The simple impulse approximation to the optical potential gives the first order
term

$$V_1(r) = - (2\pi\hbar^2/\mu_A)(1 + m_K/m_p)\ \bar{a}\ \rho_m(r)$$

where $\bar{a} = \frac{1}{2}(a_p + a_n)$ is the isospin averaged $\bar{K}N$ scattering length. (This expression
assumes $N = Z$ and $\rho_n = \rho_p$ but can easily be generalised).

There are two points to consider, namely the validity of the first order potential and the validity of using an effective scattering length. The following K N
channels are relevant.

$$K^- + p \to \begin{cases} K^- & + p \\ \bar{K}^\circ & + n \\ \Sigma & + \pi \\ \Lambda & + \pi \end{cases} \qquad K^- + n \to \begin{cases} K^- & + n \\ \Sigma & + \pi \\ \Lambda & + \pi \end{cases}$$

There have been many analyses of data using $\bar{a}$ as an adjustable parameter. A typical
result is (Koch and Sternheim, 1972)

$$\bar{a}_{eff} = - 0.44 - i0.83 \text{ fm.}$$

whereas $\frac{1}{2}(a_p + a_n) = 0.49 - i0.60$

Deloff and Law (1976) have used a K-N potential of range b folded into the nuclear density and derived the effective scattering length:

$$a_{eff} \simeq \bar{a}/(1 + 1.2\bar{a}/b) = -0.46 - i1.0 \text{ fm}$$

Qureshi (1978) has used this approach for ^{7}Li and ^{9}Be with gaussian interactions and found for the L X-rays the results shown in the table

Shift and widths in L X-rays (eV)

Nucleus	^{7}Li		^{9}Be	
	δE	Γ	δE	Γ
Experiment (Batty et al. 1978)	2 ± 26	55 ± 29	$- 79 \pm 21$	173 ± 58
$\bar{a}$ fitted to many nuclei	$- 5.3$	59	$- 59$	288
Folding model	$- 1.4$	71	$- 66$	355

The fit to ^{7}Li is good but the ^{9}Be result shows the need for better calculations. He has also calculated a second order term in the optical potential using closure approximantion and correlations corresponding to a Fermi gas. These give rise to the following equivalent local second order potential (where k_0 is imaginary).

$$V_2 = \pi^2 h^2 A(A - 1)/(ik_0\mu)[I_1(r)(\rho(r))^2 + I_2(r)\frac{1}{2}\rho(r)\nabla^2\rho(r)](\bar{a}')^2$$

where I_1 and I_2 are obtained from integrals involving the correlation function and $\bar{a}'$ in the scattering length for a kaon of energy -27 MeV. Applying this to ^{32}S gives

Shifts and widths in M X-rays in ^{32}S (keV)

	δE	Γ
Experiment (Backenstoss et al. 1974)	$- 0.55 \pm 0.06$	2.33 ± 0.20
V_1	$- 0.76$	1.67
$V_1 + V_2$	$- 0.58$	2.43

The scattering length used was a = $0.42 - i0.70$ fm

He is continuing calculations without closure or the assumption of Fermi gas correlations, and for the case of ^{12}C is evaluating the expression

$$(0|V_2|0) = \frac{-(A-1)}{A} \sum_{\alpha,\gamma<F} \langle\alpha|t|\gamma\rangle \frac{1}{E-K_0-\varepsilon_\gamma+\varepsilon_\alpha+i\varepsilon} \langle\gamma|t|\alpha\rangle$$

$$+ \frac{(A-1)}{A^2} \sum_{\alpha,\gamma<F} \langle\alpha|t|\alpha\rangle \frac{1}{E-K_0+i\varepsilon} \langle\gamma|t|\gamma\rangle$$

This involves calculating the expression

$$U_2(r) = (\bar{a})^2 \sum_{\ell_1 j_1 \ell_2 j_2} \sum_{LL'} C_{\ell_1 \ell_2 j_1 j_2 LL'} \; I^{L'}_{\ell_1 \ell_2}(r)$$

where the coefficients C represent a product of angular momentum coupling factors and the $I^{L'}_{\ell_1 \ell_2}(r)$ are integrals involving the nuclear radial wave functions.

We would expect to be able to check the K^-p interaction fairly easily by using it for a system which has very few multiple scattering corrections, namely kaonic hydrogen! There is at present some experimental interest in this system and the 8keV level is expected to have a shift and a width of a few hundred electron volts. It seems like a trivial calculation since first order perturbation theory gives for a potential of depth V_0, range b

$$\delta E_1 = \frac{4}{3} V_0 b^3 / R^3 \approx 300 \text{ eV}$$

where R = 83 fm is the Bohr radius and $V_0 b^3$ has been taken to be 200 MeV fm^3. This is much smaller than the binding energy. However the second order term is

$$\delta E_2 = \delta E_1 (1 - b^3/3k^3) \; V_0/\bar{E} \approx \delta E_1 V_0 / \bar{E} \gg \delta E_1$$

where $\bar{E}$ is the average excitation energy of the atom.

A calculation for the level shift has been carried out by Stepien-Rudzka and Wycech (1976) using a Yamaguchi separable potential and evaluating the Coulomb propagator. The shift has also been evaluated by Deloff and Law (1976) who obtained an expression for the shift from the scattering length without using a potential. For $\bar{a} = -0.885 + i0.625$ fm their results are

Shift and width in kaonic hydrogen (eV)

	δE	$\Gamma/2$
Stepien-Ruska and Wycech	356	232
Deloff and Law	349	217

In order to check the K N potentials I have considered the bound state coupled channel problem. In a pilot calculation I simulated the Coulomb force with a separable potential of range 117 fm (to give the correct atomic radius) and the separable strong interaction potential obtained by Alberg et al. (1976). The equations are then

$$(\nabla^2 + \varepsilon_i)\psi_i(\underline{r}) = \frac{2\mu_i}{\hbar^2} \sum_{nj} \lambda^n_{ij} v_n(r) \int v_n(r')\psi_j(\underline{r}')d^3r'$$

The eigenvalues are roots of the equation

$$\det (\mathfrak{U}) = 0$$

where the elements of the matrix U are

$$U_{ni,mj} = \delta_{nm}\delta_{ij} + \frac{2\mu_i}{\hbar^2} \lambda^n_{ij} \int \frac{\tilde{v}_n(k)\tilde{v}_m(k)d^3k}{k^2 - \epsilon_j}$$

If the off-diagonal terms are put equal to zero the result is

$$\delta E - i\Gamma/2 = 154 - i725 \text{ eV}$$

If the coupling is turned on slowly the real part changes sign twice and eventually reaches the value

$$\delta E - i\Gamma/2 = 1163 - i1120 \text{ eV}$$

The differences between these three results indicate that the first step needed to improve the optical potentials is to obtain a more believable kaon-nucleon potential. It is then necessary to use a coupled channel approach for the kaon-nucleus system as well. This has already been done for the case of ^{12}C by Thies (1978). Further calculations are in progress.

M. Alberg, E. Henley and L. Wilets, Ann. Phys. N.Y. 96 (1976) 43.

G. Backenstoss et al., Nuc. Phys. B 73 (1974) 184.

C.J. Batty et al., Nuc. Phys. A 282 (1977) 487.

A. Deloff and J. Law, Proc. Pittsburg Conf. on Meson-Nucleon Interactions, 1976; and Nukleonika 22 (1977) 875.

C.B. Dover and P.J. Moffa, Phys. Rev. C 16 (1977) 1087.

J.H. Koch and M.M. Sternheim, Phys. Rev. Lett. 28 (1972) 1061.

I.E. Qureshi (to be published)

W. Stepien-Rudzka and S. Wycech, Nukleonika 22 (1977) 929.

M. Thies, Nuc. Phys. A 298 (1978) 344.

SUMMARY

P. E. Hodgson
Nuclear Physics Laboratory, Oxford

1. Introduction

This meeting has been a timely one, marked by significant advances
in the accuracy of the experimental data, in the phenomenological
analyses and in the methods of calculating the microscopic potentials.

This summary begins with some general remarks about what we mean by
an optical potential, how it is defined and the ways it can be
determined, either microscopically from the nucleon-nucleon interactions, or by phenomenological analysis of the experimental data, or
in some intermediate way.

The next Section 3 is devoted to the microscopic potentials themselves, both global for a range of nuclei and particular when the
properties of particular nuclei are introduced into the calculation.
Section 4 considers the ways of adding extra theoretical or phenomenological effective potentials to the standard optical potential to
account for a recognised physical effect. Sections 5, 6 and 7 are
devoted to new experimental data and phenomenological analyses of
nucleon, light and heavy ion interactions, and Section 8 to some
concluding remarks.

2. Definition of the Optical Potential

By an optical potential we mean a potential that represents the
interaction between a nucleon or group of nucleons and a nucleus.
When inserted into the Schrödinger equation it gives the differential
cross-section and polarisation for elastic scattering, the reaction
cross-section and some other less important observable quantities.

These optical potentials can be obtained in several ways. Most
fundamentally, they can be calculated from the nucleon-nucleon interaction; this is difficult and has only recently been brought to the
stage of quantitative success. Alternatively they may be found phenomenologically by postulating a form of potential and adjusting its
parameters to optimize the fit to the experimental data. Such phenomenological analyses may be put on a firmer physical basis by using
additional information from nuclear models.

Optical potentials may also be classified by the range of experimental data they are designed to fit. Global potentials give good

overall fits to the scattering from many nuclei over a range of
energies. More precise fits can be obtained for particular interactions either by incorporating nuclear structure and nuclear reaction
information into the calculation of the potential or by adjusting the
parameters of a phenomenological potential to fit particular sets of
data.

There are many ambiguities in optical model potentials. These are
familiar in phenomenological analyses, where it is often found that
several potentials fit the same data equally well. It is usually
thought that one of these is the 'physical' potential, namely the one
that is given by a microscopic calculation. It is important to identify the physical potential, for this can be used with more confidence
in situations different from those from which it was obtained.

It should however be noticed that the 'physical' potential may have
a form that is significantly different from any of the phenomenological potentials. Furthermore, a potential is of its nature a theoretical construct, so care is necessary in describing it as 'physical'. It
is possible that even among microscopic potentials there are ambiguities in the sense that different types of calculations could conceivab-
ly give different potentials that nevertheless give equally good fits
to the data.

This possibility follows most clearly from the different ways of
expressing the potentials. We know from general considerations that
the 'physical' is non-local, and this non-locality can be expressed in
different ways. How can we compare these potentials with each other?
It is of course possible to define local potentials that are in some
specified sense equivalent to the corresponding non-local potential.
But this at once opens up further possibilities of ambiguity. The
local potential necessarily carries less information than the non-
local potential and potentials equivalent in one sense may not be
equivalent in another, and then which do we consider the more
'physical'?

These considerations concerning the ambiguities in the optical
potentials suggest that it can be misleading to compare a theoretical
microscopic potential with a phenomenological potential, since they
may differ significantly from each other and yet both be consistent
with experiment. It is therefore preferable to test theoretical
microscopic potentials by comparing them directly with the experimental data.

The relations between some of the topics discussed at this meeting
are summarised in the figure.

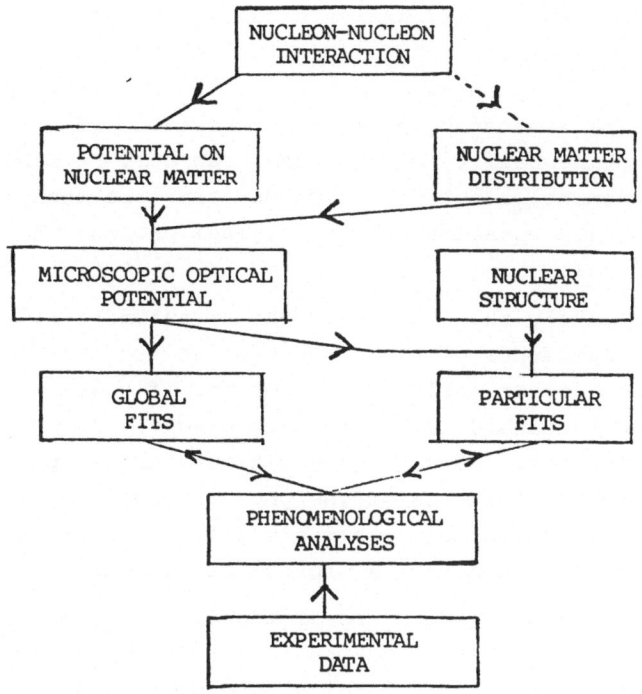

3. Microscopic Optical Potentials

By a microscopic optical potential we mean a potential calculated
from the nucleon-nucleon interaction and some nuclear properties,
usually the density distribution. Such potentials are distinguished
from phenomenological potentials which are obtained by direct fitting
to experimental elastic scattering data. These phenomenological
potentials usually fit the data much better than microscopic potentials, so what is the use of the microscopic potentials?

We are interested in microscopic potentials primarily because we
want to unify our understanding of nuclear physics. This understanding
develops in several stages: for example if we find a bump in a cross-
section we can fit it with some analytical curve. Then perhaps we can
show that the bump is a true resonance, and analysing it with the
Breit-Wigner theory can give its energy, width and quantum numbers.
Later on we might be able to show that these properties follow automatically from a potential that also correlates other nuclear properties. And finally we may succeed in calculating that potential from
the nucleon-nucleon interaction.

There is another reason for calculating microscopic optical potentials. Every phenomenological analysis relies on an analytical form for the potential and although our general physical intuition is sufficient to fix its general form the details may be wrong, and in a way that cannot be removed by parameter variation. If in such a case we insist on a high-quality fit then this can only be achieved by adding more phenomenological terms. These terms are then present solely to remove the inadequacies of our original choice, and correspond to nothing physical.

It is very unlikely that such a potential automatically gives the correct energy dependence of the cross-section, so this has to be rectified by yet another arbitrary parameter. If however the potential is physically well-founded it has greater predictive power and can account for a wide range of phenomena without further parametrisation. Furthermore there are features of the phenomenological analysis, such as the Coulomb correction term, that are not obtainable phenomenologically but must be obtained from a microscopic calculation.

It is important at the outset to distinguish between global potentials that give a good overall fit to a wide range of data for different nuclei at different energies, and particular potentials that give a precise fit to a narrow range of data. Both types of potentials have their value: the global potentials are useful for calculations that are relatively insensitive to the details of the potentials used, while the particular potentials are essential for detailed studies of specific reactions.

The calculation of microscopic optical potentials is fraught with many difficulties, and it is only in the last few years that for nucleons they have been developed to a degree of precision that approaches and in some cases even exceeds the phenomenological analyses.

These new methods have not yet been extended to light and heavy ions and so this review concentrates on nucleon potentials, with a briefer account of the phenomenological and semi-phenomenological analyses of the scattering of heavier particles that were also presented at the meeting.

Global Potentials

At this meeting we have heard of the nuclear matter approaches from Mahaux (1978) and Brieva (1978). Physically they are very similar: they start from nucleon-nucleon interactions that are equivalent in many respects and then solve the Bethe-Goldstone equation to give the correlated nucleon wave function in co-ordinate space and hence the mass operator and the t-matrix elements in momentum space. The principal difference arises in the application to finite nuclei: Mahaux applies the local density approximation directly to the mass operator whereas Brieva transforms the t-matrix to a co-ordinate representation before applying the local density approximation; this enables him to improve the geometrical aspect of the calculation and make explicit the range of the force. Both methods give complex energy-dependent microscopic potentials that already give a fair fit to the experimental data, a fit that is greatly improved by slight parameter adjustment. Mahaux has calculated many detailed features of the potential, including the volume integrals, the isospin terms and the Coulomb correction. The method of Brieva easily gives the spin dependence of the potential, and this is found to be similar to the phenomenological form. Both methods can be extended by improving the mathematical approximations, studying the convergence properties and evaluating higher order terms.

These nuclear matter calculations use information on nuclear structure in the form of nuclear density distributions. It is not certain that these contain all the nuclear structure information required even for global potentials, but it seems likely to be a good approximation. In principle it is possible to calculate the required nuclear structure information from the same nucleon-nucleon interaction, but this is generally regarded as a separate and very difficult problem. For the present purpose it is simpler and more accurate to use nuclear density distributions obtained either from Hartree-Fock calculations (Negele, 1970; Sick et al., 1975) or from the summation of squares of single-particle wave functions (Hodgson, 1976, Malaguti et al., 1978). The nuclear matter calculations are in general not too sensitive to the precise form of the density distribution, provided it has the correct RMS radius. An exception to this is those potentials that are sensitive to particular features of the density distribution, for example heavy ion potentials that are sensitive to its extreme tail.

The work described by Mahaux and by Brieva marks an important stage in the development of calculations of microscopic optical potentials from the nucleon-nucleon interaction: for the first time it is possible to obtain potentials that fit the data almost as well as phenomenological potentials with many adjustable parameters. Much work remains to be done to exploit this success. In particular, the spin-orbit and isospin terms can be studied with more realistic form

factors obtained from the microscopic calculations. It is now possible
to provide a much firmer physical basis for phenomenological calculations, and this should make it possible to obtain potentials that may
be extrapolated in energy and from one nucleus to another with much
greater confidence than before.

An alternative method of calculating the optical potential was
described by McCarthy (1978) in the context of his studies of atomic
potentials. The principle differences from the nuclear matter approach
are that the momentum distribution of the struck particle is given by
the momentum-space wave function, not by the Thomas-Fermi approximation, and the imaginary potential starts at the physical threshold for
Q-space. The spherical averaging is carried out only in the projection
operator Q and the multi-dimensional integration makes it possible to
perform co-ordinate transformations exactly. For incident momentum k,
the optical potential depends on K, p and $\underline{K}.\underline{p}$ so that a non-local
potential in co-ordinate space is non-spherical in momentum space.

A study of the three-body Bethe-Faddeev equations and single-
particle potentials in nuclei was presented by Zabolitzky (1978). A
comparison of the single-particle potential felt by a real A+1st
nucleon added to an A-nucleon system with the potential felt by one of
the A nucleons in the system when virtually excited shows that the
former quantity is of interest in optical model potential calculations
while the latter is relevant for the calculation of the ground-state
properties of the A-nucleon system.

The general theory of the optical potential was reviewed by Weigel
(1978). He used the procedure of Bell and Squires to obtain the
generalised optical potential from the Green's function formalism of
many-particle theory. This approach includes the Pauli principle and
the potential can easily be divided into an energy-independent folding
term and an energy-dependent resonant term. In this formalism it is
also easy to establish the connection with theories of nuclear structure and to calculate the contributions of inelastic and reaction
processes to the imaginary potentials. The analytic properties of
this potential are uniquely defined so that dispersion relations can
be obtained. At higher energies, where particle-particle correlations
in nuclear matter are important, the main part of the optical potential can be related to properties of nuclear ground state correlations.
Numerical results were presented for the energy dependence of the
optical potential and for the calculation of the optical potential
from dispersion relations.

A study of the kaon-nucleus optical potential was presented by Barrett (1978). To first order, the potential is proportional to the nuclear matter density. The K^+-N interaction is rather weak and can be used to study $\rho_m(r)$, the K^--N interaction is considerably stronger and is needed for the energies and widths of states in k^- atoms. The measured shifts are analysed to give values of the scattering length. Calculations of the potential to second order were also presented.

A microscopic method of calculating the imaginary part of the inelastic scattering interaction was described by Madsen, Osterfeld and Wambach (1978). This calculation is based on the expression

$$\text{Im } F(r,r') = \text{Im } \sum_N \langle \psi_f | V | \psi_N \rangle \langle \psi_N | V | \psi_i \rangle G_N(r,r')$$

The imaginary part of the projectile Green's function $G_N(r,r')$ is always negative, so the terms add constructively for elastic scattering, whereas for inelastic scattering they need not have the same sign and may be random or destructive. This expression is evaluated for inelastic scattering by assuming (1) that the initial and final nuclear states are related by creation and annihilation operators according to the expression

$$\psi_f = \sum_{\bar{m}\underline{n}} A_{\bar{m}\underline{n}} a_{\bar{m}}^+ a_{\underline{n}} \psi_i$$

where $\bar{m}$ refers to an unoccupied particle state and $\underline{n}$ to an occupied hole state based on the target nucleus. This is true for TDA and RPA collective states as well as for simple single-particle transitions. Also, (2) the intermediate states are taken as simple particle-hole doorway states based on the ground state, and (3) the identity of the projectile and target nucleons is neglected.

Evaluation of the expression then shows that there is indeed destructive interference but that it occurs in a systematic way so that the imaginary form factor is related to the strength of the real form factor and is sharply peaked at or beyond the nuclear surface due to the subtraction of the intermediate hole-scattering terms in the nuclear interior. This result provides a microscopic understanding of the phenomenological collective model prescription that relates the inelastic form factor to the derivative of the optical potential.

The odd-even dependence of the potential has been studied by Tang (1978) using the resonating group formalism. He shows that the anti-symmetrisation effects that arise when the Pauli principle is applied are represented by various nucleon-exchange terms in the exchange-normalisation and exchange-Hamiltonian kernel functions. The effects of the corresponding integral term in the Schrödinger equation can be represented by an equivalent local energy-dependent potential with a Majorana exchange term whose range decreases as the mass difference of the interacting nuclei increases. It is found that the amount of odd-even L-dependence is quite strong for the interaction of two S-shell nuclei, but is weak for $\alpha + {}^{16}O$ and for $n + {}^{40}Ca$. Exchange effects are also important for $\alpha + {}^{6}Li$, ${}^{12}C + {}^{13}C$, ${}^{12}C + {}^{16}O$ and ${}^{28}Si + {}^{16}O$ scattering (Bachelier et al., 1972; von Oertzen, 1970; Baye, 1976; Dehnhard et al., 1978).

This analysis leads one to expect that there are no odd-even effects when the masses of the interacting particles are widely different. However Kondo et al. (1975) have fitted the scattering of alpha-particles by ${}^{40}Ca$ using an L-dependent potential. Fits to this system over a wide range of energies have also been obtained with parity-independent potentials by Michel and Vanderpoorten (1977) and by Gubler et al. (1978). This shows that it is sometimes possible to fit the same data with L-dependent and with L-independent potentials, thus highlighting a pitfall in the phenomenological determination of L-dependent potentials. It should, however, be noted that it is possible to observe the effects of very small transfer amplitudes whenever there is high cancellation and this often occurs for backward scattering at high energies (Greenlees et al., 1972).

Explicit calculations were made for the $h + \alpha$ system at 60 MeV to study the importance of the various nucleon exchange terms, and these showed that the one-particle exchange and the core-exchange (three-particle exchange) terms are important, while the two-particle exchange term is negligible. Examination of the phase-shifts shows that the one-particle exchange effective potential is essentially an attractive Wigner potential, while the three-particle exchange effective potential is essentially a Majorana potential. In the case of the $\alpha + {}^{16}O$ system at 18 MeV it is sufficient to include the one and two-particle exchange terms.

In general, it is found that the one-particle and core-exchange are the most important exchange terms, the latter being significant only when the nuclei are similar in mass. The effective local potentials corresponding to these two terms are of the Majorana and Wigner types respectively. Thus in phenomenological analyses the effective potential must in general include an odd-even L-dependent or parity-dependent term. However its effect is small when the interacting nuclei are dissimilar in mass, and this explains why it is frequently possible to get the data quite well with an L-independent potential. Further discussions of L-dependent potentials are summarised in the next section.

Ioannides and Johnston (1978) have analysed the propagation of energetic deuterons through nuclear matter, and have shown that the binding energy of the deuteron in nuclear matter depends strongly on the relative orientation of the deuteron spin and centre-of-mass momentum, when both the Pauli exclusion principle and the tensor force component of the neutron-proton interaction are included in the calculation. As a result, the tensor component of the duetron-nucleus optical potential is sensitive to the high momentum and tensor components of the neutron-proton interaction. In principle this could be used to obtain more information about the neutron-proton interaction, but at present the necessary experimental data on the elastic scattering of energetic deuterons is lacking.

Particular Potentials

Particular microscopic potentials can be calculated by making use of information on the structure of the nucleus concerned. If the nuclear matter distribution is used, it can be improved by adjusting for each state the parameters of the potential used to calculate the wavefunction so as to fit experimental binding energies and the model-independent charge distribution. Other information can also be included, such as the excitation energies and other properties of the excited states. The nuclear matter and nuclear structure methods of calculating microscopic optical potentials were compared by Vinh Mau (1978). They are complementary to each other because the former neglects the contributions of the collective states, which is only important at low energies, while the later uses a weak effective nucleon-nucleon interaction which is valid at higher energies.

Detailed calculations with the Green function formalism show that the strength of the imaginary potential calculated with the nuclear structure method depends strongly on the choice of effective interaction. The method was applied to calculate the n-^{40}Ca potential at low energy (Vinh Mau and Bouyssy, 1976). This work used information on the energies and collective properties of the low-lying states to give the contributions to the absorbing potential from each of the individual excitations. It is found that the mean square radii of both the real and the imaginary parts of the optical potential are well represented for energies less than 50 MeV. The volume integrals are too small for the real potential, but for the Reichstein-Tang nucleon-nucleon interaction the imaginary potential is in accord with the empirical value, and their variation with energy is well represented. These potentials have been used by Ngô to calculate cross-sections, and it is found that at 30.3 MeV the angular positions of the oscillations are correct but that the potential has to be renormalised to reproduce the amplitudes at large angles. Calculations of the n-α and α-^{40}Ca potentials were also presented.

A microscopic analysis of p-^{40}Ca elastic scattering at 40 MeV was made by Ngô et al. (1978). They obtained the non-local, energy-dependent complex potential, as well as the local equivalent potential, for several nucleon-nucleon interactions. This potential is in good overall agreement with the experimental data, but the fit is greatly improved by a 10 % increase in the real potential. The effects of particle-vibration coupling on the Hartree-Fock potential have been studied for ^{208}Pb by Van Giai and Bernard (1978). They include the effects of long-range particle-hole correlations and calculate the real and imaginary parts of the optical potential by a formalism in which the Hartree-Fock term and the correction terms are derived from a Skyrme effective interaction. For each L-multipole all the most important collective states are included, and in their calculations for ^{208}Pb they took account of the natural parity states corresponding to L=0,2,3,4,5 for isoscalar modes and L=0,1,2 for isovector modes. The resulting potential has a volume form for the real part and a surface-peaked form for the imaginary part, and strengths rather less than those found phenomenologically. These potentials are being used to calculate elastic scattering, and preliminary results are encouraging.

A shell model description of the optical model potential was presented by Micklinghoff (1978), who solved the Bethe-Goldstone integral equation for a finite system using an approximation in the framework of the collective model and the hole line expansion. This allows the absorption due to collective excitation to be included in the optical potential.

4. Effective Optical Potentials

There are several cases where the global potential does not fit the elastic scattering data at all well, for a reason that is physically evident. It is then possible either to make a more detailed calculation that includes this effect explicitly, or to devise a new term in

the optical potential that reproduces its effects on the elastic
scattering (Hodgson, 1974). Such effective potentials can be obtained
either theoretically or phenomenologically. Notable examples of this
are provided by the Coulomb excitation of low-lying collective states
in heavy ion scattering and the coupling to giant resonances and to
specific reaction channels (von Geramb, 1972, 1973).

The long-range absorption potential that gives the effect of the
Coulomb excitation of low-lying collective states in heavy ion
scattering has been studied by Baltz et al. (1978). In many of these
reactions, especially at energies around the Coulomb barrier, it is
found that the coupling to low-lying collective states is so strong
that the elastic scattering cross-section is considerably perturbed
from its usual Fresnel form and cannot be fitted by a standard optical potential. It was found possible to account for the coupling by
adding a phenomenological absorbing potential falling off as $r^{-6.3}$
(Thorn et al., 1977). The coupling can be taken into account explicitly by a full coupled-channels calculation but this is very lengthy
due to the long range of the Coulomb forces and so efforts have been
made to calculate a potential that, when added to standard optical
potential, gives the same effect as the coupling. This has been done
by Love, Terasawa and Satchler (1977) using the Feshbach projection
operator formalism. Their potential is almost eintirely imaginary and
has a long range, falling off as r^{-5} at large distances.

A more exact expression was obtained by Baltz et al. (1977, 1978)
using Coulomb-distorted scattering states and a Coulomb-distorted
Green's function. Their potential has a different radial dependence
from that of Love et al., and is L-dependent, but the contribution of
greatest range again varies as r^{-5}. They obtained an analytical
closed-form expression for the differential elastic scattering cross-
section that is valid below the Coulomb barrier where the absorption
is weak.

Baltz et al. also calculated for the case of 50 MeV alpha-particles
on ^{154}Sm the component of the optical potential due to the direct
rotational coupling of the ground and first 2^{+} excited state. This
potential is highly L-dependent, repulsive in the low partial waves,
increasing in magnitude to the surface, changing sign and becoming
attractive, and then decreasing in magnitude for high partial waves.
The empirical optical model component obtained as the difference
between the optical model and coupled-channels parameters is also L-
dependent and repulsive. The calculated imaginary potential is L-
dependent and absorptive for all partial waves, while the empirical

imaginary potential is of small magnitude but relatively diffuse in its small absorption outside the nuclear surface. Similar results were obtained for 60 MeV ^{16}O on ^{40}Ca. In both cases the L-dependence of the imaginary potential reflects the L-window of a direct reaction in the presence of a strongly absorptive background potential; flux is lost from the elastic channel into the inelastic channel primarily in the surface partial waves. This shows that it may well be insufficiently accurate to use an L-independent potential for heavy ion reactions in which a substantial amount of flux goes into direct channels.

Calculations of the effects of coupling to excited states have also been made by Frahn and Hill (1978) who show that Frahn's generalised Fresnel model for very heavy ion scattering can describe dynamic polarisation effects due to the Coulomb excitation of low-lying collective states. They use the Coulomb-distorted eikonal approximation to calculate the reflection functions corresponding to the potentials of Love et al. and of Baltz et al., and the resulting explicit formulae are in good accord with the experimental elastic scattering data. Measurements of dynamic polarisation effects in heavy ion scattering provide a sensitive way of determining B(EL) transition probabilities and deformations.

Several adiabatic and dynamic polarisation effects in sub-Coulomb elastic scattering were considered by Baur, Rösel and Trautmann (1978). The mutual electric polarisation of two nuclei can be interpreted as transitions to excited intermediate states, characterised by an adiabacity parameter $\xi_{if} = \eta_i - \eta_f$, where η_i and η_f are the Coulomb parameters. If $\xi_{if} > 1$, only virtual excitations are likely to occur, and the most important intermediate states are the strongly collective EL transitions to the giant dipole states. In this case the effects on the elastic scattering can be represented by a real potential. If however $\xi_{if} < 1$ then flux can be removed from the elastic channel due to the excitation of low-lying collective states. In this case the cross-section can be reliably calculated by the semiclassical coupled-channels method. Some explicit calculations for ^{40}Ar on ^{238}U at 340 MeV showed that the effect of reorientation in the second state and the coupling to the 4^+ state have little effect on the elastic scattering for angles less than the grazing angle.

L-dependence

The elastic scattering of particles by nuclei is fixed by the complex phase-shifts δ_L corresponding to several values of the orbital

angular momentum L from zero to L max, the highest value of L for
which the phase-shift contributes significantly to observed quantities.
These phase-shifts can in principle be found by analysing the cross-
section, but in practice this is only possible if very few partial
waves contribute significantly. Solving the radial wave equation for
each value of L gives the phase-shifts for a particular potential V(r).
It is obviously remarkable if a potential of a particular analytic
form gives all the phase-shifts correctly even if it has several
adjustable parameters and δ_L is a smooth function of L, as is often
found to be the case. In practice when there are many partial waves
some errors in one may compensate some errors in another, so even if
the fit is very good it does not necessarily mean that all the phase-
shifts are correct. In principle any cross-section, and thus any set
of phase-shifts, can be fitted by an L-independent potential, but this
could be highly singular and not a smooth function of L. The high
precision of the fits to alpha-particle cross-sections obtained with
the spline method by Michel and Vanderpoorten shows that in this case
an L-independent potential is adequate. In other cases it is necessary
to assume some L-dependence. There is, of course, no guarantee that
this L-independence will have a simple analytic form as a function of
L, but this is the simplest initial hypothesis, and it seems to work
quite well.

There is no need to consider L-dependent potentials for most nuclei
as the scattering can be fitted very well by L-independent potentials.
Even in these cases, however, it is not certain that there is not some
L-dependence that is masked by parameter variation. The case for
trying the effect of an L-dependent potential is stronger if no amount
of parameter variation is able to give a fit to the data. There does
still remain the possibility that a potential with a different analytical form would be able to fit the data without recourse to L-dependence. Every form of potential gives some L-dependence of the phase-
shifts and it may be surmised that if this is not correct the potential
has to be supplemented by an explicit L-dependence.

Leaving aside these considerations, and remaining in the frame of
Saxon-Woods potential, it is interesting to see if data that cannot be
fitted with an L-independent potential can be fitted with the addition
of a simple L-dependent term. Mackintosh and Kobos (1978) have described their analyses using a potential with an L-dependent term of
the form

$$V = f(\ell^2, L^2, \Delta^2)\{U^L \ g(r, R_R, a_R) + i \ W^L \ g(r, R_I, a_I)\}$$

where f is the standard Saxon-Woods form and
$g(r,R,a) = -4a(d/dr)f(r,R,a)$. They found that very precise fits could
be obtained to many sets of data, including some, such as p + ^{16}O,
that cannot be well-fitted by a standard optical potential. These
potentials are valuable for distorted wave calculations of reaction
processes. The L-dependent terms are similar in all cases and have the
following properties: (1) The potential for L < L_c is substantially
different from that for L > L_c, where L_c defines the transition region,
(2) the new terms are always repulsive and absorptive for low partial
waves, (3) the repulsive real term is almost always narrowly peaked
within the nuclear surface, (4) the absorptive term is always peaked
in the far surface and is usually narrow, (5) the L-dependent terms
are particularly strong for energies around 20-40 MeV, and (6) the
parameters of the L-dependent potential vary much more smoothly with
energy than those of the L-independent potentials.

Another series of calculations by Mackintosh (1973, 1974) and
Mackintosh and Kobos (1976) showed that the data can also be fitted by
explicit inclusion of the coupling to (p,d) pick-up channels. This
suggests that the L-dependence simulates the effect of this coupling,
although the L-dependent term is much smaller than the part of the
optical potential generated by the pick-up couplings. Some similar
work has been carried out by Coulter and Satchler (1977).

The spline interpolation method is being used to obtain model-
independent potentials, and this shows the connection between the L-
dependent potentials and the corresponding L-independent potentials.

A special case of L-dependence is the odd-even staggering of the
phase-shifts that may be represented by a $(-)^L$ term in the imaginary
potential (Gelbke, Bock and Richter, 1974; Robson, 1971). In the case
of elastic scattering of particles of nearly equal mass this is due to
the constructive and destructive interference of the direct and transfer amplitudes; the matrix element of the transfer process changes
sign depending on whether the orbital angular momentum L is even or
odd, and this produces an odd-even staggering of the scattering S-
matrix elements. However the staggering in the L-dependent model comes
from the nuclear size resonances whereas in the transfer process it is
due to the symmetry of the reaction and the bound state wave function
of the exchanged particle. The energy dependence of the odd-even
staggering is given naturally by the exchange model, whereas it can be
reproduced with an L-dependent potential only by a special choice of
parameters in each case. As described in the previous section, Tang

(1978) has used the resonating group formalism to show that such terms come from antisymmetrisation of the particle wave functions.

An L-dependent absorbing potential has been frequently used in analyses of heavy-ion scattering, and it can be justified physically by simple qualitative considerations concerning energy and angular momentum conservation in the various reaction channels (Hodgson, 1978).

5. Nucleon Analyses

Three detailed semi-phenomenological analyses of a wide range of nuclear data were described by Giannini (1978), Tarrats and Escudié (1978) and Leeb and Eder (1978). They obtain formulae for the potentials appropriate for many nuclei over a range of energies. It would be useful to make a critical comparison of these formulae, and also to study the simplifications that can be obtained by using information from microscopic optical potentials.

The microscopic optical potentials should make it possible to extract improved isospin potentials from the experimental data, in particular by providing more realistic form factors. At present it is usual to assume that the isospin potential has the same form factor as the bulk of the real potential, and to express the potential depth in the form

$$V_{p,n} = V_0 \pm V_1 \frac{N-Z}{A}$$

The values of V_0 and V_1 can then be obtained by analysing scattering data for a range of nuclei. For a particular choice for the dependence of the radius parameter on A, the asymmetry potential V_1 is in general partly geometrical and partly isospin, and it is important to separate these. For bound states this may be done using the energies of the $T_<$ and $T_>$ states, and for scattering by comparing the scattering of protons and neutrons from the same target. This comparison must be made for energies differing by the Coulomb energy difference Δ_c so that the compound states are analogues. An alternative method of separating the geometrical and isospin dependence was presented by Leeb and Eder (1978). It is also necessary to take into account the Coulomb-energy correction term V_c (sometimes taken as 0.4 $Z/A^{1/3}$. As Jeukenne et al. (1977) have shown, this cannot be obtained phenomeno-logically but must be obtained from a microscopic calculation.

A survey of the application of the folding model to elastic and inelastic scattering was presented by Petrovich (1978). He showed that with a proper choice of the effective interaction and inclusion of exchange it is possible to obtain a good overall fit to the experimental data.

It is found that precision fits to analysing powers give rather small values of the spin-orbit radius parameters, around 0.9 fm, and this is consistent with some analyses of bound state data. This does not seem to be consistent with the microscopic calculations of Rook and Brieva (1978) who found that the nucleon spin-orbit term is quite sharply peaked at the maximum of the derivative of the nuclear density. This merits further study.

Roman (1978) suggested that the low value of the spin-orbit diffuseness parameter for helions compared with the normal value for tritons may be attributable to a corresponding difference for nucleons. Reanalysis of neutron polarisation data showed that it can be quite well fitted by a potential with abnormally low spin-orbit diffuseness parameter. However the neutron polarisation data is rather imprecise, so it does not determine the parameter accurately. Improved neutron polarisation data is thus very desirable.

Another intriguing result was obtained by Bendiscioli et al. (1976). They found that the differential cross-sections and anti-symmetries for 36.2 MeV proton scattering at small angles around 1-7° from several nuclei differs markedly from the optical model values. An independent measurement is called for and if the result is confirmed its theoretical implications require examination.

A new technique for obtaining ratios of cross-sections by scattering protons from mixed targets was described by Austin (1978). These cross-section ratios are free of many of the uncertainties of absolute cross-sections, and thus may provide useful data for testing optical potentials. Some results were presented for 30.3 MeV proton elastic scattering from the calcium isotopes ^{40}Ca, ^{42}Ca and ^{48}Ca. These are well fitted by a standard optical potential, and give differences between the neutron RMS radii of the isotopes that are in accord with analyses of 800 MeV proton scattering data.

Extensive experimental data on the elastic scattering of 10-40 MeV protons by a range of light nuclei at closely-spaced energies was presented by the Milano-Bari group (Fabrici et al., 1978). It is found that the differential cross-sections for deformed nuclei like ^{12}C, ^{24}Mg and ^{28}Si can be quite well fitted by standard optical potentials, while those for spherical nuclei like ^{16}O, ^{40}Ar and ^{40}Ca are much harder to fit, even with non-standard potentials. These nuclei also show an anomalous peak in the differential cross-section for energies from 30 to 36 MeV. The magnitude of this peak is for many nuclei inversely proportional to the quadrupole deformation parameters β_2, suggesting that the anomaly is connected with collective excitation. The peak values of the cross-sections for excitation of 2_1^+ states are similarly proportional to $(R\beta_2)^2$ times the corresponding elastic peak cross-section. A phase-shift analysis shows that the partial waves involved in the enhanced cross-sections are centred on the grazing partial wave with L=kR. These features provide a severe test of microscopic optical potentials, which should be able to fit data for both spherical and deformed nuclei. It is important to test a calculated potential by comparing it with an extensive range of data; comparison with one cross-section could well be misleading.

6. Light-Ion Analyses

An impressive amount of precise data on the elastic scattering of deuterons, helions and alpha-particles by nuclei was presented by several speakers, together with detailed phenomenological analyses. This provides some knowledge of the interactions, which remain to be understood in a fully microscopic way.

The deuteron work was described by Djaloeis (1978), and referred to several nuclei at energies from 59 to 130 MeV. Data for 33 MeV helion scattering by a range of nuclei was presented by Karban (1978), together with the rather surprising result that the best optical model fits required a spin-orbit term with the abnormally-small diffuseness parameter of 0.2 fm. No satisfactory explanation of this has yet been proposed. Similar analyses of triton scattering gave a normal value for this parameter, and Roman (1978) suggested that the difference between the helion and triton potentials may be traceable to corresponding differences between neutron and proton potentials. This has been discussed in the previous section.

An analysis of the scattering of 104 MeV alpha-particles by ^{40}Ca and ^{48}Ca was presented by Gils et al. (1978). They used a potential consisting of the standard optical model term augmented by a Fourier-Bessel expansion to give greater flexibility and to remove the unwanted correlations between the different regions of the potential that are implicit in the usual analytic forms. The fits obtained with this potential are substantially better than those found with standard analyses using Saxon-Woods and Saxon-Woods-squared potentials. The analysis was repeated using a folding model potential obtained from an assumed nuclear density distribution and an effective alpha-nucleon interaction with a Gaussian radial dependence and a density-dependent factor to give the required saturation effects. The parameters of the interaction were first fixed by requiring a fit to the ^{40}Ca data assuming a known density distribution, and then determining the parameters of the neutron density distribution of ^{40}Ca from a fit to the ^{48}Ca data. This gives the following results for the RMS radii differences of ^{40}Ca and ^{48}Ca: $r_m(48)-r_m(40) = 0.12 \pm 0.07$ fm, and $r_n(48)-r_p(48) = 0.08 \pm 0.10$ fm.

The results of an analysis of the elastic scattering of 40-142 MeV alpha-particles by ^{90}Zr were presented by Put (1978). He found that the data above 80 MeV is well fitted by an optical potential with constant form factors and depths that vary smoothly with energy. A different potential is needed for lower energies.

The uniqueness and shape of the alpha-particle optical potential was studied by Wiktor et al. (1978) by using a Saxon-Woods potential raised to a variable power to analyse a wide range of experimental data from 120-172.5 MeV and from ^{12}C to ^{64}Ni. It is found that the best form of the potential depends on the target mass and projects the energy.

The validity of various folding models for the analysis of 104 MeV alpha-particle and 156 MeV ^{6}Li scattering by ^{40}Ca and ^{48}Ca has been studied by Majka, Gils and Rebel (1978). The folding models used were the simple double folding, and single-folding with target and projectile folding. The density-dependence of the nucleon-nucleon interaction was included in the sudden and in the intermediate approximations. The potentials used in the single folding were obtained in a semi-microscopic way including one-nucleon exchange effects and the density-dependence of the nucleon-nucleon interaction. In each case the real potential obtained from the folding model was supplemented by a phenomenological imaginary potential, and during the fitting process the real potential was normalised and the imaginary potential optimised. All the models give quite good fits, the best corresponding to projectile folding, and it is found that antisymmetrisation effects are of minor importance compared with the density-dependence of the nucleon-nucleon interaction. With the intermediate approximation for the density-dependence it is not necessary to normalise the real potential. Calculations were also made with various phenomenological form factors for the real part of the potentiala and it was found that the Saxon-Woods potential gives a poor fit, the Saxon-Woods-squared a much better fit and the modified Fermi parametrisation the best fit of all.

The results of an extensive analysis of alpha-particle bound and scattering data with a hard-core potential were presented by Gridnev et al. (1978) (Baz et al. 1977). The hard-core takes account of the re-

pulsion due to the Pauli principle, and gives a total potential with a
surface pocket that sustains unbound alpha-cluster states. It is found
that with a particular choice of the parameters of the potential the
energies and widths of many resonances in the $^{12}C + \alpha$ system are well
reproduced, and similarly for the $^{16}O + \alpha$ system. The same potentials
also give good overall fits to the alpha-particle elastic scattering
cross-sections over a range of energies.

D.F. Jackson (1978) summarised some of her recent work on the theory of alpha-decay, and showed that the half-life of this process is
very sensitive to the barrier penetrability and hence to the alpha-particle optical potential. The restriction that such potentials must be
consistent with alpha-decay should be useful in resolving at least some
of the ambiguities among the potentials that fit elastic scattering.
Since the potentials may be calculated from a folding model, this should
lead to improved nuclear matter distributions.

7. Heavy-Ion Analyses

After reviewing the folding model calculations of the optical potential, Brink (1978) presented a method for calculating the imaginary part
of the heavy-ion optical potential based on Feynman's path integral
method. This provides a natural link between a complete quantal theory
and the semi-classical approximations. The method consists in equating
two expressions for the scattering amplitude, one in terms of an optical potential and the other in terms of a detailed coupled-channels
description. This gives an expression for the optical potential in terms
of the time evolution operation $U[r(t), t_1, t_2]$,

$$\int_{t_0}^{t_i} V_{opt}(r(t))dt = i\hbar \ln\left[<0|U[r(t), t_1, t_2]|0>\right]$$

The right hand side of this equation is approximated by the leading term
in time-dependent perturbation theory, and separating out the imaginary
part of the potential gives

$$\int_{-\infty}^{\infty} Im\ V_{opt}[r(t)]dt = -\frac{1}{2\hbar}\sum_{n\neq0}\left|\int_{-\infty}^{\infty}dt<n|v(r(t),\xi)|0>\exp[i(\varepsilon_n-\varepsilon_o)t\hbar]\right|^2$$

This expression can be used to calculate the long-range absorption potential studied by Love et al. (1977) and by Baltz et al. (1977). The
path integral method can also be used to find a relation between the
imaginary part of the optical potential describing the scattering of
heavy ions in excited states and the friction coefficient for deep inelastic scattering.

The elastic scattering and vector and tensor analysing powers of 14.22 MeV ^{6}Li and ^{7}Li ions elastically scattered by ^{58}Ni have been measured by Tungate and Fick (1978). The data can be fitted quite well by spin-orbit potentials based on the folding model, and some of the marked differences between ^{6}Li and ^{7}Li scattering can be attributed to the differences in shape of the two projectiles.

The theoretical basis of the folding model of the optical potential was reviewed by Love (1978) and applied to calculate heavy-ion potentials. Both single- and double-folding methods were used, with realistic internucleon interactions based on G-matrices in a double-folding model. Many different internucleon interactions were used, and evaluated by comparing with data on heavy-ion scattering. It is usual in folding-model calculations to optimise the normalisation of the potential, and this provides a way of taking into account to some extent some of the higher-order corrections that are present in a more fully microscopic theory. While it is now possible to devise prescriptions for folding model calculations that fit the data quite well, progress at a more fundamental level can only come from the explicit introduction of these higher-order effects.

An extensive review of molecular states in heavy ion potentials was contributed by Terlecki et al. (1978). Particular attention was devoted to the conditions for the existance and excitation of nuclear molecules, and the calculation of heavy-ion potentials. A theoretical analysis using a rotating co-ordinate system clarifies the excitation mechanisms of molecular collective and single-particle states. A relatively new development is the application of parastatistics to nucleus-nucleus scattering. This is relevant to the scattering of nuclei with integral spin: when they are well separated they are two bosons and yet when they interact strongly they form a compound fermion system. To take account of this the relative wave function is written with a variable parameter that allows a continuous transition from Bose statistics to Fermi statistics. Analysis of the cross-section for ^{12}C-^{12}C elastic scattering shows that this parameter depends linearly on the incident energy. It can be used to simulate effects due to nuclear contributions to the potential, and from the internal structure and Coulomb excitation of the nuclei.

The folding model has also been used by Sinha (1978) to calculate the heavy-ion potential. The first-order real potential is obtained by folding the two nuclear densities with a Skyrme interaction, and higher order terms are also estimated.

Reinhard and Goeke (1978) described their work on the definition of the collective Hamiltonian for heavy-ion scattering. Using the concepts of the adiabatic time-dependent Hartree-Fock theory, they extend the collective path $|q\rangle$ to a dynamic path $|q,p\rangle$ and hence obtain the Hamiltonian function $k(q,p)$. This function is quantised by subtracting terms in $\hbar^2$, giving collective Hamiltonian identical to the one obtained in a Gaussian-overlap-expansion of the generator co-ordinate method.

8. Conclusion

The present time is a very lively and exciting period in the development of the optical model. Microscopic theories of the nucleon optical potential have now been developed to the stage where they can give without parameter adjustment potentials that are in good overall agreement with the experimental data. The way is now open to study particular features of the potential and to extend the microscopic theory to light and heavy ions.

Phenomenological anylses of the experimental data have greatly improved, and the spline technique now gives very precise fits. This shows that the usual Saxon-Woods form factor is in some respects inadequate, which had already been indicated by microscopic calculations. We can look forward to convergence of two approaches. It should be possible to use this as the basis of extensive and accurate global potentials that fit the data for many nuclei over a range of energies.

Work in the area of semi-phenomenological theories, particularly the folding model, is proving very successful and should facilitate the unification of the microscopic and phenomenological theories.

These developments in the theory of the optical potential have shown the need for more extensive and accurate experimental data in several areas, and this should stimulate further experimental work. Measurements of elastic scattering may seem somewhat tame, but in fact the attainment of high precision cross-sections and polarisations over a wide range of angles and energies presents a challenging technical problem.

In the brief compass of this summary it has not been possible to do justice to all the valuable and interesting contributions to the meeting. In particular, little attempt has been made to summarise several excellent papers that were themselves summaries of some specialised topic.

I thank all those who kindly made available copies of their papers and also Professor H.V. von Geramb and his colleagues for organising such a valuable and stimulating meeting.

REFERENCES

S.M. Austin, This Meeting, 1978.

D. Bachelier, M. Bernas, J.L. Boyard, H.L. Harney, J.C. Jourdain, P. Radvanyi, M. Roy-Stephen and R. De Vries, Nucl. Phys. $\underline{A195}$,361,1972.

A.J. Baltz, N.K. Glendenning, S.K. Kauffmann and K. Pruess, This Meeting, 1987.
A.J. Baltz, S.K. Kauffmann, N.K. Glendenning and K. Pruess, Phys. Rev. Lett. $\underline{40}$,22,1978; Symposium on Nuclear Direct Reaction Mechanisms, Hosei University, Tokyo, 5-10 September 1977.

R.C. Barrett, This Meeting, 1978.

G. Baur, F. Rösel and D. Trautmann, This Meeting, 1978; Phys. Rev. $\underline{C17}$, 2256,1978.

D. Baye, Nucl. Phys. $\underline{A272}$,445,1976.

A.I. Baz, V.Z. Goldberg, N.Z. Darwisch, K.A. Gridnev, V.M. Semjonov and E.F. Hefter, Lett. Nuovo Cim. $\underline{18}$,227,1977.

G. Bendiscioli, E. Lodi-Rizzini, A. Rotondi, M.L. Stanga and A. Venaglioni, Lecture Notes in Physics Vol. 55 (Pavia Conference), Edited by S. Boffi and G. Passatore, Springer-Verlag, 1976.

F.A. Brieva, This Meeting, 1978. See also Brieva and Rook, 1977, 1978.

F.A. Brieva and J.R. Rook, Nucl. Phys. $\underline{A291}$,299,317,1977; $\underline{A297}$,206,1978.

D.M. Brink, This Meeting, 1978.

P.W. Coulter and G.R. Satchler, Nucl. Phys. $\underline{A293}$,269,1977.

D. Dehnhard, V. Shkolnik and M.A. Franey, Phys. Rev. Lett. $\underline{40}$,1549,1978.

A. Djaloeis, This Meeting, 1978.

E. Fabrici, S. Micheletti, M. Pignanelli, F. Resmini, R. De Leo, G. D'Erasmo and A. Pantaleo, This Meeting, 1978.

W.E. Frahn, Preprint 1978.

W.E. Frahn and T.F. Hill, Preprint 1978.

H.V. von Geramb, Nucl. Phys. $\underline{A199}$,545,1973; Nuovo Cim. Lett. $\underline{5}$,333,1972.

C.K. Gelbke, R. Bock and A. Richter, Phys. Rev. $\underline{C9}$,852,1974.

M. Giannini, This Meeting, 1978.

H.J. Gils, E. Friedman, H. Rebel and Z. Majka, This Meeting, 1978.

G.W. Greenlees, W. Makofske, Y.C. Tang and D.R. Thompson, Phys. Rev. $\underline{C6}$, 2057,1972.

K.A. Gridnev, V.M. Semjonov, V.B. Subbotin, E.F. Hefter and H.V. von Geramb, This Meeting, 1978.

H.P. Grubler, U. Kiebele, H.O. Meyer, G.R. Plattner and I. Sick, Phys. Lett. $\underline{74B}$,202,1978.

P.E. Hodgson, Nature $\underline{249}$,412,1974; Lecture Notes in Physics Vol. 55 (Pavia Conference), Edited by S. Boffi and G. Passatore, Springer-Verlag 1976, p. 88; Nuclear Heavy-Ion Reactions Ch.3.4 (Oxford) 1978.

A.A. Ioannides and R.C. Johnson, Phys. Rev. $\underline{C17}$,1331,1978; This Meeting, 1978.
D.F. Jackson, This Meeting, 1978.

J.P. Jeukenne, A. Lejeune and C. Mahaux, Phys. Rep. $\underline{25C}$,83,1976; Phys. Rev. $\underline{C15}$,10,1977.

O. Karban, This Meeting, 1978.

Y. Kondo, S. Nagaka, S. Ohkubo and O. Tanimura, Prog. Theor. Phys. 53, 1006,1975.

H. Leeb and G. Eder, This Meeting, 1978.

W.G. Love, This Meeting, 1978.

W.G. Love, T. Terasawa and G. Satchler, Phys. Rev. Lett. 39,6,1977; Nucl. Phys. A291,183,1977.

R.S. Mackintosh, Phys. Lett. 44B,437,1973; Nucl. Phys. A230,195,1974; Phys. Lett. 59B,431,1975.

R.S. Mackintosh and A.M. Kobos, Phys. Lett. 62B,127,1976, This Meeting 1978.

V.A. Madsen, F. Osterfeld and J. Wambach, This Meeting, 1978.

C. Mahaux, This Meeting, 1978. See also Jeukenne, Lejeune and Mahaux, 1976, 1977.

Z. Majka, H.J. Gils and H. Rebel, This Meeting, 1978.

F. Malaguti, A. Uguzzoni, E. Verondini and P.E. Hodgson, Nucl. Phys. A297,287,1978.

I.E. McCarthy, This Meeting, 1978.

F. Michel and R. Vanderpoorten, Phys. Rev. C16,142,1977; This Meeting, 1978.

M. Micklinghoff, This Meeting, 1978.

J.W. Negele, Phys. Rev. C1,1260,1970.

H. Ngô, A. Bouyssy and N. Vinh Mau, This Meeting, 1978.

F. Petrovich, This Meeting, 1978.

L. Put, This Meeting, 1978.

P.G. Reinhard and K. Goeke, This Meeting, 1978.

D. Robson, Argonne National Laboratory Report 7837,239,1971.

S. Roman, This Meeting, 1978.

I. Sick, J.B. Bellicard, M. Bernheim, B. Frois, M. Huet, P.L. Leconte, I. Mongey, Phan Xuan-Ho, D. Royer and S. Turck, Phys. Rev. Lett. 35, 910,1975.

B.K. Sinha, This Meeting, 1978.

Y.C. Tang, This Meeting, 1978.

A. Tarrats and J.L. Escudié, This Meeting, 1978.

G. Terlecki, D. Hahn, W. Scheid, R. Koennecke and W. Greiner, This Meeting, 1978.

C.E. Thorn, M.J. LeVine, J.J. Kolata, C. Flaum, P.D. Bond and J.C. Sens, Phys. Rev. Lett. 38,384,1977.

A. Tungate and D. Fick, This Meeting, 1978.

N. Van Giai and V. Bernard, This Meeting, 1978.

N. Vinh Mau and A. Bouyssy, Nucl. Phys. A257,189,1976.

W. von Oertzen, Nucl. Phys. A148,529,1970.

M. Weigel, This Meeting, 1978.

S. Wiktor, C. Mayer-Böricke, A. Kiss, M. Rogge and P. Turek, This Meeting, 1978.

J.G. Zabolitzky, This Meeting, 1978.